Undergraduate Texts in Mathematics

Series Editors
Pamela Gorkin, Mathematics Department, Bucknell University, Lewisburg, PA, USA
Jessica Sidman, Mathematics and Statistics, Amherst College, Amherst, MA, USA

Advisory Editors
Colin Adams, Williams College, Williamstown, MA, USA
Jayadev S. Athreya, University of Washington, Seattle, WA, USA
Nathan Kaplan, University of California, Irvine, CA, USA
Lisette G. de Pillis, Harvey Mudd College, Claremont, CA, USA
Jill Pipher, Brown University, Providence, RI, USA
Jeremy Tyson, University of Illinois at Urbana-Champaign, Urbana, IL, USA

Undergraduate Texts in Mathematics are generally aimed at third- and fourth-year undergraduate mathematics students at North American universities. These texts strive to provide students and teachers with new perspectives and novel approaches. The books include motivation that guides the reader to an appreciation of interrelations among different aspects of the subject. They feature examples that illustrate key concepts as well as exercises that strengthen understanding.

Emily Clader • Dustin Ross

Beginning in Algebraic Geometry

 Springer

Emily Clader
Department of Mathematics
San Francisco State University
San Francisco, CA, USA

Dustin Ross
Department of Mathematics
San Francisco State University
San Francisco, CA, USA

ISSN 0172-6056 ISSN 2197-5604 (electronic)
Undergraduate Texts in Mathematics
ISBN 978-3-031-88818-2 ISBN 978-3-031-88819-9 (eBook)
https://doi.org/10.1007/978-3-031-88819-9

This work was supported by Emily Clader and Dustin Ross.

Mathematics Subject Classification (2020): 14-XX, 14-01

© The Editor(s) (if applicable) and The Author(s) 2025. This book is an open access publication.

Open Access This book is licensed under the terms of the Creative Commons Attribution-NonCommercial 4.0 International License (http://creativecommons.org/licenses/by-nc/4.0/), which permits any noncommercial use, sharing, adaptation, distribution and reproduction in any medium or format, as long as you give appropriate credit to the original author(s) and the source, provide a link to the Creative Commons license and indicate if changes were made.
The images or other third party material in this book are included in the book's Creative Commons license, unless indicated otherwise in a credit line to the material. If material is not included in the book's Creative Commons license and your intended use is not permitted by statutory regulation or exceeds the permitted use, you will need to obtain permission directly from the copyright holder.
This work is subject to copyright. All commercial rights are reserved by the author(s), whether the whole or part of the material is concerned, specifically the rights of translation, reprinting, reuse of illustrations, recitation, broadcasting, reproduction on microfilms or in any other physical way, and transmission or information storage and retrieval, electronic adaptation, computer software, or by similar or dissimilar methodology now known or hereafter developed. Regarding these commercial rights a non-exclusive license has been granted to the publisher.
The use of general descriptive names, registered names, trademarks, service marks, etc. in this publication does not imply, even in the absence of a specific statement, that such names are exempt from the relevant protective laws and regulations and therefore free for general use.
The publisher, the authors and the editors are safe to assume that the advice and information in this book are believed to be true and accurate at the date of publication. Neither the publisher nor the authors or the editors give a warranty, expressed or implied, with respect to the material contained herein or for any errors or omissions that may have been made. The publisher remains neutral with regard to jurisdictional claims in published maps and institutional affiliations.

This Springer imprint is published by the registered company Springer Nature Switzerland AG
The registered company address is: Gewerbestrasse 11, 6330 Cham, Switzerland

If disposing of this product, please recycle the paper.

Contents

Preface	ix
Acknowledgements	xii
Notation and Conventions	xiii

0 Polynomial Rings   1
- 0.1 Polynomials 2
- 0.2 Irreducible polynomials 7
- 0.3 Ideals and quotients 12
- 0.4 Prime and maximal ideals 18
- 0.5 Single-variable polynomials 23
- 0.6 Unique factorization in polynomial rings 27
- 0.7 Irreducibility criteria 32

I Affine Algebraic Geometry   35

1 Varieties and Ideals   37
- 1.1 The $\mathcal{V}$-operator 38
- 1.2 Affine varieties 43
- 1.3 The $\mathcal{I}$-operator 47
- 1.4 Radical ideals 51
- 1.5 The Nullstellensatz 56

2 Irreducibility of Affine Varieties   61
- 2.1 Inclusions, intersections, and unions 62
- 2.2 Finite generation 68
- 2.3 Irreducible affine varieties 73
- 2.4 Irreducible decompositions 79

3 Coordinate Rings   85
- 3.1 Polynomial functions on affine varieties 86
- 3.2 K-algebras 90
- 3.3 Generators of K-algebras 95
- 3.4 Nilpotents and reduced rings 99

4 Polynomial Maps   105
- 4.1 Polynomial maps between affine varieties 106
- 4.2 Pullback homomorphisms 112
- 4.3 Pulling back is a bijection 119
- 4.4 The equivalence of algebra and geometry 124

5 Proof of the Nullstellensatz 127
- 5.1 Modules 128
- 5.2 Module generators 134
- 5.3 Integrality 139
- 5.4 Noether normalization 145
- 5.5 Proof of the Nullstellensatz 151

6 Dimension 157
- 6.1 Motivating ideas 158
- 6.2 Function fields 163
- 6.3 Transcendence bases 167
- 6.4 Transcendence degree 173
- 6.5 Dimension: definition and first properties 178
- 6.6 The Fundamental Theorem 182

7 Smoothness 191
- 7.1 Linearizations and tangent spaces 192
- 7.2 Duality of vector spaces 197
- 7.3 Tangent spaces from coordinate rings 201
- 7.4 Tangent spaces and dimension 206
- 7.5 Smooth and singular points 210

8 Products 215
- 8.1 Products of affine varieties 216
- 8.2 Attributes of products 220
- 8.3 Tensor products of modules and algebras 225
- 8.4 Tensor products and bilinearity 231
- 8.5 The coordinate ring of a product 238

II Projective Algebraic Geometry 243

9 Projective Varieties 245
- 9.1 Projective space 246
- 9.2 The projective $\mathcal{V}$-operator 251
- 9.3 The projective $\mathcal{I}$-operator 257
- 9.4 Affine restrictions 262
- 9.5 Projective closures 266
- 9.6 The projective Nullstellensatz 272

10 Maps of Projective Varieties 279
- 10.1 Regular maps of projective varieties 280
- 10.2 Isomorphisms of projective varieties 286
- 10.3 Veronese maps 292
- 10.4 Products and Segre maps 298
- 10.5 Grassmannians and Plücker maps 305

11 Quasiprojective Varieties — 313
- 11.1 Quasiprojective varieties — 314
- 11.2 Closures and irreducibility — 320
- 11.3 Regular maps and regular functions — 325
- 11.4 Affine varieties revisited — 331
- 11.5 Affine opens and local properties — 337
- 11.6 Continuity of regular maps — 342
- 11.7 Products and graphs — 348
- 11.8 Images of projective varieties — 352

12 Culminating Topics — 359
- 12.1 Rational functions — 360
- 12.2 Dimension of quasiprojective varieties — 367
- 12.3 Local rings and tangent spaces — 372
- 12.4 Blow-ups — 379
- 12.5 Theorem on Fiber Dimensions — 384
- 12.6 Lines on surfaces — 392
- 12.7 Lines on smooth cubic surfaces — 400

Coda: Where to from here? — 409

Index — 415

Bibliography — 421

Preface

*The first fear
being drowning, the
ship's first shape
was a raft, which
was hard to unflatten
after that didn't
happen...*

—*Kay Ryan, "We're Building the Ship as We Sail It"*

Algebraic geometry, broadly defined as the geometric study of polynomial equations, is a subject that has now permeated nearly all of modern mathematics. After the subject was revolutionized in the second half of the twentieth century, in large part due to the work of Alexander Grothendieck, algebraic geometry began to develop a reputation for being an almost mythically powerful machine that was impenetrable to all but the initiated. Many beautiful texts were written to chip away at this barrier to entry, helping to make the geometric intuition behind algebraic geometry's abstractions visible to students with only some grounding in commutative algebra and topology.

There remains a population of learners to whom the gates of this subject are not yet fully open, though: those for whom the concepts of abstract algebra are a new language, one that has been encountered but not yet entirely internalized. While some may insist that comfort with such notions as quotient rings, modules, and algebras is a prerequisite to learning algebraic geometry, it is our belief that these ideas can be reinforced—in some cases even introduced—by seeing their manifestation in the geometric context. In addition to affording more students entry into algebraic geometry, this perspective has the advantage of allowing the algebra learner to restrict attention from the vast world of arbitrary rings to the more focused setting of polynomial rings, in which many notions are better-behaved and more intuitively motivated.

It is largely for these learners that we wrote this book. Our intended audience has taken a first course in abstract algebra, so words such as "ring" and "ideal" are meaningful. However, our intended audience is also human, so having been exposed, for example, to the definition of a quotient ring does not necessarily mean that moving between viewing it as a collection of cosets and as an operation of "setting to zero" is entirely fluid. We strive to build facility in working with algebraic notions gradually and organically as it becomes relevant to the geometric narrative.

On the other hand, we also anticipate that this book will be valuable to students with a more robust algebraic foundation who are eager to learn the foundations of algebraic geometry. While many of the algebraic ideas introduced in this book may not be new for such a student, we expect that revisiting and reinforcing those ideas in the geometric context will be a worthwhile endeavor.

Prerequisites

If you have taken undergraduate courses in linear algebra and abstract algebra, and you are willing to think deeply and read slowly, then you are equipped to learn from this book. Some algebraic words that we will use without defining them are *ring*, *integral domain*, *field*, and *ring homomorphism*. Other words from ring theory that play an especially central role in our development of algebraic geometry, such as *polynomial ring* and *ideal*, are defined carefully in Chapter 0; we recommend that all readers, especially those unsure whether they have sufficient background in algebra, begin by perusing that chapter. With regard to linear algebra, there are moments when we require some computational familiarity with matrices—including basic properties of *determinants*, *rank*, and *nullity*—and we also expect familiarity with *finite-dimensional vector spaces*, including the notions of *spans*, *linear independence*, *bases*, *dimension*, and *linear maps*. More advanced topics from linear algebra, such as dual spaces and tensor products, will be developed as needed.

One notable prerequisite that we have not assumed is background in topology. Readers who have not yet encountered topology should find that they are at no disadvantage in Part I of the book, on affine algebraic geometry, though we occasionally mention connections to topology for those who do have some exposure. In Part II, on projective algebraic geometry, topological terminology becomes unavoidable, so we introduce the relevant background where needed. Even this part is intended to be accessible to students without prior background in topology, though we confess, having taken a course in topology would certainly make the last two chapters of the book easier to digest.

How to use this book

This is a narrative-driven book, and as such, it is meant to be read in order. That being said, there are certain subsets that could naturally be broken off in order to create a course of the desired length. In particular, a course focusing exclusively on affine algebraic geometry could reasonably cover most of Chapters 0–8 in one semester. Alternatively, for a course that reaches projective varieties more quickly, one could forego the advanced topics in affine algebraic geometry (Chapters 6–8, on dimension, smoothness, and products) and skip directly to Chapters 9–11. For students with more algebraic background, it is conceivable that the entire contents of the book could be covered in one semester by breezing through some of the purely algebraic content.

Despite the narrative style of the book, students should not be fooled into believing it can be read like a novel. You should expect to read with pencil in hand, pausing frequently to work out details and examples. Your goal is to achieve a delicate balance between small-scale and large-scale comprehension, gaining comfort with detailed computations while holding the conceptual storyline in mind. The best way to gauge one's success in this task is to attempt to explain the material clearly but succinctly to a peer.

On that note, we should mention that this book is intended to be amenable to a number of different learning contexts. We have seen it used successfully for independent studies, especially ones in which multiple students work together, meeting regularly to discuss the material and solve problems. It could also be used in a flipped classroom, where students are expected to learn at least some portion of the material through out-of-class reading. Even for entirely independent self-study, we expect that this text could be useful, especially for students preparing to enroll in a more advanced course in algebraic geometry that assumes familiarity with varieties.

We have striven to write a book that is accessible to undergraduate students, as well as to working mathematicians without prior training in algebraic geometry. Our hope is that this text will lower the bar of entry into this beautiful subject, opening the gates for a broad range of students and practitioners at every level to be exposed to the rigorous foundations of algebraic geometry.

Stylistic choices

In order to produce a text that we believe to be as readable as possible, we have made many choices regarding style and content. We are almost painfully consistent with notation; for instance, we use letters at the end of the alphabet for variables and letters at the beginning of the alphabet for values of those variables, so the parabola in $\mathbb{R}^2$ defined by $y = x^2$ consists of pairs (a, b) of real numbers satisfying $b = a^2$. This choice is based on our personal experience that notational consistency is often helpful for students with less background in the particular style of abstraction at hand. However, we recognize that rigid consistency must eventually be abandoned in favor of the flexibility of intuition; we have attempted to make the transition between these extremes happen smoothly over the course of the book.

Regarding content, one notable decision that we have made differentiates this book from many other introductory texts in algebraic geometry: we give a rather complete development of affine varieties before ever introducing the projective context. This approach centers the interplay of algebra and geometry, which is most salient in the affine setting, disentangling the most foundational algebraic developments from the more geometric and topological notions that are necessary in order to study projective and quasiprojective varieties. However, as mentioned previously, one could easily choose to skip ahead to Part II sooner if they were especially eager to familiarize themselves with projective varieties.

For the sake of creating an undergraduate text of a reasonable length, we have resisted the temptation to stuff all of our favorite topics into these pages, and we have chosen to conclude the book with quasiprojective varieties, just shy of sheaves, abstract varieties, and scheme theory. This text is by no means a comprehensive reference in algebraic geometry, and while some experts may disagree with our choices of omissions, we have made every attempt to create a cohesive narrative that includes what we believe to be the core of a rigorous foundation in algebraic varieties. After working their way through this book, students will have built a strong foundation in the setting of quasiprojective varieties, from which they can draw tools and intuition for further study in special topics or the more abstract settings of modern algebraic geometry. To help guide future study, we have concluded the book with a coda that suggests several of the most natural next steps.

Acknowledgements

We have had many teachers in our own study of algebraic geometry, and more generally in mathematics and its exposition, to whom we are deeply grateful. Particular debt is due to Bill Fulton and Jeff Achter, with whom we took our first steps as algebraic geometers, and also to our advisors Yongbin Ruan and Renzo Cavalieri, who guided us as we learned to run and play in the wild with these notions. We were inspired by many existing books, particularly those by Shafarevich [8], Smith et al [10], and Vakil [11]. We learned an immense amount about writing, typesetting, and publishing a book from Sheldon Axler; we are grateful to him for this and for providing detailed feedback on early drafts. Thanks, also, to Elizabeth Loew at Springer and Brian Treadway for their editorial and copyediting work. Many of our early readers were students at San Francisco State University, and their input was invaluable; special thanks to Nitan Avivi-Stuhl, Tracy Camacho, Paul Carmody, Alvaro Cornejo, Ernestina DaCosta, Jupiter Davis, Jason Meintjes, Cesar Meza, Anastasia Nathanson, Lauren Nowak, Patrick O'Melveny, and Ian Wallace for their sharp eye and generous feedback. We also thank Linda Chen and her students at Swarthmore, who piloted a draft of the early chapters of the book and caught many typos. We acknowledge the generous support of the National Science Foundation; the creation of this book was partially supported by NSF grants DMS–1810969, DMS–2137060, DMS–2001439, and DMS–2302024.

Finally, thanks to Maya Ross Clader for tolerating the indignity of having two algebraic geometers as parents; you are by far our greatest collaboration.

Emily Clader
Dustin Ross
San Francisco, 2025

Notation and Conventions

- All rings are commutative with unity, denoted 1.
- R and S denote rings.
- All ring homomorphisms $\varphi : R \to S$ satisfy $\varphi(1) = 1$.
- K denotes a field, which is assumed to be algebraically closed after Chapter 1.
- It is not assumed that $1 \neq 0$ for general rings, but it is assumed that $1 \neq 0$ for integral domains (and fields).
- $\mathbb{Z}$, $\mathbb{Q}$, $\mathbb{R}$, and $\mathbb{C}$ denote the sets of integers, rational numbers, real numbers, and complex numbers, respectively.
- $\mathbb{N} = \{0, 1, 2, 3, \dots\}$ denotes the set of natural numbers.
- $\mathbb{Z}_+ = \{1, 2, 3, \dots\}$ denotes the set of positive integers.

Chapter 0

Polynomial Rings

> LEARNING OBJECTIVES FOR CHAPTER 0
> - Define and work with polynomials and polynomial rings.
> - Define and give examples of ideals and quotients, especially in the context of polynomial rings.
> - Describe properties of polynomials over fields, such as the existence and uniqueness of irreducible factorizations.
> - Determine if a given polynomial is irreducible.

Algebraic geometry studies solutions of polynomial equations by building a dictionary between the geometry of the solution sets and the algebra of the defining polynomial equations. In this preliminary chapter, we develop the algebraic notions of polynomial rings that are prerequisite to the study of algebraic geometry. The chapter culminates with a proof of the important fact that every polynomial over a field factors uniquely into irreducible polynomials.

The reader is assumed to have taken a first course in ring theory and to be familiar with the notions of rings, integral domains, and fields. However, knowledge beyond the most fundamental definitions and results is not expected. Surely, this chapter will read more quickly for those students with a more robust algebraic background, while a student newer to abstract algebra may choose to devote a significant amount of time to mastering the contents of these pages.

The purpose of this chapter is to establish the algebraic foundation on which the rest of the book is built, focusing on the fundamental properties of polynomial rings that will be most useful for later developments. As such, the choice has often been made to forego generality for the sake of brevity; for example, we will never consider noncommutative rings, so every ideal will be a two-sided ideal. Nearly all of the examples in this chapter illustrate concepts in the specific setting of polynomial rings, though we hope that the intuition developed in this setting might help the interested student study these concepts more generally.

Section 0.1 Polynomials

Polynomials and their solutions are some of the first objects we encounter in our mathematical lives. For example, you may even remember the first time you learned that the solutions in $\mathbb{R}^2$ of the two-variable polynomial equation

$$x^2 + y^2 - 1 = 0$$

describe the unit circle. However, if our goal is to study the unit circle, then there are many other polynomials that one might choose to describe it; for example, the unit circle is also equal to the solutions of either

$$2x^2 + 2y^2 - 2 = 0 \quad \text{or} \quad (x^2 + y^2 - 1)^2 = 0.$$

This leads to a natural question: is there a *best* polynomial that describes the unit circle? The answer proposed by algebraic geometry is, in some sense, the most egalitarian: all of the polynomials that describe the unit circle are equally important, and we should study them together as a *set*. What does it mean, then, to study a *set* of polynomials?

As we will soon learn, the set of polynomials describing the unit circle is much more than just a set; it has important algebraic structure that reflects the geometry of the circle. To be able to describe this algebraic structure in this example and beyond, we must first establish precise notation and terminology regarding sets of polynomials and their algebraic structure. To begin our formal discussion of polynomials, we start with the notion of a monomial.

0.1 DEFINITION *Monomials*

A *monomial* in the variables $x_1, \ldots, x_n$ is an expression of the form

$$x^\alpha = x_1^{\alpha_1} \cdots x_n^{\alpha_n},$$

where $\alpha = (\alpha_1, \ldots, \alpha_n) \in \mathbb{N}^n$ is the *exponent vector*. Two monomials are equal if and only if they have the same exponent vector.

The variables $x_1, \ldots, x_n$ should be viewed as formal symbols, and their role is simply to serve as placeholders for $\alpha_1, \ldots, \alpha_n$. The data of a monomial in $x_1, \ldots, x_n$ is equivalent to the data

> When there are only a few variables, they are often represented with distinct letters such as x, y, and z.

of an exponent vector $(\alpha_1, \ldots, \alpha_n)$; however, placing each α_i as the exponent of x_i will prove useful when multiplication of monomials is defined below. Variables that appear with an exponent of 0 are typically omitted; for example,

$$x^2 y^3 z^0 = x^2 y^3 \quad \text{and} \quad x^0 y^0 z^0 = 1.$$

As in the case of $f(x, y) = x^2 + y^2 - 1$, polynomials are built by taking linear combinations of monomials. In the most general setting, the coefficients of these linear combinations belong to an arbitrary ring R, as in the following definition.

0.2 DEFINITION *Polynomials over R*

A *polynomial* in the variables $x_1, \ldots, x_n$ over R is an expression of the form

$$f = f(x_1, \ldots, x_n) = \sum_{\alpha \in \mathbb{N}^n} a_\alpha x^\alpha,$$

where $a_\alpha \in R$ for each $\alpha \in \mathbb{N}^n$ and $a_\alpha = 0$ for all but finitely many α. Two polynomials

$$f = \sum_\alpha a_\alpha x^\alpha \quad \text{and} \quad g = \sum_\alpha b_\alpha x^\alpha$$

are equal if and only if $a_\alpha = b_\alpha$ for all $\alpha \in \mathbb{N}^n$.

0.3 EXAMPLE *Polynomials*

The following are two examples of polynomials in the variables x and y over the ring of integers $\mathbb{Z}$:

Polynomials are typically written as finite sums, omitting all summands that have a coefficient of zero.

$$f = xy^2 + 3xy + 2 \quad \text{and} \quad g = -xy + 4.$$

The polynomials f and g can also be viewed as having coefficients in $\mathbb{Q}$, $\mathbb{R}$, $\mathbb{C}$ or any other ring containing $\mathbb{Z}$. Observe that we can create new polynomials from f and g by adding them and multiplying them in the familiar way:

$$f + g = xy^2 + 2xy + 6,$$
$$fg = -x^2y^3 - 3x^2y^2 + 4xy^2 + 10xy + 8.$$

As the reader is encouraged to verify in Exercise 0.1.1, the operations of addition and multiplication, formalized in the next definition, endow the set of polynomials with the structure of a ring.

0.4 DEFINITION *Polynomial rings*

The *polynomial ring* $R[x_1, \ldots, x_n]$ is the set of all polynomials in variables $x_1, \ldots, x_n$ over R. Polynomial addition and multiplication are defined by

$$\left(\sum_\alpha a_\alpha x^\alpha \right) + \left(\sum_\alpha b_\alpha x^\alpha \right) = \sum_\alpha (a_\alpha + b_\alpha) x^\alpha$$

and

$$\left(\sum_\alpha a_\alpha x^\alpha \right) \left(\sum_\alpha b_\alpha x^\alpha \right) = \sum_\alpha \left(\sum_{\alpha_1 + \alpha_2 = \alpha} a_{\alpha_1} b_{\alpha_2} \right) x^\alpha.$$

The additive identity $0 \in R[x_1, \ldots, x_n]$ is the polynomial for which $a_\alpha = 0$ for all α, and the multiplicative identity $1 \in R[x_1, \ldots, x_n]$ is the polynomial for which

$$a_\alpha = \begin{cases} 1 & \text{if } (\alpha_1, \ldots, \alpha_n) = (0, \ldots, 0), \\ 0 & \text{if } (\alpha_1, \ldots, \alpha_n) \neq (0, \ldots, 0). \end{cases}$$

When working with polynomial rings, it can be useful to leverage their recursive nature in order to utilize proofs by induction on the number of variables. The next result is somewhat self-evident, but we state it carefully as it will be used often.

0.5 PROPOSITION *Recursive nature of polynomial rings*

There is a canonical isomorphism of rings

$$R[x_1,\ldots,x_n] = R[x_1,\ldots,x_{n-1}][x_n],$$

where the right-hand side is the ring of polynomials in one variable x_n with coefficients in the ring $R[x_1,\ldots,x_{n-1}]$.

The next example illustrates the main idea behind Proposition 0.5.

0.6 EXAMPLE $\mathbb{Z}[x,y]$ versus $\mathbb{Z}[x][y]$

Consider the polynomial

$$f = x^3 y^2 + xy^2 - 2xy - x + 1 \in \mathbb{Z}[x,y].$$

We can view f as an element of $\mathbb{Z}[x][y]$ by grouping all terms that have the same exponent in y. In doing so, we write

$$f = (x^3 + x)y^2 + (-2x)y + (-x+1) \in \mathbb{Z}[x][y].$$

As a polynomial in y, the coefficients of f are $x^3 + x$, $-2x$, and $-x+1$, all of which are elements of $\mathbb{Z}[x]$.

PROOF OF PROPOSITION 0.5

To prove the proposition, we describe the canonical ring isomorphism, which, as illustrated in Example 0.6, simply groups all terms of a polynomial in $R[x_1,\ldots,x_n]$ for which x_n appears with the same exponent. To make this more precise, consider the following function:

> When two rings are isomorphic, there will often be a multitude of possible isomorphisms. The word "canonical" means that there is one natural choice among all possible isomorphisms. The symbol $\cong$ is used to denote isomorphisms, while $=$ is used for canonical isomorphisms.

$$\varphi : R[x_1,\ldots,x_n] \to R[x_1,\ldots,x_{n-1}][x_n]$$

$$\sum_{\alpha \in \mathbb{N}^n} a_\alpha x_1^{\alpha_1} \cdots x_n^{\alpha_n} \mapsto \sum_{d \geq 0} \left(\sum_{\substack{\alpha \in \mathbb{N}^n \\ \alpha_n = d}} a_\alpha x_1^{\alpha_1} \cdots x_{n-1}^{\alpha_{n-1}} \right) x_n^d.$$

The verification that φ is a ring isomorphism is Exercise 0.1.3. □

The following result concerning polynomial rings over integral domains is a first application of Proposition 0.5.

0.1. POLYNOMIALS

> **0.7 PROPOSITION** *Polynomials over integral domains*
>
> If R is an integral domain, then $R[x_1, \ldots, x_n]$ is an integral domain.

PROOF   We proceed by induction on the number of variables.
(Base case) Let $f, g \in R[x_1]$ be nonzero. We must prove that $fg \neq 0$. Write

$$f = a_d x_1^d + a_{d-1} x_1^{d-1} + \cdots + a_0 \quad \text{and} \quad g = b_e x_1^e + b_{e-1} x_1^{e-1} + \cdots + b_0,$$

where $a_d \neq 0$ and $b_e \neq 0$. By definition of multiplication in $R[x_1]$,

$$fg = (a_d b_e) x_1^{d+e} + (a_d b_{e-1} + a_{d-1} b_e) x_1^{d+e-1} + \cdots + a_0 b_0.$$

Since R is an integral domain, $a_d b_e \neq 0$. Since fg has at least one nonzero coefficient, we conclude that $fg \neq 0$.

(Induction step) Assume $S = R[x_1, \ldots, x_{n-1}]$ is an integral domain; we must show that $R[x_1, \ldots, x_n]$ is an integral domain. By Proposition 0.5,

$$R[x_1, \ldots, x_n] = S[x_n].$$

Since S is an integral domain, the argument used in the base case immediately implies that $S[x_n]$, and thus $R[x_1, \ldots, x_n]$, is an integral domain. □

The numbers d and e appearing in the proof of Proposition 0.7 are important attributes of the polynomials f and g, called their degrees. The next definition generalizes the notion of degree to polynomials with any number of variables.

> **0.8 DEFINITION** *Monomial and polynomial degree*
>
> The *degree* of a monomial $x^\alpha = x_1^{\alpha_1} \cdots x_n^{\alpha_n}$ is
>
> $$\deg(x^\alpha) = \alpha_1 + \cdots + \alpha_n \in \mathbb{N}.$$
>
> The *degree* of a nonzero polynomial $f = \sum a_\alpha x^\alpha \in R[x_1, \ldots, x_n]$ is
>
> $$\deg(f) = \max\{\deg(x^\alpha) \mid a_\alpha \neq 0\}.$$

0.9 EXAMPLE   Degree
The monomials $x^2 yz$, z^4, x, and 1 have degrees 4, 4, 1, and 0, respectively, and

$$\deg(x^2 yz + z^4 + x + 1) = 4.$$

The reader may have noticed that the zero polynomial has not been assigned a degree, which is intentional. One of the most useful properties of degree is additivity, described in the next result, which fails for any choice of $\deg(0) \in \mathbb{N}$.

> **0.10 PROPOSITION** *Additivity of degree*
>
> If R is an integral domain and $f, g \in R[x_1, \ldots, x_n]$ are nonzero, then
> $$\deg(fg) = \deg(f) + \deg(g).$$

PROOF Let $f, g \in R[x_1, \ldots, x_n]$ be nonzero polynomials of degrees d and e, respectively. Write
$$f = f_d + f_{d-1} + \cdots + f_0 \quad \text{and} \quad g = g_e + g_{e-1} + \cdots + g_0,$$
where f_i comprises all terms in f of degree i, and similarly for g_j. By assumption, $f_d \neq 0$ and $g_e \neq 0$. Some reflection should convince the reader that degree is additive on monomials, so the highest-degree monomials that could possibly appear with nonzero coefficients in fg have degree $d + e$ and arise in the product $f_d g_e$. Since $R[x_1, \ldots, x_n]$ is an integral domain, we see that $f_d g_e \neq 0$, from which we conclude that $\deg(fg) = d + e = \deg(f) + \deg(g)$. $\square$

Exercises for Section 0.1

0.1.1 Review the definition of a ring (commutative with unity) and prove that, for any ring R, addition and multiplication of polynomials endows $R[x_1, \ldots, x_n]$ with the structure of a ring (commutative with unity).

0.1.2 Group the terms of the polynomial
$$f = xyz^2 + xyz + z^3 + x^2 z^2 + yz^2 + z + x + 1 \in R[x, y, z]$$
to view it as an element of $R[x, y][z]$, $R[x][y, z]$, and $R[y, z][x]$.

0.1.3 Prove that the function
$$\varphi : R[x_1, \ldots, x_n] \to R[x_1, \ldots, x_{n-1}][x_n]$$
defined in the proof of Proposition 0.5 is a ring isomorphism.

0.1.4 A polynomial $f = \sum_\alpha a_\alpha x^\alpha \in R[x_1, \ldots, x_n]$ is called a *constant polynomial* if $a_\alpha = 0$ whenever $\alpha \neq (0, \ldots, 0)$. Prove that the set of constant polynomials form a subring of $R[x_1, \ldots, x_n]$ that is isomorphic to R.

0.1.5 Let $a = (a_1, \ldots, a_n) \in R^n$. Prove that there is a unique ring homomorphism
$$\varphi_a : R[x_1, \ldots, x_n] \to R$$
such that $\varphi_a(r) = r$ for all $r \in R$ and $\varphi_a(x_i) = a_i$ for all i. (This homomorphism is called *evaluation at a* and $\varphi_a(f)$ is often written as $f(a_1, \ldots, a_n)$.)

0.1.6 Show that Propositions 0.7 and 0.10 fail if R is not assumed to be an integral domain.

0.1.7 Show that Proposition 0.10 does not extend to any choice of $\deg(0) \in \mathbb{N}$. (Some mathematicians take the convention that $\deg(0) = -\infty$.)

0.1.8 Prove that $\deg(f + g) \leq \max\{\deg(f), \deg(g)\}$.

Section 0.2 Irreducible polynomials

Algebraic geometry is primarily interested in polynomial rings $K[x_1, \ldots, x_n]$ with coefficients in a field K. (Throughout this section—and this book—we always use K to denote a field.) In order to study these polynomial rings, it is useful to have an understanding of their "atomic" elements and how each polynomial decomposes into atomic ones. As motivation for these ideas, it is instructive to first consider the more familiar ring $\mathbb{Z}$, where the atomic elements are the *prime* numbers.

Recall that an integer $p \in \mathbb{Z}_{\geq 2}$ is *prime* if for all $m, n \in \mathbb{Z}$,

(0.11) $$p = mn \implies m = \pm 1 \text{ or } n = \pm 1.$$

In other words, a prime integer is one that cannot be factored in a nontrivial way. One of the central results in number theory (and in all of mathematics, for that matter) is the existence and uniqueness of prime factorizations: every integer $n \in \mathbb{Z}_{\geq 2}$ can be written as a product of prime numbers in a unique way, up to reordering the factors.

We would like to study these ideas more generally, especially in the context of polynomial rings. To do so, we begin with the ring-theoretic definition of "atomic," including the notion of a *unit*, which generalizes the $\pm 1 \in \mathbb{Z}$ appearing in (0.11).

0.12 DEFINITION *Units and (ir)reducible elements*

An element $u \in R$ is called a *unit* if it has a multiplicative inverse. The set of units in R is denoted $R^* \subseteq R$. An element $p \in R$ is called *irreducible* if it is neither zero nor a unit, and

$$p = ab \implies a \in R^* \text{ or } b \in R^*.$$

An element is *reducible* if it is neither zero, a unit, nor irreducible.

In other words, a nonzero element is irreducible if it cannot be factored into a product of two nonunits. In the case of polynomial rings $K[x_1, \ldots, x_n]$ over a field K, the units are the nonzero constant polynomials (Exercise 0.2.2):

> The distinction between units and nonunits is necessary because every element factors if we allow units:
> $$a = u(u^{-1}a).$$

$$K[x_1, \ldots, x_n]^* = K^* = K \setminus \{0\}.$$

It follows that a polynomial $f \in K[x_1, \ldots, x_n]$ is irreducible if and only if

(i) f is nonconstant, and
(ii) f cannot be written as a product of two nonconstant polynomials.

0.13 EXAMPLE Linear polynomials in $K[x_1, \ldots, x_n]$ are irreducible

We say that a polynomial $f \in K[x_1, \ldots, x_n]$ is *linear* if $\deg(f) = 1$. If f is linear and $f = gh$, then by additivity of degree,

$$1 = \deg(f) = \deg(g) + \deg(h).$$

It follows that either $\deg(g) = 0$ or $\deg(h) = 0$, implying that g or h is constant.

0.14 EXAMPLE $y - x^2$ is irreducible in $K[x,y]$

Suppose that
$$y - x^2 = gh$$
for some $g, h \in K[x,y]$. As an element of the ring $K[x][y]$, the polynomial $y - x^2$ has degree 1, implying that (as polynomials in y) one of g or h must also have degree 1 and the other must have degree 0. Without loss of generality, we can write
$$g = ay + b \quad \text{and} \quad h = c$$
where $a, b, c \in K[x]$. This implies that $y - x^2 = acy + bc$. By matching coefficients of y, we see that $ac = 1$, which implies that $h = c \in K^*$. Thus, $y - x^2$ is irreducible.

0.15 EXAMPLE $x^2 + 1 \in \mathbb{C}[x]$ versus $x^2 + 1 \in \mathbb{R}[x]$

In $\mathbb{C}[x]$, we have a factorization
$$x^2 + 1 = (x - i)(x + i),$$
which shows that $x^2 + 1$ is reducible in $\mathbb{C}[x]$. In $\mathbb{R}[x]$, on the other hand, it is not possible to factor $x^2 + 1$ into two linear factors (Exercise 0.2.3), implying that $x^2 + 1$ is irreducible in $\mathbb{R}[x]$. This example illustrates how the behavior of polynomial rings heavily depends on the choice of coefficient ring.

Since prime factorization is such a fundamental property of the ring of integers, it is useful to have a generalization of this property to the setting of integral domains. The following definition captures the essence of unique prime factorization in $\mathbb{Z}$.

0.16 DEFINITION   *Unique factorization domain*

An integral domain R is called a *factorization domain* (FD) if

  (i) for every nonzero, nonunit $a \in R$, there exist irreducible elements $p_1, \ldots, p_\ell \in R$ such that $a = p_1 \cdots p_\ell$.

It is called a *unique factorization domain* (UFD) if, in addition,

  (ii) whenever $p_1 \cdots p_\ell = q_1 \cdots q_m$ for some irreducible elements p_i and q_j, then $\ell = m$ and, after possibly reordering, there exist units u_i such that $p_i = u_i q_i$ for all i.

Unique prime factorization in the ring of integers implies that $\mathbb{Z}$ is a UFD. One of the fundamental properties of $K[x_1, \ldots, x_n]$ is that it is also a UFD, as we will see over the course of this chapter.

To prove that $K[x_1, \ldots, x_n]$ is a UFD, we must prove that irreducible factorizations exist and that they are unique. In the current section, we content ourselves with proving existence of irreducible factorizations.

For examples of an integral domain that is not a FD and a FD that is not a UFD, see Exercises 0.2.10 and 0.3.15, respectively.

0.2. IRREDUCIBLE POLYNOMIALS

0.17 PROPOSITION $K[x_1, \ldots, x_n]$ *is a FD*

If $f \in K[x_1, \ldots, x_n]$ is a nonconstant polynomial, then there exist irreducible polynomials $p_1, \ldots, p_\ell \in K[x_1, \ldots, x_n]$ such that $f = p_1 \cdots p_\ell$.

PROOF We proceed by induction on the degree of f.

(Base case) If $\deg(f) = 1$, then f is irreducible by Example 0.13. In particular, f has an irreducible factorization (with $\ell = 1$).

(Induction step) Assume that every polynomial of degree less than d can be factored into irreducible polynomials, and suppose $f \in K[x_1, \ldots, x_n]$ has degree d. If f is irreducible, then f has an irreducible factorization with $\ell = 1$. If f is not irreducible, then $f = gh$ with $\deg(g) < d$ and $\deg(h) < d$. By the induction hypothesis, there are irreducible factorizations

$$g = p_1 \cdots p_\ell \quad \text{and} \quad h = q_1 \cdots q_k.$$

Thus, f admits an irreducible factorization

$$f = p_1 \cdots p_\ell \cdot q_1 \cdots q_k. \qquad \square$$

0.18 EXAMPLE *An irreducible factorization*

It follows from Examples 0.13 and 0.15 that

$$x^2 + 1 = (x - i)(x + i)$$

is an irreducible factorization of $x^2 + 1 \in \mathbb{C}[x]$.

For inspiration on how one might prove uniqueness of irreducible factorizations, we return to the familiar case of $\mathbb{Z}$. The key to proving uniqueness of prime factorizations in the integers is Euclid's Lemma, which says that $p \in \mathbb{Z}_{\geq 2}$ is prime if and only if, for all $m, n \in \mathbb{Z}$,

$$p \mid mn \implies p \mid m \text{ or } p \mid n.$$

This second characterization of prime integers naturally generalizes to rings.

> We use the standard notation $a \mid b$ as shorthand for "a divides b," which means that $b = ca$ for some c.

0.19 DEFINITION *Prime element*

An element $p \in R$ is *prime* if it is neither zero nor a unit and, for all $a, b \in R$,

$$p \mid ab \implies p \mid a \text{ or } p \mid b.$$

As we will see below, the question of whether irreducible factorizations are unique can be reduced to proving the ring-theoretic analogue of Euclid's Lemma. More specifically, given an integral domain R, Euclid's Lemma translates to a statement equating prime elements in R with irreducible elements. The following result verifies one of the implications: in integral domains, all primes are irreducible.

0.20 PROPOSITION *Prime implies irreducible*

In an integral domain, every prime element is irreducible.

PROOF   Let R be an integral domain and let $p \in R$ be prime. Toward proving that p is irreducible, suppose that

(0.21) $$p = ab \quad \text{for some} \quad a, b \in R;$$

we must prove that $a \in R^*$ or $b \in R^*$. Notice that (0.21) implies, in particular, that $p \mid ab$. By primeness of p, we have $p \mid a$ or $p \mid b$. Without loss of generality, assume $p \mid a$ and write $a = pc$ for some $c \in R$. Substituting this expression into (0.21), we obtain

$$p = pcb.$$

Since R is an integral domain and $p \neq 0$, we can cancel p from both sides to obtain $1 = cb$, implying that $b \in R^*$. □

The converse of Proposition 0.20 is not true in general. In fact, the converse is, in some sense, equivalent to uniqueness of factorizations, which is the content of the next result.

Look ahead to Exercise 0.3.15 for an example where the converse of Proposition 0.20 fails.

0.22 PROPOSITION *FDs versus UFDs*

Let R be a factorization domain. Then R is a unique factorization domain if and only if every irreducible element of R is prime.

PROOF   We prove both implications.

($\Rightarrow$) Suppose that R is a UFD and let $p \in R$ be irreducible. Toward proving that p is prime, suppose that $p \mid ab$ for some $a, b \in R$, and choose $c \in R$ such that $ab = pc$. Since R is a FD, everything admits an irreducible factorization:

$$a = p_1 \cdots p_k, \quad b = q_1 \cdots q_\ell, \quad \text{and} \quad c = r_1 \cdots r_m.$$

Thus, we have two irreducible factorizations

$$p_1 \cdots p_k \cdot q_1 \cdots q_\ell = p \cdot r_1 \cdots r_m.$$

Since R is a UFD, there exists a unit u such that $p = up_i$ or $p = uq_j$ for some i or j. It follows that $p \mid a$ or $p \mid b$, so p is prime.

($\Leftarrow$) Suppose that every irreducible element of R is prime, and let

(0.23) $$p_1 \cdots p_\ell = q_1 \cdots q_m$$

be two irreducible (hence, prime) factorizations. Assume without loss of generality that $\ell \leq m$. Since p_1 is prime and $p_1 \mid q_1 \cdots q_m$, it follows from Exercise 0.2.7 that $p_1 \mid q_j$ for some j. After possibly reordering, assume $j = 1$ and write $q_1 = u_1 p_1$ for some $u_1 \in R$. Since q_1 is irreducible, u_1 or p_1 must be a unit, but p_1 cannot be

0.2. IRREDUCIBLE POLYNOMIALS

a unit because it is irreducible. Thus, u_1 is a unit. Canceling p_1 from both sides of (0.23), we obtain
$$p_2 \cdots p_\ell = u_1 q_2 \cdots q_m.$$

Since $p_2 \nmid u_1$, as otherwise p_2 would be a unit itself, we can repeat the above argument to see that, after possibly reordering, there is a unit $u_2 \in R$ such that $q_2 = u_2 p_2$ and
$$p_3 \cdots p_\ell = u_1 u_2 q_3 \cdots q_m.$$

Continuing this process for ℓ steps, we see that there are units $u_1, \ldots, u_\ell$ such that $q_i = u_i p_i$ and
$$1 = u_1 \cdots u_\ell q_{\ell+1} \cdots q_m.$$

Since each q_j is irreducible, and thus not a unit, we conclude that $\ell = m$, finishing the proof that R is a UFD. □

Since we know that $K[x_1, \ldots, x_n]$ is a FD, Proposition 0.22 provides a strategy for proving that $K[x_1, \ldots, x_n]$ is a UFD: it suffices to show that every irreducible polynomial is prime. In order to accomplish this, we need a more robust algebraic foundation upon which we can work with these ideas. In order to build this foundation, we first turn to a discussion of ideals and quotients (Section 0.3), prime and maximal ideals (Section 0.4), and the special case of single-variable polynomials (Section 0.5). We then return to unique factorization of polynomials in Section 0.6.

Exercises for Section 0.2

0.2.1 Prove that the set of units $R^* \subseteq R$ is a group under multiplication.

0.2.2 Prove that $K[x_1, \ldots, x_n]^* = K^* = K \setminus \{0\}$.

0.2.3 Using the fact that $a^2 + 1 \neq 0$ for any real number $a \in \mathbb{R}$, prove that $x^2 + 1$ does not factor into linear terms in $\mathbb{R}[x]$.

0.2.4 Let $f \in K[x]$ be a polynomial of degree 2 or 3. Prove that f is irreducible if and only if there does not exist an element $a \in K$ such that $f(a) = 0$.

0.2.5 Show that the previous problem fails for polynomials of degree 4 by giving an explicit example of a reducible polynomial in $\mathbb{R}[x]$ that has no zeros.

0.2.6 Using degree arguments, prove that $x^2 + y^2 - 1$ is irreducible in $\mathbb{C}[x, y]$.

0.2.7 Suppose that $p \in R$ is prime. If $p \mid a_1 \cdots a_n$, prove that $p \mid a_i$ for some i.

0.2.8 (a) Describe the units in $\mathbb{Z}[x]$.
    (b) Give an example of a linear polynomial in $\mathbb{Z}[x]$ that is reducible.

0.2.9 Explain why every field is a UFD.

0.2.10 Consider the ring $R = \{f \in \mathbb{Q}[x] \mid f(0) \in \mathbb{Z}\} \subseteq \mathbb{Q}[x]$.
    (a) Prove that R is an integral domain.
    (b) Prove that the element $x \in R$ is neither a unit nor irreducible.
    (c) Prove that $x \in R$ cannot be factored into irreducibles, so R is not a FD.

Section 0.3 Ideals and quotients

One of the central constructions in ring theory is that of taking quotients by ideals. In this section, we review the quotient construction, along with the most fundamental result regarding quotients: the First Isomorphism Theorem. Ideals and quotient rings are standard topics in a first course in ring theory, so the proofs in this section are left as exercises. However, we include a number of instructive examples to illustrate how to work with ideals and quotients in the context of polynomial rings.

Our discussion of quotient rings begins with the notion of an ideal.

0.24 DEFINITION *Ideals*

An *ideal* of R is a nonempty subset $I \subseteq R$ satisfying two properties:

(i) $a, b \in I \implies a - b \in I$, and

(ii) $a \in I$ and $r \in R \implies ra \in I$.

In words, a nonempty subset of a ring is an ideal if (i) it is closed under subtraction and (ii) it absorbs multiplication. In many contexts, the easiest way to describe an ideal is by specifying a *generating set*. This method of describing ideals is made precise in the next definition.

0.25 DEFINITION *Generating sets of ideals*

If $A \subseteq R$ is a subset, then the *ideal generated by* A, denoted $\langle A \rangle$, is the set of all R-linear combinations of elements of A:

$$\langle A \rangle = \left\{ \sum_{i=1}^{n} r_i a_i \;\middle|\; n \in \mathbb{N},\, r_1, \ldots, r_n \in R,\, a_1, \ldots, a_n \in A \right\} \subseteq R.$$

If $A = \{a_1, \ldots, a_n\}$ is finite, we write $\langle A \rangle = \langle a_1, \ldots, a_n \rangle$. An ideal that can be generated by a single element is called *principal*.

The reader is encouraged to verify that the set $\langle A \rangle$ is, in fact, an ideal. Moreover, it is the smallest ideal of R that contains the set A (Exercise 0.3.4). Let us consider an explicit example of an ideal in $R[x_1, \ldots, x_n]$, along with a set of generators.

0.26 EXAMPLE Generators for an ideal in $R[x_1, \ldots, x_n]$

Let $I \subseteq R[x_1, \ldots, x_n]$ be the subset of all polynomials whose constant term is zero. The set of such polynomials is closed under subtraction and absorbs multiplication, so I is an ideal. Moreover, a polynomial is in I if and only if you can factor out at least one variable from each term, and this implies that $I = \langle x_1, \ldots, x_n \rangle$.

By definition, a principal ideal $\langle a \rangle$ consists of all multiples of its generator:

$$\langle a \rangle = \{ ra \mid r \in R \} = \{ b \in R \mid a \text{ divides } b \}.$$

Principal ideals are especially nice, but not all ideals in polynomial rings are principal; for instance, the ideal in Example 0.26 is not principal (Exercise 0.3.5).

0.3. IDEALS AND QUOTIENTS

Given an ideal $I \subseteq R$, define a relation on R by

$$r \sim s \iff r - s \in I.$$

Using Condition (i) in Definition 0.24, it can be shown that $\sim$ is an equivalence relation (Exercise 0.3.6). The equivalence class of an element $r \in R$ under this equivalence relation is called a *coset*, denoted

$$[r] = r + I = \{s \in R \mid s \sim r\} \subseteq R.$$

We typically prefer the notation $[r]$ when the ideal I is understood from context, but use the notation $r + I$ when it is useful to emphasize the role of I. Note that

$$[r] = [s] \iff r - s \in I.$$

0.27 EXAMPLE Cosets

Consider the principal ideal $I = \langle x \rangle \subseteq R[x]$. Notice that $[x + 2] = [x^2 + 2]$ because

$$(x + 2) - (x^2 + 2) = x - x^2 \in \langle x \rangle.$$

More generally, $[f(x)] = [g(x)]$ if and only if $f(0) = g(0) \in R$. In other words, the collection of cosets is in natural bijection with the ring R via the identification

$$[f(x)] \longmapsto f(0) \in R.$$

In the previous example, we saw that the collection of cosets is in bijection with the coefficient ring R, and can therefore be given the structure of a ring. The next definition describes how to endow the set of cosets with a ring structure for any ideal.

> ### 0.28 DEFINITION Quotient rings
>
> Let $I \subseteq R$ be an ideal. The *quotient ring* R/I is the set of cosets
>
> $$R/I = \{[r] \mid r \in R\}.$$
>
> Coset addition and multiplication are defined by
>
> $$[r] + [s] = [r + s] \quad \text{and} \quad [r][s] = [rs].$$

It is not obvious that addition and multiplication in R/I are well-defined. In particular, since different elements can be chosen to represent the same coset, it is necessary to verify that the operations are independent of the choice of coset representatives. This verification follows from Conditions (i)

> In the quotient R/I, the additive identity is $0 = [0]$ and the multiplicative identity is $1 = [1]$. Since $[a] = 0$ if and only if $a \in I$, the quotienting process can be thought of as "setting elements of I equal to zero."

and (ii) in Definition 0.24; we leave the computation to the reader (Exercise 0.3.8).

Notice that, for any ideal $I \subseteq R$, there is a ring homomorphism

$$\pi : R \to R/I$$
$$a \mapsto [a].$$

This homomorphism is called the *quotient homomorphism*.

0.29 EXAMPLE   Quotient ring computations

Consider the principal ideal $\langle y - x^2 \rangle \subseteq R[x,y]$. By definition of addition and multiplication in the quotient ring, we see that

$$[y] - [x]^2 = [y - x^2] = 0 \in \frac{R[x,y]}{\langle y - x^2 \rangle}.$$

In particular, this implies that $[y] = [x]^2$. Taking this logic a step farther, we see, for example, that

$$[y]^2 = [x]^4 \text{ and } [x]^2[y]^3 = [x]^8.$$

In general, for any polynomial $f(x,y) \in R[x,y]$, observe that

$$[f(x,y)] = f([x],[y]) = f([x],[x]^2) = [f(x,x^2)].$$

In other words, when we form the quotient by the ideal $\langle y - x^2 \rangle$, we are able to treat $y - x^2$ as the zero element and replace every occurrence of y with x^2. In particular, every element of the quotient can be represented in the variable x alone. As we will see in Example 0.32 below, the quotient ring in this example is isomorphic to $R[x]$.

The most important application of the quotient construction is that it provides a tool for turning homomorphisms into isomorphisms. The proofs of the three points in the following fundamental result are left to the reader (Exercise 0.3.9).

0.30 THEOREM   *First Isomorphism Theorem for rings*

If $\varphi : R \to S$ is a ring homomorphism, then

(i) $\mathrm{im}(\varphi) = \{s \in S \mid s = \varphi(r) \text{ for some } r \in R\}$ is a subring of S,

(ii) $\ker(\varphi) = \{r \in R \mid \varphi(r) = 0\}$ is an ideal of R, and

(iii) the function

$$[\varphi] : \frac{R}{\ker(\varphi)} \to \mathrm{im}(\varphi)$$
$$[r] \mapsto [\varphi(r)]$$

is a well-defined ring isomorphism.

We close this section with a few detailed examples that demonstrate applications of the First Isomorphism Theorem in the context of polynomial rings. For more examples, we direct the reader to the exercises.

0.3. IDEALS AND QUOTIENTS

0.31 EXAMPLE $\langle x_1, \ldots, x_n \rangle \subseteq R[x_1, \ldots, x_n]$

Consider the ring homomorphism

$$\varphi : R[x_1, \ldots, x_n] \to R$$
$$f(x_1, \ldots, x_n) \mapsto f(0, \ldots, 0).$$

Notice that $f(0, \ldots, 0)$ is simply the constant term of f. Some reflection should convince the reader that φ is surjective and that the kernel of φ consists of all polynomials with a vanishing constant term. Thus, by Example 0.26,

$$\ker(\varphi) = \langle x_1, \ldots, x_n \rangle \subseteq R[x_1, \ldots, x_n],$$

and by the First Isomorphism Theorem, we conclude that $[\varphi]$ gives an isomorphism:

$$\frac{R[x_1, \ldots, x_n]}{\langle x_1, \ldots, x_n \rangle} \cong R.$$

Using similar arguments, this example can be generalized in a number of ways. See, for example, Exercises 0.3.11 and 0.3.12.

0.32 EXAMPLE $\langle y - x^2 \rangle \subseteq R[x, y]$

This example verifies that

$$\frac{R[x, y]}{\langle y - x^2 \rangle} \cong R[x].$$

Based on the computations in Example 0.29, this should make sense: we can replace every occurrence of y with x^2 and write every coset in terms of x alone. To make this argument precise using the First Isomorphism Theorem, it suffices to construct a surjective homomorphism $\varphi : R[x, y] \to R[x]$ with kernel $\langle y - x^2 \rangle$.

Define

$$\varphi : R[x, y] \to R[x]$$
$$f(x, y) \mapsto f(x, x^2),$$

which the reader can verify is a surjective ring homomorphism. Thus, it remains to prove that $\langle y - x^2 \rangle = \ker(\varphi)$. We prove both inclusions.

($\subseteq$) Suppose $f(x, y) \in \langle y - x^2 \rangle$. This means that there exists $g(x, y) \in R[x, y]$ such that

$$f(x, y) = (y - x^2) g(x, y).$$

Evaluating φ at $f(x, y)$, we see that

$$\varphi(f(x, y)) = (x^2 - x^2) g(x, x^2) = 0,$$

so $f(x, y) \in \ker(\varphi)$.

($\supseteq$) Suppose $f(x, y) \in \ker(\varphi)$. By the computations in Example 0.29, we see that $[f(x, y)] = [f(x, x^2)]$ in the quotient ring $R[x, y] / \langle y - x^2 \rangle$, implying that

(0.33) $$f(x, y) - f(x, x^2) \in \langle y - x^2 \rangle.$$

Moreover, since $f(x,y) \in \ker(\varphi)$, we know that

(0.34) $\qquad 0 = \varphi(f(x,y)) = f(x,x^2) \in R[x] \subseteq R[x,y].$

Combining (0.34) and (0.33), we conclude that

$$f(x,y) = f(x,y) - f(x,x^2) \in \langle y - x^2 \rangle.$$

Exercise 0.3.13 provides a useful generalization of this result.

Exercises for Section 0.3

0.3.1 Let $I \subseteq R$ be an ideal. Prove that $I = R$ if and only if I contains a unit.

0.3.2 Prove that ideals contain 0 and are closed under addition.

0.3.3 Prove that the only ideals of a field K are $\{0\}$ and K.

0.3.4 Let $A \subseteq R$ be a subset. Prove the following.
   (a) The set $\langle A \rangle$ is an ideal of R.
   (b) If $I \subseteq R$ is any ideal containing A, then $\langle A \rangle \subseteq I$.

0.3.5 Prove that $\langle x_1, \ldots, x_n \rangle \subseteq K[x_1, \ldots, x_n]$ is not a principal ideal if $n > 1$.

0.3.6 Let $I \subseteq R$ be an ideal and consider the relation on R given by

$$r \sim s \iff r - s \in I.$$

   (a) Prove that $\sim$ is *reflexive*: $r \sim r$ for all $r \in R$.
   (b) Prove that $\sim$ is *symmetric*: $r \sim s$ implies $s \sim r$.
   (c) Prove that $\sim$ is *transitive*: if $r \sim s$ and $s \sim t$, then $r \sim t$.
   Thus, $\sim$ is an equivalence relation.

0.3.7 Given an ideal $I \subseteq R$, prove that $[r] = \{r + a \mid a \in I\}$, justifying the notation $[r] = r + I$.

0.3.8 Let $I \subseteq R$ be an ideal and let $r_1, r_2, s \in R$. Prove the following.
   (a) If $r_1 \sim r_2$, then $r_1 + s \sim r_2 + s$.
   (b) If $r_1 \sim r_2$, then $r_1 s \sim r_2 s$.
   Thus, addition and multiplication in the quotient ring R/I are well-defined.

0.3.9 Prove the First Isomorphism Theorem for rings.

0.3.10 Let $\varphi : R \to S$ be a ring homomorphism.
   (a) If $I \subseteq S$ is an ideal, prove that $\varphi^{-1}(I) \subseteq R$ is an ideal.
   (b) If φ is surjective and $I \subseteq R$ is an ideal, prove that $\varphi(I) \subseteq S$ is an ideal.
   (c) Give an example of a nonsurjective ring homomorphism $\varphi : R \to S$ and an ideal $I \subseteq R$ such that $\varphi(I)$ is not an ideal.

0.3.11 Prove that
$$\frac{R[x_1, \ldots, x_n]}{\langle x_{k+1}, \ldots, x_n \rangle} \cong R[x_1, \ldots, x_k].$$

0.3. IDEALS AND QUOTIENTS

0.3.12 Let $a_1, \ldots, a_n$ be elements of R. Prove that

$$\frac{R[x_1, \ldots, x_n]}{\langle x_1 - a_1, \ldots, x_n - a_n \rangle} \cong R.$$

0.3.13 Let $f_1, \ldots, f_k \in R[x_1, \ldots, x_n]$ and $g \in R[x_1, \ldots, x_{n-1}]$. Consider the ring homomorphism

$$\varphi : R[x_1, \ldots, x_n] \to R[x_1, \ldots, x_{n-1}]$$
$$f(x_1, \ldots, x_{n-1}, x_n) \mapsto f(x_1, \ldots, x_{n-1}, g).$$

Use the First Isomorphism Theorem to prove that

$$\frac{R[x_1, \ldots, x_n]}{\langle f_1, \ldots, f_k, x_n - g \rangle} \cong \frac{R[x_1, \ldots, x_{n-1}]}{\langle \varphi(f_1), \ldots, \varphi(f_k) \rangle}.$$

(Notice that Examples 0.31 and 0.32 and Exercise 0.3.11 and 0.3.12 all follow from this result. This result essentially says that quotienting by $x_n - g$ is equivalent to replacing all occurrences of x_n with g.)

0.3.14 Prove that every element of the quotient ring

$$\frac{R[x, y]}{\langle x^2 + y^2 - 1 \rangle}$$

can be represented uniquely by a polynomial of the form $f(x) + yg(x)$ where $f(x), g(x) \in R[x]$.

0.3.15 Consider the quotient ring

$$R = \frac{K[x, y]}{\langle x^2 - y^3 \rangle}.$$

(a) Prove that every element of R can be represented by a polynomial of the form $f(y) + xg(y)$ where $f(y), g(y) \in K[y]$.
(b) Prove that R is an integral domain.
(c) Prove that R is a factorization domain.
(d) Prove that $[x], [y] \in R$ are irreducible.
(e) Prove that $[x], [y] \in R$ are not prime.
(f) Find an element in R that has two distinct irreducible factorizations.

Section 0.4 Prime and maximal ideals

Given an ideal $I \subseteq R$, how do the ring-theoretic properties of R/I translate to properties of the ideal I? For example, if R/I is an integral domain or a field, what does this tell us about I? In this section, we discuss these questions through the introduction of two special types of ideals—prime and maximal ideals—that are central to ring theory and algebraic geometry. We provide several examples of prime and maximal ideals in the context of polynomial rings, and we close with an application of how these notions can be used to study irreducible factorizations.

> **0.35 DEFINITION** *Prime and maximal ideals*
>
> An ideal $I \subseteq R$ is *prime* if $I \neq R$ and, for all $a, b \in R$,
>
> $$ab \in I \implies a \in I \text{ or } b \in I.$$
>
> An ideal $I \subseteq R$ is *maximal* if $I \neq R$ and there does not exist an ideal $J \subseteq R$ such that
>
> $$I \subsetneq J \subsetneq R.$$

0.36 EXAMPLE *Eponymous example of prime ideals*

It follows from the definitions (Exercise 0.4.1) that a nonzero element $p \in R$ is prime if and only if the principal ideal $\langle p \rangle \subseteq R$ is a prime ideal. In particular, $\langle 2 \rangle$, $\langle 3 \rangle$, $\langle 5 \rangle$, and $\langle 7 \rangle$ are all prime ideals in $\mathbb{Z}$, but $\langle 1 \rangle$, $\langle 4 \rangle$, $\langle 6 \rangle$, and $\langle 9 \rangle$ are not.

0.37 EXAMPLE $\langle x \rangle \subseteq K[x]$ *is maximal*

To prove that $\langle x \rangle$ is maximal, suppose that $J \subseteq K[x]$ is an ideal and $\langle x \rangle \subsetneq J$; we prove that $J = K[x]$.

Since $\langle x \rangle$ consists of all polynomials without a constant term, J must contain at least one polynomial with a nonzero constant term:

$$f = a_0 + a_1 x + a_2 x^2 + \cdots + a_n x^n \in J \quad \text{and} \quad a_0 \neq 0.$$

As $\langle x \rangle$ is a subset of J, all polynomials without a constant term are elements of J. This implies that

$$g = a_1 x + a_2 x^2 + \cdots + a_n x^n \in J.$$

Since ideals are closed under subtraction, we obtain $a_0 = f - g \in J$, and since a_0 is a unit in $K[x_1, \ldots, x_n]$, this implies that $J = K[x_1, \ldots, x_n]$.

The next result is a useful characterization of prime and maximal ideals in terms of the ring-theoretic properties of their quotients.

0.4. PRIME AND MAXIMAL IDEALS

0.38 PROPOSITION *Quotients by prime and maximal ideals*

Let $I \subseteq R$ be an ideal.

(i) The ideal I is prime if and only if R/I is an integral domain.

(ii) The ideal I is maximal if and only if R/I is a field.

PROOF We prove the forward implication for each property, and we leave the reverse implications to Exercises 0.4.4 and 0.4.5.

We prove the forward implication in (i) by contrapositive. Assume that R/I is not an integral domain, so that there exist nonzero cosets $[a], [b] \in R/I$ such that $[a][b] = 0$. By definition of the quotient ring, this implies that $a, b \notin I$ but $ab \in I$. In other words, I is not a prime ideal.

To prove the forward implication of (ii), suppose I is maximal and let $[a] \in R/I$ be a nonzero coset, meaning that $a \notin I$. We must show that $[a]$ has a multiplicative inverse. Consider the ideal generated by I and a:

$$J = \langle I, a \rangle.$$

Since $a \notin I$, $I \subsetneq J$. By maximality of I, this implies that $J = R$, so $1 \in J$. By definition of generating sets of ideals, we can write 1 as

$$1 = r_1 b_1 + \cdots + r_n b_n + sa$$

for some $b_1, \ldots, b_n \in I$ and $r_1, \ldots, r_n, s \in R$. Since I is itself an ideal, this implies that $1 = b + sa$ where $b = r_1 b_1 + \cdots + r_n b_n \in I$ and $s \in R$. Therefore, in the quotient ring R/I,

$$1 = [1] = [b + sa] = [b] + [s][a] = [s][a].$$

Thus, $[a]$ has a multiplicative inverse. □

Since every field is an integral domain, we obtain the following immediate consequence of Proposition 0.38.

0.39 COROLLARY *Maximal implies prime*

Every maximal ideal is a prime ideal.

0.40 EXAMPLE $\langle x^2 \rangle \subseteq R[x]$ *is not prime*

Notice that

$$[x] \in \frac{R[x]}{\langle x^2 \rangle}$$

is a nonzero element of the quotient ring, but $[x][x] = [x^2] = 0$. Thus, $[x]$ is a zero divisor. Since the quotient ring $R[x]/\langle x^2 \rangle$ contains a zero divisor, it is not an integral domain. From Proposition 0.38, we conclude that $\langle x^2 \rangle$ is not a prime ideal. In particular, this implies that $\langle x^2 \rangle$ is not maximal, which can also be verified directly:

$$\langle x^2 \rangle \subsetneq \langle x \rangle \subsetneq K[x].$$

0.41 EXAMPLE $\langle y - x^2 \rangle \subseteq K[x, y]$ is prime but not maximal

By Example 0.32,
$$K[x,y]/\langle y - x^2 \rangle \cong K[x].$$
Since $K[x]$ is an integral domain, but not a field, we conclude that $\langle y - x^2 \rangle$ is a prime ideal, but not a maximal ideal.

0.42 EXAMPLE $\langle x_1, \ldots, x_n \rangle \subseteq K[x_1, \ldots, x_n]$ is maximal

By Example 0.31,
$$K[x_1, \ldots, x_n]/\langle x_1, \ldots, x_n \rangle \cong K.$$
Since K is a field, we conclude that $\langle x_1, \ldots, x_n \rangle \subseteq K[x_1, \ldots, x_n]$ is a maximal ideal.

0.43 EXAMPLE $\langle x^2 + 1 \rangle \subseteq \mathbb{R}[x]$ versus $\langle x^2 + 1 \rangle \subseteq \mathbb{C}[x]$

Consider the quotient
$$\frac{\mathbb{R}[x]}{\langle x^2 + 1 \rangle}.$$
Observe that $[x]$ satisfies $[x]^2 = -[1] = -1$. In other words, $[x]$ is a square root of -1. We know of another ring that has a square root of -1; namely, the field of complex numbers $\mathbb{C}$.

Consider the function
$$\varphi : \mathbb{R}[x] \to \mathbb{C}$$
$$f(x) \mapsto f(i).$$
It can be shown (Exercise 0.4.6) that φ is a surjective ring homomorphism and that $\ker(\varphi) = \langle x^2 + 1 \rangle$. Thus, by the First Isomorphism Theorem, we conclude that the quotient ring is isomorphic to the field of complex numbers:
$$\frac{\mathbb{R}[x]}{\langle x^2 + 1 \rangle} \cong \mathbb{C}.$$
Therefore, $\langle x^2 + 1 \rangle \subseteq \mathbb{R}[x]$ is a maximal ideal.

Now, consider the quotient ring
$$\frac{\mathbb{C}[x]}{\langle x^2 + 1 \rangle}.$$
In this case, neither of the elements $[x - i]$ nor $[x + i]$ is zero, but their product is:
$$[x - i][x + i] = [x^2 + 1] = 0 \in \frac{\mathbb{C}[x]}{\langle x^2 + 1 \rangle}.$$
Thus, the quotient ring over $\mathbb{C}$ contains zero divisors—it is not an integral domain. Therefore, $\langle x^2 + 1 \rangle \subseteq \mathbb{C}[x]$ is neither prime nor maximal.

We close this section with an application that illustrates how the notions of prime and maximal ideals can be used to help study questions concerning irreducible and prime elements. We begin with a definition of a particularly nice type of ring.

0.4. PRIME AND MAXIMAL IDEALS

> **0.44 DEFINITION** *Principal ideal domain*
>
> An integral domain R is called a *principal ideal domain* (*PID*) if every ideal in R is principal.

For example, the ring of integers $\mathbb{Z}$ is a PID, as the reader is encouraged to verify in Exercise 0.4.7. As Exercise 0.3.5 shows, the polynomial ring $K[x_1, \ldots, x_n]$ is not a PID for $n > 1$ because $\langle x_1, \ldots, x_n \rangle$ is not a principal ideal. However, as we will see in the next section, the single-variable polynomial ring $K[x]$ is a PID.

The next result uses the notions of prime and maximal ideals to prove that every PID is a UFD. In particular, along with Exercise 0.4.7, this provides a self-contained proof of the uniqueness of prime factorization in $\mathbb{Z}$.

> **0.45 PROPOSITION** *PIDs are UFDs*
>
> Every principal ideal domain is a unique factorization domain.

PROOF Suppose that R is a PID. We begin by proving that R is a FD. Suppose, toward a contradiction, that there exists a nonzero, nonunit $a \in R$ that does not factor as a product of irreducible elements. This implies that a is not irreducible so we can factor it as $a = a_1 b_1$ where neither a_1 nor b_1 is a unit. If both a_1 and b_1 factored as products of irreducible elements, then so would a. Therefore, without loss of generality, assume a_1 does not factor as a product of irreducible elements and write $a_1 = a_2 b_2$ where neither a_2 nor b_2 is a unit. As before, we can assume, without loss of generality, that a_2 does not factor as a product of irreducible elements.

Continuing the above process, we recursively construct a sequence

$$(a = a_0, a_1, a_2, a_3, \ldots)$$

where, for every $i \geq 0$, we have $a_i = a_{i+1} b_{i+1}$ for some nonunit b_{i+1}. It follows from Exercise 0.4.8 that the ideals $\langle a_i \rangle$ fit into a chain of strict containment:

$$\langle a_0 \rangle \subsetneq \langle a_1 \rangle \subsetneq \langle a_2 \rangle \subsetneq \langle a_3 \rangle \subsetneq \cdots.$$

The union of the above ideals is an ideal by Exercise 0.4.9. Since R is a PID, choose $c \in R$ such that

$$\langle c \rangle = \bigcup_{i=0}^{\infty} \langle a_i \rangle.$$

Then c must be in $\langle a_n \rangle$ for some n, which implies that $\langle c \rangle = \langle a_k \rangle$ for all $k \geq n$. This contradicts the strict containment $\langle a_n \rangle \subsetneq \langle a_{n+1} \rangle$.

Now, to prove that R is a UFD, it suffices, by Proposition 0.22, to prove that every irreducible element of R is prime. Let $p \in R$ be irreducible. To prove that p is prime, it suffices to prove that $\langle p \rangle$ is maximal. Toward this end, suppose $I \subseteq R$ is an ideal such that $\langle p \rangle \subseteq I \subseteq R$. We must prove that either $I = \langle p \rangle$ or $I = R$.

Since R is a PID, $I = \langle r \rangle$ for some $r \in R$. Since $p \in \langle p \rangle \subseteq \langle r \rangle$, it follows that $p = rs$ for some $s \in R$. By the irreducibility of p, either r or s is a unit. But r being a unit implies that $I = R$ and s being a unit implies that $I = \langle p \rangle$. □

Exercises for Section 0.4

0.4.1 Prove that a nonzero element $p \in R$ is prime if and only if the principal ideal $\langle p \rangle \subseteq R$ is a prime ideal.

0.4.2 Prove that $\langle y, y - x^2 \rangle \subseteq K[x, y]$ is not prime, even though both generators are prime elements.

0.4.3 Prove that the zero ideal $\langle 0 \rangle \subseteq R$ is prime if and only if R is an integral domain.

0.4.4 Let $I \subseteq R$ be an ideal such that R/I is an integral domain. Prove that I is a prime ideal.

0.4.5 Let $I \subseteq R$ be an ideal such that R/I is a field. Prove that I is a maximal ideal.

0.4.6 Prove that
$$\frac{\mathbb{R}[x]}{\langle x^2 + 1 \rangle} \cong \mathbb{C}.$$

0.4.7 Prove that $\mathbb{Z}$ is a PID. (Hint: If $I \subseteq \mathbb{Z}$ is an ideal, let n be the smallest positive integer in I. Prove that $I = \langle n \rangle$.)

0.4.8 Let R be an integral domain. If $a = bc$ and $c \notin R^*$, prove that $\langle a \rangle \subsetneq \langle b \rangle$.

0.4.9 Let $I_1 \subseteq I_2 \subseteq I_3 \subseteq \cdots$ be an ascending chain of ideals of a ring R. Prove that
$$I = \bigcup_{k=1}^{\infty} I_k$$
is an ideal of R.

0.4.10 Let R be a PID. Prove that every nonzero prime ideal in R is maximal.

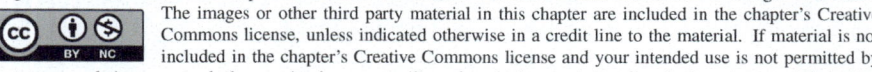

Open Access This chapter is licensed under the terms of the Creative Commons Attribution-NonCommercial 4.0 International License (http://creativecommons.org/licenses/by-nc/4.0/), which permits any noncommercial use, sharing, adaptation, distribution and reproduction in any medium or format, as long as you give appropriate credit to the original author(s) and the source, provide a link to the Creative Commons license and indicate if changes were made.

The images or other third party material in this chapter are included in the chapter's Creative Commons license, unless indicated otherwise in a credit line to the material. If material is not included in the chapter's Creative Commons license and your intended use is not permitted by statutory regulation or exceeds the permitted use, you will need to obtain permission directly from the copyright holder.

Section 0.5 Single-variable polynomials

In this section, we consider rings of single-variable polynomials with coefficients in a field. We introduce a number of fundamental results, concluding with the fact that $K[x]$ is a UFD, which will serve as the starting place to prove that $K[x_1, \ldots, x_n]$ is a UFD in the next section. The results in this section are direct consequences of the *polynomial division algorithm*, which is the polynomial analogue of the long division algorithm that many students learn in grade school.

> **0.46 Theorem** *Polynomial division algorithm*
>
> For any $f, g \in K[x]$ with $g \neq 0$, there exist unique polynomials $q, r \in K[x]$ such that
> $$f = qg + r$$
> with $r = 0$ or $\deg(r) < \deg(g)$.

The polynomials q and r satisfying the conditions in the division algorithm are called the *quotient* and *remainder* of f divided by g. Notice that the remainder is zero if and only if $g \mid f$.

Before presenting a proof of the polynomial division algorithm, we work through a detailed example that illustrates the step-by-step process for computing the quotient and remainder. The reader is encouraged to work out several additional examples in Exercise 0.5.1.

0.47 Example Polynomial long division

Consider polynomials $f = x^3 + x^2 + 1$ and $g = x - 2$ in $\mathbb{Q}[x]$. We compute the quotient and remainder of f divided by g.

Step 1: Subtract the unique multiple of g that cancels the leading term of f:
$$f - x^2 g = 3x^2 + 1.$$

Step 2: Subtract another multiple of g to cancel the leading term of $3x^2 + 1$:
$$f - x^2 g - 3xg = 6x + 1.$$

Step 3: Subtract another multiple of g to cancel the leading term of $6x + 1$:
$$f - x^2 g - 3xg - 6g = 13.$$

Final step: Since the polynomial 13 has degree strictly smaller than g, we stop here. Rearranging terms, we see that
$$f = qg + r$$
where $q = x^2 + 3x + 6$ and $r = 13$.

Notice that each step decreased the degree of the polynomial appearing in the right-hand side, ensuring that the process would eventually terminate.

PROOF OF THEOREM 0.46   We begin by proving that quotients and remainders exist, then we prove that they are unique.

Fix a nonzero polynomial g. In order to show that quotients and remainders exist for any f divided by this particular g, we proceed by induction on $\deg(f)$ (in the case where $f = 0$, set $q = r = 0$).

(**Base case**) Suppose $\deg(f) = 0$. If $\deg(g) > 0$, set $q = 0$ and $r = f$. If $\deg(g) = 0$, then g is a nonzero constant. Since K is a field, g has a multiplicative inverse. Set $q = fg^{-1}$ and $r = 0$. The reader can directly verify that these choices of q and r satisfy the conditions in the division algorithm.

(**Induction step**) Let $f \in K[x]$ be a polynomial of degree $d > 0$. If $\deg(g) > d$, set $q = 0$ and $r = f$. If $\deg(g) \leq d$, set $k = \deg(g)$ and let a_d and b_k be the leading coefficients of f and g, respectively. By construction, the polynomial

$$\tilde{f} = f - a_d b_k^{-1} x^{d-k} g$$

is zero or $\deg(\tilde{f}) < d$. If $\tilde{f} = 0$, set $q = a_d b_k^{-1} x^{d-k}$ and $r = 0$. Otherwise, by the induction hypothesis, choose $\tilde{q}$ and $\tilde{r}$ such that $\tilde{r} = 0$ or $\deg(\tilde{r}) < \deg(g)$ and

$$\tilde{f} = \tilde{q} g + \tilde{r},$$

and set $q = \tilde{q} + a_d b_k^{-1} x^{d-k}$ and $r = \tilde{r}$. In each case, the reader can check that q and r satisfy the conditions in the division algorithm, completing the induction step and the proof of existence.

It remains to prove uniqueness. If q, r and $\tilde{q}, \tilde{r}$ both satisfy the conclusion of the division algorithm, then

$$f = qg + r = \tilde{q}g + \tilde{r} \implies g(\tilde{q} - q) = (r - \tilde{r}).$$

By assumption, either $r - \tilde{r} = 0$ or $\deg(r - \tilde{r}) < \deg(g)$. In the latter case, additivity of degree implies that $\deg(\tilde{q} - q) < 0$, a contradiction. Therefore, it must be the case that $r = \tilde{r}$. Since $g \neq 0$ and $K[x]$ is an integral domain, it then follows that $q = \tilde{q}$. □

With the division algorithm in hand, we now prove a slew of important consequences. The first two applications concern zeros of single-variable polynomials.

0.48 COROLLARY   *Factor theorem*

If $f(x) \in K[x]$ and $a \in K$, then $f(a) = 0$ if and only if $(x - a) \mid f(x)$.

PROOF   Using the division algorithm, write

$$f(x) = (x - a)q + r$$

for some $q, r \in K[x]$ such that $r = 0$ or $\deg(r) < \deg(x - a) = 1$. In either case, the remainder must be a constant: $r \in K$. Evaluating at $x = a$, we see that

$$f(a) = (a - a)q(a) + r = r.$$

The result then follows from the observation that $(x - a) \mid f(x)$ if and only if the remainder of $f(x)$ divided by $x - a$, which we just proved is $f(a)$, is zero. □

0.49 EXAMPLE $x^n - 1 \in K[x]$

Consider the polynomial $f(x) = x^n - 1 \in K[x]$. Since $f(1) = 1^n - 1 = 0$, Corollary 0.48 implies that $(x-1) \mid (x^n - 1)$. Indeed, by multiplying out the right-hand side of the following equation, one checks that

$$x^n - 1 = (x-1)(x^{n-1} + x^{n-2} + \cdots + x + 1).$$

0.50 COROLLARY *Finite zeros theorem*

If $f(x) \in K[x]$ is a nonzero polynomial of degree d, then there are at most d values $a \in K$ such that $f(a) = 0$.

PROOF We proceed by induction on d.

(**Base case**) Suppose $d = 0$. Then $f = b$ for some nonzero constant $b \in K$. Thus, $f(a) = b \neq 0$ for all $a \in K$, so f does not have any zeros.

(**Induction step**) Let f be a polynomial of degree $d > 0$. If f does not have any zeros, then we are done. If f has at least one zero $a \in K$, then Corollary 0.48 implies that

$$f = (x-a)g$$

for some $g \in K[x]$. By additivity of degree, $\deg(g) = d - 1$, so the induction hypothesis implies that g has at most $d - 1$ zeros. Since every zero of f other than a must also be a zero of g, we conclude that f has at most d zeros. $\square$

0.51 EXAMPLE $x^n - 1 \in \mathbb{C}[x]$

By Corollary 0.50, the polynomial $x^n - 1 \in \mathbb{C}[x]$ has at most n zeros. For $j = 1, \ldots, n$, consider the complex number

> It may be helpful for this example to recall Euler's formula:
> $$e^{ix} = \cos(x) + i\sin(x).$$

$$a_j = e^{\frac{2\pi i}{n} j}.$$

Since

$$(a_j)^n = e^{2\pi i j} = 1^j = 1,$$

we see that a_j is a zero of $x^n - 1$ for every j. Since $\{a_1, \ldots, a_n\}$ is a set of n distinct zeros, we conclude that these must be all of the zeros of $x^n - 1$.

The following result is another important consequence of the division algorithm.

0.52 COROLLARY *$K[x]$ is a PID*

The single-variable polynomial ring $K[x]$ is a principal ideal domain.

PROOF Let $I \subseteq K[x]$ be an ideal and define the set

$$D = \{\deg(f) \mid f \in I \text{ and } f \neq 0\} \subseteq \mathbb{N}.$$

If $D = \emptyset$, then I is the zero ideal $\langle 0 \rangle$, thus principal. If $D \neq \emptyset$, then, by the well-ordering principle, D contains a least element, call it d. Let $f \in I$ be a nonzero element such that $\deg(f) = d$. To prove that I is principal, we show that $I = \langle f \rangle$.

Since $f \in I$, we obtain the inclusion $\langle f \rangle \subseteq I$ for free. To prove the other inclusion, suppose $g \in I$. Applying the division algorithm, we have

$$g = qf + r$$

with $r = 0$ or $\deg(r) < \deg(f) = d$. Since $f, g \in I$, it follows that $r \in I$. If $r \neq 0$, then $\deg(r) \in D$ and $\deg(r) < d$, contradicting that $d = \min(D)$. Thus, $r = 0$, from which we conclude that $g \in \langle f \rangle$, completing the proof. □

We close this section with the important result that $K[x]$ is a UFD, which is an immediate consequence of Corollary 0.52 and Proposition 0.45.

0.53 COROLLARY $K[x]$ *is a UFD*

The single-variable polynomial ring $K[x]$ is a unique factorization domain.

Exercises for Section 0.5

0.5.1 Compute the quotient and remainder for the following pairs in $\mathbb{Q}[x]$.
   (a) $f = 2x^3 + 7x^2 + 2x + 9$ and $g = 2x + 3$
   (b) $f = 3x^3 - 2x^2 + 5$ and $g = x^2 - 1$
   (c) $f = x^3 + 3x^2 - 4x - 12$ and $g = x^2 + x - 6$

0.5.2 Consider $f = x^3 - x^2 + x - 1 \in \mathbb{Q}[x]$.
   (a) Use Corollary 0.48 to show that $x - 1$ divides f.
   (b) Compute the quotient of f divided by $x - 1$.

0.5.3 Show by example that the polynomial division algorithm fails in $\mathbb{Z}[x]$.

0.5.4 Prove that Corollary 0.48 holds in $R[x]$ for any ring R. (Hint: Try to write $f(x) - f(a)$ as a multiple of $x - a$.)

0.5.5 Prove that Corollary 0.50 holds in $R[x]$ if and only if R is an integral domain.

0.5.6 Give an example of a ring R and a nonzero polynomial $f \in R[x]$ with infinitely many zeros.

0.5.7 Compute the unique irreducible factorization of $x^n - 1 \in \mathbb{C}[x]$.

Section 0.6 Unique factorization in polynomial rings

In this section, we conclude the proof that $K[x_1,\ldots,x_n]$ is a unique factorization domain. By Propositions 0.17 and 0.22, all that remains to be proved is the following analogue of Euclid's Lemma.

0.54 PROPOSITION *Euclid's Lemma for polynomials*

Every irreducible polynomial in $K[x_1,\ldots,x_n]$ is prime.

The proof of Proposition 0.54 involves an induction argument on the number of variables, starting with the base case of $K[x]$. Because the proof is rather involved, we start with a brief overview of the main ideas, then we develop each of those ideas in turn, finally merging them into a formal proof at the end of this section.

To motivate the ideas that follow, recall that we can view the polynomial ring $K[x_1,\ldots,x_n]$ as a polynomial ring in $n-1$ variables:

$$K[x_1,\ldots,x_n] = R[x_1,\ldots,x_{n-1}],$$

where $R = K[x_n]$. One of the important new ideas that we introduce in this section is that of the *fraction field*, which associates a field $\text{Frac}(R)$ to any integral domain R, along with a canonical inclusion $R \subseteq \text{Frac}(R)$. In particular, if $K' = \text{Frac}(K[x_n])$, we obtain an inclusion

$$K[x_1,\ldots,x_n] \subseteq K'[x_1,\ldots,x_{n-1}].$$

Using this inclusion, we can begin to see an induction argument coming together. In particular, if our induction hypothesis is that irreducible polynomials in $n-1$ variables over any field are prime, then we can proceed to prove that irreducible polynomials in n variables are prime using the following two steps (Lemmas 0.59 and 0.60, respectively):

1. Prove that every irreducible polynomial in $K[x_1,\ldots,x_n]$ remains irreducible in $K'[x_1,\ldots,x_{n-1}]$ (hence prime, by the induction hypothesis).
2. Prove that every irreducible polynomial in $K[x_1,\ldots,x_n]$ that happens to be prime in $K'[x_1,\ldots,x_{n-1}]$ is also prime in $K[x_1,\ldots,x_n]$.

Now that we have outlined the road ahead, we begin in earnest by developing the notions of fraction fields, starting with the definition of fractions.

0.55 PROPOSITION/DEFINITION *Fractions*

Let R be an integral domain. A *fraction* of elements in R is an expression of the form a/b where $a, b \in R$ and $b \neq 0$. Equality of fractions is defined by

$$\frac{a}{b} = \frac{c}{d} \iff ad = bc \in R.$$

Equality of fractions is an equivalence relation, and the set of equivalence classes is denoted $\text{Frac}(R)$.

0.56 EXAMPLE $\mathbb{Q} = \text{Frac}(\mathbb{Z})$

While the definition of equality of fractions might be confusing at first glance, it is modeled on the familiar way that rational numbers are constructed from the integers. In particular, as we learn in grade school, to check an equality of rational numbers, such as

$$\frac{3}{4} = \frac{6}{8},$$

we cross-multiply: $3 \cdot 8 = 4 \cdot 6 \in \mathbb{Z}$.

0.57 EXAMPLE Rational functions

As the polynomial rings $K[x_1, \ldots, x_n]$ play such a central role in our story, fractions of polynomials have a special name: they are called *rational functions*. We denote the set of rational functions using the following notation:

$$K(x_1, \ldots, x_n) = \text{Frac}(K[x_1, \ldots, x_n]).$$

Consider, for example, the following two elements of $K(x, y)$:

$$\frac{2x^2 + x - 2xy - y}{x^2 - y^2} \quad \text{and} \quad \frac{2x + 1}{x + y}.$$

In fact, these rational functions are equal because, as the reader can verify,

$$(2x^2 + x - 2xy - y)(x + y) = (2x + 1)(x^2 - y^2).$$

Another way to view this equality is by canceling like factors in the numerator and denominator:

$$\frac{2x^2 + x - 2xy - y}{x^2 - y^2} = \frac{(x - y)(2x + 1)}{(x - y)(x + y)} = \frac{2x + 1}{x + y}.$$

The term "function" here is standard but misleading; a rational function should not necessarily be thought of as a function, per se, with a domain, a range, and a rule. Rather, it is simply a formal quotient of polynomials.

In fact, the set of all fractions is more than just a set; it forms a field under the familiar operations of addition and multiplication of quotients. Since the same fraction can be represented in multiple ways, it needs to be verified that the operations are well-defined, meaning that they are independent of the choice of representatives. In addition, one should verify that, with these operations, the set of fractions satisfies the field axioms. We formalize the addition and multiplication operations below and leave the needed verifications to Exercise 0.6.2.

0.6. UNIQUE FACTORIZATION IN POLYNOMIAL RINGS

> **0.58 PROPOSITION/DEFINITION** *Fraction field*
>
> Let R be an integral domain. The operations of addition and multiplication defined by
> $$\frac{a}{b} + \frac{c}{d} = \frac{ad+bc}{bd} \quad \text{and} \quad \frac{a}{b} \cdot \frac{c}{d} = \frac{ac}{bd}.$$
> are well-defined on equivalence classes of fractions and endow $\text{Frac}(R)$ with the structure of a field, called the *fraction field* of R.

One of the most important properties of the fraction field is that it canonically contains R as a subring (Exercise 0.6.3):
$$R = \{a/1 \mid a \in R\} \subseteq \text{Frac}(R).$$

In particular, $K[x_n] \subseteq K(x_n)$, and thus, $K[x_1, \ldots, x_n] \subseteq K(x_n)[x_1, \ldots, x_{n-1}]$. Moreover, given any element $f \in K(x_n)[x_1, \ldots, x_{n-1}]$, we can always find a polynomial $r \in K[x_n]$ such that $rf \in K[x_1, \ldots, x_n]$ (Exercise 0.6.5). Multiplying f by such an r is called *clearing the denominators in f*, and is used frequently in the proofs of this section.

We are now ready to prove the two lemmas required for Proposition 0.54.

> **0.59 LEMMA**
>
> If f is irreducible in $K[x_1, \ldots, x_n]$ and $f \notin K[x_n]$, then f is irreducible as an element of $K(x_n)[x_1, \ldots, x_{n-1}]$.

PROOF We prove the contrapositive. Assume that $f \in K[x_1, \ldots, x_n]$ is reducible in $K(x_n)[x_1, \ldots, x_{n-1}]$; our goal is to prove that f is reducible in $K[x_1, \ldots, x_n]$. Since f is reducible in $K(x_n)[x_1, \ldots, x_{n-1}]$, we can write $f = gh$ where neither g nor h is an element of $K(x_n)$. By clearing the denominators in both g and h, we can write
$$rf = g_0 h_0,$$
where $r \in K[x_n]$ and $g_0, h_0 \in K[x_1, \ldots, x_n] \setminus K[x_n]$. Since the single-variable polynomial ring $K[x_n]$ is a UFD (Corollary 0.53), write $r = p_1 \cdots p_\ell$ where each p_i is irreducible, and thus prime. We have
$$p_1 \cdots p_\ell f = g_0 h_0.$$

Since p_1 is prime in $K[x_n]$, it follows from Exercise 0.6.6 that p_1 is also prime in $K[x_n][x_1, \ldots, x_{n-1}] = K[x_1, \ldots, x_n]$. Therefore, by definition of prime elements, $p_1 \mid g_0$ or $p_1 \mid h_0$. Without loss of generality, suppose $p_1 \mid g_0$ and write $g_0 = p_1 g_1$ and $h_1 = h_0$ so that
$$p_2 \cdots p_\ell f = g_1 h_1.$$

Repeating the above procedure with $p_2, \ldots, p_\ell$, we conclude that $f = g_\ell h_\ell$. Since neither g_0 nor h_0 were elements of $K[x_n]$, it follows that neither g_ℓ nor h_ℓ are elements of K. Thus, f is reducible in $K[x_1, \ldots, x_n]$. □

0.60 LEMMA

If $f \in K[x_1,\ldots,x_n]$ is irreducible as an element of $K[x_1,\ldots,x_n]$ and prime as an element of $K(x_n)[x_1,\ldots,x_{n-1}]$, then f is prime in $K[x_1,\ldots,x_n]$.

PROOF Suppose that $f \in K[x_1,\ldots,x_n]$ is irreducible, and that f is prime in $K(x_n)[x_1,\ldots,x_{n-1}]$. To prove that f is prime in $K[x_1,\ldots,x_n]$, suppose $f \mid gh$ in $K[x_1,\ldots,x_n]$. Since f is prime in $K(x_n)[x_1,\ldots,x_{n-1}]$, we know that $f \mid g$ or $f \mid h$ in $K(x_n)[x_1,\ldots,x_{n-1}]$. Without loss of generality, assume $f \mid g$ and write

$$af = g$$

for some $a \in K(x_n)[x_1,\ldots,x_{n-1}]$. By clearing denominators in a, write

$$a_0 f = rg \in K[x_1,\ldots,x_n]$$

for some $a_0 \in K[x_1,\ldots,x_n]$ and $r \in K[x_n]$. As in the proof of Lemma 0.59, let $r = p_1 \cdots p_\ell$ be a prime factorization of r, from which it follows that p_1 must divide a_0 or f. Since f is irreducible in $K[x_1,\ldots,x_n]$, the only way p_1 could divide f is if they differed by a constant, which would imply that $f \in K[x_n]$. This would mean that f is a unit in $K(x_n)[x_1,\ldots,x_{n-1}]$, which contradicts the assumption that f is prime in $K(x_n)[x_1,\ldots,x_{n-1}]$. Thus, p_1 must divide a_0. Write $a_0 = a_1 p_1$ so that

$$a_1 f = p_2 \cdots p_\ell g.$$

Repeating this procedure for $p_2,\ldots,p_\ell$, we conclude that

$$a_\ell f = g \in K[x_1,\ldots,x_n].$$

Therefore, $f \mid g$ in $K[x_1,\ldots,x_n]$ and it follows that f is prime. □

Combining the previous two lemmas, we now prove Proposition 0.54.

PROOF OF PROPOSITION 0.54   Proceeding by induction on n, the base case $n = 1$ is Corollary 0.53. To prove the induction step, suppose that, for some $n \geq 2$ and for any field K, every irreducible polynomial in $K[x_1,\ldots,x_{n-1}]$ is prime. Let K be a field and let $f \in K[x_1,\ldots,x_n]$ be irreducible; we must show that f is prime.

Since f is not a unit, it has positive degree in at least one variable; suppose without loss of generality that it has positive degree in a variable other than x_n. Then f is irreducible and not an element of $K[x_n]$, so Lemma 0.59 implies that f is irreducible as an element of $K(x_n)[x_1,\ldots,x_{n-1}]$. Because $K(x_n)$ is a field, the induction hypothesis implies that f is prime in $K(x_n)[x_1,\ldots,x_{n-1}]$. Therefore, upon applying Lemma 0.60, we conclude that f is prime in $K[x_1,\ldots,x_n]$. □

As an immediate consequence of Propositions 0.17, 0.22, and 0.54, we now conclude that $K[x_1,\ldots,x_n]$ is a unique factorization domain. For ease of reference, we close this section with the precise statement of this result.

It can be proved, more generally, that $R[x_1,\ldots,x_n]$ is a UFD whenever R is a UFD. This level of generality is not necessary for our purposes.

0.61 THEOREM $K[x_1,\ldots,x_n]$ *is a UFD*

If $f \in K[x_1,\ldots,x_n]$ is nonconstant, then there exist irreducible polynomials $p_1,\ldots,p_\ell \in K[x_1,\ldots,x_n]$ such that

$$f = p_1 \cdots p_\ell.$$

If $f = q_1 \cdots q_m$ is another irreducible factorization, then $\ell = m$ and, after possibly reordering terms, q_i is a constant multiple of p_i for every i.

Exercises for Section 0.6

0.6.1 Prove that equality of fractions is an equivalence relation.

0.6.2 Let R be an integral domain and $a/b, c/d, r/s \in \text{Frac}(R)$ with $a/b = c/d$.
   (a) Prove that
   $$\frac{a}{b} + \frac{r}{s} = \frac{c}{d} + \frac{r}{s} \quad \text{and} \quad \frac{a}{b} \cdot \frac{r}{s} = \frac{c}{d} \cdot \frac{r}{s}.$$
   (b) Prove that $\text{Frac}(R)$ satisfies the field axioms.

0.6.3 Let R be an integral domain.
   (a) Prove that the function
   $$\varphi : R \to \text{Frac}(R)$$
   $$a \mapsto a/1$$
   is an injective ring homomorphism.
   (b) Let K be a field with $R \subseteq K$. Prove that $\text{Frac}(R) \subseteq K$.

0.6.4 Suppose that R is not an integral domain. Explain what might go wrong if we try to construct the fraction field of R.

0.6.5 Let $f \in K(x_n)[x_1,\ldots,x_{n-1}]$. Prove that there is a polynomial $r \in K[x_n]$ such that $rf \in K[x_1,\ldots,x_n]$.

0.6.6 Let $a \in R$ and consider the surjective homomorphism
$$\pi : R[x_1,\ldots,x_n] \to (R/\langle a \rangle)[x_1,\ldots,x_n].$$
   (a) Prove that $\ker(\pi) = \langle a \rangle \subseteq R[x_1,\ldots,x_n]$ and conclude that
   $$\frac{R[x_1,\ldots,x_n]}{\langle a \rangle} \cong (R/\langle a \rangle)[x_1,\ldots,x_n].$$
   (b) Prove that a is prime in R if and only if a is prime in $R[x_1,\ldots,x_n]$.

Section 0.7 Irreducibility criteria

In order to have a large bank of concrete examples in algebraic geometry, it is useful to have methods at our disposal for studying specific polynomials. For example, if we have a particular polynomial in mind, such as

$$x^2 + y^2 + z^2 - 1 \in \mathbb{R}[x, y, z],$$

it might be helpful to be able to determine quickly whether or not this polynomial is irreducible. In this final section of the chapter, we discuss two criteria for determining whether a given polynomial is irreducible. The first result follows quickly from our prior developments and we leave its proof to Exercise 0.7.1.

0.62 PROPOSITION *Characterizations of irreducible polynomials*

Let $f \in K[x_1, \ldots, x_n]$. The following are equivalent.

1. f is irreducible;
2. f is prime;
3. $\langle f \rangle$ is a prime ideal;
4. $K[x_1, \ldots, x_n]/\langle f \rangle$ is an integral domain.

0.63 EXAMPLE $y - x^2$ is irreducible in $K[x, y]$, revisited

We already proved directly that $y - x^2$ is irreducible. In light of Proposition 0.62, this can also be seen from the fact that $K[x]$ is an integral domain and

$$\frac{K[x, y]}{\langle y - x^2 \rangle} \cong K[x].$$

It is not always helpful to apply Proposition 0.62 in practice, because the problem of showing that an ideal is prime or that a quotient is an integral domain is typically just as difficult as showing directly that a polynomial is irreducible—the proposition translates the problem but does not necessarily simplify it. The next test, called Eisenstein's Criterion, while a bit more complicated to state, is much more useful in practice, as will be illustrated in the subsequent examples.

0.64 PROPOSITION *Eisenstein's Criterion*

Let R be an integral domain and let $f = a_n x^n + \cdots + a_1 x + a_0 \in R[x]$ be a polynomial satisfying the following conditions.

1. There does not exist a nonunit $b \in R$ such that $b \mid f$.
2. There exists a prime element $p \in R$ such that
   - $p \mid a_i$ for $i < n$, and
   - $p^2 \nmid a_0$.

Then f is irreducible.

0.7. IRREDUCIBILITY CRITERIA

Unlike Proposition 0.62, Eisenstein's Criterion is not an if-and-only-if statement. In particular, Eisenstein's Criterion can never be used to determine whether a single-variable polynomial $f \in K[x]$ is irreducible, simply because fields do not contain prime elements. Before proving Eisenstein's Criterion, we provide a few example applications to demonstrate how to use it in the context of multivariable polynomials over fields. We include a number of further examples in the exercises.

0.65 EXAMPLE $f = x^2 + y^2 - 1$ is irreducible in $\mathbb{R}[x, y]$

If we view f as an element of $R[y]$ where $R = \mathbb{R}[x]$, then

$$f = a_2 y^2 + a_1 y + a_0$$

where $a_2 = 1$, $a_1 = 0$, and $a_0 = x^2 - 1 = (x-1)(x+1)$. Notice that these coefficients do not have any common nonconstant divisors in $\mathbb{R}[x]$, so f meets the first condition in Eisenstein's Criterion. Since the quotient

$$\frac{\mathbb{R}[x]}{\langle x - 1 \rangle} \cong \mathbb{R},$$

is an integral domain, $p = x - 1$ is prime in $\mathbb{R}[x]$ and satisfies the second condition of Eisenstein's Criterion. Therefore, we conclude that f is an irreducible polynomial.

> Working over $\mathbb{R}$ is not essential for these two examples; the same argument works for any field for which $1 \neq -1$ (when $\text{char}(K) \neq 2$).

0.66 EXAMPLE $f = x^2 + y^2 + z^2 - 1$ is irreducible in $\mathbb{R}[x, y, z]$

As in the previous example, write

$$f = a_2 z^2 + a_1 z + a_0$$

where $a_2 = 1$, $a_1 = 0$ and $a_0 = x^2 + y^2 - 1$. These coefficients do not have any common nonconstant divisors in $\mathbb{R}[x, y]$, so f meets the first condition in Eisenstein's Criterion. Set $p = x^2 + y^2 - 1$, which is irreducible by the previous example, and thus prime because $\mathbb{R}[x, y]$ is a UFD. The polynomial p satisfies the second condition of Eisenstein's Criterion, from which we conclude that f is irreducible.

0.67 EXAMPLE $f = x^2 y^2 + yz^2 + x^3 z^2 \in K[x, y, z]$ is irreducible

Write

$$f = a_3 x^3 + a_2 x^2 + a_1 x + a_0 \in R[x]$$

where $R = K[y, z]$ and $a_3 = z^2$, $a_2 = y^2$, $a_1 = 0$, and $a_0 = yz^2$. Notice that these coefficients do not have any common nonconstant divisors in $K[y, z]$, so f meets the first condition in Eisenstein's Criterion. Let $p = y \in R$. Since

$$\frac{R}{\langle p \rangle} = \frac{K[y, z]}{\langle y \rangle} \cong K[z]$$

is an integral domain, we see that $p = y$ is prime in R. Because p satisfies the second condition of Eisenstein's Criterion, we conclude that f is irreducible.

PROOF OF PROPOSITION 0.64   Let f and p be as in the statement of the proposition, and suppose $f = gh$ for some $g, h \in R[x]$. We must show that g or h is a unit in R. By the first condition, f is not divisible by any nonunits in R, so it suffices to prove that either g or h is an element of R.

Toward a contradiction, suppose $g = b_k x^k + \cdots + b_0$ and $h = c_\ell x^\ell + \cdots + c_0$ both have positive degree. Consider the ring homomorphism

$$\varphi : R[x] \to (R/\langle p \rangle)[x]$$
$$\sum_{i \geq 0} r_i x^i \mapsto \sum_{i \geq 0} [r_i] x^i.$$

By our assumptions on f in the statement of Proposition 0.64, it follows that

$$\varphi(f) = [a_n] x^n = \varphi(g)\varphi(h),$$

with $[a_n] \neq 0$. Since $R/\langle p \rangle$ is an integral domain, $\varphi(g)$ and $\varphi(h)$ must each consist of a single nonzero term (Exercise 0.7.2), and by additivity of degree, it follows that $\varphi(g) = [b_k] x^k$ and $\varphi(h) = [c_\ell] x^\ell$. In particular, this implies that $p \mid b_0$ and $p \mid c_0$, so that $p^2 \mid b_0 c_0 = a_0$, which contradicts the assumptions on p. □

Exercises for Section 0.7

0.7.1 Prove Proposition 0.62 by citing relevant results from previous sections.

0.7.2 Suppose R is an integral domain, and let $g, h \in R[x]$ be such that gh has a single nonzero term—that is, $gh = ax^n$ for some nonzero $a \in R$ and $n \in \mathbb{N}$. Prove that each of g and h has a single nonzero term.

0.7.3 Prove that $wx - yz \in K[w, x, y, z]$ is irreducible.

0.7.4 Prove that $xyz + x^2 z^2 + yz^3 + x \in K[x, y, z]$ is irreducible.

0.7.5 Assume $\text{char}(K) \neq 2$ and $n \geq 2$. Prove that $x_1^2 + \cdots + x_n^2 - 1$ is irreducible in $K[x_1, \ldots, x_n]$.

0.7.6 Assume $n \geq 3$ and $m \geq 1$. Prove that $x_1^m + \cdots + x_n^m \in \mathbb{C}[x_1, \ldots, x_n]$ is irreducible.

Part I

Affine Algebraic Geometry

Chapter 1

Varieties and Ideals

> LEARNING OBJECTIVES FOR CHAPTER 1
> - Define and work with affine space $\mathbb{A}_K^n$.
> - Use the $\mathcal{V}$- and $\mathcal{I}$-operators to move between subsets of $K[x_1, \ldots, x_n]$ and subsets of $\mathbb{A}_K^n$.
> - Define and give examples of affine varieties in $\mathbb{A}_K^n$ and radical ideals in $K[x_1, \ldots, x_n]$.
> - State the Nullstellensatz and use it to describe the bijection between affine varieties and radical ideals.

Algebraic geometry is, at its heart, a dictionary for translating between different languages: the language of algebra and the language of geometry. As in any dual-language dictionary, this involves translation in both directions. Given an algebraic object, such as a polynomial, we produce a geometric object by determining the vanishing set of the polynomial; the vanishing set of the polynomial $x^2 + y^2 - 1$, for instance, is the unit circle in the plane. Conversely, given a geometric object, we produce an algebraic object by determining the set of all polynomials that vanish on the given geometric set.

In this chapter, we begin our study of algebraic geometry in earnest by making these two operations precise by way of the $\mathcal{V}$- and $\mathcal{I}$-operators. Crucially, we find that these operators are not surjective. Not every geometric set is the vanishing set of some collection of polynomials; those geometric sets that are obtained in this way are called *affine varieties*. Conversely, not every set of polynomials is obtainable by starting from a geometric set and calculating the polynomials that vanish on it; the study of algebraic sets that are obtained in this way will lead us to develop the algebraic notion of a *radical ideal*.

The chapter culminates with the statement of a result that might properly be termed the "Fundamental Theorem of Algebraic Geometry," though it instead traditionally goes by the German name *Nullstellensatz*. Under one key hypothesis—that the ground field is *algebraically closed*—the Nullstellensatz asserts that when one restricts attention to affine varieties on the geometric side and to radical ideals on the algebraic side, the $\mathcal{V}$- and $\mathcal{I}$-operators provide a true dictionary—a bijection, to put it mathematically—between algebra and geometry.

Section 1.1 The $\mathcal{V}$-operator

In order to study the vanishing sets of polynomials, we must begin by specifying where those polynomials and their solutions live. Choose a field K, referred to as the *ground field*, and consider the polynomial ring $K[x_1, \ldots, x_n]$. Elements of this ring, in addition to being abstract polynomials, can also be used to define functions that take elements of K^n as input, and output elements of K. For example, the polynomial
$$f = 2xy + 4z^2 \in \mathbb{R}[x, y, z]$$
defines a function from $\mathbb{R}^3$ to $\mathbb{R}$. If one inputs the element $(1, -1, 3)$, then the output of f is the single real number
$$f(1, -1, 3) = 2 \cdot 1 \cdot (-1) + 4 \cdot 3^2 = 34.$$
The domain of a polynomial function is referred to as *affine space*.

1.1 DEFINITION *Affine space*

The *n-dimensional affine space over K*, denoted $\mathbb{A}_K^n$, is the set of n-tuples of elements of K:
$$\mathbb{A}_K^n = \{(a_1, \ldots, a_n) \mid a_i \in K\}.$$

As a set, $\mathbb{A}_K^n$ is the same as the vector space K^n. So why give it a new name and a new notation? When we write K^n, we are viewing this set as an algebraic object with addition and scalar multiplication operations—that is, as a vector space. When we write $\mathbb{A}_K^n$, on the other hand, we forget the algebraic structure on this set: we view its elements not as vectors that can be added to one another, but rather as inputs to polynomial functions. In particular, the element $(0, \ldots, 0)$ is very special in the vector space K^n—it is the additive identity—but it is essentially the same as any other element in the affine space $\mathbb{A}_K^n$.

> While elements of K^n are typically referred to as "vectors," elements of $\mathbb{A}_K^n$ are called "points" to highlight their geometric significance.

> Often, when the ground field K is understood from context, we simply write $\mathbb{A}^n$ instead of $\mathbb{A}_K^n$.

To distinguish between polynomials and their evaluations, we use letters at the end of the alphabet (x, y, and z) to denote variables, and letters at the beginning of the alphabet (a, b, and c) to denote elements of K. So, for example, $f = x^2 y$ denotes an element of the ring $K[x, y]$, whereas $f(a, b) = a^2 b$ denotes the element of K obtained by evaluating f at the point $(a, b) \in \mathbb{A}^2$. If $f \in K[x_1, \ldots, x_n]$ satisfies
$$f(a_1, \ldots, a_n) = 0$$
for some point $(a_1, \ldots, a_n) \in \mathbb{A}^n$, we say that f *vanishes* at $(a_1, \ldots, a_n)$.

Given a polynomial $f \in K[x_1, \ldots, x_n]$, the set of all points at which f vanishes is a subset of $\mathbb{A}^n$. For example, if
$$f = y - x^2 \in \mathbb{R}[x, y],$$

1.1. THE $\mathcal{V}$-OPERATOR

then f vanishes at the point $(0,0)$, as well as at the point $(1,1)$, the point $(2,4)$, the point $(-1,1)$, and so on. The set of all points in $\mathbb{A}_\mathbb{R}^2$ at which $f = y - x^2$ vanishes forms the familiar parabola. This is our first taste of algebraic geometry: an algebraic object (namely, the element $y - x^2$ of the ring $\mathbb{R}[x,y]$) led us to a geometric object (the parabola).

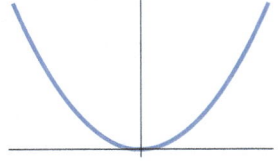

More generally, one can study the set of points in $\mathbb{A}^n$ at which every element of a (possibly infinite) set of polynomials vanishes.

1.2 DEFINITION *Vanishing set*

Let $\mathcal{S} \subseteq K[x_1, \ldots, x_n]$ be a set of polynomials. The *vanishing set of* $\mathcal{S}$ is

$$\mathcal{V}(\mathcal{S}) = \{(a_1, \ldots, a_n) \in \mathbb{A}^n \mid f(a_1, \ldots, a_n) = 0 \text{ for all } f \in \mathcal{S}\} \subseteq \mathbb{A}^n.$$

It is common to say that points of $\mathcal{V}(\mathcal{S})$ are *solutions* of the *polynomials* in $\mathcal{S}$. When $\mathcal{S} = \{f_1, \ldots, f_r\}$ is finite, we write $\mathcal{V}(f_1, \ldots, f_r)$ instead of $\mathcal{V}(\{f_1, \ldots, f_r\})$.

1.3 EXAMPLE Curves in $\mathbb{A}_\mathbb{R}^2$

Consider $\mathcal{S} = \{y - x^2\} \subseteq K[x,y]$. Then $\mathcal{V}(\mathcal{S}) = \{(a, a^2) \mid a \in K\} \subseteq \mathbb{A}^2$. When $K = \mathbb{R}$, this is the parabola above. Similarly, $\mathcal{V}(x^2 + y^2 - 1)$, $\mathcal{V}(y^2 - x^3 - x^2)$, and $\mathcal{V}(y^2 - (x+1)^3)$ are the plane curves depicted in $\mathbb{A}_\mathbb{R}^2$ below.

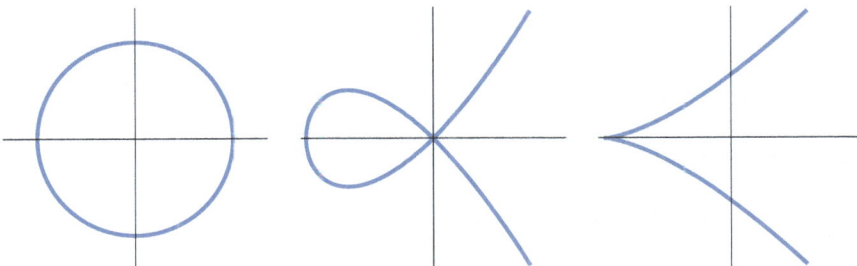

1.4 EXAMPLE Surfaces in $\mathbb{A}_\mathbb{R}^3$

The vanishing sets $\mathcal{V}(x^2 + y^2 + z^2 - 1)$, $\mathcal{V}(x^2 + y^2 - z^2)$, and $\mathcal{V}(x^2 + y^2 z)$ are the unit sphere, the cylindrical cone, and the "Whitney umbrella," respectively, pictured below in $\mathbb{A}_\mathbb{R}^3$.

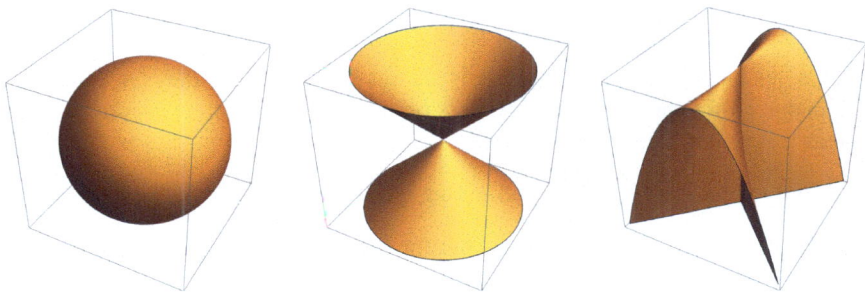

1.5 EXAMPLE The coordinate axes

If $S = \{xy\} \subseteq K[x,y]$, then $\mathcal{V}(S) = \{(a,b) \in \mathbb{A}^2 \mid ab = 0\} \subseteq \mathbb{A}^2$. Since K is a field, $ab = 0$ if and only if either $a = 0$ or $b = 0$ (or both), so $\mathcal{V}(S)$ is the union of the points where $a = 0$ and those where $b = 0$. When $K = \mathbb{R}$, this is the union of the x-axis and the y-axis in the real plane $\mathbb{A}^2_{\mathbb{R}}$.

1.6 EXAMPLE Single points

Let $S = \{x,y\} \subseteq K[x,y]$. Then
$$\mathcal{V}(S) = \{(a,b) \in \mathbb{A}^2 \mid a = 0 \text{ and } b = 0\} = \{(0,0)\} \subseteq \mathbb{A}^2.$$
That is, $\mathcal{V}(S)$ consists of a single point: the origin.

Similarly, if $S = \{x - i, y - 1 - i, z - 5\} \subseteq \mathbb{C}[x,y,z]$, then
$$\mathcal{V}(S) = \{(i, 1+i, 5)\} \subseteq \mathbb{A}^3_{\mathbb{C}},$$
which, again, is a single point.

1.7 EXAMPLE A curve in $\mathbb{A}^3$

The vanishing set of
$$S = \{x^2 + y^2 - 1,\ x^2 - y^2 - z\} \subseteq \mathbb{R}[x,y,z]$$
consists of $(a,b,c) \in \mathbb{A}^3_{\mathbb{R}}$ where $a^2 + b^2 - 1$ and $a^2 - b^2 - c$ both vanish. This curve is de-

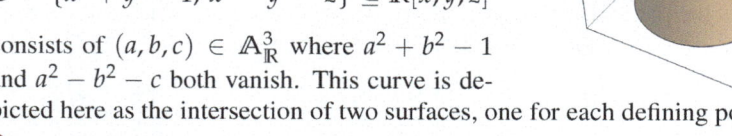

picted here as the intersection of two surfaces, one for each defining polynomial.

The ground field for the images above is $K = \mathbb{R}$. Over $\mathbb{R}$, we have a geometric intuition; for example, we have an idea about what it means to be a "curve" or "surface." Algebraic geometry aims to make this intuition precise for general fields.

Given that we are now viewing polynomials as functions $\mathbb{A}^n \to K$, it is worth pointing out that there is a subtle but important difference between *polynomials* and *polynomial functions*. In particular, it is a somewhat unsettling fact that different polynomials in $K[x_1, \ldots, x_n]$ can give rise to the same function $\mathbb{A}^n \to K$.

For example, let $K = \mathbb{F}_2 = \{0,1\}$, the field with two elements; recall that addition and multiplication in this field are both carried out modulo 2. Let

(1.8) $$f(x) = 1 + x + x^2 \quad \text{and} \quad g(x) = 1.$$

As elements of $\mathbb{F}_2[x]$, these polynomials are not equal, because they have different coefficients on the monomial x as well as on the monomial x^2. However, viewing them as functions $\mathbb{A}^1 \to \mathbb{F}_2$, we see that

$$f(0) = 1 + 0 + 0^2 = 1 = g(0) \quad \text{and} \quad f(1) = 1 + 1 + 1^2 = 1 = g(1).$$

Thus, since f and g give the same output for every input in their domain, they are equal as functions, even though they are different as polynomials.

1.1. THE $\mathcal{V}$-OPERATOR

To describe the difference between polynomials and their corresponding functions more generally, let $K[\mathbb{A}^n]$ denote the set of polynomial functions $\mathbb{A}^n \to K$. That is, an element of $K[\mathbb{A}^n]$ is a function of the form

$$(a_1, \ldots, a_n) \mapsto f(a_1, \ldots, a_n)$$

for some $f \in K[x_1, \ldots, x_n]$. Some reflection should convince the reader that $K[\mathbb{A}^n]$ is a ring under addition and multiplication of functions to K, and taking a polynomial to its corresponding function defines a surjective ring homomorphism $K[x_1, \ldots, x_n] \to K[\mathbb{A}^n]$. This homomorphism fails to be injective exactly when different polynomials give rise to the same function. For example,

$$\mathbb{F}_2[x] \to \mathbb{F}_2[\mathbb{A}^1]$$

is not an injection, because the different polynomials $f, g \in \mathbb{F}_2[x]$ defined in (1.8) give rise to the same polynomial function in $\mathbb{F}_2[\mathbb{A}^1]$.

The next result shows that the above phenomenon is unique to finite fields, so we need not worry about this subtlety as long as we assume that K is infinite.

1.9 PROPOSITION *Polynomials versus functions*

The ring homomorphism $K[x_1, \ldots, x_n] \to K[\mathbb{A}^n]$ that takes polynomials to their corresponding functions is an isomorphism if and only if K is infinite.

PROOF We prove both directions of the if-and-only-if statement.

($\Rightarrow$) By contrapositive, assume that $K = \{a_1, \ldots, a_n\}$ is finite, and consider the nonzero polynomial

$$f = \prod_{i=1}^{n}(x_1 - a_i) \in K[x_1] \subseteq K[x_1, \ldots, x_n].$$

Plugging in any value of K for x_1 produces a factor of zero, so f defines the zero function $\mathbb{A}^n \to K$. This shows that $K[x_1, \ldots, x_n] \to K[\mathbb{A}^n]$ is not an isomorphism.

($\Leftarrow$) Suppose K is infinite. By definition, $K[x_1, \ldots, x_n] \to K[\mathbb{A}^n]$ is surjective. Thus, it remains to prove injectivity, or equivalently, to prove that the kernel is the zero polynomial. We accomplish this by induction on n.

(Base case) Suppose $h \in K[x]$ defines the zero function $\mathbb{A}^1 \to K$. Since K is an infinite field, this implies that h has infinitely many zeros. By Corollary 0.50, nonzero polynomials have finitely many zeros, so h must be the zero polynomial.

(Induction step) Suppose $h \in K[x_1, \ldots, x_n]$ defines the zero function, and write

$$h = \sum_{i=0}^{d} h_i x_n^i$$

where $h_i \in K[x_1, \ldots, x_{n-1}]$. For each $(a_1, \ldots, a_{n-1}) \in \mathbb{A}^{n-1}$, the single-variable polynomial $h(a_1, \ldots, a_{n-1}, x_n) \in K[x_n]$ defines the zero function $\mathbb{A}^1 \to K$. Thus, by the argument in the base case, $h(a_1, \ldots, a_{n-1}, x_n)$ is the zero polynomial. In particular, this implies that $h_i(a_1, \ldots, a_{n-1}) = 0$ for all i. Since this is true for every $(a_1, \ldots, a_{n-1}) \in \mathbb{A}^{n-1}$, it follows that $h_i : \mathbb{A}^{n-1} \to K$ is the zero function for all i. By the induction hypothesis, $h_i \in K[x_1, \ldots, x_{n-1}]$ is the zero polynomial for every i, and it follows that h is the zero polynomial. $\square$

Exercises for Section 1.1

1.1.1 Sketch the following vanishing sets.
   (a) $\mathcal{V}(x^2 - 1) \subseteq \mathbb{A}^1_\mathbb{R}$
   (b) $\mathcal{V}(x^2 - y^2) \subseteq \mathbb{A}^2_\mathbb{R}$
   (c) $\mathcal{V}(y - x^2, y - x) \subseteq \mathbb{A}^2_\mathbb{R}$
   (d) $\mathcal{V}(x^2 + y^2 + z^2 - 1, z) \subseteq \mathbb{A}^3_\mathbb{R}$
   (e) $\mathcal{V}(x^2 + y^2 - z^2, x) \subseteq \mathbb{A}^3_\mathbb{R}$
   (f) $\mathcal{V}(x^2 + y^2 - z^2, z) \subseteq \mathbb{A}^3_\mathbb{R}$
   (g) $\mathcal{V}(xy - y^2, x^2 - xy - x + y) \subseteq \mathbb{A}^2_\mathbb{R}$

1.1.2 Express the following sets as $\mathcal{V}(\mathcal{S})$ for some $\mathcal{S}$.
   (a) $\{0, \pi, -1\} \subseteq \mathbb{A}^1_\mathbb{R}$
   (b) the x-axis in $\mathbb{A}^3_\mathbb{R}$
   (c) $\{(4, -1, 3)\} \subseteq \mathbb{A}^3_\mathbb{R}$
   (d) $\{(-1, 0), (1, 0)\} \subseteq \mathbb{A}^2_\mathbb{R}$
   (e) $\{(a, a, a) \mid a \in \mathbb{R}\} \subseteq \mathbb{A}^3_\mathbb{R}$
   (f) $\{(\cos(a), \sin(a), \cos^2(a) - \sin^2(a)) \mid a \in [0, 2\pi)\} \subseteq \mathbb{A}^3_\mathbb{R}$

1.1.3 Show that the origin $\{(0,0)\} \subseteq \mathbb{A}^2_K$ is the vanishing set of a single polynomial if $K = \mathbb{R}$, but not if $K = \mathbb{C}$.

1.1.4 Let $f, g \in K[x_1, \ldots, x_n]$. Prove that
   (a) $\mathcal{V}(fg) = \mathcal{V}(f) \cup \mathcal{V}(g)$;
   (b) $\mathcal{V}(f, g) = \mathcal{V}(f) \cap \mathcal{V}(g)$.

1.1.5 Let
$$X = \mathcal{V}(x^2 - yz, xz - x) \subseteq \mathbb{A}^3_K.$$

   (a) Prove that
$$X = \mathcal{V}(x, y) \cup \mathcal{V}(x, z) \cup \mathcal{V}(x^2 - y, z - 1).$$

   (b) Use part (a) to sketch X in the case where $K = \mathbb{R}$.

1.1.6 For any field K, prove that every finite set in $\mathbb{A}^1_K$ can be expressed as the vanishing set of one polynomial.

1.1.7 For any field K, prove that every finite set in $\mathbb{A}^2_K$ can be expressed as the vanishing set of two polynomials.

1.1.8 Generalizing the previous two exercises, prove that every finite set in $\mathbb{A}^n_K$ can be expressed as the vanishing set of a collection of n polynomials.
   (Hint: Consider the map $\mathbb{A}^n \to \mathbb{A}^{n-1}$ that forgets the last coordinate, and use induction on n.)

Section 1.2 Affine varieties

Not every subset of $\mathbb{A}^n$ is the vanishing set of some collection of polynomials. For example, let $X \subseteq \mathbb{A}^1_\mathbb{R}$ be the set of all nonzero real numbers:

$$X = \{a \in \mathbb{R} \mid a \neq 0\} \subsetneq \mathbb{A}^1_\mathbb{R}.$$

Let us argue that X cannot be the vanishing set of any set of polynomials. Suppose, toward a contradiction, that $X = \mathcal{V}(\mathcal{S})$ for some set $\mathcal{S} \subseteq \mathbb{R}[x]$. Then $\mathcal{S}$ must contain at least one nonzero element, since if $\mathcal{S}$ were either $\emptyset$ or $\{0\}$, its vanishing set would be all of $\mathbb{A}^1_\mathbb{R}$. Let $f \in \mathcal{S}$ be any nonzero element. Since $X = \mathcal{V}(\mathcal{S})$, we have $f(a) = 0$ for all $a \in \mathbb{R} \setminus \{0\}$. This means that $f \in \mathbb{R}[x]$ is a nonzero polynomial with infinitely many zeros, contradicting Corollary 0.50.

In light of examples such as this, we give a name to those special subsets of affine space that can be defined as the vanishing set of a set of polynomials.

1.10 DEFINITION *Affine variety*

A subset $X \subseteq \mathbb{A}^n$ is called an *affine variety* if $X = \mathcal{V}(\mathcal{S})$ for some set of polynomials $\mathcal{S} \subseteq K[x_1, \ldots, x_n]$.

We have already met a number of affine varieties in the examples presented in Section 1.1. Additionally, for each affine variety in Section 1.1, we specified a set of polynomials that realized it as a vanishing set. Two more examples that are perhaps more basic than any of the previous ones, but nevertheless crucial, are the empty set and the entirety of affine space.

1.11 EXAMPLE The empty set and affine space are affine varieties

The constant polynomial $1 \in K[x_1, \ldots, x_n]$ does not vanish at any point of $\mathbb{A}^n$, so

$$\mathcal{V}(1) = \emptyset \subseteq \mathbb{A}^n.$$

The zero polynomial $0 \in K[x_1, \ldots, x_n]$ vanishes at every point of $\mathbb{A}^n$, so

$$\mathcal{V}(0) = \mathbb{A}^n.$$

Thus, both $\emptyset$ and $\mathbb{A}^n$ are affine varieties.

It is quite easy, and not very enlightening, to come up with an endless list of examples of affine varieties by simply writing down sets of polynomials and considering their vanishing sets. A more subtle task is to understand what makes affine varieties special among subsets of affine space. In other words, what sorts of subsets of affine space are not affine varieties? We have already seen one example: the set $\mathbb{R} \setminus \{0\}$ is not an affine variety in $\mathbb{A}^1_\mathbb{R}$. In fact, the only affine varieties in $\mathbb{A}^1$ are finite collections of points (including $\emptyset$) and all of $\mathbb{A}^1$ (Exercise 1.2.1).

In $\mathbb{A}^2$, on the other hand, an infinite proper subset can certainly be an affine variety; the parabola in $\mathbb{A}^2_\mathbb{R}$ is an example. What a proper subset of $\mathbb{A}^2_\mathbb{R}$ cannot have, however, if it is to be an affine variety, is a nonempty interior. The following example illustrates this phenomenon.

1.12 EXAMPLE A solid square is not an affine variety

Let X be the filled-in square in $\mathbb{A}_\mathbb{R}^2$ defined as

$$X = \{(a,b) \in \mathbb{A}_\mathbb{R}^2 \mid -1 \leq a \leq 1 \text{ and } -1 \leq b \leq 1\}.$$

Suppose, toward a contradiction, that $X = \mathcal{V}(\mathcal{S})$ for some set $\mathcal{S} \subseteq \mathbb{R}[x,y]$. Consider any $f \in \mathcal{S}$; we will argue that f is the zero polynomial. Write

$$f = \sum_{i=0}^{d} f_i y^i$$

where $f_i \in \mathbb{R}[x]$ for each i. Since $f \in \mathcal{S}$, it vanishes at all values of X. In other words, for any value $a \in [-1,1]$, the single-variable polynomial

$$f(a,y) = \sum_{i=0}^{d} f_i(a) y^i \in \mathbb{R}[y]$$

vanishes at all values $b \in [-1,1]$. Since a nonzero single-variable polynomial can only have finitely many zeros, it follows that $f(a,y)$ must be the zero polynomial, implying that $f_i(a) = 0$ for all i. In other words, we have argued that the single-variable polynomials $f_i \in \mathbb{R}[x]$ vanish at all values $a \in [-1,1]$. Again, using the fact that nonzero single-variable polynomials have finitely many zeros, this implies that f_i is the zero polynomial for each i, so f is the zero polynomial. This argument shows that the only polynomial that can be in $\mathcal{S}$ is the zero polynomial, from which it follows that $X = \mathcal{V}(\mathcal{S}) = \mathbb{A}_\mathbb{R}^2$, a contradiction.

Students with a background in topology are encouraged to prove, more generally, that if one gives $\mathbb{A}_\mathbb{R}^n$ the Euclidean topology, then the only affine variety in $\mathbb{A}_\mathbb{R}^n$ that has a nonempty topological interior is the entirety of $\mathbb{A}_\mathbb{R}^n$ (Exercise 1.2.8).

1.13 EXAMPLE The graph of e^x in $\mathbb{A}_\mathbb{R}^2$ is not an affine variety

Let X be the graph of the exponential function on the real numbers:

(1.14) $$X = \{(a,b) \in \mathbb{A}_\mathbb{R}^2 \mid b = e^a\} \subseteq \mathbb{A}_\mathbb{R}^2.$$

A careful proof that X is not an affine variety is outlined in Exercise 1.2.7. Intuitively, it should seem reasonable that X is not an affine variety: it is defined by the vanishing of the expression $y - e^x$, which is not a polynomial in x and y. One should be careful with this sort of reasoning, however. For example, the expression $\sin^2(x) + \cos^2(x) + y$ is not a polynomial either, but the set

$$\{(a,b) \in \mathbb{A}_\mathbb{R}^2 \mid \sin^2(a) + \cos^2(a) + b = 0\} \subseteq \mathbb{A}_\mathbb{R}^2$$

is the same as the set

$$\{(a,b) \in \mathbb{A}_\mathbb{R}^2 \mid 1 + b = 0\} = \mathcal{V}(1+y) \subseteq \mathbb{A}_\mathbb{R}^2,$$

so it is an affine variety.

1.2. AFFINE VARIETIES

It is certainly possible for the same affine variety to arise as $\mathcal{V}(\mathcal{S})$ for different sets $\mathcal{S}$. For example,

$$\mathcal{V}(x,y) = \mathcal{V}(x+y, x-y) = \{(0,0)\} \subseteq \mathbb{A}^2,$$

as the reader can readily verify. In fact, the set $\mathcal{S}$ can be replaced by the entire ideal $\langle \mathcal{S} \rangle \subseteq K[x_1, \ldots, x_n]$ that is generated by $\mathcal{S}$ without affecting its vanishing set.

> **1.15 PROPOSITION** *Affine varieties are defined by ideals*
> If $\mathcal{S} \subseteq K[x_1, \ldots, x_n]$ is a set of polynomials, then
> $$\mathcal{V}(\mathcal{S}) = \mathcal{V}(\langle \mathcal{S} \rangle).$$

PROOF Exercise 1.2.2. □

To put the discussion of this section schematically, we view the $\mathcal{V}$-operator as a function

$$\mathcal{V} : \{\text{subsets of } K[x_1, \ldots, x_n]\} \longrightarrow \{\text{subsets of } \mathbb{A}^n\}.$$

This function is not surjective, since not every subset of $\mathbb{A}^n$ is an affine variety, but it becomes surjective, by definition, if we restrict the codomain to affine varieties:

$$\mathcal{V} : \{\text{subsets of } K[x_1, \ldots, x_n]\} \longrightarrow \{\text{affine varieties in } \mathbb{A}^n\}.$$

The $\mathcal{V}$-operator is also not injective, because different subsets of $K[x_1, \ldots, x_n]$ can define the same affine variety. As a first pass toward making it bijective, Proposition 1.15 shows that we can restrict the domain to ideals and maintain a surjective function:

$$\mathcal{V} : \{\text{ideals of } K[x_1, \ldots, x_n]\} \longrightarrow \{\text{affine varieties in } \mathbb{A}^n\}.$$

In fact, this operator is still not injective (see Exercise 1.2.3). It will take a further restriction on the domain and an assumption on the ground field K in order to finally obtain a bijective version of the $\mathcal{V}$-operator.

Exercises for Section 1.2

1.2.1 Prove that the only affine varieties in $\mathbb{A}^1$ are finite collections of points and all of $\mathbb{A}^1$.

1.2.2 Prove Proposition 1.15.

1.2.3 Give an example of different ideals $I, J \in K[x]$ such that $\mathcal{V}(I) = \mathcal{V}(J)$.

1.2.4 Prove that the set

$$X = \{(a, a^2, a^3) \mid a \in K\} \subseteq \mathbb{A}^3$$

is an affine variety by finding a set of polynomials $\mathcal{S}$ for which $X = \mathcal{V}(\mathcal{S})$. (The variety X is called the *affine twisted cubic curve*.)

1.2.5 Let K be an infinite field. A *plane* in $\mathbb{A}^3_K$ is an affine variety P that can be defined as the vanishing set of a nonconstant linear polynomial:

$$P = \mathcal{V}(Ax + By + Cz + D),$$

where $A, B, C, D \in K$ and A, B, C are not all zero. Prove the following.

(a) For any polynomials $f_1, f_2, f_3 \in K[x]$ of degree at most two, the set

$$X = \{(f_1(a), f_2(a), f_3(a)) \mid a \in K\} \subseteq \mathbb{A}^3$$

is contained in at least one plane. (Hint: Set up linear equations for A, B, C, D and argue that there is at least one solution.)

(b) The affine twisted cubic of Exercise 1.2.4 is not contained in any plane. (Thus, (a) fails with "two" replaced by "three.")

1.2.6 Let K be an infinite field. Prove that the set

$$X = \{(a, ab) \mid a, b \in K\} \subseteq \mathbb{A}^2$$

is not an affine variety.

1.2.7 Prove that the vanishing set X of the expression $y - e^x$ inside $\mathbb{A}^2_{\mathbb{R}}$ is not an affine variety, possible using the following strategy.

(a) Suppose, toward a contradiction, that $X = \mathcal{V}(S)$ for some $S \subseteq \mathbb{R}[x, y]$. Explain why there exists a nonzero polynomial $f \in \mathbb{R}[x, y]$ such that $f(a, e^a) = 0$ for all $a \in \mathbb{R}$.

(b) For any polynomial f as above, write

$$f(x, y) = p_0(x) + p_1(x)y + p_2(x)y^2 + \cdots + p_d(x)y^d,$$

where $p_d \neq 0 \in \mathbb{R}[x]$. Show that

$$\frac{p_0(a)}{e^{da}} + \frac{p_1(a)}{e^{(d-1)a}} + \frac{p_2(a)}{e^{(d-2)a}} + \cdots + \frac{p_{d-1}(a)}{e^a} + p_d(a) = 0$$

for all $a \in \mathbb{R}$.

(c) Take the limit of the above expression as $a \to \infty$ to conclude that

$$\lim_{a \to \infty} p_d(a) = 0.$$

By arguing that $\lim_{a \to \infty} g(a) \neq 0$ for any nonzero $g \in \mathbb{R}[x]$, deduce a contradiction.

1.2.8 (For students with some knowledge of topology) View $\mathbb{A}^n_{\mathbb{R}} = \mathbb{R}^n$ as a topological space with the Euclidean topology. Let $X \subseteq \mathbb{A}^n_{\mathbb{R}}$ be any affine variety other than $\mathbb{A}^n_{\mathbb{R}}$ itself. Prove that, as a topological subspace of $\mathbb{A}^n_{\mathbb{R}}$, the interior of X is empty. (The same result also holds, by a similar proof, with $\mathbb{R}$ replaced by $\mathbb{C}$.)

Section 1.3 The $\mathcal{I}$-operator

In Section 1.1, we learned how to associate, to any subset of $K[x_1, \ldots, x_n]$, a subset of $\mathbb{A}^n$ via the $\mathcal{V}$-operator. In this section, we reverse the procedure, describing an operator that associates a subset of $K[x_1, \ldots, x_n]$ to any subset of $\mathbb{A}^n$.

1.16 DEFINITION *Vanishing ideal*

Let $X \subseteq \mathbb{A}^n$ be a subset. The *vanishing ideal* of X is defined by

$$\mathcal{I}(X) = \{f \in K[x_1, \ldots, x_n] \mid f(a_1, \ldots, a_n) = 0 \text{ for all } (a_1, \ldots, a_n) \in X\}.$$

In other words, the vanishing ideal of X is the set of all polynomials that vanish on all of X. As the name suggests, the set $\mathcal{I}(X)$ is more than just a subset of the polynomial ring $K[x_1, \ldots, x_n]$, it is an ideal (Exercise 1.3.1).

1.17 EXAMPLE Polynomials vanishing at $(0,0)$

Let $X = \{(0,0)\} \subseteq \mathbb{A}^2$. A polynomial $f \in K[x,y]$ vanishes at $(0,0)$ if and only if the constant term of f is zero. As explained in Example 0.26, the set of all such polynomials comprises the ideal $\langle x,y \rangle$. Thus, $\mathcal{I}(X) = \langle x,y \rangle$.

1.18 EXAMPLE Vanishing ideals in one variable

Let $X = \{1,3\} \subseteq \mathbb{A}^1_\mathbb{R}$. By Corollary 0.48, a polynomial $f \in \mathbb{R}[x]$ vanishes at 1 if and only if $x - 1$ divides f, and f vanishes at 3 if and only if $x - 3$ divides f. Since $x - 1$ and $x - 3$ are irreducible, it follows from unique factorization in $\mathbb{R}[x]$ that

$$\mathcal{I}(X) = \{f \in \mathbb{R}[x] \mid (x-1)(x-3) \text{ divides } f\} = \langle x^2 - 4x + 3 \rangle.$$

A similar procedure computes the vanishing ideal of any $X \subseteq \mathbb{A}^1$ (Exercise 1.3.3).

1.19 EXAMPLE Vanishing ideal of the parabola

Let K be an infinite field and let X be the affine variety defined by

$$X = \mathcal{V}(y - x^2) = \{(a, a^2) \mid a \in K\} \subseteq \mathbb{A}^2.$$

Which polynomials vanish at every point of X? Certainly any polynomial of the form $(y - x^2)f(x,y)$ vanishes at every point of X, so $\langle y - x^2 \rangle \subseteq \mathcal{I}(X)$. Let's show that the reverse containment also holds.

Let $f \in \mathcal{I}(X)$. The arguments in Example 0.29 show that $[f(x,y)] = [f(x,x^2)]$ in the quotient ring $K[x,y]/\langle y - x^2 \rangle$. In particular, this means that

(1.20) $$f(x,y) - f(x,x^2) \in \langle y - x^2 \rangle.$$

Using our assumption that f vanishes on $X = \{(a, a^2) \mid a \in K\}$, we see that $f(a, a^2) = 0$ for every $a \in K$, which implies that $f(x, x^2) \in K[x]$ is a single-variable polynomial with infinitely many zeros. This is only possible if $f(x, x^2)$ is the zero polynomial. Substituting $f(x, x^2) = 0$ into equation (1.20), we see that $f \in \langle y - x^2 \rangle$, implying that $\mathcal{I}(X) \subseteq \langle y - x^2 \rangle$.

Having argued both inclusions, we conclude that $\mathcal{I}(X) = \langle y - x^2 \rangle$.

Even in the simple case of the parabola in Example 1.19, it was already somewhat involved to show that the vanishing ideal was $\langle y - x^2 \rangle$. Indeed, computing the vanishing ideals of affine varieties in general is a nontrivial task that usually requires ad hoc methods in each case (see Exercises 1.3.3–1.3.8 for more examples of such computations). As we will see in Section 1.5, the Nullstellensatz greatly simplifies the task of computing vanishing ideals.

Now that we have introduced both the $\mathcal{V}$- and $\mathcal{I}$-operators, which pass between subsets of $K[x_1, \ldots, x_n]$ and subsets of $\mathbb{A}^n$, it is natural to wonder to what extent these operators are inverse to each other. The next result provides a first answer, showing that they are not generally inverse, but that they become inverse upon being restricted to vanishing ideals and affine varieties.

1.21 PROPOSITION *Composing $\mathcal{V}$- and $\mathcal{I}$-operators*

1. Let $X \subseteq \mathbb{A}^n$. Then
$$\mathcal{V}(\mathcal{I}(X)) \supseteq X,$$
with equality if and only if $X = \mathcal{V}(\mathcal{S})$ for some $\mathcal{S} \subseteq K[x_1, \ldots, x_n]$.

2. Let $\mathcal{S} \subseteq K[x_1, \ldots, x_n]$. Then
$$\mathcal{I}(\mathcal{V}(\mathcal{S})) \supseteq \mathcal{S},$$
with equality if and only if $\mathcal{S} = \mathcal{I}(X)$ for some $X \subseteq \mathbb{A}^n$.

PROOF We prove Part 1 and leave Part 2 to Exercise 1.3.2.

Let $a = (a_1, \ldots, a_n) \in X$. To prove that $a \in \mathcal{V}(\mathcal{I}(X))$, consider any polynomial $f \in \mathcal{I}(X)$. By definition of the vanishing ideal, we have that $f(b) = 0$ for all $b \in X$. In particular, $f(a) = 0$. Thus, we have proved that, for every $f \in \mathcal{I}(X)$, $f(a) = 0$. This implies that $a \in \mathcal{V}(\mathcal{I}(X))$.

If equality holds, then X is the vanishing set of $\mathcal{S} = \mathcal{I}(X)$, proving one direction of the if-and-only-if statement. To prove the converse, suppose that $X = \mathcal{V}(\mathcal{S})$ for some set $\mathcal{S} \subseteq K[x_1, \ldots, x_n]$; we must prove that $\mathcal{V}(\mathcal{I}(X)) \subseteq X$. This is equivalent to proving that $a \notin X$ implies $a \notin \mathcal{V}(\mathcal{I}(X))$, so suppose the former. Since $X = \mathcal{V}(\mathcal{S})$ and $a \notin X$, there exists some $f \in \mathcal{S}$ such that $f(a) \neq 0$. Since $f \in \mathcal{S}$ and $X = \mathcal{V}(\mathcal{S})$, it follows that f vanishes on X, implying that $f \in \mathcal{I}(X)$. Because $f \in \mathcal{I}(X)$ and $f(a) \neq 0$, we conclude that $a \notin \mathcal{V}(\mathcal{I}(X))$. □

> If $X \subseteq \mathbb{A}^n$ is an affine variety, Part 1 of the proposition says that there is a distinguished ideal for which X is the vanishing set, namely $\mathcal{I}(X)$.

If $X \subseteq \mathbb{A}^n$ is a subset, then the set $\mathcal{V}(\mathcal{I}(X))$ appearing in the first part of Proposition 1.21 has another interpretation: it is the smallest affine variety containing X (Exercise 1.3.10). In this sense, it is analogous to the ideal generated by a set $\mathcal{S}$, which is the smallest ideal containing $\mathcal{S}$; we might even call $\mathcal{V}(\mathcal{I}(X))$ the "affine variety generated by X" and denote it by $\langle X \rangle$. With this notation, the last part of Proposition 1.21 becomes analogous to Proposition 1.15:

$$\mathcal{I}(X) = \mathcal{I}(\langle X \rangle).$$

1.3. THE $\mathcal{I}$-OPERATOR

From the if-and-only-if statements in Proposition 1.21, we see that the containments can certainly be strict. For example, if X is not an affine variety, then $X \neq \mathcal{V}(\mathcal{S})$ for any $\mathcal{S}$, so $X \neq \mathcal{V}(\mathcal{I}(X))$. Similarly, if $\mathcal{S}$ is not an ideal, then it cannot be the case that $\mathcal{S} = \mathcal{I}(\mathcal{V}(\mathcal{S}))$, because the latter is an ideal. In fact, even when $\mathcal{S}$ *is* an ideal, equality still need not hold, as illustrated in the next example.

1.22 EXAMPLE $\mathcal{I}(\mathcal{V}(I)) \neq I$

Let $I = \langle x^2 \rangle \subseteq K[x]$. Then

$$\mathcal{V}(I) = \{a \in K \mid a^2 = 0\}.$$

Since K is a field, $a^2 = 0$ if and only if $a = 0$. Thus, $\mathcal{V}(I) = \{0\}$. The same reasoning as in Example 1.18 shows that $\mathcal{I}(\{0\}) = \langle x \rangle$, so

$$\mathcal{I}(\mathcal{V}(I)) = \mathcal{I}(\{0\}) = \langle x \rangle.$$

The vanishing ideal $\mathcal{I}(\mathcal{V}(I)) = \langle x \rangle$ contains but is not equal to $I = \langle x^2 \rangle$. Thus, as a consequence of the if-and-only-if statement in the second part of Proposition 1.21, we conclude that I is not the vanishing ideal of any $X \subseteq \mathbb{A}^1$.

At the end of Section 1.2, we saw that the $\mathcal{V}$-operator gives a surjection

$$\mathcal{V} : \{\text{ideals in } K[x_1, \ldots, x_n]\} \longrightarrow \{\text{affine varieties in } \mathbb{A}^n\}.$$

It follows from Proposition 1.21 that the $\mathcal{V}$-operator becomes a bijection (with inverse given by the $\mathcal{I}$-operator) if we restrict the domain to the set of *vanishing ideals*—those ideals that arise as vanishing ideals of some set. Our goal then, if we want to understand the dictionary between algebra and geometry, is to obtain a better understanding of the vanishing ideals in $K[x_1, \ldots, x_n]$. Motivated by the observations in Example 1.22, we take a first step in this direction in the next section, where we introduce the algebraic notion of a radical ideal.

Exercises for Section 1.3

1.3.1 For any subset $X \subseteq \mathbb{A}^n$, prove that $\mathcal{I}(X) \subseteq K[x_1, \ldots, x_n]$ is an ideal.

1.3.2 Prove Proposition 1.21, Part 2.

1.3.3 Let $X \subseteq \mathbb{A}^1$.
  (a) Suppose that $X = \{a_1, \ldots, a_r\}$ is finite. Use unique factorization in $K[x]$ to prove that
  $$\mathcal{I}(X) = \langle (x - a_1) \cdots (x - a_r) \rangle \subseteq K[x].$$
  (b) Suppose that X is infinite. Prove that $\mathcal{I}(X) = \langle 0 \rangle$.

1.3.4 Compute the vanishing ideal of $\mathcal{V}(x^2 + 1) \subseteq \mathbb{A}^1$
  (a) over $\mathbb{R}$, and
  (b) over $\mathbb{C}$.

1.3.5 Let $X = \{(a_1, \ldots, a_n)\} \in \mathbb{A}^n$ be a single point. Prove that
$$\mathcal{I}(X) = \langle x_1 - a_1, \ldots, x_n - a_n \rangle.$$

1.3.6 Let $X = \mathcal{V}(x^2 + y^2 - 1) \subseteq \mathbb{A}^2_\mathbb{R}$ be the unit circle. This exercise proves that $\mathcal{I}(X) = \langle x^2 + y^2 - 1 \rangle$.
   (a) Prove that $\mathcal{I}(X) \supseteq \langle x^2 + y^2 - 1 \rangle$.
   (b) Prove that $\mathcal{I}(X) \subseteq \langle x^2 + y^2 - 1 \rangle$, possibly using the following proof outline. Suppose $f \in \mathcal{I}(X)$.
      i. Prove that
      $$f - g_1 - yg_2 \in \langle x^2 + y^2 - 1 \rangle$$
      for some $g_1, g_2 \in \mathbb{R}[x]$.
      ii. Use the fact that $f \in \mathcal{I}(X)$ to prove that
      $$g_1(a)^2 = (1 - a^2)g_2(a)^2$$
      for all $a \in [-1, 1]$, and conclude that $g_1(x)^2 = (1 - x^2)g_2(x)^2$.
      iii. Use unique factorization to prove that g_1 and g_2 are both the zero polynomial, and thereby conclude that $f \in \langle x^2 + y^2 - 1 \rangle$.

1.3.7 Let $X = \mathcal{V}(x_1^2 + \cdots + x_n^2 - 1) \subseteq \mathbb{A}^n_\mathbb{R}$ be the unit n-sphere. Generalize the previous exercise to prove that $\mathcal{I}(X) = \langle x_1^2 + \cdots + x_n^2 - 1 \rangle$.

1.3.8 Let K be an infinite field with $1 \neq -1$ and let $X = \mathcal{V}(x^2 + y^2 - 1) \subseteq \mathbb{A}^2_K$.
   (a) For any $c \in K$ with $c^2 \neq -1$, prove that
   $$\left(\frac{c^2 - 1}{c^2 + 1}, \frac{2c}{c^2 + 1} \right) \in X.$$
   (b) Prove that there are infinitely many values $a \in K$ such that $(a, b) \in X$ for some $b \in K$.
   (c) Prove $\mathcal{I}(X) = \langle x^2 + y^2 - 1 \rangle$, possibly by adapting the strategy of Exercise 1.3.6.

1.3.9 Let $K = \mathbb{F}_2$, the finite field with two elements. Let
$$X = \mathcal{V}(y - x^2) \subseteq \mathbb{A}^2_{\mathbb{F}_2}.$$
Prove that $\mathcal{I}(X) \neq \langle y - x^2 \rangle$, in contrast to the analogous case over $\mathbb{R}$ studied in Example 1.19. What is $\mathcal{I}(X)$?

1.3.10 Let $X \subseteq \mathbb{A}^n$ be any subset.
   (a) Prove that $\mathcal{V}(\mathcal{I}(X))$ is the smallest affine variety containing X, in the following sense: if $Y \subseteq \mathbb{A}^n$ is any affine variety and $X \subseteq Y$, then $\mathcal{V}(\mathcal{I}(X)) \subseteq Y$.
   (b) Demonstrate part (a) by choosing any set $X \subseteq \mathbb{A}^2_\mathbb{R}$ that is not an affine variety and calculating $\mathcal{V}(\mathcal{I}(X))$.

Section 1.4 Radical ideals

In the previous section, we learned that not every ideal in $K[x_1,\ldots,x_n]$ arises as a vanishing ideal of some subset in $\mathbb{A}^n$. In particular, we noticed in Example 1.22 that the ideal $\langle x^2 \rangle \subseteq K[x]$ is not a vanishing ideal. How, then, can we recognize whether a given ideal is $\mathcal{I}(X)$ for some X? We investigate one important property of vanishing ideals in this section: vanishing ideals are *radical*.

1.23 DEFINITION *Radical ideal*

An ideal $I \subseteq R$ is *radical* if, for all $a \in R$,

$$a^m \in I \text{ for some integer } m > 0 \implies a \in I.$$

In other words, an ideal is radical if it is closed under taking roots: whenever a power of an element is in the ideal, the element itself must be in the ideal. Notice that the ideal $\langle x^2 \rangle \subseteq K[x]$ from Example 1.22 is not radical, because $x^2 \in \langle x^2 \rangle$ but $x \notin \langle x^2 \rangle$. The next result says that the property of radical-ness is an attribute of all vanishing ideals.

1.24 PROPOSITION *Vanishing ideals are radical*

For any set $X \subseteq \mathbb{A}^n$, the vanishing ideal $\mathcal{I}(X)$ is a radical ideal.

In particular, Proposition 1.24 and the observation that $\langle x^2 \rangle$ is not a radical ideal together imply that $\langle x^2 \rangle$ is not a vanishing ideal, as we observed in Example 1.22.

PROOF OF PROPOSITION 1.24   Let $X \subseteq \mathbb{A}^n$ be a subset and suppose there exists a positive integer m such that $f^m \in \mathcal{I}(X)$, or in other words, such that $f^m(a) = 0$ for all $a \in X$. Then

$$0 = f^m(a) = \big(f(a)\big)^m \in K.$$

Since K is a field, it has no zero divisors; in particular, $\big(f(a)\big)^m = 0$ if and only if $f(a) = 0$. Thus, $f(a) = 0$ for all $a \in X$, so $f \in \mathcal{I}(X)$. $\qquad\square$

Proposition 1.24 suggests, in particular, that radical ideals play a central role in algebraic geometry. Consequently, we devote the rest of this section to collecting some of the fundamental notions pertaining to radical ideals.

By definition, an ideal fails to be radical if it is not closed under taking roots. You might suspect, then, that you could construct a radical ideal simply by adding in the missing roots. The next definition makes this construction precise.

1.25 DEFINITION *Radical of an ideal*

Let $I \subseteq R$ be an ideal. The *radical* of I is

$$\sqrt{I} = \{a \in R \mid a^m \in I \text{ for some } m > 0\}.$$

This process of "adding in the missing roots" indeed yields a radical ideal, as the next result shows.

1.26 PROPOSITION *Radicals are radical*

If $I \subseteq R$ is an ideal, then $\sqrt{I}$ is a radical ideal.

PROOF The proof that $\sqrt{I}$ is an ideal is Exercise 1.4.2. To prove that $\sqrt{I}$ is radical, assume $a^k \in \sqrt{I}$ for some integer $k > 0$; we must prove that $a \in \sqrt{I}$. By the definition of $\sqrt{I}$, we have

$$a^k \in \sqrt{I} \implies (a^k)^m \in I \text{ for some integer } m > 0.$$

In other words, $a^{(km)} \in I$, from which it follows that $a \in \sqrt{I}$. □

If I is a radical ideal, then it is already closed under taking roots, so $\sqrt{I} = I$. Conversely, if $\sqrt{I} = I$, then Proposition 1.26 implies that I is a radical ideal. Thus, we have proved the following useful characterization of radical ideals.

1.27 COROLLARY *Characterization of radical ideals*

The ideal $I \subseteq R$ is radical if and only if $I = \sqrt{I}$.

We now provide a few examples to help familiarize ourselves with radical ideals.

1.28 EXAMPLE

If $R = K[x]$ and $I = \langle x^2 \rangle$, then we have already seen that I is not radical. If we want to enlarge I by adding all possible roots of elements in I, what should we add? Let's show that

$$\sqrt{I} = \langle x \rangle.$$

To see this, first note that $x \in \sqrt{I}$ because $x^2 \in I$. Since $\sqrt{I}$ is an ideal, the fact that $x \in \sqrt{I}$ then implies that $\langle x \rangle \subseteq \sqrt{I}$, proving one containment.

Conversely, suppose that $f \notin \langle x \rangle$. This means that f has a nonzero constant term. It follows that f^m has a nonzero constant term for all integers $m > 0$, so $f^m \notin I$. Hence, $f \notin \sqrt{I}$, which proves by contrapositive that $\sqrt{I} \subseteq \langle x \rangle$.

1.29 EXAMPLE

If $R = \mathbb{Z}$ and $I = \langle 12 \rangle$, then I is not radical because $6^2 \in I$ but $6 \notin I$. In fact,

$$\sqrt{I} = \langle 6 \rangle.$$

To prove the containment $\sqrt{I} \subseteq \langle 6 \rangle$, let $r \in \sqrt{I}$. Then $r^m = 12k$ for some positive integer m and some integer k. Since 2 is prime and $2 \mid r^m$, it follows from Euclid's Lemma that $2 \mid r$. The same reasoning shows that $3 \mid r$. Since r is divisible by both 2 and 3, it is divisible by 6, which means that $r \in \langle 6 \rangle$, proving one containment.

Conversely, if $r \in \langle 6 \rangle$, then $r = 6k$ for some integer k. Thus,

$$r^2 = 36k^2 = 12(3k^2),$$

so $r^2 \in I$. This implies that $r \in \sqrt{I}$, verifying that $\langle 6 \rangle \subseteq \sqrt{I}$.

1.4. RADICAL IDEALS

Comparing the radical ideal $\langle 6 \rangle$ and the nonradical ideal $\langle 12 \rangle$ in the previous example, we see that 6 is *square-free*—in other words, not divisible by the square of a prime—but 12 is not square-free; it is divisible by 2^2. Arguments similar to those in Example 1.29 show that $\langle n \rangle \subseteq \mathbb{Z}$ is radical if and only if n is square-free.

In the context of algebraic geometry, it would be useful to be able to determine when an ideal $I \subseteq K[x_1, \ldots, x_n]$ is radical. For general ideals, this can be quite difficult. However, the situation is analogous to the case of the integers when I is a principal ideal. To state the result precisely, let $f \in K[x_1, \ldots, x_n]$ be a nonconstant polynomial, and consider an irreducible factorization:

$$f = p_1 \cdots p_m.$$

Because of irreducibility, $p_i \mid p_j$ if and only if they differ by a constant. By collecting all of the terms that differ by a constant, we can write

$$(1.30) \qquad f = a q_1^{k_1} \cdots q_\ell^{k_\ell}$$

where $a \in K$ is nonzero, each q_i is irreducible, and $q_i \nmid q_j$ whenever $i \neq j$. We say that (1.30) is a *distinct irreducible factorization* of f and that the q_i are the *distinct irreducible factors* of f. It follows from unique factorization that the distinct irreducible factors are unique up to reordering and multiplying by constants.

The next result describes the radical of a principal ideal in terms of these factors.

1.31 PROPOSITION *Radicals of principal ideals*

If $f \in K[x_1, \ldots, x_n]$ is a nonconstant polynomial with distinct irreducible factors $q_1, \ldots, q_\ell$, then

$$\sqrt{\langle f \rangle} = \langle q_1 \cdots q_\ell \rangle.$$

In particular, $\langle f \rangle$ is radical if and only if f is not divisible by the square of a nonconstant polynomial.

PROOF We prove both inclusions in the equality and leave the deduction of the if-and-only-if assertion to Exercise 1.4.3.

($\subseteq$) Suppose that $g \in \sqrt{\langle f \rangle}$. Then $g^m = hf$ for some $m > 0$ and some $h \in K[x_1, \ldots, x_n]$. In particular, it follows that $q_i \mid g^m$ for every i. By uniqueness of distinct irreducible factors, each q_i must be one of the distinct irreducible factors of g^m. Since the distinct irreducible factors of g^m are the same as those of g, then each q_i must also be one of the distinct irreducible factors of g. It then follows that $q_1 \cdots q_\ell \mid g$, so $g \in \langle q_1 \cdots q_\ell \rangle$.

($\supseteq$) Suppose that $g \in \langle q_1 \cdots q_\ell \rangle$ and write $g = hq_1 \cdots q_\ell$ for some polynomial h. Regarding the distinct irreducible factorization

$$f = a q_1^{k_1} \cdots q_\ell^{k_\ell},$$

set $m = \max\{k_1, \ldots, k_\ell\}$. It follows that

$$g^m = h^m q_1^m \cdots q_\ell^m = \left(h^m q_1^{m-k_1} \cdots q_\ell^{m-k_\ell} a^{-1}\right) f \in \langle f \rangle,$$

which implies that $g \in \sqrt{\langle f \rangle}$. $\square$

1.32 EXAMPLE $\langle x^2 + y^2 \rangle \subseteq \mathbb{C}[x,y]$ is radical

The irreducible factorization of $x^2 + y^2$ in $\mathbb{C}[x,y]$ is

$$x^2 + y^2 = (x - iy)(x + iy).$$

Since the two irreducible factors are distinct, $x^2 + y^2$ is square-free. Thus, the ideal $\langle x^2 + y^2 \rangle \subseteq \mathbb{C}[x,y]$ is radical.

1.33 EXAMPLE A radical ideal that is not a vanishing ideal

Over the real numbers, the polynomial $x^2 + y^2 \in \mathbb{R}[x,y]$ is irreducible. Thus, by Proposition 1.31, the ideal $\langle x^2 + y^2 \rangle$ is radical. However, since the origin is the only point at which $x^2 + y^2$ vanishes, it follows that

$$\mathcal{I}(\mathcal{V}(\langle x^2+y^2\rangle)) = \mathcal{I}(\{(0,0)\}) = \langle x,y \rangle \supsetneq \langle x^2+y^2 \rangle.$$

Thus, Proposition 1.21 implies that $\langle x^2 + y^2 \rangle$ is not a vanishing ideal. In particular, this example shows that the converse of Proposition 1.24 does not hold over $\mathbb{R}$: radical ideals need not be vanishing ideals.

In our study of rings, we have now had the opportunity to meet three special types of ideals: maximal, prime, and radical. We already know that maximal ideals are prime. The next result adds radical ideals to this hierarchy.

1.34 PROPOSITION *Prime and maximal ideals are radical*

Every prime ideal is a radical ideal.

PROOF Toward proving the contrapositive, assume that I is an ideal that is not radical. Choose an element $a \notin I$ such that $a^m \in I$ for some $m > 1$. If m_0 is the smallest integer greater than 1 such that $a^{m_0} \in I$, then neither a nor a^{m_0-1} are elements of I, but their product a^{m_0} is an element of I. Thus, I is not prime. $\square$

Schematically, for any ring R, we have the following hierarchy of ideals:

$$\{\text{ideals}\} \supseteq \{\text{radical ideals}\} \supseteq \{\text{prime ideals}\} \supseteq \{\text{maximal ideals}\}.$$

While these inclusions are not strict for every ring, they are strict for multivariable polynomial rings over fields (Exercise 1.4.4).

To conclude this section, we return to our overarching goal of obtaining a bijection between algebraic objects and geometric objects. Since $\mathcal{V}(\mathcal{I}(X)) = X$ for all affine varieties $X \subseteq \mathbb{A}^n$ (Proposition 1.21) and $\mathcal{I}(X)$ is radical (Proposition 1.24), the $\mathcal{V}$-operator remains a surjection upon restricting the domain:

(1.35) $\mathcal{V} : \{\text{radical ideals in } K[x_1, \ldots, x_n]\} \longrightarrow \{\text{affine varieties in } \mathbb{A}^n\}.$

One might be so optimistic as to hope that (1.35) is our sought-after bijection between algebraic objects and geometric objects. Unfortunately, Example 1.33 provides a counterexample: $\langle x^2 + y^2 \rangle$ and $\langle x, y \rangle$ are distinct radical ideals in $\mathbb{R}[x,y]$

1.4. RADICAL IDEALS

with the same vanishing set $\{(0,0)\} \in \mathbb{A}_\mathbb{R}^2$. Nonetheless, if we make one additional assumption on the ground field K—that it is *algebraically closed*—then (1.35) is, indeed, the bijection we desire. This result is a consequence of the Nullstellensatz, to which we turn in the next section.

Exercises for Section 1.4

1.4.1 Determine which of the following ideals are radical. For those that are not radical, compute their radical.
   (a) $\langle 4 \rangle \subseteq \mathbb{Z}$
   (b) $\langle 6 \rangle \subseteq \mathbb{Z}$
   (c) $\langle 18 \rangle \subseteq \mathbb{Z}$
   (d) $\langle x^2 + y^2 - 1 \rangle \subseteq \mathbb{R}[x,y]$
   (e) $\langle x^2, y^3 \rangle \subseteq \mathbb{R}[x,y]$
   (f) $\langle y - x^2, y \rangle \subseteq \mathbb{R}[x,y]$
   (g) $\langle x^2 - y^2 \rangle \subseteq \mathbb{R}[x,y]$

1.4.2 Let $I \subseteq R$ be an ideal. Prove that $\sqrt{I} \subseteq R$ is an ideal. (Hint: Use the binomial theorem to prove that $\sqrt{I}$ is closed under addition/subtraction.)

1.4.3 Prove that a principal ideal $\langle f \rangle \subseteq K[x_1, \ldots, x_n]$ is radical if and only if f is not divisible by the square of a nonconstant polynomial.

1.4.4 (a) Give an example of an ideal $I \subseteq K[x,y]$ that is not radical.
   (b) Give an example of a radical ideal $I \subseteq K[x,y]$ that is not prime.
   (c) Give an example of a prime ideal $I \subseteq K[x,y]$ that is not maximal.

1.4.5 Prove, by example, that an ideal $\langle f_1, f_2 \rangle \subseteq K[x_1, \ldots, x_n]$ need not be radical even if f_1 and f_2 are both square-free.

1.4.6 Prove that the set of all zero divisors in a ring R is a radical ideal.

1.4.7 Let R be a ring and let $I \subseteq R$ be an ideal. Prove that $\sqrt{I} = R$ if and only if $I = R$.

1.4.8 Let R be a ring and let $I, J \subseteq R$ be ideals. Prove that
$$\sqrt{I \cap J} = \sqrt{I} \cap \sqrt{J}.$$

1.4.9 Let R be a ring and let $I \subseteq R$ be an ideal. Prove that $\sqrt{I}$ is the intersection of all prime ideals in R that contain I.

Open Access This chapter is licensed under the terms of the Creative Commons Attribution-NonCommercial 4.0 International License (http://creativecommons.org/licenses/by-nc/4.0/), which permits any noncommercial use, sharing, adaptation, distribution and reproduction in any medium or format, as long as you give appropriate credit to the original author(s) and the source, provide a link to the Creative Commons license and indicate if changes were made.

The images or other third party material in this chapter are included in the chapter's Creative Commons license, unless indicated otherwise in a credit line to the material. If material is not included in the chapter's Creative Commons license and your intended use is not permitted by statutory regulation or exceeds the permitted use, you will need to obtain permission directly from the copyright holder.

Section 1.5 The Nullstellensatz

In this section, we state the theorem that forms the foundation of much of algebraic geometry: the *Nullstellensatz*. This result allows us to set up a powerful correspondence between affine varieties and radical ideals in polynomial rings, which is the backbone of the dictionary between the realms of geometry and algebra.

> *"Nullstellensatz" is a German word composed of "Nullstellen" (zeros) and "Satz" (theorem).*

To state the theorem, we require a key assumption on the ground field.

1.36 DEFINITION *Algebraically closed field*

A field K is said to be *algebraically closed* if every nonconstant polynomial in $K[x]$ has at least one zero in K.

For example, the fields $\mathbb{Q}$ and $\mathbb{R}$ are not algebraically closed because $x^2 + 1$ does not have any rational or real zeros. However, $x^2 + 1$ does have zeros in $\mathbb{C}$, namely i and $-$i. In fact, every nonconstant polynomial in $\mathbb{C}[x]$ has at least one zero, which is the statement of the Fundamental Theorem of Algebra.

1.37 THEOREM *Fundamental Theorem of Algebra*

The field $\mathbb{C}$ is algebraically closed.

Although the Fundamental Theorem of Algebra is often taught at an early stage, it is by no means obvious. There are many proofs, including arguments via Galois theory, complex analysis, and topology. None of these falls within the scope of this book, and the fact that $\mathbb{C}$ is algebraically closed is not necessary for the logical development of the material. The Fundamental Theorem of Algebra is introduced here simply to emphasize that there is at least one familiar and concrete example of an algebraically closed field, and we encourage the reader to accept it without proof.

As we observed in Section 1.1, much of algebraic geometry is more straightforward when the ground field is infinite, because one no longer needs to draw a distinction between polynomials and the functions they define. One advantage of working with algebraically closed fields is that they are automatically infinite.

1.38 PROPOSITION *Algebraically closed fields are infinite*

If K is an algebraically closed field, then K is infinite.

PROOF Exercise 1.5.1. □

In particular, none of the finite fields $\mathbb{F}_p = \mathbb{Z}/\langle p \rangle$ for any prime $p \in \mathbb{Z}$ are algebraically closed. What fields are algebraically closed, then, besides $\mathbb{C}$? There are perhaps no other familiar examples of algebraically closed fields, but there is a procedure by which one can construct, from any field K, an *algebraic closure* $\overline{K}$, which is the smallest algebraically closed field in which K is contained. The most

1.5. THE NULLSTELLENSATZ

familiar application of this construction says that $\overline{\mathbb{R}} = \mathbb{C}$. Applying this procedure to any field at all yields a host of new examples of algebraically closed fields, albeit not particularly familiar ones: $\overline{\mathbb{Q}}$, $\overline{\mathbb{F}_p}$, and so on. Readers unfamiliar with this material are encouraged, whenever K is assumed to be algebraically closed, to think of the more familiar setting $K = \mathbb{C}$.

Having discussed what it means for a field to be algebraically closed, we can now state the Nullstellensatz, the proof of which is deferred to Chapter 5.

1.39 THEOREM *Nullstellensatz*

Let K be an algebraically closed field. Then, for any ideal $I \subseteq K[x_1, \ldots, x_n]$,

$$\mathcal{I}(\mathcal{V}(I)) = \sqrt{I}.$$

The containment $\sqrt{I} \subseteq \mathcal{I}(\mathcal{V}(I))$ is true over any field, and can be proved directly from the definitions (Exercise 1.5.3). The other inclusion requires a good deal of work, for which K being algebraically closed is essential.

The Nullstellensatz helps us answer the motivating question posed at the end of Section 1.2: on what domain and codomain does the $\mathcal{V}$-operator become a bijection? Over algebraically closed fields, the answer is that $\mathcal{V}$ is a bijection between radical ideals and affine varieties. This is the first key instance of the precise dictionary between algebra and geometry.

1.40 COROLLARY *Radical ideals and affine varieties*

If K is algebraically closed, then

$$\mathcal{V} : \{\text{radical ideals of } K[x_1, \ldots, x_n]\} \longrightarrow \{\text{affine varieties in } \mathbb{A}^n\}$$

is a bijection with inverse $\mathcal{I}$.

PROOF Since a function is bijective if and only if it has an inverse, it suffices to prove that $\mathcal{I}$ is the inverse of $\mathcal{V}$. In other words, we must show that, for any affine variety $X \subseteq \mathbb{A}^n$,

(1.41) $$\mathcal{V}(\mathcal{I}(X)) = X$$

and, for any radical ideal $I \subseteq K[x_1, \ldots, x_n]$,

(1.42) $$\mathcal{I}(\mathcal{V}(I)) = I.$$

The equality (1.41) is the first part of Proposition 1.21, so it is true without any assumptions on K. To prove (1.42), notice that

$$\mathcal{I}(\mathcal{V}(I)) = \sqrt{I} = I,$$

where the first equality is the Nullstellensatz and the second is the characterization of radical ideals as those ideals that are equal to their radical (Corollary 1.27). □

In addition to providing a bijection between affine varieties and radical ideals, the Nullstellensatz is also a useful tool for computing vanishing ideals. In particular, it is often much easier to determine whether an ideal is radical than it is to determine whether it is a vanishing ideal. The following examples illustrate this point.

1.43 EXAMPLE $\mathcal{I}(\mathcal{V}(y - x^2)) = \langle y - x^2 \rangle$

Let K be algebraically closed and consider the affine variety

$$X = \mathcal{V}(y - x^2) \subseteq \mathbb{A}^2.$$

Since every algebraically closed field is infinite, we know from Example 1.19 that $\mathcal{I}(X) = \langle y - x^2 \rangle$. However, the argument in that example was somewhat involved and special to the particular polynomial $y - x^2$. For algebraically closed fields, there is a much simpler argument that applies to all irreducible polynomials.

In particular, knowing that $y - x^2 \in K[x, y]$ is an irreducible polynomial, it follows that $\langle y - x^2 \rangle \subseteq K[x, y]$ is a prime ideal (Proposition 0.62) and therefore radical (Proposition 1.34). Thus, by the Nullstellensatz,

$$\mathcal{I}(X) = \mathcal{I}(\mathcal{V}(y - x^2)) = \sqrt{\langle y - x^2 \rangle} = \langle y - x^2 \rangle.$$

1.44 EXAMPLE $\mathcal{I}(\mathcal{V}(f)) = \langle f \rangle$ when f is square-free

Generalizing the previous example, we see that, whenever K is algebraically closed and $f \in K[x_1, \ldots, x_n]$ is not divisible by the square of a nonzero polynomial, we have

$$\mathcal{I}(\mathcal{V}(f)) = \sqrt{\langle f \rangle} = \langle f \rangle.$$

The first equality is the Nullstellensatz and the second follows from Proposition 1.31. In particular, if f is irreducible, then the vanishing ideal of $\mathcal{V}(f)$ is simply $\langle f \rangle$.

Notice that the conclusion of this example fails when K is not algebraically closed. For instance, consider the irreducible polynomial $x^2 + 1 \in \mathbb{R}[x]$. Since this polynomial does not have any zeros,

$$\mathcal{I}(\mathcal{V}(x^2 + 1)) = \mathcal{I}(\emptyset) = K[x] \supsetneq \langle x^2 + 1 \rangle.$$

The bijection between radical ideals and affine varieties merely scratches the surface of the rich dictionary that we will continue to build between algebra and geometry. To draw an analogy with languages, the bijection in Corollary 1.40 should be thought of as a translation of nouns between two languages; such a translation might allow us to have very simple conversations, but if we want to take full advantage of the richness of language, we should also translate the verbs, the adjectives, the adverbs, and so on.

Over the course of the next three chapters, we will continue to build the dictionary between algebra and geometry, assuming throughout that the Nullstellensatz holds. As we do so, we will have the opportunity to introduce a number of new algebraic notions that are useful along the way. In Chapter 5, once we have developed a more robust algebraic foundation and a fuller appreciation of the dictionary between algebra and geometry, we will return to the proof of the Nullstellensatz.

1.5. THE NULLSTELLENSATZ

ASSUMPTIONS REGARDING THE GROUND FIELD K

The Nullstellensatz is the backbone of algebraic geometry, and as such, **we assume for the remainder of this book, unless otherwise stated, that K is an algebraically closed field.** Many of the definitions and results that we develop remain valid over general fields. Others, however, require slight modifications, and some are just outright wrong in the non-algebraically-closed setting. To help the reader appreciate our assumptions, we regularly turn to the setting of $K = \mathbb{R}$ to illustrate nonexamples of results where being algebraically closed is essential.

Even though the central results of algebraic geometry do not hold over the field of real numbers—because it is not algebraically closed—much of our geometric intuition for affine varieties arises from viewing solutions of polynomials over $\mathbb{R}$. Indeed, every geometric image of a vanishing set in Section 1.1 depicts an affine variety over $\mathbb{R}$. As algebraic geometers, it is important to develop the skill of using our knowledge and intuition over $\mathbb{R}$ as a source of insight, while at the same time not being misled by phenomena that may occur in that special setting as a result of the fact that $\mathbb{R}$ is not algebraically closed.

As we move forward, even though our ground field will always be assumed to be algebraically closed, we will continue to discuss and depict examples of varieties by looking at their solutions over $\mathbb{R}$, and we will continue to use familiar words from our years of experience working with these sets. For example, we refer to the variety $\mathcal{V}(y - x^2) \subseteq \mathbb{A}^2$ as a *parabola* and $\mathcal{V}(x^2 + y^2 + z^2 - 1) \subseteq \mathbb{A}^3$ as the *unit sphere*, even though, over general fields, these varieties may not closely resemble the geometric picture in our mind that the words *parabola* and *sphere* connote. Since $\mathbb{R}$ is a subset of the algebraically closed field $\mathbb{C}$, the reader is welcome to assume $K = \mathbb{C}$ throughout, in which case the images over $\mathbb{R}$ depicted in the examples are a subset of the full solution set over $\mathbb{C}$. The images do not give us the whole picture, but they at least provide a glimpse—a slice, if you will—into the nature of the variety.

Another important attribute of a field is its *characteristic*. Recall that the characteristic of K, denoted $\text{char}(K)$, is the smallest positive integer p such that

$$\underbrace{1 + \cdots + 1}_{p} = 0 \in K.$$

If no such p exists, as is the case for $\mathbb{Q}$, $\mathbb{R}$, and $\mathbb{C}$, then we say that the field has characteristic 0. All of the results in this book hold for a general algebraically closed field. However, in examples, we may want to avoid a finite list of characteristics because they might exhibit unusual behavior with particular types of polynomials. For example, when $\text{char}(K) \neq 2$, the polynomial $x^2 - y^2 - 1$ is irreducible, but if $1 + 1 = 0 \in K$, then

$$x^2 + y^2 - 1 = (x + y + 1)^2.$$

Rather than mentioning the exceptional characteristics, we often assume for simplicity in specific examples that $K = \mathbb{C}$, even though the conclusions being drawn in the examples usually extend to general algebraically closed fields with only finitely many exceptions on the characteristic.

Exercises for Section 1.5

1.5.1 Prove that any algebraically closed field is infinite.
(Hint: If $K = \{a_1, \ldots, a_r\}$ is finite, can you construct a polynomial in $K[x]$ with no zeros in K?)

1.5.2 Let K be algebraically closed and let $f \in K[x]$ be a polynomial of degree d. Prove that there exist $a_0, a_1, \ldots, a_d \in K$, not necessarily distinct, such that
$$f = a_0(x - a_1) \cdots (x - a_d).$$

1.5.3 Let K be any field and $I \subseteq K[x_1, \ldots, x_n]$ an ideal. Prove that one inclusion of the Nullstellensatz holds without assuming that K is algebraically closed:
$$\sqrt{I} \subseteq \mathcal{I}(\mathcal{V}(I)).$$

1.5.4 Prove that the Nullstellensatz fails for any field that is not algebraically closed.

1.5.5 Let K be algebraically closed and let $I \subseteq K[x_1, \ldots, x_n]$ be an ideal. Assuming the Nullstellensatz, prove that $\mathcal{V}(I) = \emptyset$ if and only if $I = K[x_1, \ldots, x_n]$. (This result is often called the *Weak Nullstellensatz*.)

1.5.6 Assuming the Nullstellensatz, calculate $\mathcal{I}(X)$ for the following varieties.
  (a) $X = \mathcal{V}(x^2 - y^3, x^2 + y^3) \subseteq \mathbb{A}_\mathbb{C}^2$
  (b) $X = \mathcal{V}(x) \cup \mathcal{V}(y - z) \subseteq \mathbb{A}_\mathbb{C}^3$
  (c) $X = \mathcal{V}((x^2y^2 + yz^2 + x^3z^2)(x^2 + y^2 + 1)) \subseteq \mathbb{A}_\mathbb{C}^3$

1.5.7 Let K be algebraically closed. For each $a \in K$, define
$$X_a = \mathcal{V}(y - x^2, y - a) \subseteq \mathbb{A}_K^2.$$

  (a) For what values of a do we have $\mathcal{I}(X_a) = \langle y - x^2, y - a \rangle$? When this is not the case, what is $\mathcal{I}(X_a)$?
  (b) Draw a picture over $\mathbb{R}$ of the affine varieties X_a for several representative values of $a \in \mathbb{R}$. Can you explain, geometrically, the difference between the values of a for which the equality in part (a) holds and the values of a for which it does not hold?

1.5.8 Assume $\text{char}(K) = 2$. Prove that $x^2 + y^2 - 1 = (x + y + 1)^2 \in K[x]$.

1.5.9 Generalizing the previous example, assume $\text{char}(K) = p$ for some prime $p \in \mathbb{Z}$ and show that
$$\left(\sum_{i=1}^m f_i\right)^p = \sum_{i=1}^m f_i^p \in K[x_1, \ldots, x_n]$$
for any $f_1, \ldots, f_m \in K[x_1, \ldots, x_n]$.

Chapter 2

Irreducibility of Affine Varieties

> LEARNING OBJECTIVES FOR CHAPTER 2
> - Give examples of inclusions, intersections, and unions of affine varieties, and describe each in terms of the corresponding ideals.
> - Prove that every affine variety can be written as the vanishing set of a finite set of polynomials.
> - Define what it means for an affine variety to be irreducible, and detect irreducibility in examples.
> - Compute irreducible decompositions of affine varieties.
> - Describe how the dictionary between radical ideals and affine varieties behaves when restricted to prime or maximal ideals.

When studying the integers, a key tool is the existence of prime factorizations. There is an analogue when studying polynomials (factorization into irreducibles) or when studying finite abelian groups (decomposition as a direct sum of cyclic groups). These settings all demonstrate the way in which one can understand a class of mathematical objects by specifying the atomic, indecomposable objects as well as how a general object decomposes into its atomic pieces. In this chapter, we apply this philosophy to affine varieties, introducing the notion of an *irreducible* affine variety and describing how every affine variety decomposes uniquely as a finite union of irreducible affine varieties—its *irreducible components*.

In order to get there, it is necessary to lay some preliminary groundwork. First, since our goal is to prove that every affine variety can be written as a union of its irreducible components, we need a general understanding of how affine varieties behave with respect to set-theoretic notions like inclusions, intersections, and unions, which we discuss in Section 2.1. Furthermore, just as an integer can be factored into primes by a process of repeated factorization that eventually terminates, we need to be sure that the analogous process for affine varieties cannot produce an infinitely nested chain of smaller varieties. This condition translates to a purely algebraic property of $K[x_1, \ldots, x_n]$—that it is a *Noetherian ring*—which is the topic of Section 2.2. Once the groundwork has been laid, we introduce the notion of irreducibility in Section 2.3 and we prove that every affine variety uniquely decomposes as a finite union of irreducible affine varieties in Section 2.4.

Section 2.1 Inclusions, intersections, and unions

In this section, we discuss the ways in which the $\mathcal{V}$- and $\mathcal{I}$-operators interact with set-theoretic notions, beginning with their behavior with respect to inclusions.

> **2.1 PROPOSITION** $\mathcal{V}$ and $\mathcal{I}$ are inclusion-reversing
>
> Let $\mathcal{S}, \mathcal{T} \subseteq K[x_1, \ldots, x_n]$ and $X, Y \subseteq \mathbb{A}^n$ be subsets.
> 1. If $\mathcal{S} \subseteq \mathcal{T}$, then $\mathcal{V}(\mathcal{S}) \supseteq \mathcal{V}(\mathcal{T})$.
> 2. If $X \subseteq Y$, then $\mathcal{I}(X) \supseteq \mathcal{I}(Y)$.
>
> Furthermore, if X and Y are affine varieties, then
>
> $$X \subseteq Y \text{ if and only if } \mathcal{I}(X) \supseteq \mathcal{I}(Y).$$

In words, the first item says that a larger set of polynomials has fewer common solutions than a smaller one, while the second item says that a larger set of points in $\mathbb{A}^n$ has fewer polynomials that vanish on it than a smaller one. The reader is encouraged to take a moment to convince themselves of these statements on an intuitive level before attempting a formal proof.

PROOF OF PROPOSITION 2.1   Items 1 and 2 are left to Exercises 2.1.1 and 2.1.2, respectively, where it is also shown that the converse of each of these statements can fail. For the final if-and-only-if statement, the "only-if" direction is the statement of Item 2, so it remains to prove the "if" direction. Assume, then, that $\mathcal{I}(X) \supseteq \mathcal{I}(Y)$, which implies by Item 1 that $\mathcal{V}(\mathcal{I}(X)) \subseteq \mathcal{V}(\mathcal{I}(Y))$. Using the assumption that X and Y are affine varieties, we apply Proposition 1.21 to see that $\mathcal{V}(\mathcal{I}(X)) = X$ and $\mathcal{V}(\mathcal{I}(Y)) = Y$, from which we conclude that $X \subseteq Y$. □

2.2 EXAMPLE   A line on a hyperboloid

Consider the ideals

$$I = \langle x^2 + y^2 - z^2 - 1 \rangle \text{ and } J = \langle x - z, y - 1 \rangle.$$

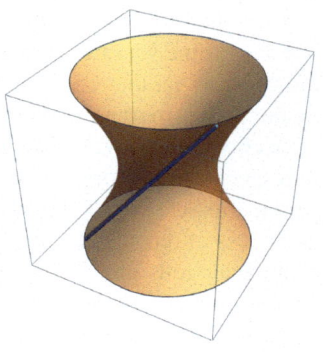

Then $\mathcal{V}(I)$ is the one-sheeted hyperboloid depicted to the right over $\mathbb{R}$, and the variety $\mathcal{V}(J)$ is the line contained in the hyperboloid, whose points are of the form $\{(a, 1, a) \mid a \in K\}$. The containment $\mathcal{V}(I) \supseteq \mathcal{V}(J)$ follows from the containment of ideals $I \subseteq J$, which is verified by noting that the generator of I lies in J:

$$x^2 + y^2 - z^2 - 1 = (x+z)(x-z) + (y+1)(y-1) \in J.$$

Having discussed inclusions, we now turn our attention to intersections and unions. Is the intersection or union of a set of affine varieties itself an affine variety? If so, and if we happen to know a set of defining equations for the original collection of varieties, can we find defining equations for the intersection and union? We explore these questions, beginning with a familiar example.

2.1. INCLUSIONS, INTERSECTIONS, AND UNIONS

2.3 EXAMPLE *Intersection and union of coordinate axes*

Consider the affine varieties $\mathcal{V}(x) \subseteq \mathbb{A}^2$ and $\mathcal{V}(y) \subseteq \mathbb{A}^2$, which are the y-axis and the x-axis, respectively. The intersection of the two coordinate axes is simply the origin, which we saw in Example 1.6 is defined by the vanishing of the set $\{x, y\}$. Thus,

(2.4) $$\mathcal{V}(x) \cap \mathcal{V}(y) = \mathcal{V}(x, y).$$

Their union, on the other hand, is the affine variety of Example 1.5, which is defined by the vanishing of the polynomial xy. Thus,

(2.5) $$\mathcal{V}(x) \cup \mathcal{V}(y) = \mathcal{V}(xy).$$

If we interpret equations (2.4) and (2.5) in terms of ideals, then the ideal $\langle xy \rangle$ appearing in equation (2.5) is the intersection of the ideals $\langle x \rangle$ and $\langle y \rangle$. The ideal $\langle x, y \rangle$ in equation (2.4) is not quite the union of $\langle x \rangle$ and $\langle y \rangle$—since this union is not an ideal—but is the ideal generated by the union (Exercise 2.1.3). In this way, Example 2.3 illustrates the following general result.

2.6 PROPOSITION *Intersections and unions of vanishing sets*

For any ideals $I, J \subseteq K[x_1, \ldots, x_n]$,

$$\mathcal{V}(I) \cap \mathcal{V}(J) = \mathcal{V}(I \cup J),$$

$$\mathcal{V}(I) \cup \mathcal{V}(J) = \mathcal{V}(I \cap J).$$

PROOF We prove the second equality and leave the first to Exercise 2.1.4.

($\subseteq$): Since $I \cap J \subseteq I$, Proposition 2.1 implies that $\mathcal{V}(I) \subseteq \mathcal{V}(I \cap J)$. By the same token, we have $\mathcal{V}(J) \subseteq \mathcal{V}(I \cap J)$. Taking these together, we conclude that

$$\mathcal{V}(I) \cup \mathcal{V}(J) \subseteq \mathcal{V}(I \cap J).$$

($\supseteq$): Suppose $a = (a_1, \ldots, a_n) \notin \mathcal{V}(I) \cup \mathcal{V}(J)$. Since $a \notin \mathcal{V}(I)$, there exists $f \in I$ such that $f(a) \neq 0$. Similarly, there exists $g \in J$ such that $g(a) \neq 0$. Because ideals absorb multiplication, the product fg lies in I and J, so $fg \in I \cap J$. Since

$$(fg)(a) = f(a)g(a) \neq 0,$$

we conclude that $a \notin \mathcal{V}(I \cap J)$, completing the proof. □

Since every affine variety X is the vanishing set of some ideal (namely, the ideal $\mathcal{I}(X)$), Proposition 2.6 implies that the intersection and union of any two affine varieties are, themselves, affine varieties. Each of the equations in Proposition 2.6 has a downside, however. In the first equation, the issue is that the union $I \cup J$ of ideals is not, in general, an ideal (Exercise 2.1.5). In the second equation, although $I \cap J$ is an ideal, it can be inconvenient to work with in practice; for example, if one knows generators for I and J, it is not obvious how to deduce generators for $I \cap J$.

Both of these issues can be rectified by rephrasing Proposition 2.6 in terms of the following pair of algebraic operations on ideals.

2.7 DEFINITION   *Sums and products of ideals*

Let I and J be ideals in a ring R. The *sum* of I and J is the ideal

$$I + J = \{r + s \mid r \in I,\, s \in J\},$$

and the *product* of I and J is the ideal

$$I \cdot J = \left\{ \sum_{i=1}^{m} r_i s_i \;\middle|\; m \in \mathbb{N},\, r_i \in I,\, s_i \in J \right\}.$$

The definition of the sum of two ideals is what you might expect: it is the set consisting of pairwise sums, which happens to be an ideal. The product, however, requires an additional step: since the set of pairwise products is not closed under addition, one needs to include all finite sums of pairwise products in order to obtain an ideal. The verification that the sum and product of two ideals are, in fact, ideals is left to Exercise 2.1.7, where a number of other useful properties are developed.

An important aspect of working with sums and products of ideals is that, if we have generators for I and J, say $I = \langle a_1, \ldots, a_k \rangle$ and $J = \langle b_1, \ldots, b_\ell \rangle$, then we can immediately write down generators for the sum and product ideals (Exercise 2.1.7):

$$I + J = \langle a_1, \ldots, a_k, b_1, \ldots, b_\ell \rangle,$$
$$I \cdot J = \langle a_i b_j \mid i = 1, \ldots, k \text{ and } j = 1, \ldots, \ell \rangle.$$

For example, in the ring $K[x,y]$, we have

$$\langle x \rangle + \langle y \rangle = \langle x, y \rangle,$$
$$\langle x \rangle \cdot \langle y \rangle = \langle xy \rangle.$$

Utilizing sums and products, we have the following modification of Proposition 2.6.

2.8 PROPOSITION   *Intersections and unions revisited*

For any ideals $I, J \subseteq K[x_1, \ldots, x_n]$,

$$\mathcal{V}(I) \cap \mathcal{V}(J) = \mathcal{V}(I + J),$$
$$\mathcal{V}(I) \cup \mathcal{V}(J) = \mathcal{V}(I \cdot J).$$

PROOF   For the first equation, we need only observe that $I + J$ is the ideal generated by $I \cup J$ (Exercise 2.1.7), and then the first equation of Proposition 2.8 follows from the first equation of Proposition 2.6.

To prove the second equation, notice that $I \cdot J$ is contained in both I and J (Exercise 2.1.7). Therefore, the proof that $\mathcal{V}(I) \cup \mathcal{V}(J) \subseteq \mathcal{V}(I \cdot J)$ carries over verbatim from Proposition 2.6. Similarly, because $fg \in I \cdot J$ whenever $f \in I$ and $g \in J$, the proof that $\mathcal{V}(I) \cup \mathcal{V}(J) \supseteq \mathcal{V}(I \cdot J)$ also applies unchanged.   □

2.1. INCLUSIONS, INTERSECTIONS, AND UNIONS

2.9 EXAMPLE Intersection and union via ideals

Consider the affine varieties $\mathcal{V}(x,y)$ and $\mathcal{V}(x-y)$ in $\mathbb{A}^2$, which are the origin and a line through the origin. This example computes their intersection and union algebraically, verifying what one would expect.

Applying Proposition 2.8 and the description of the ideal sum in terms of generators, we have

$$\mathcal{V}(x,y) \cap \mathcal{V}(x-y) = \mathcal{V}(\langle x,y \rangle + \langle x-y \rangle) = \mathcal{V}(x,y,x-y).$$

Note that $\langle x, y, x-y \rangle = \langle x, y \rangle$, since $x - y$ is already in the ideal that x and y generate. Thus, the above can be expressed as

$$\mathcal{V}(x,y) \cap \mathcal{V}(x-y) = \mathcal{V}(x,y),$$

which captures the geometric observation that the intersection of these two affine varieties is the origin.

As for their union, Proposition 2.8 implies that

$$\mathcal{V}(x,y) \cup \mathcal{V}(x-y) = \mathcal{V}(\langle x,y \rangle \cdot \langle x-y \rangle) = \mathcal{V}(x(x-y), y(x-y)).$$

An element $(a,b) \in \mathcal{V}(x(x-y), y(x-y))$ must satisfy the equations

$$a(a-b) = 0 \quad \text{and} \quad b(a-b) = 0.$$

The first of these implies that either $a = 0$ or $a = b$. In case $a = 0$, the second equation implies that $b^2 = 0$ and hence $b = 0$, and in case $a = b$, the second equation is automatically satisfied. In this way, one confirms that

$$\mathcal{V}(x(x-y), y(x-y)) = \{(a,b) \in \mathbb{A}^2 \mid a = b\}.$$

In other words, we have verified algebraically that the union of the origin and the line $y = x$ is, as expected, just the line.

Thus far, we have considered only pairwise unions and intersections, but the astute reader may realize that everything generalizes to unions and intersections of finitely many affine varieties $\mathcal{V}(I_1), \ldots, \mathcal{V}(I_k)$. In fact, intersections can be pushed even further, to collections of infinitely many affine varieties $\mathcal{V}(I_1), \mathcal{V}(I_2), \mathcal{V}(I_3), \ldots$ or even collections of uncountably many affine varieties.

Notationally, in order to speak of arbitrary collections of ideals, we consider sets $\{I_\alpha\}_{\alpha \in A}$, where A is an arbitrary set (the *indexing set*) and $I_\alpha \subseteq K[x_1, \ldots, x_n]$ is an ideal for each $\alpha \in A$. For example, if $A = \{1, 2, 3\}$, this would be a collection $\{I_1, I_2, I_3\}$. If $A = \mathbb{N}$, it would be a collection $\{I_0, I_1, I_2, \ldots\}$ of countably-infinitely many ideals. We could even have $A = \mathbb{R}$, meaning the collection contains not just ideals $I_0, I_1, I_2, \ldots$ but also ideals $I_{-1}, I_{1/2}, I_{\sqrt{2}}, I_\pi$, and so on.

With this notation established, the general result is the following.

> **2.10 PROPOSITION** *General intersections and unions*
>
> For any collection $\{I_\alpha\}_{\alpha \in A}$ of ideals $I_\alpha \subseteq K[x_1, \ldots, x_n]$,
>
> $$\bigcap_{\alpha \in A} \mathcal{V}(I_\alpha) = \mathcal{V}\left(\bigcup_{\alpha \in A} I_\alpha\right).$$
>
> For any finite collection $\{I_1, \ldots, I_k\}$ of ideals $I_i \subseteq K[x_1, \ldots, x_n]$,
>
> $$\bigcup_{i=1}^{k} \mathcal{V}(I_i) = \mathcal{V}\left(\bigcap_{i=1}^{k} I_i\right).$$

PROOF The proof mimics the proof of Proposition 2.6. The reason finiteness is required in the second equation is that the product fg that appears in the proof of Proposition 2.6 is replaced here by a product of one f_i from each I_i, and infinite products of polynomials are not polynomials.   □

> *Proposition 2.10 can also be stated in terms of ideal sums and products, but a bit of care must be taken in defining infinite sums of ideals.*

Finiteness is essential in order for the union of affine varieties to be an affine variety. For example, $\mathbb{Z} \subseteq \mathbb{C} = \mathbb{A}_{\mathbb{C}}^1$ is an infinite union of its points, each of which is an affine variety, but we know that $\mathbb{Z} \subseteq \mathbb{A}_{\mathbb{C}}^1$ is not an affine variety, because it is neither finite nor all of $\mathbb{A}_{\mathbb{C}}^1$.

Proposition 2.10 implies that arbitrary intersections and finite unions of affine varieties are affine varieties. Readers familiar with topology may recognize these conditions: along with the property that $\emptyset$ and $\mathbb{A}^n$ are affine varieties, these form the defining conditions on the closed sets of a topology, so their complements form the open sets. This topology on $\mathbb{A}^n$ is called the *Zariski topology*, named in honor of Oscar Zariski (1899–1986), who made foundational contributions to modern algebraic geometry by placing the classical Italian approach, in which he was trained, on a more rigorous algebraic footing.

Though familiarity with topology will not be assumed in this book, the terminology of Zariski-open and Zariski-closed sets permeates throughout algebraic geometry, so we present the definition here for future reference.

> **2.11 DEFINITION** *Zariski topology on $\mathbb{A}^n$*
>
> A subset $X \subseteq \mathbb{A}^n$ is called *Zariski-closed* if X is an affine variety.
>
> A subset $U \subseteq \mathbb{A}^n$ is called *Zariski-open* if $\mathbb{A}^n \setminus U$ is an affine variety.

> *The modifier "Zariski" distinguishes this topology from other natural topologies on $\mathbb{A}_K^n$, such as the Euclidean topology for $K = \mathbb{R}$ or $\mathbb{C}$.*

The interested reader with a background in topology is directed to Exercise 2.1.9 to explore some basic properties of the Zariski topology and how it compares to the Euclidean topology.

Exercises for Section 2.1

2.1.1 Let $\mathcal{S}, \mathcal{T} \subseteq K[x_1, \ldots, x_n]$ be subsets.
   (a) Prove that $\mathcal{S} \subseteq \mathcal{T}$ implies that $\mathcal{V}(\mathcal{S}) \supseteq \mathcal{V}(\mathcal{T})$.
   (b) Prove, by example, that the converse of (a) can fail.

2.1.2 Let $X, Y \subseteq \mathbb{A}^n$ be subsets.
   (a) Prove that $X \subseteq Y$ implies that $\mathcal{I}(X) \supseteq \mathcal{I}(Y)$.
   (b) Prove, by example, that the converse of (a) can fail.

2.1.3 This exercise concerns the ideals $\langle x \rangle \subseteq K[x,y]$ and $\langle y \rangle \subseteq K[x,y]$.
   (a) Prove that $\langle x \rangle \cap \langle y \rangle = \langle xy \rangle$.
   (b) Prove that $\langle \langle x \rangle \cup \langle y \rangle \rangle = \langle x, y \rangle$.

2.1.4 Complete the proof of Proposition 2.6 by proving that

$$\mathcal{V}(I) \cap \mathcal{V}(J) = \mathcal{V}(I \cup J)$$

for any ideals $I, J \subseteq K[x_1, \ldots, x_n]$.

2.1.5 Prove that a union of two ideals in a ring is an ideal if and only if one of the ideals is contained in the other.

2.1.6 Let I and J be ideals of a ring R. Prove, by example, that $\{ab \mid a \in I, b \in J\}$ is not necessarily an ideal of R.

2.1.7 Let I and J be ideals of a ring R.
   (a) Prove that $I + J$ and $I \cdot J$ are both ideals.
   (b) Suppose that $I = \langle a_1, \ldots, a_k \rangle$ and $J = \langle b_1, \ldots, b_\ell \rangle$. Prove that

   $$I + J = \langle a_1, \ldots, a_k, b_1, \ldots, b_\ell \rangle$$

   and

   $$I \cdot J = \langle \{a_i b_j \mid i = 1, \ldots, k,\ j = 1, \ldots, \ell\} \rangle.$$

   (c) Prove that $I + J$ is the ideal generated by $I \cup J$.
   (d) Prove that $I \cdot J \subseteq I \cap J$.

2.1.8 Assume that K is infinite. Prove that any two nonempty Zariski-open sets in $\mathbb{A}_K^n$ have nonempty intersection. (For students with some background in topology, this says that the Zariski topology on $\mathbb{A}_K^n$ is *not* Hausdorff.)

2.1.9 (For students with some background in topology) Compare the Zariski topology on $\mathbb{A}_\mathbb{R}^n = \mathbb{R}^n$ to the Euclidean topology. Is one of these topologies coarser than the other?

Section 2.2 Finite generation

The notion of a vanishing set $\mathcal{V}(\mathcal{S})$ makes sense whether $\mathcal{S}$ is finite or infinite, but often an infinite set can be replaced by a finite one without affecting the corresponding vanishing set. For example, the ideal $\langle y - x^2 \rangle \subseteq K[x,y]$ contains infinitely many polynomials, but $\mathcal{V}(\langle y - x^2 \rangle)$ is equal to $\mathcal{V}(y - x^2)$, the vanishing set of just a single polynomial.

A natural question, then, is whether *every* affine variety is equal to $\mathcal{V}(\mathcal{S})$ for some *finite* set $\mathcal{S}$. The answer to this question is "yes," and the algebraic proof of this fact is the primary aim of this section. We begin our discussion with a bit of algebraic terminology.

2.12 DEFINITION *Finitely-generated ideals*

An ideal I of a ring R is said to be *finitely generated* if $I = \langle r_1, \ldots, r_k \rangle$ for finitely many elements $r_1, \ldots, r_k \in R$.

The ideal $\langle y - x^2 \rangle \subseteq K[x,y]$, for example, is finitely generated, as is the ideal $\langle x, y \rangle \subseteq K[x,y]$. In fact, one must look to a ring that is rather less familiar to find an example of an ideal that is not finitely generated.

2.13 EXAMPLE An ideal that is not finitely generated

Consider the polynomial ring $K[x_1, x_2, x_3, \ldots]$ in infinitely many variables. Explicitly, monomials in $K[x_1, x_2, x_3, \ldots]$ are expressions of the form

$$x^\alpha = x_1^{\alpha_1} x_2^{\alpha_2} x_3^{\alpha_3} \cdots,$$

where $\alpha = (\alpha_1, \alpha_2, \alpha_3, \ldots)$ is an exponent vector satisfying

(i) $\alpha_i \in \mathbb{N}$ for all i, and

(ii) $\alpha_i = 0$ for all but finitely many i.

An element of R is a finite K-linear combination of monomials:

$$f = \sum_\alpha a_\alpha x^\alpha,$$

where the sum is over all exponent vectors α satisfying (i) and (ii), the coefficients a_α are elements of K, and $a_\alpha = 0$ for all but finitely many α.

In $K[x_1, x_2, x_3, \ldots]$, the ideal generated by all of the variables,

$$I = \langle x_1, x_2, x_3, \ldots \rangle,$$

is not finitely generated, as the reader is encouraged to verify (Exercise 2.2.1).

Despite the previous example, many of the rings with which we are familiar have the property that all of their ideals are finitely generated. These rings are given a special name, in honor of Emmy Noether (1882–1935). In addition to her pioneering work in abstract algebra, Noether also guided the development of modern physics by discovering the connection between symmetries and conservation laws.

2.2. FINITE GENERATION

2.14 DEFINITION *Noetherian ring*

A ring is said to be *Noetherian* if all of its ideals are finitely generated.

Our primary goal in this section is to prove that $K[x_1, \ldots, x_n]$ is Noetherian. First, let us discuss a few familiar examples that we already know to be Noetherian.

2.15 EXAMPLE Fields are Noetherian

If K is any field, then the only ideals of K are $\{0\}$ and K. Both of these are finitely generated, because $\{0\} = \langle 0 \rangle$ and $K = \langle 1 \rangle$. Thus, K is Noetherian.

2.16 EXAMPLE PIDs are Noetherian

By definition, every ideal in a principal ideal domain is generated by a single element, and is thus finitely generated. Therefore, the rings $\mathbb{Z}$ and $K[x]$ are both examples of Noetherian rings.

There is an alternative way to characterize what it means for a ring to be Noetherian, using nested sequences of ideals. Although this second characterization is not as simple to state, it can be very useful in practice, as we will see below.

2.17 PROPOSITION *The ascending chain condition*

A ring R is Noetherian if and only if, given any ideals $I_1, I_2, I_3, \ldots$ of R with

$$I_1 \subseteq I_2 \subseteq I_3 \subseteq \cdots,$$

there exists a positive integer k such that $I_d = I_k$ for all $d \geq k$.

In other words, Proposition 2.17 says that Noetherian rings are characterized by the *ascending chain condition*: every ascending chain of ideals must eventually stabilize. The ascending chain condition is not satisfied for the (non-Noetherian) ring $R = K[x_1, x_2, x_3, \ldots]$; for example, the chain of ideals

$$\langle x_1 \rangle \subsetneq \langle x_1, x_2 \rangle \subsetneq \langle x_1, x_2, x_3 \rangle \subsetneq \cdots$$

continues to grow at each step.

PROOF OF PROPOSITION 2.17 We prove both implications.
 ($\Rightarrow$) Suppose R is Noetherian, and let

$$I_1 \subseteq I_2 \subseteq I_3 \subseteq \cdots$$

be an ascending chain of ideals of R. Consider the union of these nested ideals

$$I = \bigcup_{k=1}^{\infty} I_k,$$

which is an ideal of R by Exercise 0.4.9. Since R is Noetherian, $I = \langle a_1, \ldots, a_r \rangle$ for some $a_1, \ldots, a_r \in R$. Each a_i is in the union of the I_k, so it must lie in one of them; say $a_i \in I_{k_i}$. Since the ideals are nested, it follows that $a_i \in I_d$ for all $d \geq k_i$.

In particular, if we set $k = \max\{k_1, \ldots, k_r\}$, then $\{a_1, \ldots, a_r\} \subseteq I_d$ for all $d \geq k$, implying that $I = \langle a_1, \ldots, a_r \rangle \subseteq I_d$ for all $d \geq k$. However, since I is the union of the I_k, we also have $I \supseteq I_d$, from which we conclude that $I = I_d$ for all $d \geq k$, verifying the ascending chain condition.

($\Leftarrow$) We prove this direction by proving the contrapositive. Suppose R is not Noetherian, and choose an ideal I of R that is not finitely generated. Choose any element $a_1 \in I$ and define $I_1 = \langle a_1 \rangle$. Then $I_1 \subseteq I$, but since I is not finitely generated, the inclusion must be strict. Thus, choose an element $a_2 \in I \setminus I_1$ and define $I_2 = \langle a_1, a_2 \rangle$. We now have

$$I_1 \subsetneq I_2 \subsetneq I;$$

the first inclusion is strict because $a_2 \notin I_1$, and the second inclusion is strict because I is not finitely generated. We can continue this process indefinitely by choosing $a_{n+1} \in I \setminus \langle a_1, \ldots, a_n \rangle$ and defining $I_{n+1} = \langle a_1, \ldots, a_{n+1} \rangle$. This process recursively produces an ascending chain of ideals that never stabilizes, so the ascending chain condition fails. $\square$

Equipped with the ascending chain characterization of the Noetherian property, we are now ready to show that the polynomial rings $K[x_1, \ldots, x_n]$ are Noetherian. The proof uses induction, adding one variable at a time. The induction step follows from the following key algebraic result that goes by the name of *Hilbert's Basis Theorem*,

> *The word "basis" is a somewhat outdated artifact: in Hilbert's time, a set of ideal generators was referred to as a basis. This terminology persists today, to some extent—for example, in the term "Gröbner basis"—but is relatively uncommon.*

in honor of David Hilbert (1862–1943), a prolific mathematician who was the first to prove this result in 1890 as part of his work on invariant theory.

2.18 THEOREM *Hilbert's Basis Theorem*

If R is a Noetherian ring, then $R[x]$ is a Noetherian ring.

PROOF   Suppose that R is a Noetherian ring and, toward a contradiction, suppose that $R[x]$ is not Noetherian. Let $I \subseteq R[x]$ be an ideal that is not finitely generated. Define an infinite sequence of polynomials in $R[x]$ by the following recursion:

1. Choose $f_1 \in I$ to be a nonzero polynomial of minimum degree.
2. Having chosen $f_1, \ldots, f_j \in I$, choose $f_{j+1} \in I \setminus \langle f_1, \ldots, f_j \rangle$ to be a nonzero polynomial of minimum degree.

It cannot be the case that $\deg(f_j) > \deg(f_{j+1})$, as this would contradict the choice of f_j having minimum degree in $I \setminus \langle f_1, \ldots, f_{j-1} \rangle$. Therefore, the degrees of the polynomials in the sequence $(f_1, f_2, f_3, \ldots)$ are nondecreasing.

For each j, let a_j be the leading coefficient of f_j and consider the following ascending chain of ideals in R:

$$\langle a_1 \rangle \subseteq \langle a_1, a_2 \rangle \subseteq \langle a_1, a_2, a_3 \rangle \subseteq \cdots.$$

2.2. FINITE GENERATION

Since R is Noetherian, this chain must eventually stabilize; suppose it stabilizes at the kth step. Then $a_{k+1} \in \langle a_1, \ldots, a_k \rangle$. Choose elements $r_1, \ldots, r_k \in R$ such that

$$a_{k+1} = r_1 a_1 + \cdots + r_k a_k.$$

Using the fact that $\deg(f_i) \leq \deg(f_{k+1})$ for all $i \leq k$, define the polynomial

$$g = x^{\deg(f_{k+1}) - \deg(f_1)} r_1 f_1 + \cdots + x^{\deg(f_{k+1}) - \deg(f_k)} r_k f_k \in R[x].$$

By design, g has the same leading term as f_{k+1}, which implies that

$$\deg(f_{k+1} - g) < \deg(f_{k+1}).$$

Since f_{k+1} and g are both elements of I, it follows that $f_{k+1} - g \in I$. However, since g is an element of $\langle f_1, \ldots, f_k \rangle$ and f_{k+1} is not, it follows that $f_{k+1} - g$ cannot be in $\langle f_1, \ldots, f_k \rangle$. Thus, $f_{k+1} - g \in I \setminus \langle f_1, \ldots, f_k \rangle$ and $\deg(f_{k+1} - g) < \deg(f_{k+1})$, contradicting the minimality of degree in the choice of f_{k+1}. □

This finally brings us to the following fundamental property of polynomial rings.

2.19 COROLLARY $K[x_1, \ldots, x_n]$ *is Noetherian*

For any field K, the polynomial ring $K[x_1, \ldots, x_n]$ is Noetherian.

PROOF   The proof is by induction on n.

(**Base case**) As was noted in Example 2.15, any field K is Noetherian, proving the base case $n = 0$.

(**Induction step**) Suppose $K[x_1, \ldots, x_{n-1}]$ is Noetherian. Using the canonical isomorphism

$$K[x_1, \ldots, x_n] = K[x_1, \ldots, x_{n-1}][x_n]$$

and applying Hilbert's Basis Theorem for $R = K[x_1, \ldots, x_{n-1}]$, we conclude that $K[x_1, \ldots, x_n]$ is Noetherian. □

The geometric interpretation of the fact that $K[x_1, \ldots, x_n]$ is Noetherian is that every affine variety can be defined as the vanishing set of finitely many polynomials. In fact, a somewhat stronger statement is true, as stated in the next result.

2.20 COROLLARY *Affine varieties are finitely generated*

If $\mathcal{S} \subseteq K[x_1, \ldots, x_n]$ is any subset, then there is a finite subset $\mathcal{T} \subseteq \mathcal{S}$ such that

$$\mathcal{V}(\mathcal{S}) = \mathcal{V}(\mathcal{T}).$$

PROOF   Let $\mathcal{S} \subseteq K[x_1, \ldots, x_n]$ be a (possibly infinite) set. By Corollary 2.19, there exist elements $f_1, \ldots, f_k \in K[x_1, \ldots, x_n]$ such that

$$\langle \mathcal{S} \rangle = \langle f_1, \ldots, f_k \rangle.$$

> This result is stronger than simply saying that $\mathcal{V}(\mathcal{S})$ can be defined by a finite set of polynomials; it also asserts that the finite set can be chosen to be a subset of $\mathcal{S}$.

The polynomials $f_1, \ldots, f_k$ may not themselves belong to the set $\mathcal{S}$, but by definition of $\langle \mathcal{S} \rangle$, we can write each f_i as a linear combination of elements in $\mathcal{S}$:

(2.21) $$f_i = \sum_{j=1}^{\ell_i} g_{i,j} h_{i,j}$$

where $g_{i,j} \in K[x_1, \ldots, x_n]$ and $h_{i,j} \in \mathcal{S}$. Define the finite subset

$$\mathcal{T} = \{h_{i,j} \mid 1 \leq i \leq k, \ 1 \leq j \leq \ell_i\} \subseteq \mathcal{S}.$$

Equation (2.21) implies that $f_i \in \langle \mathcal{T} \rangle$ for all i, so $\langle \mathcal{S} \rangle = \langle f_1, \ldots, f_k \rangle \subseteq \langle \mathcal{T} \rangle$. Conversely, since $\mathcal{T} \subseteq \mathcal{S}$, we obtain $\langle \mathcal{T} \rangle \subseteq \langle \mathcal{S} \rangle$. Thus, $\langle \mathcal{S} \rangle = \langle \mathcal{T} \rangle$, and we conclude that

$$\mathcal{V}(\mathcal{S}) = \mathcal{V}(\langle \mathcal{S} \rangle) = \mathcal{V}(\langle \mathcal{T} \rangle) = \mathcal{V}(\mathcal{T}). \qquad \square$$

Exercises for Section 2.2

2.2.1 Prove that $\langle x_1, x_2, x_3, \ldots \rangle \subseteq K[x_1, x_2, x_3, \ldots]$ is not finitely generated.

2.2.2 Because $\mathbb{Z}$ is a principal ideal domain (thus, Noetherian), Proposition 2.17 implies that any ascending chain of ideals $I_1 \subseteq I_2 \subseteq I_3 \subseteq \cdots$ in $\mathbb{Z}$ must stabilize. Explain this phenomenon concretely: namely, expressing each $I_i = \langle a_i \rangle$ for some integer a_i, what is the relationship between a_i and a_{i+1}? Why must there exist $k \in \mathbb{N}$ such that $a_k = a_{k+1} = a_{k+2} = \cdots$?

2.2.3 Let R be a Noetherian ring and $I \subseteq R$ an ideal. Prove that R/I is Noetherian.

2.2.4 Prove that every Noetherian ring is a factorization domain.

2.2.5 This exercise shows that a subring of a Noetherian ring need not be Noetherian. Let $K[yz, yz^2, yz^3, \ldots]$ denote the image of the ring homomorphism

$$\varphi : K[x_1, x_2, x_3, \ldots] \to K[y, z]$$
$$f(x_1, x_2, x_3, \ldots) \mapsto f(yz, yz^2, yz^3, \ldots).$$

Prove that $K[yz, yz^2, yz^3, \ldots]$ is a non-Noetherian subring of $K[y, z]$.

2.2.6 Let R be a Noetherian ring and let $\varphi : R \to R$ be a ring homomorphism. Prove that φ is an isomorphism if and only if φ is surjective. (Hint: Consider the ideals $I_1 = \ker(\varphi)$, $I_2 = \ker(\varphi \circ \varphi)$, $I_3 = \ker(\varphi \circ \varphi \circ \varphi)$, and so on.)

2.2.7 Suppose
$$\mathbb{A}^n = \bigcup_{\alpha \in A} U_\alpha$$
where each $U_\alpha \subseteq \mathbb{A}^n$ is Zariski-open. Prove that there is a finite subset $B \subseteq A$ such that
$$\mathbb{A}^n = \bigcup_{\alpha \in B} U_\alpha.$$
(For students with some background in topology, this says that the Zariski topology on $\mathbb{A}^n$ is compact.)

Section 2.3 Irreducible affine varieties

We now come to the heart of this chapter and a central concept in algebraic geometry: the notion of irreducibility. To motivate the idea, consider the affine varieties

$$X_1 = \mathcal{V}(y - x^2), \quad X_2 = \mathcal{V}(y - 2), \quad \text{and} \quad X_3 = \mathcal{V}(x - 1, y - 3).$$

Over $\mathbb{R}$, the affine variety $X = X_1 \cup X_2 \cup X_3$ is shown to the right. Imagine now that you were given this image without being told that X was a union of three affine varieties. You could probably still tell, visually, that X was equal to such a union, and by studying the picture carefully you might even be able to determine the varieties.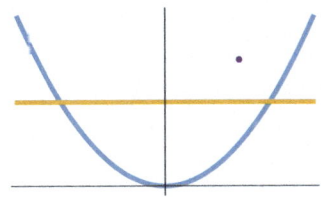
These varieties are the atomic pieces, or irreducible components, of X.

As this discussion suggests, the way we decompose affine varieties into their constituent pieces is by breaking them up into a finite union of smaller affine varieties. As such, the atomic ones are those that cannot be written as a union of two smaller affine varieties. We make this notion precise in the next definition.

2.22 DEFINITION *Reducible and irreducible affine variety*

An affine variety $X \subseteq \mathbb{A}^n$ is *reducible* if $X = X_1 \cup X_2$ for some affine varieties $X_1, X_2 \subsetneq X$, and X is *irreducible* if it is neither empty nor reducible.

2.23 EXAMPLE

The affine variety $X = \mathcal{V}(x^2 - y^2) \subseteq \mathbb{A}_{\mathbb{C}}^2$ is reducible. To see this, notice that

$$X = \mathcal{V}((x-y)(x-y)) = \mathcal{V}(x+y) \cup \mathcal{V}(x-y).$$

Therefore, the two affine varieties

$$X_1 = \mathcal{V}(x+y) \subsetneq X$$

and

$$X_2 = \mathcal{V}(x-y) \subsetneq X$$

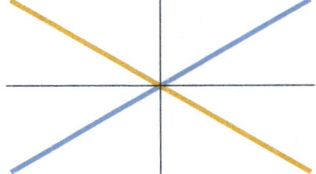

satisfy $X = X_1 \cup X_2$. Visually, upon restricting to the real numbers, X_1 and X_2 are the two lines that constitute X in the image above. In fact, as we will see in the next section, the two varieties X_1 and X_2 are the unique irreducible components of X.

2.24 EXAMPLE

By contrast, the parabola $X = \mathcal{V}(y - x^2) \subseteq \mathbb{A}^2$ is irreducible. This assertion should be geometrically believable: unlike the affine variety of Example 2.23, the parabola consists of just a single "piece." While this intuition is not yet a proof, the irreducibility of X will follow from Proposition 2.25 below.

As the previous examples illustrate, proving that an affine variety X is reducible is straightforward: one must find a pair of affine varieties $X_1 \subsetneq X$ and $X_2 \subsetneq X$ whose union is X. It is less clear how to prove that an affine variety is irreducible. The following algebraic characterization of irreducibility provides a key tool.

2.25 PROPOSITION *Irreducibility algebraically*

An affine variety $X \subseteq \mathbb{A}^n$ is irreducible if and only if $\mathcal{I}(X)$ is a prime ideal.

PROOF   We prove both implications by proving their contrapositives.

($\Leftarrow$) Suppose that X is reducible and choose affine varieties $X_1, X_2 \subsetneq X$ such that
$$X = X_1 \cup X_2.$$
Since $X_1 \subsetneq X$, it follows from Proposition 2.1 that $\mathcal{I}(X_1) \supsetneq \mathcal{I}(X)$. Thus, there exists $f \in \mathcal{I}(X_1)$ with $f \notin \mathcal{I}(X)$, and similarly, there exists $g \in \mathcal{I}(X_2)$ with $g \notin \mathcal{I}(X)$. For any $a \in X$, we either have $a \in X_1$ (and hence $f(a) = 0$) or $a \in X_2$ (and hence $g(a) = 0$), so $(fg)(a) = f(a)g(a) = 0$, implying that $fg \in \mathcal{I}(X)$. We have thus argued the existence of a pair of elements $f, g \in K[x_1, \ldots, x_n]$ with

(2.26) $\quad f \notin \mathcal{I}(X), \quad g \notin \mathcal{I}(X), \quad \text{and} \quad fg \in \mathcal{I}(X),$

which proves that $\mathcal{I}(X)$ is not prime.

($\Rightarrow$) Suppose that $\mathcal{I}(X)$ is not prime and choose $f, g \in K[x_1, \ldots, x_n]$ satisfying the conditions in (2.26). Define
$$X_1 = \mathcal{V}(f) \cap X \quad \text{and} \quad X_2 = \mathcal{V}(g) \cap X.$$
Proposition 2.6 implies that X_1 and X_2 are both affine varieties, and both are contained in X. Furthermore, the containments must be strict; if $X = X_1$, for example, then $X = \mathcal{V}(f) \cap X$, which means that $X \subseteq \mathcal{V}(f)$. If this were the case, then $f(a) = 0$ for all $a \in X$, meaning that $f \in \mathcal{I}(X)$, contradicting our assumptions.

By construction, we have $X_1 \cup X_2 \subseteq X$, but the other containment also holds. To see this, let $a \in X$. The fact that $fg \in \mathcal{I}(X)$ implies that $(fg)(a) = f(a)g(a) = 0$. It follows that either $f(a) = 0$ or $g(a) = 0$, implying that either $a \in \mathcal{V}(f)$ or $a \in \mathcal{V}(g)$. Since $a \in X$ by assumption, we conclude that either $a \in \mathcal{V}(f) \cap X = X_1$ or $a \in \mathcal{V}(g) \cap X = X_2$, so $a \in X_1 \cup X_2$. We have thus found two affine varieties $X_1, X_2 \subsetneq X$ such that $X = X_1 \cup X_2$, so X is reducible.   □

2.27 EXAMPLE   *The parabola is irreducible*

Consider the affine variety $X = \mathcal{V}(y - x^2) \subseteq \mathbb{A}^2$, whose vanishing ideal we computed in Example 1.19 to be $\mathcal{I}(X) = \langle y - x^2 \rangle$. Since $y - x^2$ is irreducible, $\mathcal{I}(X)$ is prime (Proposition 0.62), which proves that X is irreducible.

It is often the case that an affine variety is described in terms of defining equations, or equivalently, an ideal $I = \langle f_1, \ldots, f_k \rangle$. However, this defining ideal may not be equal to the vanishing ideal. This raises the question: given an ideal I, is there a way to determine if $\mathcal{V}(I)$ is irreducible, without drawing upon knowledge of the vanishing ideal? The Nullstellensatz provides the following useful answer.

2.3. IRREDUCIBLE AFFINE VARIETIES

2.28 PROPOSITION *Irreducibility of* $\mathcal{V}(I)$

If $I \subseteq K[x_1, \ldots, x_n]$ is a prime ideal, then $\mathcal{V}(I)$ is irreducible. In particular, if $f \in K[x_1, \ldots, x_n]$ is an irreducible polynomial, then $\mathcal{V}(f)$ is irreducible.

PROOF Suppose that I is a prime ideal. Then I is radical by Proposition 1.34. Thus, by the Nullstellensatz,
$$\mathcal{I}(\mathcal{V}(I)) = \sqrt{I} = I.$$
Since $\mathcal{I}(\mathcal{V}(I))$ is prime, we conclude from Proposition 2.25 that $\mathcal{V}(I)$ is irreducible.

To prove the second assertion, assume that f is an irreducible polynomial. Then $\langle f \rangle$ is a prime ideal by Proposition 0.62. Therefore, the first statement in the proposition implies that $\mathcal{V}(f) = \mathcal{V}(\langle f \rangle)$ is irreducible. □

We point out that Proposition 2.28 fails when K is not algebraically closed. For example, $x^2 + 1 \in \mathbb{R}[x]$ is irreducible but
$$\mathcal{V}(x^2 + 1) = \emptyset \in \mathbb{A}^1_{\mathbb{R}}$$
and, by definition, the empty set is not irreducible. For an example of an irreducible polynomial over $\mathbb{R}$ that defines a *nonempty* reducible variety, see Exercise 2.3.6.

Still, as we have previously mentioned, it is often useful to use our intuition over $\mathbb{R}$ to glean information about varieties more generally. For example, it should be somewhat intuitively clear that the parabola is irreducible over $\mathbb{R}$ because it is comprised of a single "piece," and this intuition extends to more general fields: the variety $\mathcal{V}(y - x^2)$ is irreducible over any infinite field K. However, one should be careful with this sort of reasoning over $\mathbb{R}$, as the next example illustrates.

2.29 EXAMPLE The hyperbola is irreducible

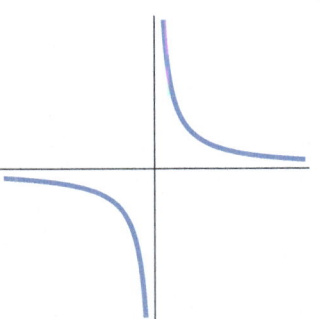

Consider the hyperbola $X = \mathcal{V}(xy - 1) \subseteq \mathbb{A}^2$, which is pictured to the right over $\mathbb{R}$. At a glance, our geometric intuition tells us that X consists of two "pieces," one in the first quadrant and one in the third. It would be natural to guess, then, that X is reducible. To the contrary, X is actually irreducible. Over an algebraically closed field, this follows from the fact that $xy - 1$ is irreducible. However, X can also be shown to be irreducible over any infinite field, including $\mathbb{R}$ (Exercise 2.3.7).

How, then, did our intuition fail us in this example? The answer is that the solutions over $\mathbb{R}$ do not capture the entire picture. If we expand our viewpoint and consider the zeros of $xy - 1$ over the algebraic closure of $\mathbb{R}$, namely $\mathbb{C}$, then we see that the two "pieces" are actually connected to each other via complex solutions. For example, we can get from the point $(1, 1)$ in the upper piece to the point $(-1, -1)$ in the lower piece by walking along the set of complex solutions
$$\{(e^{\pi i \theta}, e^{-\pi i \theta}) \mid \theta \in [0, 1]\}.$$

Motivated by this observation, let us attempt to draw $\mathcal{V}(xy - 1) \subseteq \mathbb{A}_\mathbb{C}^2$. This is a bit challenging because $\mathbb{A}_\mathbb{C}^2$ is 4-dimensional over the real numbers: $\mathbb{A}_\mathbb{C}^2 = \mathbb{C}^2 = \mathbb{R}^4$. However, by mapping down to $\mathbb{R}^3$, one can show (Exercise 2.3.8) that the complex solutions can be identified with the surface to the right, where we have included a depiction of the real hyperbola and the path connecting $(1, 1)$ to $(-1, -1)$. Thus, we see that the complex picture is much more consistent with what we might expect an

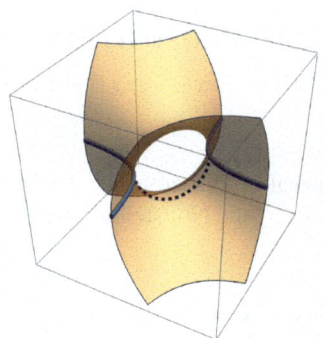

irreducible variety to look like intuitively: it is comprised of just a single "piece."

This example illustrates that, while it is important in algebraic geometry to use our geometric intuition over $\mathbb{R}$, we may not see the whole picture when we do so, and this intuition should only be trusted insofar as it can be justified algebraically.

A special type of irreducible affine variety is one that consists of a single point: $X = \{a\} \subseteq \mathbb{A}^n$. Since single points are minimal among varieties (with respect to inclusion), then the inclusion-reversing nature of the $\mathcal{V}$- and $\mathcal{I}$-operators suggest that their vanishing ideals should be maximal ideals, which, as we know, are special types of prime ideals. This is true, and even more can be said.

2.30 PROPOSITION *Single points and maximal ideals*

Let $I \subseteq K[x_1, \ldots, x_n]$ be an ideal. The following are equivalent.
  (i) $I = \mathcal{I}(\{a\})$ for some point $a = (a_1, \ldots, a_n) \in \mathbb{A}^n$.
  (ii) $I = \langle x_1 - a_1, \ldots, x_n - a_n \rangle$ for some $a_1, \ldots, a_n \in K$.
  (iii) $K[x_1, \ldots, x_n]/I \cong K$.
  (iv) I is a maximal ideal.

PROOF   That (i) implies (ii) is Exercise 1.3.5. That (ii) implies (iii) is Exercise 0.3.12. That (iii) implies (iv) follows from Proposition 0.38. Thus, it remains to prove that (iv) implies (i). Suppose that I is maximal. Since maximal ideals are radical, the Nullstellensatz implies that $I = \mathcal{I}(\mathcal{V}(I))$. To prove (i), we must prove that $\mathcal{V}(I)$ is a single point. Toward a contradiction, suppose that $\mathcal{V}(I)$ is not a single point. There are two cases: $\mathcal{V}(I) = \emptyset$ or $\mathcal{V}(I)$ has more than one point. If $\mathcal{V}(I) = \emptyset$, then
$$I = \mathcal{I}(\mathcal{V}(I)) = K[x_1, \ldots, x_n],$$
contradicting the maximality of I. If, on the other hand, $\mathcal{V}(I)$ has more than one point, let $a \in \mathcal{V}(I)$ be any point and notice that $\emptyset \subsetneq \{a\} \subsetneq \mathcal{V}(I)$ is a strict containment of affine varieties. Applying Proposition 2.1, we obtain
$$K[x_1, \ldots, x_n] \supsetneq \mathcal{I}(\{a\}) \supsetneq \mathcal{I}(\mathcal{V}(I)) = I,$$
which, again, contradicts the maximality of I. $\square$

2.3. IRREDUCIBLE AFFINE VARIETIES

Notice that the equivalence of (iii) and (iv) is a purely algebraic statement that does not hold over non-algebraically-closed fields; for example, the ideal $\langle x^2 + 1 \rangle$ is maximal in $\mathbb{R}[x]$ even though

$$\mathbb{R}[x]/\langle x^2 + 1 \rangle \cong \mathbb{C} \not\cong \mathbb{R}.$$

This observation reflects that $\langle x^2 + 1 \rangle \subseteq \mathbb{R}[x]$ is not a vanishing ideal.

We close this section on irreducibility by describing a refined dictionary between ideals and varieties. As we have already seen, the Nullstellensatz implies that the $\mathcal{V}$-operator is a bijection between radical ideals in $K[x_1, \ldots, x_n]$ and affine varieties in $\mathbb{A}^n$, with inverse given by the $\mathcal{I}$-operator (Corollary 1.40). We now introduce a refinement of this bijection that adds prime and maximal ideals to the mix.

2.31 PROPOSITION *Refined dictionary between ideals and varieties*

The $\mathcal{V}$- and $\mathcal{I}$-operators are mutually inverse, inclusion-reversing bijections that translate between the following hierarchies of ideals and varieties:

$\{\text{radical ideals in } K[x_1, \ldots, x_n]\} \longleftrightarrow \{\text{affine varieties in } \mathbb{A}^n\}$

$\cup\vert \qquad\qquad\qquad\qquad\qquad\qquad\qquad \cup\vert$

$\{\text{prime ideals in } K[x_1, \ldots, x_n]\} \longleftrightarrow \{\text{irreducible varieties in } \mathbb{A}^n\}$

$\cup\vert \qquad\qquad\qquad\qquad\qquad\qquad\qquad \cup\vert$

$\{\text{maximal ideals in } K[x_1, \ldots, x_n]\} \longleftrightarrow \{\text{points in } \mathbb{A}^n\}.$

PROOF If I is a radical ideal and $X = \mathcal{V}(I)$, then the Nullstellensatz implies that $I = \mathcal{I}(X)$. Thus, that the bijection between radical ideals and affine varieties is inclusion-reversing is the final statement in Proposition 2.1. To show that $\mathcal{V}$ and $\mathcal{I}$ restrict to a bijection between prime ideals and irreducible varieties, it suffices to observe that $I = \mathcal{I}(X)$ is prime if and only if $X = \mathcal{V}(I)$ is irreducible (Proposition 2.25). To show that $\mathcal{V}$ and $\mathcal{I}$ restrict to a bijection between maximal ideals and single points, it suffices to observe that $I = \mathcal{I}(X)$ is maximal if and only if $X = \mathcal{V}(I)$ is a single point (Proposition 2.30). $\square$

Exercises for Section 2.3

2.3.1 Prove that affine space is irreducible over any infinite field.

2.3.2 Explain why an affine variety over a finite field (which cannot be algebraically closed) is irreducible if and only if it consists of a single point.

2.3.3 Prove that $\mathcal{V}(xy) \subseteq \mathbb{A}^2$ is reducible.

2.3.4 Let X be an irreducible affine variety, and suppose that

$$X = \bigcup_{i=1}^{r} X_i,$$

where each X_i is an affine variety. Prove that $X = X_i$ for some i.

2.3.5 Prove that the affine variety $\mathcal{V}(x^2 + y^2) \subseteq \mathbb{A}^2$ is irreducible over $\mathbb{R}$ but reducible over $\mathbb{C}$.

2.3.6 Consider the function $f = (x^2 + 1)(x^2 - 1)^2 + y^2 \in \mathbb{R}[x,y]$.
   (a) Use Eisenstein's criterion to prove that f is irreducible.
   (b) Prove that $\mathcal{V}(f)$ is reducible, consisting of two distinct points.

2.3.7 Let K be an infinite field. This exercise proves the irreducibility of
$$X = \mathcal{V}(xy - 1) \subseteq \mathbb{A}_K^2.$$
   (a) Use properties of $\mathcal{I}$ and $\mathcal{V}$ to prove that $\mathcal{I}(X) \supseteq \langle xy - 1 \rangle$.
   (b) Prove that $\mathcal{I}(X) \subseteq \langle xy - 1 \rangle$, possibly using the following strategy.
      i. Let $f \in \mathcal{I}(X)$. Prove that
      $$x^k f - g \in \langle xy - 1 \rangle$$
      for some $k \in \mathbb{N}$ and $g \in K[x]$.
      ii. Using that $f \in \mathcal{I}(X)$, prove that g is the zero polynomial.
      iii. Using that $x^k f \in \langle xy - 1 \rangle$, prove that $f \in \langle xy - 1 \rangle$.
   (c) Prove that $\langle xy - 1 \rangle$ is a prime ideal, and use this to explain why X is irreducible.

2.3.8 The surface pictured in Example 2.29 is the real affine variety
$$Y = \mathcal{V}(uv + w^2 - 1) \subseteq \mathbb{A}_\mathbb{R}^3.$$
Let $X = \mathcal{V}(xy - 1) \subseteq \mathbb{A}_\mathbb{C}^2$.
   (a) Prove that $X = \{(re^{i\theta}, (re^{i\theta})^{-1}) \mid r \in \mathbb{R}_{>0},\ 0 \leq \theta < 2\pi\}$.
   (b) Define a function $F : X \to \mathbb{A}_\mathbb{R}^3$ by taking $(re^{i\theta}, (re^{i\theta})^{-1}) \in X$ to
   $$\left(\frac{r^2 - 1}{2r} + \frac{r^2 + 1}{2r} \cos\theta,\ \frac{1 - r^2}{2r} + \frac{r^2 + 1}{2r} \cos\theta,\ \frac{r^2 + 1}{2r} \sin\theta \right).$$
   Prove that F is a bijection onto Y.

Section 2.4 Irreducible decompositions

In the previous section, we were introduced to the notion of irreducibility for affine varieties. In this section, we prove the fundamental fact that every affine variety is the finite union of a unique set of irreducible affine varieties. This is one of the most important consequences of the algebraic fact that $K[x_1, \ldots, x_n]$ is Noetherian.

If X is reducible, then we can write X as a union of affine varieties $X_1, X_2 \subsetneq X$, which may themselves be reducible. By further decomposing X_1 and X_2, one can split X into a union comprised of more and more affine varieties, stopping only when the constituent pieces are irreducible. This process is analogous to the way in which one gradually factors an integer into primes or factors a polynomial into irreducibles. The following proposition says that, just like for prime factorizations of integers or irreducible factorizations of polynomials, this process eventually terminates, and the decomposition obtained in this way is unique.

2.32 Proposition/Definition *Irreducible decomposition*

Let $X \subseteq \mathbb{A}^n$ be a nonempty affine variety. Then there exist irreducible affine varieties $X_1, \ldots, X_r \subseteq X$ such that $X_i \not\subseteq X_j$ for any $i \neq j$ and

(2.33) $$X = \bigcup_{i=1}^{r} X_i.$$

Moreover, the irreducible affine varieties $X_1, \ldots, X_r$ are unique up to reordering; we call these the *irreducible components* of X, and refer to (2.33) as the *irreducible decomposition* of X.

We should stress here that both the finiteness of the number of irreducible components and the fact that $X_i \not\subseteq X_j$ for all $i \neq j$ are crucial features in order for the irreducible decomposition to be unique. To see why, consider the parabola $X = \mathcal{V}(y - x^2) \subseteq \mathbb{A}^2$. Because X is already irreducible, its irreducible decomposition has just a single component. On the other hand, if we did not insist on the finiteness of the number of X_i, then expressing X as the union of all of its points,

$$X = \bigcup_{p \in X} \{p\},$$

would be a different irreducible decomposition. If we did not insist that $X_i \not\subseteq X_j$ for all $i \neq j$, then we could obtain a different irreducible decomposition as, for example,

$$X = \{(0,0)\} \cup X.$$

Proof of Proposition 2.32 We first prove existence and then uniqueness.

(**Existence**) Suppose, toward a contradiction, that $X \subseteq \mathbb{A}^n$ is a nonempty affine variety that does not have a finite irreducible decomposition. In particular, this implies that X is not irreducible, so write

$$X = X_1 \cup X_1' \quad \text{for some} \quad X_1, X_1' \subsetneq X.$$

If both X_1 and X_1' have finite irreducible decompositions, then the union of these would be a finite irreducible decomposition of X, which goes against our supposition. Thus, it must be the case that either X_1 or X_1' does not have a finite irreducible decomposition. Without loss of generality, suppose X_1 does not have a finite irreducible decomposition, and write

$$X_1 = X_2 \cup X_2' \quad \text{for some} \quad X_2, X_2' \subsetneq X.$$

Again, since X_1 does not have a finite irreducible decomposition, then at least one of X_2 or X_2' does not have a finite irreducible decomposition. Suppose that X_2 does not have a finite irreducible decomposition, and write

$$X_2 = X_3 \cup X_3' \quad \text{for some} \quad X_3, X_3' \subsetneq X.'$$

Continuing in this way, we construct an infinite chain of nested affine varieties in $\mathbb{A}^n$:

$$X \supsetneq X_1 \supsetneq X_2 \supsetneq X_3 \supsetneq \cdots.$$

By Proposition 2.1, this yields an infinite chain of nested ideals in $K[x_1, \ldots, x_n]$:

$$\mathcal{I}(X) \subsetneq \mathcal{I}(X_1) \subsetneq \mathcal{I}(X_2) \subsetneq \mathcal{I}(X_3) \subsetneq \cdots,$$

which contradicts that $K[x_1, \ldots, x_n]$ is Noetherian (Corollary 2.19). The contradiction implies that every affine variety $X \subseteq \mathbb{A}^n$ must have at least one finite irreducible decomposition.

(**Uniqueness**) Let

$$X = \bigcup_{i=1}^{r} X_i = \bigcup_{j=1}^{s} Y_j$$

be two irreducible decompositions of X. We must prove that $r = s$ and that, after possibly reordering, we have $X_i = Y_i$ for each i. Without loss of generality, assume that $r \geq s$.

Since $X_1 \subseteq X$, we have

$$X_1 = X_1 \cap X = X_1 \cap \left(\bigcup_{j=1}^{s} Y_j \right) = \bigcup_{j=1}^{s} (X_1 \cap Y_j).$$

Given that X_1 is irreducible, it follows (Exercise 2.3.4) that $X_1 = X_1 \cap Y_j$ for some j. Reordering $Y_1, \ldots, Y_s$, assume that $X_1 = X_1 \cap Y_1$, which implies that $X_1 \subseteq Y_1$.

By the same token, since $Y_1 \subseteq X$, we have

$$Y_1 = Y_1 \cap X = Y_1 \cap \left(\bigcup_{i=1}^{r} X_i \right) = \bigcup_{i=1}^{r} (Y_1 \cap X_i),$$

so $Y_1 = Y_1 \cap X_i$ for some i, and $Y_1 \subseteq X_i$. It follows that $X_1 \subseteq Y_1 \subseteq X_i$. Since, by the definition of an irreducible decomposition, $X_1 \not\subseteq X_i$ for any $i \neq 1$, it must be the case that $i = 1$, and the containment $X_1 \subseteq Y_1 \subseteq X_1$ implies $X_1 = Y_1$.

Repeating this argument with X_2 in place of X_1 shows that $X_2 = Y_j$ for some j. Since $X_2 \neq X_1$, it cannot be the case that $j = 1$. Thus, after reordering $Y_2, \ldots, Y_s$, we may assume that $X_2 = Y_2$. We can continue in this way, showing that $X_i = Y_i$ for each $i \in \{1, \ldots, r\}$. In particular, this proves that $r \leq s$. Since we assumed that $r \geq s$, we conclude that $r = s$ and $X_i = Y_i$ for all i. $\square$

2.4. IRREDUCIBLE DECOMPOSITIONS

2.34 EXAMPLE Irreducible components of an intersection

Consider the affine variety

$$X = \mathcal{V}(2x^2 + 2y^2 - z^2 - 1, x^2 + y^2 - 1) \subseteq \mathbb{A}_\mathbb{C}^3.$$

Notice that $X = Y_1 \cap Y_2$ where

$$Y_1 = \mathcal{V}(2x^2 + 2y^2 - z^2 - 1)$$

and

$$Y_2 = \mathcal{V}(x^2 + y^2 - 1).$$

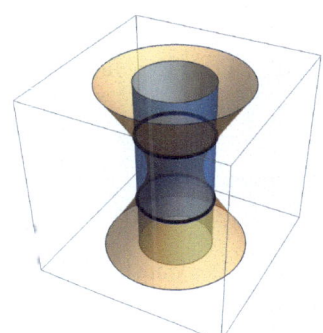

To gain some intuition for the variety X, let us consider the picture for $K = \mathbb{R}$. In that case, Y_1 is a one-sheeted hyperboloid and Y_2 is a circular cylinder, depicted to the right. From this image, we can see that their intersection, X, consists of two circles. One might naturally guess that these two circles are the irreducible components of X, so let us take this intuition and verify it algebraically.

If $(a, b, c) \in X$, then the coordinates satisfy

$$2a^2 + 2b^2 - c^2 = 1 \quad \text{and} \quad a^2 + b^2 = 1.$$

Subtracting twice the second equation from the first, we see that these equations are satisfied if and only if

$$c^2 = 1 \quad \text{and} \quad a^2 + b^2 = 1.$$

Since $c^2 = 1$ if and only if $c = \pm 1$, we then see that

$$X = \{(a, b, 1) \mid a^2 + b^2 = 1\} \cup \{(a, b, -1) \mid a^2 + b^2 = 1\}$$
$$= \underbrace{\mathcal{V}(x^2 + y^2 - 1, z - 1)}_{X_1} \cup \underbrace{\mathcal{V}(x^2 + y^2 - 1, z + 1)}_{X_2}.$$

Over the real numbers, X_1 and X_2 are precisely the circles depicted above.

To prove that X_1 and X_2 are, in fact, the irreducible components of X, it remains to prove that they are each irreducible. While this fact can be proved over $\mathbb{R}$, it is much simpler to prove over an algebraically closed field, like $\mathbb{C}$. In particular, over $\mathbb{C}$, it suffices (by Proposition 2.28) to observe that

$$\langle x^2 + y^2 - 1, z + 1 \rangle, \langle x^2 + y^2 - 1, z - 1 \rangle \subseteq \mathbb{C}[x, y, z]$$

are prime ideals, which the reader is encouraged to verify (see Exercise 2.4.2).

In general, it is not easy to determine the irreducible components of a variety, especially if the variety is described by many polynomials in a lot of variables, making it impossible to draw a picture and use our geometric intuition. However, in the special case that the variety is defined by a single polynomial f, the irreducible decomposition of $\mathcal{V}(f)$ is closely related to the irreducible factorization of f.

> **2.35 PROPOSITION** *Irreducible decomposition of $\mathcal{V}(f)$*
>
> If $f \in K[x_1,\ldots,x_n]$ has distinct irreducible factors $q_1,\ldots,q_m$, then the irreducible decomposition of $\mathcal{V}(f)$ is
>
> $$\mathcal{V}(f) = \mathcal{V}(q_1) \cup \cdots \cup \mathcal{V}(q_m).$$

PROOF   From the Nullstellensatz, it follows that

$$\mathcal{V}(f) = \mathcal{V}\left(\sqrt{\langle f \rangle}\right).$$

Applying Proposition 1.31, we see that $\sqrt{\langle f \rangle} = \langle q_1 \cdots q_m \rangle$, so

$$\mathcal{V}(f) = \mathcal{V}(q_1 \cdots q_m).$$

Since $q_1 \cdots q_m$ vanishes at a point if and only if one of the q_i vanishes at that point, we have

(2.36) $$\mathcal{V}(f) = \mathcal{V}(q_1) \cup \cdots \cup \mathcal{V}(q_m).$$

As each q_i is irreducible, Proposition 2.28 implies that each $\mathcal{V}(q_i)$ is irreducible.

To finish the proof, we must verify that $\mathcal{V}(q_i) \not\subseteq \mathcal{V}(q_j)$ for any $i \neq j$. By definition of distinct irreducible factors, we know that $q_i \nmid q_j$ for any $i \neq j$. This implies that $\langle q_i \rangle \not\supseteq \langle q_j \rangle$ for any $i \neq j$. Since each q_i is irreducible, then $\langle q_i \rangle$ is a prime ideal, and the Nullstellensatz implies that

$$\mathcal{I}(\mathcal{V}(q_i)) = \mathcal{I}(\mathcal{V}(\langle q_i \rangle)) = \sqrt{\langle q_i \rangle} = \langle q_i \rangle.$$

Thus, since $\mathcal{I}(\mathcal{V}(q_i)) \not\supseteq \mathcal{I}(\mathcal{V}(q_j))$ for any $i \neq j$, the final statement in Proposition 2.1 implies that $\mathcal{V}(q_i) \not\subseteq \mathcal{V}(q_j)$ for any $i \neq j$. □

Not every ideal in $K[x_1,\ldots,x_n]$ is generated by a single polynomial, and one might naturally wonder how to compute an irreducible decomposition of $\mathcal{V}(I)$ for nonprincipal ideals I. While this is a difficult task to do by hand, it is accomplishable with the aid of a computer. In particular, given a set of generators $I = \langle f_1,\ldots,f_k \rangle$, there are effective algorithms utilizing Gröbner bases for computing the irreducible decomposition of $\mathcal{V}(I)$. We have chosen not to describe those computational tools in this book, focusing instead on the more theoretical aspects of the dictionary between algebra and geometry.

> For an introductory treatment of algebraic geometry that focuses on the more computational and algorithmic aspects of the theory, see the book of Cox, Little, and O'Shea [2].

Exercises for Section 2.4

2.4.1 Let K be an infinite field of characteristic not equal to 2. Prove that the irreducible decomposition of $\mathcal{V}(x^2 - y^2) \subseteq \mathbb{A}^2$ is

$$\mathcal{V}(x^2 - y^2) = \mathcal{V}(x+y) \cup \mathcal{V}(x-y).$$

What changes if the characteristic is equal to 2?

2.4. IRREDUCIBLE DECOMPOSITIONS

2.4.2 Prove that
$$\frac{\mathbb{C}[x,y,z]}{\langle x^2+y^2-1, z-1\rangle} \cong \frac{\mathbb{C}[x,y]}{\langle x^2+y^2-1\rangle}.$$
Conclude that $\langle x^2+y^2-1, z-1\rangle$ is prime.

2.4.3 What are the irreducible components of
$$\mathcal{V}(x^2+y^2+z^2-2z, x^2+y^2-z^2) \subseteq \mathbb{A}_\mathbb{R}^3?$$
(Hint: Graph the surfaces to see how they intersect.)

2.4.4 Calculate the irreducible decomposition of
$$\mathcal{V}(x^2+y^2+z^2-2z, \ x^2+y^2-z^2) \subseteq \mathbb{A}_\mathbb{C}^3.$$
(Hint: There are three irreducible components. How does the complex picture differ from the real picture in the previous exercise?)

2.4.5 Calculate the irreducible decomposition of
$$\mathcal{V}(xy+z, \ x^2-x+y^2+yz) \subseteq \mathbb{A}_\mathbb{C}^3.$$
(Hint: There are two irreducible components.)

Chapter 3

Coordinate Rings

> LEARNING OBJECTIVES FOR CHAPTER 3
> - Define the coordinate ring $K[X]$ of an affine variety X, both in terms of polynomial functions and as a quotient ring.
> - Define the notion of a K-algebra, and describe how to identify finitely-generated K-algebras with quotients of polynomial rings.
> - Determine whether a ring is reduced, and using quotients, whether an ideal is radical.
> - Characterize coordinate rings algebraically as those rings that are finitely-generated reduced K-algebras.
> - Find, in specific examples, an affine variety whose coordinate ring is a given finitely-generated reduced K-algebra.

The work we did in Chapter 1 provides us with a method for moving back and forth between the worlds of algebra and geometry, using the $\mathcal{V}$- and $\mathcal{I}$-operators as inverse bijections between affine varieties in $\mathbb{A}^n$ and radical ideals in $K[x_1, \ldots, x_n]$. But on the algebraic side, ideals play a special role that we have not yet invoked: they are precisely the subsets of $K[x_1, \ldots, x_n]$ by which one can take a quotient to produce a ring. If $X \subseteq \mathbb{A}^n$ is an affine variety, then, how should we interpret the quotient of $K[x_1, \ldots, x_n]$ by $\mathcal{I}(X)$ in terms of the affine variety X?

The answer, as we will see in this chapter, is that this quotient is naturally isomorphic to the *coordinate ring* of X, a ring $K[X]$ whose elements are *polynomial functions* from X to the ground field K. Once we define these objects precisely in Section 3.1, we will have a new way to pass from the world of geometry to the world of algebra:

$$\{\text{affine varieties}\} \to \{\text{rings}\}$$
$$X \mapsto K[X].$$

As always, then, we ask whether this association is a two-way dictionary. It is not, at the outset, because not every ring is $K[X]$ for some X. Our search for an algebraic characterization of the rings that arise as coordinate rings will lead us to define the notion of a *K-algebra*—a special class of rings into which polynomial rings and their quotients fall—and to study their key algebraic properties. The culminating result of the chapter is that it is precisely when a ring is a *finitely-generated reduced K-algebra* that it is the coordinate ring of some affine variety.

Section 3.1 Polynomial functions on affine varieties

We learned in Section 1.1 that every polynomial $f \in K[x_1, \ldots, x_n]$ can be used to define a function $\mathbb{A}^n \to K$, obtained by mapping $(a_1, \ldots, a_n)$ to $f(a_1, \ldots, a_n)$. If $X \subseteq \mathbb{A}^n$ is an affine variety, then we can restrict the domain of such a function to X, yielding a new function

$$f|_X : X \to K$$
$$(a_1, \ldots, a_n) \mapsto f(a_1, \ldots, a_n).$$

A function $F : X \to K$ that arises in this way is referred to as a *polynomial function*.

> **3.1 DEFINITION** *Polynomial function*
>
> Let $X \subseteq \mathbb{A}^n$ be an affine variety. A *polynomial function on X* is a function $F : X \to K$ such that $F = f|_X$ for some $f \in K[x_1, \ldots, x_n]$.

3.2 EXAMPLE Polynomial functions on the parabola

Let $X = \mathcal{V}(y - x^2) \subseteq \mathbb{A}^2$. The function

$$F : X \to K$$
$$(a, b) \mapsto a + b$$

is a polynomial function, since

$$F = f|_X \quad \text{where} \quad f = x + y \in K[x, y].$$

Note that $f = x + y$ is not the only polynomial that gives rise to F. For example, since $a^2 = b$ for all $(a, b) \in X$, it follows that $F = g|_X$ where $g = x + x^2$ and $F = h|_X$ where $h = x + 2y - x^2$.

3.3 EXAMPLE Coordinate functions

Let $X \subseteq \mathbb{A}^n$ be an affine variety. Then, for each $i \in \{1, \ldots, n\}$, the *ith coordinate function* on X is the function

$$C_i : X \to K$$
$$(a_1, \ldots, a_n) \mapsto a_i.$$

The coordinate functions are polynomial because C_i is the restriction of the function associated to the polynomial $x_i \in K[x_1, \ldots, x_n]$.

3.4 EXAMPLE The empty function

If $X = \emptyset \subseteq \mathbb{A}^n$, then there is only one function $F : \emptyset \to K$, the *empty function*. Moreover, upon restricting the domain to the empty set, every function $\mathbb{A}^n \to K$ gives rise to the empty function. In particular, this implies that the empty function is the unique polynomial function on the affine variety $\emptyset \subseteq \mathbb{A}^n$.

3.1. POLYNOMIAL FUNCTIONS ON AFFINE VARIETIES

It should be clear from this discussion that one can concoct polynomial functions on $X \subseteq \mathbb{A}^n$ simply by choosing any polynomial $f \in K[x_1, \ldots, x_n]$, considering the corresponding function on $\mathbb{A}^n$, and then restricting its domain to X. Why, then, do we define polynomial functions on X in what appears to be the opposite way: starting from F and then searching for an f that restricts to it?

The primary reason we take this approach is that, for a given polynomial function $F : X \to K$, there may be many polynomials $f \in K[x_1, \ldots, x_n]$ such that $F = f|_X$, and we do not wish to view these as different polynomial functions on X. This is already apparent in Example 3.2, where the distinct polynomials $f, g, h \in K[x, y]$ all define the same function $F : X \to K$. It is not the polynomials $f \in K[x_1, \ldots, x_n]$ that are important here; rather, it is the function $F : X \to K$ that we intend to study.

Starting from F has its drawbacks, however, because depending on how the definition of F is presented, it may not be immediately clear whether it is the restriction of a polynomial. The next example illustrates this phenomenon, and serves as a caution against making quick judgments about polynomiality.

3.5 EXAMPLE A nonobviously polynomial function

Let $X = \mathcal{V}(xy - 1) \subseteq \mathbb{A}^2$. Since $a \neq 0$ for any $(a, b) \in X$, we can consider the function defined by

$$F : X \to K$$
$$(a, b) \mapsto \frac{1}{a}.$$

The output $\frac{1}{a}$ is not a polynomial in a and b, which may lead one to guess that F is not a polynomial function. However, the fact that $ab - 1 = 0$ for all $(a, b) \in X$ means that $\frac{1}{a} = b$, and therefore, $F = f|_X$ where $f = y \in K[x, y]$.

More generally, any function $F : X \to K$ of the form

$$F(a, b) = \frac{f(a, b)}{a^j b^k}$$

with $f \in K[x, y]$ and $j, k \in \mathbb{N}$ is a polynomial function on X (Exercise 3.1.3).

The set of polynomial functions on X can be endowed with the structure of a ring by adding and multiplying functions in the usual way:

$$(F + G)(a_1, \ldots, a_n) = F(a_1, \ldots, a_n) + G(a_1, \ldots, a_n),$$
$$(F \cdot G)(a_1, \ldots, a_n) = F(a_1, \ldots, a_n) \cdot G(a_1, \ldots, a_n).$$

Thus, starting with an affine variety X, we can produce a ring associated to it. This ring is central to the study of algebraic geometry.

3.6 DEFINITION Coordinate ring

Let $X \subseteq \mathbb{A}^n$ be an affine variety. The *coordinate ring* of X, denoted $K[X]$, is the ring of all polynomial functions on X.

The additive identity $0 \in K[X]$ is the constant function that takes the value $0 \in K$ for all $a \in X$, and the multiplicative identity is the constant function that takes the value $1 \in K$ for all $a \in X$. These functions arise from the polynomials $0, 1 \in K[x_1, \ldots, x_n]$, respectively: $0 = 0|_X$ and $1 = 1|_X$.

Given an affine variety X, can we compute $K[X]$? In other words, can we identify $K[X]$ with a more familiar ring? The next result provides a step in this direction by presenting the coordinate ring as a quotient.

3.7 PROPOSITION *The coordinate ring as a quotient*

If $X \subseteq \mathbb{A}^n$ is an affine variety, then there is a canonical ring isomorphism

$$K[X] = \frac{K[x_1, \ldots, x_n]}{\mathcal{I}(X)}.$$

PROOF By the First Isomorphism Theorem, it suffices to find a canonical surjective homomorphism $\varphi : K[x_1, \ldots, x_n] \to K[X]$ whose kernel is $\mathcal{I}(X)$. Define φ by

$$\varphi(f) = f|_X.$$

Noting that $(f + g)|_X = f|_X + g|_X$ and $(f \cdot g)|_X = f|_X \cdot g|_X$, we see that φ is a ring homomorphism. By definition, every polynomial function on X arises from some polynomial $f \in K[x_1, \ldots, x_n]$, so φ is surjective. Finally, $f \in \ker \varphi$ if and only if $f|_X = 0$, which is the same as saying that $f \in \mathcal{I}(X)$. This shows that $\ker \varphi = \mathcal{I}(X)$. □

3.8 EXAMPLE Coordinate ring of affine space

Proposition 1.9, and the fact that K is algebraically closed and hence infinite, implies that $K[\mathbb{A}^n] = K[x_1, \ldots, x_n]$.

3.9 EXAMPLE Coordinate ring of the parabola

In Example 3.2, where $X = \mathcal{V}(y - x^2) \subseteq \mathbb{A}^2$, we saw that the three polynomials

$$f = x + y, \quad g = x + x^2, \quad \text{and} \quad h = x - x^2 + 2y$$

all give rise to the same polynomial function $F : X \to K$. This reflects the fact that $[f] = [g] = [h]$ in the quotient ring

$$K[X] = \frac{K[x, y]}{\mathcal{I}(X)} = \frac{K[x, y]}{\langle y - x^2 \rangle},$$

as we verify: $f - g = y - x^2 \in \langle y - x^2 \rangle$ and $g - h = 2x^2 - 2y \in \langle y - x^2 \rangle$.

For an affine variety $X \subseteq \mathbb{A}^n$, Definition 3.6 and Proposition 3.7 provide two different characterizations of the coordinate ring $K[X]$. It is important to keep both interpretations in mind: elements of the coordinate ring should be simultaneously viewed as functions $F : X \to K$ and as equivalence classes of polynomials in $K[x_1, \ldots, x_n]$. The canonical isomorphism of Proposition 3.7 identifies the polynomial function $f|_X$ with the equivalence class $[f]$ for any $f \in K[x_1, \ldots, x_n]$.

3.1. POLYNOMIAL FUNCTIONS ON AFFINE VARIETIES

As advertised in the introduction to this chapter, we have now introduced a new way of passing from geometry to algebra, by associating to an affine variety X its coordinate ring $K[X]$. In keeping with our philosophy that the passage from geometry to algebra should be a two-way dictionary, we now ask whether the passage from an affine variety to its coordinate ring can be reversed. More precisely, given a ring R, does there exist an affine variety X such that $R = K[X]$? The answer, we will find, is affirmative if R is a *finitely-generated reduced K-algebra*, and the next three sections are devoted to defining and studying these terms.

Exercises for Section 3.1

3.1.1 Describe the ring of polynomial functions on the empty set. What is 0? 1?

3.1.2 Let $X = \mathcal{V}(x^2 + y^2 - 2z^2) \subseteq \mathbb{A}^3$. List three distinct elements of $K[x, y, z]$ that restrict to the same polynomial function in $K[X]$, and list two elements of $K[x, y, z]$ that restrict to different polynomial functions in $K[X]$.

3.1.3 Let $X = \mathcal{V}(xy - 1) \subseteq \mathbb{A}^2$. Prove that any function $F : X \to K$ of the form

$$F(a, b) = \frac{f(a, b)}{a^j b^k}$$

with $f \in K[x, y]$ and $j, k \in \mathbb{N}$ is a polynomial function.

3.1.4 Let $X \subseteq \mathbb{A}^n$ be an affine variety.
   (a) Prove that $K[X]$ is an integral domain if and only if X is irreducible.
   (b) As an illustration of (a), let $X = \mathcal{V}(x^2 - xy) \subseteq \mathbb{A}^2$. Prove that X is reducible by finding two affine varieties $X_1 \subsetneq X$ and $X_2 \subsetneq X$ such that $X = X_1 \cup X_2$. Then, verify that $K[X]$ is not an integral domain by finding two nonzero functions in $K[X]$ whose product is zero.

3.1.5 Let $X = \{p\} \subseteq \mathbb{A}^n$ be a single point. Prove that the function $\varphi : K[X] \to K$ defined by $\varphi(F) = F(p)$ is a ring isomorphism.

3.1.6 Let $p_1, \ldots, p_m \in \mathbb{A}^n$ be distinct points in $\mathbb{A}^n$, and let $X = \{p_1, \ldots, p_m\}$. Prove that the function

$$\varphi : K[X] \to \overbrace{K \oplus \cdots \oplus K}^{m}$$
$$\varphi(F) = (F(p_1), \ldots, F(p_m))$$

is a ring isomorphism. (Recall that addition and multiplication in direct sums are defined componentwise.)

3.1.7 (a) Give an example of an infinite affine variety $X \subseteq \mathbb{A}^3$ such that the three coordinate functions are all the same polynomial function.
   (b) Prove that the solution to part (a) is unique.

Section 3.2 K-algebras

What types of rings arise as $K[X]$ for some affine variety X? The first answer to this question is given by considering the special role played by the ground field K.

To motivate our discussion, consider the case $X = \mathbb{A}^1$, for which

$$K[\mathbb{A}^1] = K[x].$$

As we have discussed at length, $K[x]$ is a ring. However, it is more than just an ordinary ring; it also has the structure of a vector space over K, since along with being able to add and multiply polynomials, we can also multiply polynomials by scalars in K (and the two operations of addition and scalar multiplication satisfy the usual vector-space axioms). Unlike the vector spaces one typically studies in a first linear algebra course, $K[x]$ is infinite-dimensional, with a basis given by

$$\mathcal{B} = \{1, x, x^2, x^3, \ldots\}.$$

Nonetheless, just like the more familiar finite-dimensional vector spaces, every element of $K[x]$ can be written uniquely as a linear combination of elements in $\mathcal{B}$.

More generally, every coordinate ring has this same enhanced structure: it is simultaneously a ring and a vector space over K. In this section, we develop the algebraic foundations of K-algebras, which formalize this structure. (We remind the reader, here, that all rings in this book are assumed commutative with unity.)

3.10 DEFINITION *K-algebra*

A K-algebra is a ring A together with a *scalar multiplication* function

$$K \times A \to A$$
$$(r, a) \mapsto r \cdot a$$

satisfying the following axioms.

1. $r \cdot (a + b) = r \cdot a + r \cdot b$ for all $r \in K$ and all $a, b \in A$.
2. $(r + s) \cdot a = r \cdot a + s \cdot a$ for all $r, s \in K$ and all $a \in A$.
3. $(rs) \cdot a = r \cdot (s \cdot a)$ for all $r, s \in K$ and all $a \in A$.
4. $1 \cdot a = a$ for all $a \in A$.
5. $r \cdot (ab) = (r \cdot a)b = a(r \cdot b)$ for all $r \in K$ and all $a, b \in A$.

> To help parse the axioms in Definition 3.10, the products within K and A have been written by concatenating the elements, reserving the symbol "·" for scalar multiplication.

The first four axioms stipulate that A forms a vector space over K, while the fifth axiom specifies how scalar multiplication interacts with multiplication operation in A. It follows from the axioms above, and is true of vector spaces in general, that scalar multiplying any $a \in A$ by $0 \in K$ gives $0 \in A$ and scalar multiplying any $a \in A$ by $-1 \in K$ gives the additive inverse $-a \in A$ (Exercise 3.2.1).

3.2. K-ALGEBRAS

3.11 EXAMPLE Polynomial rings

The prototypical example of a K-algebra, especially from the perspective of algebraic geometry, is the polynomial ring $K[x_1, \ldots, x_n]$. Indeed, along with being able to add and multiply polynomials, we can also multiply a polynomial by a scalar in K, and the axioms in Definition 3.10 are straightforward to verify.

3.12 EXAMPLE Coordinate rings

Let X be an affine variety. The coordinate ring $K[X]$ forms a K-algebra. For any $F \in K[X]$ and $r \in K$, we define $r \cdot F \in K[X]$ to be the function given by

$$(r \cdot F)(a_1, \ldots, a_n) = r \cdot F(a_1, \ldots, a_n),$$

where the multiplication on the right-hand side is the usual multiplication in K. To check that $r \cdot F$ is in fact an element of $K[X]$, notice that $F = f|_X$ for some $f \in K[x_1, \ldots, x_n]$, and it follows that $r \cdot F = (r \cdot f)|_X$. Since $r \cdot f \in K[x_1, \ldots, x_n]$, we see that $r \cdot F \in K[X]$. The axioms in Definition 3.10 are again readily verified.

3.13 EXAMPLE Extension rings of K

If A is any ring that contains K as a subring, then A is naturally a K-algebra where scalar multiplication is the usual ring multiplication in A. In fact, given our assumptions (rings are commutative with unity), every nontrivial K-algebra arises in this way. More precisely, given any K-algebra $A \neq \{0\}$, there is a canonical inclusion $K \to A$, and viewing K as a subring of A under this inclusion, scalar multiplication is identified with the usual multiplication in A (Exercise 3.2.5). In particular, every nontrivial K-algebra canonically contains a copy of K.

3.14 EXAMPLE Nonexamples of K-algebras

By Example 3.13, any nonzero ring that does not contain K is not a K-algebra. For example, since $\mathbb{Z}$ does not contain a field, it is not a K-algebra for any field K.

Our development of K-algebras is not complete until we specify the appropriate morphisms between them. Given that a K-algebra is a ring enhanced with a scalar multiplication operation, a K-algebra homomorphism is an enhanced ring homomorphism that preserves scalar multiplication.

3.15 DEFINITION *Homomorphism of K-algebras*

Let A and B be K-algebras. A *K-algebra homomorphism* $\varphi : A \to B$ is a ring homomorphism for which

$$\varphi(r \cdot a) = r \cdot \varphi(a)$$

for all $r \in K$ and $a \in A$. We say that φ is an *isomorphism of K-algebras* and write $A \cong B$ if φ has an inverse that is also a K-algebra homomorphism.

Being an isomorphism appears to be stronger than being a bijection—not only should an inverse function exist, but it must also be a K-algebra homomorphism. However, as is conveniently the case for groups, rings, and fields, if φ is a bijective homomorphism, then its inverse is automatically a homomorphism (Exercise 3.2.2).

3.16 **EXAMPLE** *K-algebra homomorphisms from polynomial rings*

Consider the evaluation function

$$\varphi : \mathbb{R}[x,y] \to \mathbb{R}$$
$$f(x,y) \mapsto f(2,3).$$

Some reflection should convince the reader that φ is an $\mathbb{R}$-algebra homomorphism. In addition, knowing that φ is an $\mathbb{R}$-algebra homomorphism, we can also see that it is completely determined by the image of x and y. For example, once we know that $\varphi(x) = 2$ and $\varphi(y) = 3$, then using the fact that φ is a ring homomorphism that preserves scalar multiplication, we obtain

$$\varphi(5x^2y + 2y + xy) = 5(2)^2(3) + 2(3) + (2)(3) = 72.$$

More generally, for any K-algebra A and subset $\{a_1, \ldots, a_n\} \subseteq A$, there is a unique K-algebra homomorphism

$$\varphi : K[x_1, \ldots, x_n] \to A$$

satisfying $\varphi(x_i) = a_i$ for all i (Exercise 3.2.3). This shows that a K-algebra homomorphism $K[x_1, \ldots, x_n] \to A$ is equivalent to a choice of $a_1, \ldots, a_n \in A$.

Just like for groups and rings, there is a First Isomorphism Theorem for K-algebras, which is a fundamental tool for proving that two K-algebras are isomorphic. In order to state it, we must first define K-algebra quotients and subalgebras.

Since a K-algebra A is a ring, we already know that we can form the quotient ring A/I for any ideal $I \subseteq A$. The next result shows that the quotient ring A/I naturally inherits a K-algebra structure from A.

3.17 **PROPOSITION** *Quotient K-algebras*

Let A be a K-algebra. If $I \subseteq A$ is an ideal, then the quotient ring A/I is a K-algebra, in which scalar multiplication is defined by

(3.18) $$r \cdot [a] = [r \cdot a].$$

PROOF We must check that the scalar multiplication given by (3.18) is well-defined. Toward this end, the key point is that ideals are automatically closed under scalar multiplication: if $a \in I$ and $r \in K$, then

$$r \cdot a = r \cdot (1a) = (r \cdot 1)a.$$

where 1 is the unity in A. Since $r \cdot 1 \in A$ and $a \in I$, the absorbing property of ideals implies that $(r \cdot 1)a \in I$. Thus, $r \cdot a \in I$, verifying that I is closed under scalar multiplication.

3.2. K-ALGEBRAS

From here, to see that scalar multiplication is well-defined, suppose $[a] = [b]$. Then $a - b \in I$, and using the fact that I absorbs scalar multiplication, we have $r \cdot a - r \cdot b = r \cdot (a - b) \in I$, which tells us that $[r \cdot a] = [r \cdot b]$.

Since I is an ideal, we already know that A/I is a ring, so all that remains to be checked is the five axioms of Definition 3.10. These all follow readily from the validity of the corresponding axioms for A. □

We now provide the final ingredient required for the First Isomorphism Theorem.

3.19 DEFINITION *Subalgebra*

Let A be a K-algebra. A *subalgebra* $B \subseteq A$ is a subring for which $r \cdot b \in B$ for all $r \in K$ and $b \in B$.

In other words, a subset of a K-algebra is a subalgebra if it is both a subring and closed under scalar multiplication, thereby forming a K-algebra in its own right. A natural example of a subalgebra is the K-algebra $K[x]$ as a subalgebra of $K[x, y]$.

3.20 THEOREM *First Isomorphism Theorem for K-algebras*

If $\varphi : A \to B$ is a K-algebra homomorphism, then

(i) $\text{im}(\varphi)$ is a subalgebra of B,

(ii) $\ker(\varphi)$ is an ideal of A, and

(iii) the function

$$[\varphi] : \frac{A}{\ker(\varphi)} \to \text{im}(\varphi)$$
$$[a] \mapsto \varphi(a)$$

is a well-defined isomorphism of K-algebras.

PROOF Exercise 3.2.7. □

3.21 EXAMPLE $\mathbb{C}$ as a quotient $\mathbb{R}$-algebra

Consider $\mathbb{C}$ as an $\mathbb{R}$-algebra, where scalar multiplication is the usual multiplication, and let φ be the function

$$\varphi : \mathbb{R}[x] \to \mathbb{C}$$
$$\varphi(f) = f(i).$$

One can check (Exercise 3.2.8) that φ is a surjective $\mathbb{R}$-algebra homomorphism with kernel $\langle x^2 + 1 \rangle$. Thus, by Theorem 3.20, we obtain an $\mathbb{R}$-algebra isomorphism

$$\mathbb{C} \cong \mathbb{R}[x]/\langle x^2 + 1 \rangle.$$

As vector spaces over $\mathbb{R}$, the set of complex numbers $\mathbb{C}$ has a basis $\{1, i\}$ and the quotient $\mathbb{R}[x]/\langle x^2 + 1 \rangle$ has a basis $\{1, [x]\}$. The isomorphism $[\varphi]$ identifies 1 with 1 and $[x]$ with i, which is motivated by the fact that $[x]^2 = i^2 = -1 \in \mathbb{R}$.

Returning to our motivating example of coordinate rings, we may now use our knowledge of K-algebras to tailor Proposition 3.7 to the setting of K-algebras. Let $X \subseteq \mathbb{A}^n$ be an affine variety and consider the ring homomorphism

$$K[x_1, \ldots, x_n] \to K[X]$$
$$f \mapsto f|_X.$$

Since $(r \cdot f)|_X = r \cdot (f|_X)$, this is a K-algebra homomorphism. Exactly as in the proof of Proposition 3.7, it is surjective with kernel $\mathcal{I}(X)$. Thus, we obtain the following result from the First Isomorphism Theorem.

3.22 PROPOSITION *The coordinate ring as a quotient K-algebra*

If $X \subseteq \mathbb{A}^n$ is an affine variety, then there is a canonical K-algebra isomorphism

$$K[X] \cong \frac{K[x_1, \ldots, x_n]}{\mathcal{I}(X)}.$$

Exercises for Section 3.2

3.2.1 Let A be a K-algebra. Prove that $0 \cdot a = 0$ and $(-1) \cdot a = -a$ for all $a \in A$.

3.2.2 Let $\varphi : A \to B$ be a homomorphism of K-algebras. Suppose that φ is a bijection, so that φ has an inverse function $\varphi^{-1} : B \to A$. Prove that this inverse function is a homomorphism of K-algebras.

3.2.3 Let A be a K-algebra and $\{a_1, \ldots, a_n\} \subseteq A$ a subset. Prove that there is a unique K-algebra homomorphism $\varphi : K[x_1, \ldots, x_n] \to A$ that satisfies $\varphi(x_i) = a_i$ for all $i = 1, \ldots, n$.

3.2.4 Give an example of a ring homomorphism $\varphi : \mathbb{C} \to \mathbb{C}$ that is not a $\mathbb{C}$-algebra homomorphism.

3.2.5 Let $A \neq \{0\}$ be a K-algebra.
   (a) Prove that the function $\varphi : K \to A$ defined by $\varphi(r) = r \cdot 1$ is an injective ring homomorphism.
   (b) Viewing K as a subring of A via the homomorphism in (a), prove that scalar multiplication is identified with the usual multiplication in A.

3.2.6 Let A and B be nontrivial K-algebras. By Exercise 3.2.5, both A and B contain a copy of K in a canonical way. Prove that any K-algebra homomorphism $\varphi : A \to B$ restricts to the identity on K.

3.2.7 Prove the First Isomorphism Theorem for K-algebras.

3.2.8 Fill in the details of the proof in Example 3.21 that $\mathbb{C} \cong \mathbb{R}[x]/\langle x^2 + 1 \rangle$ as $\mathbb{R}$-algebras.

Section 3.3 Generators of K-algebras

Our ongoing task, recall, is to determine precisely which rings arise as coordinate rings. The previous section shows that, for a ring R to be a coordinate ring, it must be a K-algebra, and Proposition 3.22 refines this statement: R must be a quotient of a polynomial ring. But it may not be immediately obvious whether a given ring is isomorphic to such a quotient; we saw in Example 3.21, for instance, that $\mathbb{C}$ is isomorphic as an $\mathbb{R}$-algebra to a quotient of $\mathbb{R}[x]$, despite not being initially presented as such.

The goal of this section is to characterize exactly which K-algebras arise as quotients of polynomial rings. The key ingredients that we require in order to do this are the notions of polynomial combinations and K-algebra generators.

3.23 DEFINITION *Polynomial combination, generators*

Let A be a K-algebra and let $\mathcal{S} \subseteq A$ be a subset. A *polynomial combination of $\mathcal{S}$* is an element of A of the form

$$f(a_1, \ldots, a_n)$$

for some polynomial $f \in K[x_1, \ldots, x_n]$ and $a_1, \ldots, a_n \in \mathcal{S}$. The set of all polynomial combinations of $\mathcal{S}$ is called the *subalgebra of A generated by $\mathcal{S}$*, and it is denoted $K[\mathcal{S}]$.

The reader is encouraged to check that $K[\mathcal{S}]$ is indeed a subalgebra of A, and it is the smallest subalgebra that contains $\mathcal{S}$ (Exercise 3.3.1). An alternative way to think of $K[\mathcal{S}]$ is that it

> *If $\mathcal{S} = \{a_1, \ldots, a_n\}$ is a finite set, we omit the set brackets and write $K[\mathcal{S}] = K[a_1, \ldots, a_n]$.*

consists of all elements in A that can be obtained from elements of $\mathcal{S}$ using the operations of addition, multiplication, and scalar multiplication by elements of K.

3.24 EXAMPLE Subalgebras of $K[x, y]$

Consider the K-algebra $A = K[x, y]$. If we set $a = x$, then we see that the subalgebra generated by a is the collection of all polynomials in x:

$$K[a] = K[x] \subseteq K[x, y].$$

On the other hand, if $b = x + y$, then $K[b]$ is the collection of all polynomials of the form

$$\sum_{i=0}^{k} r_i (x+y)^i$$

where $k \in \mathbb{N}$ and $r_i \in K$ for all i. Taking the generators a and b together, we see that $K[a, b]$ contains both $x = a$ and $y = b - a$, from which we conclude that $K[a, b]$ is the set of all polynomials in x and y:

$$K[a, b] = K[x, y].$$

3.25 **EXAMPLE** Generators for $K[x,y]/\langle xy-1\rangle$

Consider the K-algebra
$$A = \frac{K[x,y]}{\langle xy-1\rangle}.$$
Any element of A is of the form $[f(x,y)]$ for some $f(x,y) \in K[x,y]$. By definition of coset arithmetic, we have
$$[f(x,y)] = f([x],[y]).$$
Thus, any element of A can be written as a polynomial expression in $a = [x]$ and $b = [y]$. This implies that A is generated as an algebra by a and b:
$$A = K[a,b].$$

While the notation $K[a,b]$ in Example 3.25 is reminiscent of that for a polynomial ring, A is *not* the same thing as the ring of polynomials in variables a and b. In particular, there is a relation between the generators:
$$ab = 1.$$
Here and throughout, we use letters at the end of the alphabet, such as x and y, to denote variables in polynomial rings. By definition, these variables do not have any relations among themselves, meaning that two polynomials are equal if and only if they have the same coefficients. On the other hand, we use letters at the beginning of the alphabet, such as a and b, to denote generators of K-algebras, which may satisfy relations; in other words, it is possible to have $f(a,b) = g(a,b)$ even if f and g are different polynomials.

In the previous two examples, the entire K-algebra could be generated by finitely many elements. We capture this by saying that they are *finitely-generated K-algebras*.

3.26 **DEFINITION** *Finitely-generated K-algebras*

Let A be a K-algebra. We say that A is *finitely generated* if there exist $a_1, \ldots, a_n \in A$ such that $A = K[a_1, \ldots, a_n]$.

For example, the polynomial ring $K[x_1, \ldots, x_n]$ is a finitely-generated K-algebra, simply by taking $a_i = x_i$ for all $i = 1, \ldots, n$. More generally, the quotient ring $K[x_1, \ldots, x_n]/I$ is finitely generated for any ideal I, as we can take $a_i = [x_i]$ for all $i = 1, \ldots, n$, generalizing Example 3.25. In fact, up to isomorphism, these are the only examples of finitely-generated K-algebras, as we now verify.

3.27 **PROPOSITION** *Characterization of finitely-generated K-algebras*

Let A be a K-algebra. Then A is finitely generated if and only if there is an isomorphism of K-algebras
$$A \cong K[x_1, \ldots, x_n]/I$$
for some $n \in \mathbb{N}$ and some ideal $I \subseteq K[x_1, \ldots, x_n]$.

3.3. GENERATORS OF K-ALGEBRAS

PROOF Suppose that A is finitely generated. By definition, this means that there exist $a_1, \ldots, a_n \in A$ such that $A = K[a_1, \ldots, a_n]$. Define a function
$$\varphi : K[x_1, \ldots, x_n] \to A$$
$$f(x_1, \ldots, x_n) \mapsto f(a_1, \ldots, a_n).$$
It is straightforward to check that φ is a K-algebra homomorphism, and the fact that $A = K[a_1, \ldots, a_n]$ is equivalent to the statement that φ is surjective. Letting $I = \ker(\varphi)$, the First Isomorphism Theorem implies that
$$A \cong K[x_1, \ldots, x_n]/I.$$
Conversely, suppose that there exists an isomorphism
$$\psi : K[x_1, \ldots, x_n]/I \to A,$$
and let $a_i = \psi([x_i])$ for $i = 1, \ldots, n$. We aim to show that $A = K[a_1, \ldots, a_n]$. Suppose $a \in A$; we must show that a is a polynomial expression in $a_1, \ldots, a_n$. Since ψ is surjective, there exists $[f] \in K[x_1, \ldots, x_n]/I$ such that $\psi([f]) = a$. Then
$$a = \psi([f(x_1, \ldots, x_n)]) = \psi(f([x_1], \ldots, [x_n])) = f(\psi([x_1]), \ldots, \psi([x_n])),$$
where the second equality follows from the arithmetic of cosets and the third from the assumption that ψ is a K-algebra homomorphism. Since $a_i = \psi([x_i])$, we see that a is a polynomial expression in $a_1, \ldots, a_n$, and we conclude that A is finitely generated. □

An isomorphism of the form in Proposition 3.27 is often referred to as a *presentation* of the K-algebra A. The images of $x_1, \ldots, x_n$ are the *generators* of the presentation and the polynomials in I are the *relations* of the presentation.

By Proposition 3.22, coordinate rings of affine varieties are canonically isomorphic to quotients of polynomial rings. Thus, we obtain the next result as a consequence of Proposition 3.27.

> For a few examples of algebras that are not finitely generated, have a look at Exercises 3.3.7 – 3.3.9.

3.28 COROLLARY $K[X]$ is finitely generated

Let $X \subseteq \mathbb{A}^n$ be an affine variety. Then the coordinate ring $K[X]$ is a finitely-generated K-algebra.

More explicitly, the proof of Proposition 3.27 shows that $K[X]$ can be generated as a K-algebra by the elements $[x_1], \ldots, [x_n]$ in the canonical isomorphism
$$K[x_1, \ldots, x_n]/\mathcal{I}(X) = K[X]$$
of Proposition 3.22. These images are the n coordinate functions
$$c_i : X \to K$$
$$(a_1, \ldots, a_n) \mapsto a_i.$$
Thus, $K[X]$ is the K-algebra generated by the coordinate functions on X; this is precisely the reason we call it the "coordinate ring" of X.

Exercises for Section 3.3

3.3.1 Let A be a K-algebra and let $\mathcal{S} \subseteq A$ be a subset.
   (a) Prove that $K[\mathcal{S}]$ is a subalgebra of A.
   (b) Prove that $K[\mathcal{S}]$ is contained in every subalgebra of A that contains $\mathcal{S}$.

3.3.2 Give two different examples of three elements that generate $K[x,y,z]$.

3.3.3 Consider the K-algebra
$$A = K[x,y,z]/\langle y - x^2, z \rangle.$$
Find an element $a \in A$ such that $A = K[a]$.

3.3.4 Consider the K-algebra
$$A = K[x,y]/\langle xy - 1 \rangle.$$
Prove that $A \neq K[a]$ for any $a \in A$.

3.3.5 Suppose that A and B are K-algebras such that $A = K[a_1, \ldots, a_n]$ for some $a_1, \ldots, a_n \in A$. Prove that, if $\varphi, \psi : A \to B$ are K-algebra homomorphisms such that $\varphi(a_i) = \psi(a_i)$ for all $i = 1, \ldots, n$, then $\varphi(a) = \psi(a)$ for all $a \in A$.

3.3.6 Consider $\mathbb{R}$ as a $\mathbb{Q}$-algebra.
   (a) The subalgebra $\mathbb{Q}[\sqrt{2}] \subseteq \mathbb{R}$ is finitely generated, so Proposition 3.27 implies that $\mathbb{Q}[\sqrt{2}] \cong \mathbb{Q}[x_1, \ldots, x_n]/I$ for some n and I. Find an explicit n and I for which this is the case.
   (b) Repeat part (a) for the subalgebra $\mathbb{Q}[\sqrt[3]{2}] \subseteq \mathbb{R}$.
   (c) Repeat part (a) for the subalgebra $\mathbb{Q}[\pi] \subseteq \mathbb{R}$.

3.3.7 Prove that $K[x_1, x_2, x_3, \ldots]$ is not a finitely-generated K-algebra. (The ring $K[x_1, x_2, x_3, \ldots]$ was defined in Example 2.13.)

3.3.8 Prove that any finitely-generated $\mathbb{Q}$-algebra is countable and conclude that $\mathbb{R}$ is not a finitely-generated $\mathbb{Q}$-algebra.

3.3.9 Consider the subalgebra
$$A = K[x, xy, xy^2, xy^3, xy^4, \ldots] \subseteq K[x,y].$$
Prove that A is not a finitely-generated K-algebra. Thus, a subalgebra of a finitely-generated algebra need not be finitely generated.

Section 3.4 Nilpotents and reduced rings

We learned in Corollary 3.28 that the coordinate ring $K[X]$ of an affine variety X is a finitely-generated K-algebra. This follows from the interpretation of $K[X]$ as a quotient:
$$K[X] = K[x_1, \ldots, x_n]/\mathcal{I}(X).$$
However, there is still one important aspect of this quotient that we have not yet taken into account: $K[X]$ is not just a quotient by an arbitrary ideal, it is a quotient by a *radical* ideal. What, then, does the fact that $\mathcal{I}(X)$ is radical imply about the algebraic properties of the coordinate ring $K[X]$? The answer, as it turns out, can be phrased in terms the following definition.

> **3.29 DEFINITION** *Nilpotents and reduced rings*
>
> Let R be a ring. An element $a \in R$ is *nilpotent* if there exists an integer $m \geq 1$ such that $a^m = 0$. We say that R is *reduced* if it has no nonzero nilpotent elements.

3.30 EXAMPLE Reduced rings

Any integral domain is necessarily reduced, since a nonzero nilpotent element would be a zero divisor. Not all reduced rings are integral domains, however. For example, the quotient ring
$$\frac{K[x,y]}{\langle xy \rangle}$$
is not an integral domain, because $[x]$ and $[y]$ are zero divisors, but it is reduced. To see that this ring is reduced, suppose that $[f]^m = 0$; we must prove that $[f] = 0$. Since
$$[f^m] = [f]^m = 0 \in \frac{K[x,y]}{\langle xy \rangle},$$
we know that xy divides f^m, implying that x and y both divide f^m. Since x and y are irreducible in $K[x,y]$, and thus prime, it follows that x and y both divide f. Therefore, xy divides f, and $[f] = 0$. Thus, the ring does not contain any nonzero nilpotents, so it is reduced.

That every integral domain is reduced, but not vice versa, is a manifestation of the fact that every prime ideal is radical, but not vice versa (see Proposition 3.32).

3.31 EXAMPLE A nonreduced ring

The quotient ring
$$\frac{K[x]}{\langle x^2 \rangle}$$
is not reduced, because it has a nonzero nilpotent:
$$[x] \neq 0 \quad \text{satisfies} \quad [x]^2 = 0.$$

The next result is a quotient characterization of radical ideals, analogous to the quotient characterizations of prime and maximal ideals.

> **3.32 PROPOSITION** *Quotients by radical ideals*
>
> An ideal $I \subseteq R$ is radical if and only if R/I is reduced.

PROOF   Exercise 3.4.2.   □

We can now give a complete algebraic characterization of the type of rings that arise as coordinate rings of affine varieties over K; they are finitely-generated K-algebras that are reduced as rings.

> **3.33 PROPOSITION** *Characterization of coordinate rings*
>
> If $X \subseteq \mathbb{A}^n$ is an affine variety, then the coordinate ring $K[X]$ is a finitely-generated reduced K-algebra. Conversely, if A is a finitely-generated reduced K-algebra, then $A \cong K[X]$ for some affine variety $X \subseteq \mathbb{A}^n$.

PROOF   Suppose X is an affine variety. By Proposition 3.22,
$$K[X] = K[x_1, \ldots, x_n]/\mathcal{I}(X).$$
Thus, $K[X]$ is finitely generated by Proposition 3.27, and it is reduced by Proposition 3.32 and the fact that $\mathcal{I}(X)$ is a radical ideal.

Conversely, suppose that A is a finitely-generated reduced K-algebra. By Proposition 3.27, we can write
$$A \cong K[x_1, \ldots, x_n]/I$$
for some n and I, and by Proposition 3.32, we know that I is a radical ideal. Define $X = \mathcal{V}(I) \subseteq \mathbb{A}^n$. By the Nullstellensatz,
$$\mathcal{I}(X) = \mathcal{I}(\mathcal{V}(I)) = \sqrt{I} = I.$$
It then follows from Proposition 3.22 that $A \cong K[X]$.   □

To make Proposition 3.33 effective, we should be able to produce, given a finitely-generated reduced K-algebra A, an affine variety X for which $K[X] \cong A$. In the next example, we illustrate how to carry out this procedure in practice.

3.34 EXAMPLE   Determining X from $K[X]$

Let A be the subalgebra
$$A = K[u^2, uv, v^2] \subseteq K[u, v].$$
Then A is manifestly finitely generated, as it is generated by the three elements u^2, uv, and v^2, and it is reduced, because it is a subalgebra of the reduced K-algebra $K[u, v]$. Thus, there should exist an affine variety X such that $K[X] \cong A$.

To find X, we first give the three generators suggestive names:
$$x = u^2, \quad y = uv, \quad \text{and} \quad z = v^2.$$

3.4. NILPOTENTS AND REDUCED RINGS 101

Notice that these three generators satisfy the relation
$$xz - y^2 = (u^2)(v^2) - (uv)^2 = 0.$$
Consider the affine variety defined by this relation:
$$X = \mathcal{V}(xz - y^2) \subseteq \mathbb{A}^3.$$
In the rest of this example, we prove that $K[X] \cong A$.

First, observe that the polynomial $xz - y^2$ is irreducible (using, for example, Eisenstein's Criterion), so the ideal $\langle xz - y^2 \rangle$ is prime and hence radical. It follows from the Nullstellensatz that
$$\mathcal{I}(X) = \mathcal{I}(\mathcal{V}(xz - y^2)) = \sqrt{\langle xz - y^2 \rangle} = \langle xz - y^2 \rangle,$$
so
$$K[X] \cong \frac{K[x, y, z]}{\langle xz - y^2 \rangle}.$$
What remains to be shown is that

(3.35)
$$\frac{K[x, y, z]}{\langle xz - y^2 \rangle} \cong A.$$

To justify (3.35), define a K-algebra homomorphism $\varphi : K[x, y, z] \to A$ by
$$\varphi(f) = f(u^2, uv, v^2).$$
Since the three generators u^2, uv, v^2 of A are all in the image of φ, it follows that φ is surjective. Therefore, the sought-after isomorphism in (3.35) follows from the First Isomorphism Theorem for K-algebras if we can prove that $\ker(\varphi) = \langle xz - y^2 \rangle$.

Since
$$\varphi(xz - y^2) = (u^2)(v^2) - (uv)^2 = 0,$$
every element of $\langle xz - y^2 \rangle$ is sent to 0 by φ, so $\ker(\varphi) \supseteq \langle xz - y^2 \rangle$. To prove the other inclusion, suppose $f \in \ker(\varphi)$ and consider the coset
$$[f] \in \frac{K[x, y, z]}{\langle xz - y^2 \rangle}.$$
By repeated use of the equation $[y^2] = [xz]$, we see that
$$[f] = [g(x, z) + y h(x, z)]$$
for some polynomials $g, h \in K[x, z]$. In other words,
$$f = g(x, z) + y \cdot h(x, z) + \ell(x, y, z)(xz - y^2)$$
for some polynomial $\ell \in K[x, y, z]$. Applying φ, we obtain
$$0 = \varphi(f) = g(u^2, v^2) + uv \cdot h(u^2, v^2) \in K[u, v].$$
Since the term $g(u^2, v^2)$ is a polynomial with only even powers of both u and v and the term $uv \cdot h(u^2, v^2)$ is a polynomial with only odd powers of u and v, there can be no cancellation between these two terms. Therefore, we must have $g = h = 0$, implying that $f \in \langle xz - y^2 \rangle$. Thus, $\ker(\varphi) \subseteq \langle xz - y^2 \rangle$, completing the argument.

The diagram below depicts the developments of this chapter. In particular, in the category of rings, we have pinned down exactly which rings arise as coordinate rings of affine varieties over K: they are K-algebras that are both finitely generated and reduced.

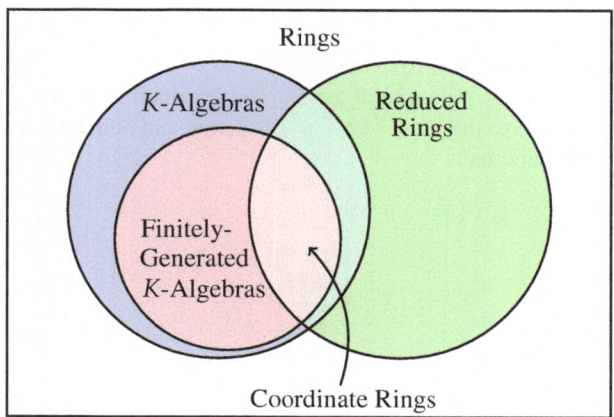

Now that we have an algebraic language in which we can characterize and discuss coordinate rings, our goal in the next chapter is to investigate what the coordinate ring $K[X]$ tells us about the affine variety X. As we will see, coordinate rings know essentially everything about their corresponding affine varieties, a powerful fact that allows us to bring all of the algebraic tools of rings and K-algebras to bear on the study of affine varieties.

Exercises for Section 3.4

3.4.1 Explain why any subring of a reduced ring is reduced.

3.4.2 Prove Proposition 3.32.

3.4.3 (a) Prove that, if R or S is reduced, then the direct sum $R \oplus S$ is reduced. (Recall that addition and multiplication in direct sums are defined componentwise.)
(b) Prove that the K-algebra $A = K[x]/\langle x^2 \rangle$ is isomorphic to $K \oplus K$ as a K-vector space but not as a ring.

3.4.4 Let R be a reduced ring. Prove that $R[x]$ is reduced, and conclude, by induction, that $R[x_1, \ldots, x_n]$ is reduced for any n.

3.4.5 Let $A = K[u^2, u^3] \subseteq K[u]$.
(a) Explain how you know that A is finitely generated and reduced.
(b) Find an affine variety X such that $K[X] \cong A$, and prove your answer.

3.4.6 Let $A = K[u+w, v+w] \subseteq K[u, v, w]$. Find an affine variety X such that $K[X] \cong A$, and prove your answer.

3.4. NILPOTENTS AND REDUCED RINGS

3.4.7 Let A be the K-vector space with basis
$$\{x^i \mid i \geq 0\} \cup \{x^i y \mid i \geq 0\}.$$
Define a (commutative) product on the elements of this basis by setting
$$x^i \cdot x^j = x^{i+j},$$
$$x^i \cdot x^j y = x^{i+j} y, \text{ and}$$
$$x^i y \cdot x^j y = x^{i+j+3},$$
and extending this product to all elements of A by linearity in K.
 (a) Prove that A is a finitely-generated reduced K-algebra.
 (b) Find an affine variety X such that $K[X] \cong A$, and prove your answer.

3.4.8 Let $a_1, \ldots, a_m \in K[x_1, \ldots, x_n]$ and consider the subalgebra
$$A = K[a_1, \ldots, a_m] \subseteq K[x_1, \ldots, x_n].$$
Let $I \subseteq K[y_1, \ldots, y_m]$ be the *ideal of relations* of $a_1, \ldots, a_m$:
$$f \in I \iff f(a_1, \ldots, a_m) = 0 \in K[x_1, \ldots, x_n].$$
Define $X = \mathcal{V}(I) \subseteq \mathbb{A}^m$. Prove that $K[X] \cong A$.

3.4.9 Give an explicit example of a ring that belongs in each region of the Venn diagram presented at the end of this section.

Chapter 4
Polynomial Maps

LEARNING OBJECTIVES FOR CHAPTER 4

- Define and give examples of polynomial maps between affine varieties, including isomorphisms.
- Define the pullback homomorphism associated to a polynomial map, and compute it in concrete examples.
- Use the bijectivity of pulling back to prove that affine varieties are isomorphic if and only if their coordinate rings are isomorphic.
- Describe the equivalence of algebra and geometry and how it translates between intrinsic properties of K-algebras and affine varieties.

The previous chapter provides us with an association

$$\{\text{affine varieties}\} \to \{\text{finitely-generated reduced } K\text{-algebras}\}$$
$$X \mapsto K[X]$$

and confirms that it is surjective. But is it injective? That is, if $K[X] = K[Y]$, is it necessarily the case that $X = Y$?

This question is more subtle than it might first appear. To answer it, one must precisely decide when two K-algebras are "the same"; is $K[x]$ the same as $K[y]$, for example? The literal answer is no, but the reader would be forgiven for finding this answer unsatisfying, given how conditioned we are to viewing isomorphic rings as identical. To capture the intuition that isomorphism is "sameness," we might rephrase our question: If $K[X] \cong K[Y]$, is it necessarily the case that $X \cong Y$?

To address this question, we must first introduce a precise notion of what it means for two affine varieties to be "isomorphic," beginning with the notion of a structure-preserving "morphism" between affine varieties. This is something one should do whenever a new type of mathematical object—groups, rings, topological spaces, et cetera—is introduced: ask which maps between those objects preserve their relevant structure. In the context of groups, the relevant maps are group homomorphisms, while for rings they are ring homomorphisms, and for topological spaces they are continuous maps. Because algebraic geometry is concerned with polynomials, it may come as no surprise that the relevant maps between affine varieties are *polynomial maps*, which we introduce in this chapter.

Equipped with the definition of polynomial maps, we can make sense of what it means to say that affine varieties X and Y are *isomorphic*, and we can prove that $X \cong Y$ if and only if $K[X] \cong K[Y]$. This statement, which is referred to as "the equivalence of algebra and geometry," is the goal of this chapter and the heart of our dictionary between algebra and geometry.

Section 4.1 Polynomial maps between affine varieties

In the previous chapter, we were introduced to the coordinate ring $K[X]$ of an affine variety $X \subseteq \mathbb{A}^m$, whose elements are polynomial functions $F : X \to K$. Identifying K with $\mathbb{A}^1$, we can view such functions as a special case of maps between affine varieties (namely, between X and $\mathbb{A}^1$), and our first goal is to extend the definition to allow for maps from any affine variety to any other.

> *There is a deliberate distinction between the words "function" and "map." A function takes values in the ground field whereas a map takes values in an affine variety.*

Let $X \subseteq \mathbb{A}^m$ be an affine variety. Since every element of the coordinate ring $K[X]$ is a function $X \to K$, we see that n elements $F_1, \ldots, F_n \in K[X]$ give rise to a map $X \to \mathbb{A}^n$ defined by $a \in X \mapsto (F_1(a), \ldots, F_n(a)) \in \mathbb{A}^n$. If $Y \subseteq \mathbb{A}^n$ is an affine variety and the image of the map $X \to \mathbb{A}^n$ happens to lie in Y, then we obtain a map $X \to Y$. Maps that arise from polynomial functions in this way are called *polynomial maps*.

4.1 DEFINITION *Polynomial map between affine varieties*

Let $X \subseteq \mathbb{A}^m$ and $Y \subseteq \mathbb{A}^n$ be affine varieties. A map $F : X \to Y$ is said to be a *polynomial map* if there exist $F_1, \ldots, F_n \in K[X]$ such that, for every $a \in X$,
$$F(a) = (F_1(a), \ldots, F_n(a)).$$

In particular, a polynomial function on X is the exact same notion as a polynomial map $F : X \to \mathbb{A}^1$. In the next example, we consider a polynomial map whose codomain is the affine space $\mathbb{A}^3$.

4.2 EXAMPLE Polynomial maps to affine space

Consider the parabola $X = \mathcal{V}(x_2 - x_1^2) \subseteq \mathbb{A}^2$. (We have named the variables x_1 and x_2, rather than x and y, in preparation for the next example.) Then
$$F : X \to \mathbb{A}^3$$
$$(a_1, a_2) \mapsto (a_1 - a_1^2, a_1 + a_2, a_1^2 - a_2^2)$$
is a polynomial map, since its three component functions
$$F_1(a_1, a_2) = a_1 - a_1^2, \quad F_2(a_1, a_2) = a_1 + a_2, \quad \text{and} \quad F_3(a_1, a_2) = a_1^2 - a_2^2$$
arise from the polynomials
$$f_1 = x_1 - x_1^2, \quad f_2 = x_1 + x_2, \quad \text{and} \quad f_3 = x_1^2 - x_2^2,$$
respectively, and are thus elements of $K[X]$. Notice that the three polynomials
$$g_1 = x_1 - x_2, \quad g_2 = x_1 + x_1^2, \quad \text{and} \quad g_3 = x_2 - x_2^2$$
give rise to the same polynomial map $F : X \to \mathbb{A}^3$, because $[f_i] = [g_i] \in K[X]$.

4.1. POLYNOMIAL MAPS BETWEEN AFFINE VARIETIES

If $X \subseteq \mathbb{A}^m$ is an affine variety, it is straightforward to produce polynomial maps $F: X \to \mathbb{A}^n$ whose codomain is an affine space. In particular, any choice of polynomials $f_1, \ldots, f_n \in K[x_1, \ldots, x_m]$ defines the coordinate functions $F_1, \ldots, F_n$ of such a map, by setting

$$F_i = [f_i] \in \frac{K[x_1, \ldots, x_m]}{\mathcal{I}(X)} = K[X],$$

and another choice $g_1, \ldots, g_n$ of polynomials produces the same polynomial map if and only if $f_i - g_i \in \mathcal{I}(X)$ for every i.

On the other hand, if $Y \subsetneq \mathbb{A}^n$ is an affine variety other than affine space itself, then not every choice of $F_1, \ldots, F_n \in K[X]$ gives a map $F: X \to Y$. In particular, in order to ensure that the image of every point of X is a point in Y, we must require that for every $a \in X$, the point

$$(F_1(a), \ldots, F_n(a)) \in \mathbb{A}^n$$

actually lies within Y, meaning that it is a zero of the defining polynomials of Y. More concretely, if $Y = \mathcal{V}(\mathcal{S})$ where $\mathcal{S} \subseteq K[x_1, \ldots, x_n]$, then we must check that, for every $a \in X$ and every $g \in \mathcal{S}$,

$$g(F_1(a), \ldots, F_n(a)) = 0.$$

4.3 EXAMPLE A polynomial map to an affine variety

As in Example 4.2, let $X = \mathcal{V}(x_2 - x_1^2) \subseteq \mathbb{A}^2$, but now let $Y = \mathcal{V}(y_1 y_2 - y_3) \subseteq \mathbb{A}^3$. Over the real numbers, Y is the hyperbolic paraboloid depicted to the right, and the function

$$F: X \to Y$$
$$(a_1, a_2) \mapsto (a_1 - a_1^2, a_1 + a_2, a_1^2 - a_2^2)$$

of Example 4.2 is a polynomial map from X to Y, whose image is illustrated as the curve on the surface. We have already checked that the three component functions of F are polynomial, but we now must also confirm that the image

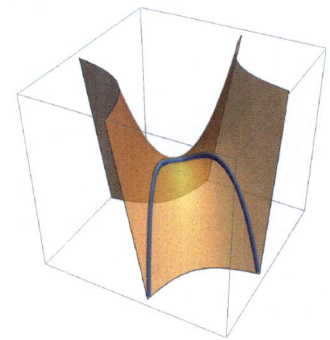

actually lies in Y. This is equivalent to the claim that $F(a_1, a_2)$ satisfies the defining equation $y_1 y_2 - y_3$ of Y, or in other words that

$$F_1(a_1, a_2) F_2(a_1, a_2) - F_3(a_1, a_2) = 0$$

whenever $(a_1, a_2) \in X$. To check this, we simply substitute in the expressions for F_1, F_2, F_3 and rearrange:

$$(a_1 - a_1^2)(a_1 + a_2) - (a_1^2 - a_2^2) = (a_1 + a_2)(a_2 - a_1^2) = (a_1 + a_2) \cdot 0 = 0,$$

where the second equality follows from $(a_1, a_2) \in X$, implying that $a_2 - a_1^2 = 0$.

> *To distinguish between coordinates on X and on Y, we often use subscripts as in the previous example: $x_1, x_2, \ldots$ for X and $y_1, y_2, \ldots$ for Y.*

From the algebraic context, the reader is already familiar with the notion that some homomorphisms are isomorphisms, and that isomorphic objects share all of their relevant properties; isomorphic groups, for example, share all group-theoretic properties. Similarly, isomorphisms of affine varieties allow us to talk about what it means for affine varieties to be essentially the same.

4.4 DEFINITION *Isomorphism of affine varietes*

Let $X \subseteq \mathbb{A}^m$ and $Y \subseteq \mathbb{A}^n$ be affine varieties. A polynomial map $F : X \to Y$ is said to be an *isomorphism* if it has an inverse function $F^{-1} : Y \to X$ that is also a polynomial map. If such an isomorphism exists, we say that X and Y are *isomorphic* and write $X \cong Y$.

4.5 EXAMPLE The parabola is isomorphic to $\mathbb{A}^1$

Let $X = \mathcal{V}(x_2 - x_1^2) \subseteq \mathbb{A}^2$. Notice that the two maps F and G defined by

$$F : X \to \mathbb{A}^1 \qquad G : \mathbb{A}^1 \to X$$
$$(a_1, a_2) \mapsto a_1 \qquad b \mapsto (b, b^2)$$

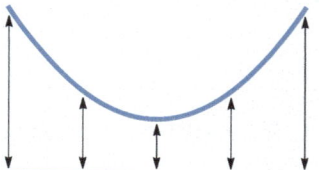

are both polynomial maps. Furthermore, it is straightforward to check that $F \circ G$ is the identity on $\mathbb{A}^1$ and, using the fact that $a_2 = a_1^2$ for every $(a_1, a_2) \in X$, it can also be checked that $G \circ F$ is the identity on X. Thus, $X \cong \mathbb{A}^1$. In the figure above, we have depicted the isomorphisms between the affine line and the parabola over $\mathbb{R}$, where F is the downward map and G is the upward map.

4.6 EXAMPLE *Isomorphic projections*

Generalizing the previous example, suppose that $X = \mathcal{V}(x_m - g) \subseteq \mathbb{A}^m$ where $g \in K[x_1, \ldots, x_{m-1}]$. Consider the two maps

$$F : X \to \mathbb{A}^{m-1}$$
$$(a_1, \ldots, a_m) \mapsto (a_1, \ldots, a_{m-1})$$

and

$$G : \mathbb{A}^{m-1} \to X$$
$$(b_1, \ldots, b_{m-1}) \mapsto (b_1, \ldots, b_{m-1}, g(b_1, \ldots, b_{m-1})).$$

Both F and G are polynomial maps, and using the fact that $a_m = g(a_1, \ldots, a_{m-1})$ for every $(a_1, \ldots, a_m) \in X$, it follows that they are inverse to each other. Thus, we conclude that $X \cong \mathbb{A}^{m-1}$. See Exercise 4.1.4 for a further generalization of this example.

4.1. POLYNOMIAL MAPS BETWEEN AFFINE VARIETIES

4.7 EXAMPLE Translations are isomorphisms

Given a reference point $c = (c_1, \ldots, c_m) \in \mathbb{A}^m$, consider the translation map

$$T_c : \mathbb{A}^m \to \mathbb{A}^m$$
$$(a_1, \ldots, a_m) \mapsto (a_1 + c_1, \ldots, a_m + c_m).$$

Then for any affine variety $X \subseteq \mathbb{A}^m$, the translation $T_c(X)$ is also an affine variety; to see this, notice that a polynomial $f(x_1, \ldots, x_m)$ vanishes on $T_c(X)$ if and only if $f(x_1 + c_1, \ldots, x_n + c_m)$ vanishes on X, so

$$T_c(X) = \mathcal{V}(\{ f \in K[x_1, \ldots, x_n] \mid f(x_1 + c_1, \ldots, x_m + c_m) \in \mathcal{I}(X) \}).$$

The map $T_c : X \to T_c(X)$ is manifestly polynomial, and in fact, it is an isomorphism. To prove this, it suffices to notice that it has a polynomial inverse defined by translating back, using the reference point $(-c_1, \ldots, -c_m) \in \mathbb{A}^m$.

Exercise 4.1.6 generalizes this example to compositions of translations with invertible linear maps; such compositions are called *affine linear transformations*.

In order for a map of affine varieties to be an isomorphism, it must be bijective, because this is necessary for an inverse function to exist. However, *not every bijective polynomial map of affine varieties is an isomorphism*, because an inverse function, even if it exists, need not be a polynomial map. The next example illustrates this phenomenon.

> *Geometry and algebra differ here: in algebra, if a homomorphism (of groups, rings, algebras, et cetera) has an inverse function, then that inverse function is automatically a homomorphism. See Exercise 3.2.2.*

4.8 EXAMPLE Bijective polynomial maps need not be isomorphisms

Consider $X = \mathcal{V}(x^2 - y^3) \subseteq \mathbb{A}^2$. Then

$$F : \mathbb{A}^1 \to X$$
$$a \mapsto (a^3, a^2)$$

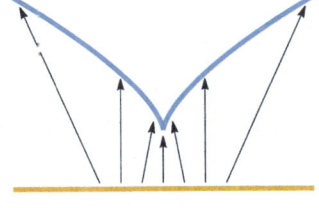

is the polynomial map depicted to the right. (Note that for every $a \in \mathbb{A}^1$, the point $(a^3, a^2) \in \mathbb{A}^2$ indeed satisfies the equation $x^2 - y^3 = 0$.) Moreover, the reader can verify that an inverse to F is given by

$$G : X \to \mathbb{A}^1$$
$$(b, c) \mapsto \begin{cases} b/c & \text{if } c \neq 0, \\ 0 & \text{if } c = 0. \end{cases}$$

However, G is not a polynomial map. That G involves a quotient of its inputs certainly hints at its non-polynomiality, but, as we saw in Example 3.5, more care is required to be sure that G is not polynomial. In the next section, we develop the necessary tools to prove that $X \not\cong \mathbb{A}^1$, from which it follows that G cannot be a polynomial map. The "cusp" point, where the curve X appears to change directions, is a visual clue that $X \not\cong \mathbb{A}^1$.

To prove that $X \cong Y$, the task at hand is somewhat straightforward: we must find an isomorphism between X and Y. But proving that $X \not\cong Y$ is more subtle. How can we rigorously prove the nonexistence of any isomorphism? In the context of algebra, we have quite a few tools for doing so: proving that two rings are not isomorphic involves finding a ring-theoretic property—like being an integral domain or a UFD—that one has but the other does not.

Therefore, if we want to be able to detect when affine varieties are not isomorphic, a natural approach is to prove that $X \cong Y$ implies $K[X] \cong K[Y]$. Once we have accomplished this, then an algebraic argument that $K[X] \not\cong K[Y]$ would imply the geometric conclusion $X \not\cong Y$, allowing us to employ algebraic methods to detect when affine varieties are not isomorphic. Thus, our goal is to develop a procedure for converting isomorphisms of affine varieties to isomorphisms of their corresponding coordinate rings. This procedure is the *pullback* and is the topic of the next section.

Exercises for Section 4.1

4.1.1 Let $X = \mathcal{V}(y^2 - z^2 + xy - z, z^2 - x^3y^2) \subseteq \mathbb{A}^3$. Prove that $F(a) = (1, a, a)$ defines a polynomial map $F : \mathbb{A}^1 \to X$. Is F an isomorphism?

4.1.2 Let $X = \mathcal{V}(x^2 + y^2 - 1) \subseteq \mathbb{A}^2$ and let $Y = \mathcal{V}(u^2 + v^2 - 2) \subseteq \mathbb{A}^2$. Prove that $F(a, b) = (a + b, a - b)$ defines a polynomial map $F : X \to Y$. Is F an isomorphism?

4.1.3 Let $X = \mathcal{V}(y^2 - x^3 - x^2) \subseteq \mathbb{A}^2$. Prove that $F(a) = (a^2 - 1, a^3 - a)$ defines a polynomial map $F : \mathbb{A}^1 \to X$. Is F an isomorphism? (Hint: Draw a picture.)

4.1.4 Let $X = \mathcal{V}(f_1, \ldots, f_k, x_m - g) \subseteq \mathbb{A}^m$ where $g \in K[x_1, \ldots, x_{m-1}]$ and $f_1, \ldots, f_k \in K[x_1, \ldots, x_m]$. For each $i = 1, \ldots, k$, define

$$\tilde{f}_i = f_i(x_1, \ldots, x_{m-1}, g) \in K[x_1, \ldots, x_{m-1}],$$

and set $Y = \mathcal{V}(\tilde{f}_1, \ldots, \tilde{f}_k) \subseteq \mathbb{A}^{m-1}$. Prove that $X \cong Y$.

(This shows that if an affine variety is such that one of its defining equations is linear in one of the variables, then it can be replaced with an isomorphic affine variety defined by fewer equations in fewer variables.)

4.1.5 (a) Prove that the image of the polynomial map

$$F : \mathbb{A}^1 \to \mathbb{A}^3$$
$$a \mapsto (a, a^2, a^3)$$

is an affine variety.

(b) Prove that the image of the polynomial map

$$G : \mathbb{A}^2 \to \mathbb{A}^2$$
$$(a, b) \mapsto (a, ab)$$

is not an affine variety.

Thus, images of polynomial maps may or may not be affine varieties.

4.1. POLYNOMIAL MAPS BETWEEN AFFINE VARIETIES

4.1.6 Let M be an invertible $m \times m$ matrix with coefficients in K and $c \in \mathbb{A}^m$. Identifying $\mathbb{A}^m$ with the vector space K^m and M with a linear transformation $\varphi_M : K^m \to K^m$, define the function

$$F : \mathbb{A}^m \to \mathbb{A}^m$$
$$a \mapsto \varphi_M(a) + c.$$

Prove the following.

(a) If $X \subseteq \mathbb{A}^m$ is an affine variety, then $F(X) \subseteq \mathbb{A}^m$ is an affine variety.
(b) If $X \subseteq \mathbb{A}^m$ is an affine variety, then $X \cong F(X)$.

4.1.7 Let $X = \mathcal{V}(\ell_1, \ldots, \ell_k) \subseteq \mathbb{A}^n$ where each ℓ_i is a linear polynomial:

$$\ell_i = a_{i1} x_1 + \cdots + a_{in} x_n + b_i \in K[x_1, \ldots, x_n].$$

Let $M = (a_{ij})$ be the $k \times n$ matrix of coefficients. Assuming that X is nonempty, prove that

$$X \cong \mathbb{A}^{n - \text{rk}(M)}.$$

(Hint: Use the Rank-Nullity Theorem.)

Section 4.2 Pullback homomorphisms

Polynomial maps form the structure-preserving maps between affine varieties in much the same way that homomorphisms form the structure-preserving maps between algebraic objects. And just as one can move from

> A collection of mathematical objects together with their structure-preserving maps is, loosely speaking, the definition of a "category."

geometry to algebra on the level of objects—sending X to $K[X]$—there is also a passage from geometry to algebra on the level of maps—sending a polynomial map between affine varieties to a corresponding K-algebra homomorphism between coordinate rings. This passage between maps is given by the *pullback homomorphism*.

4.9 DEFINITION *Pullback homomorphism*

Let $X \subseteq \mathbb{A}^m$ and $Y \subseteq \mathbb{A}^n$ be affine varieties, and let $F : X \to Y$ be a polynomial map. The *pullback homomorphism* induced by F is

$$F^* : K[Y] \to K[X]$$
$$F^*(G) = G \circ F.$$

> Notice that pulling back changes the direction of the map:
>
> $F : X \to Y \Rightarrow F^* : K[Y] \to K[X]$.

In order for the definition of the pullback homomorphism to make sense, one must verify that, for every $G \in K[Y]$, the composition $G \circ F$ is an element of $K[X]$. Since $G : Y \to K$ and $F : X \to Y$, the definition of compositions implies that $G \circ F$ is, indeed, a function from X to K; schematically:

$$X \xrightarrow{F} Y \xrightarrow{G} K \implies X \xrightarrow{G \circ F} K.$$

The fact that $G \circ F$ is, moreover, a polynomial function follows from the fact that compositions of polynomial functions are polynomial functions (Exercise 4.2.3).

4.10 EXAMPLE Pullback homomorphism

Let $X = \mathcal{V}(x_2 - x_1^2) \subseteq \mathbb{A}^2$ and let $Y = \mathcal{V}(y_1 y_2 - y_3) \subseteq \mathbb{A}^3$. Recall the polynomial map $F : X \to Y$ of Example 4.3:

$$F(a_1, a_2) = (a_1 - a_1^2, a_1 + a_2, a_1^2 - a_2^2).$$

Consider the function $G \in K[Y]$ defined by $G(b_1, b_2, b_3) = b_1^2 - b_2 b_3$. Pulling back by F, we obtain the polynomial function $F^*(G) \in K[X]$ defined by

$$(F^*G)(a_1, a_2) = (G \circ F)(a_1, a_2) = (a_1 - a_1^2)^2 - (a_1 + a_2)(a_1^2 - a_2^2).$$

Similarly, pulling back $H \in K[Y]$ defined by $H(b_1, b_2, b_3) = b_1 + b_2 - b_3$, we obtain the polynomial function $F^*(H) \in K[X]$ defined by

$$(F^*H)(a_1, a_2) = (H \circ F)(a_1, a_2) = (a_1 - a_1^2) + (a_1 + a_2) - (a_1^2 - a_2^2).$$

4.2. PULLBACK HOMOMORPHISMS

As seen in the previous example, once we have chosen polynomial expressions for $F = (F_1, \ldots, F_n)$ and G, then we obtain a polynomial expression for $F^*(G)$ simply by composing the polynomial expressions for F and G. To expand on this observation, suppose that $X \subseteq \mathbb{A}^m$ and $Y \subseteq \mathbb{A}^n$. Then

$$K[X] = \frac{K[x_1, \ldots, x_m]}{\mathcal{I}(X)} \quad \text{and} \quad K[Y] = \frac{K[y_1, \ldots, y_n]}{\mathcal{I}(Y)}.$$

If we choose polynomials $f_1, \ldots, f_n \in K[x_1, \ldots, x_m]$ and $g \in K[y_1, \ldots, y_n]$ such that $F_i = [f_i]$ and $G = [g]$, then for every $a \in X$ and $b = (b_1, \ldots, b_n) \in Y$,

$$F(a) = (f_1(a), \ldots, f_n(a)) \quad \text{and} \quad G(b_1, \ldots, b_n) = g(b_1, \ldots, b_n).$$

This implies that

$$F^*(G)(a) = G(F(a)) = g(f_1(a), \ldots, f_n(a)).$$

Thus, $F^*(G) = [g(f_1, \ldots, f_n)]$, where $g(f_1, \ldots, f_n) \in K[x_1, \ldots, x_m]$ is the polynomial obtained from g by replacing y_i with $f_i(x_1, \ldots, x_m)$.

4.11 EXAMPLE Pullback homomorphism, revisited

In the same setting as Example 4.10, the component functions of F arise from the polynomials

$$f_1(x_1, x_2) = x_1 - x_1^2, \quad f_2(x_1, x_2) = x_1 + x_2, \quad \text{and} \quad f_3(x_1, x_2) = x_1^2 - x_2^2.$$

The polynomial function $G \in K[Y]$ arises from the polynomial

$$g = y_1^2 - y_2 y_3.$$

By visual inspection, one can see that the pullback function $F^*(G)$ arises from the polynomial

$$g(f_1, f_2, f_3) = (x_1 - x_1^2)^2 - (x_1 + x_2)(x_1^2 - x_2^2).$$

As the name suggests, the pullback homomorphism is more than just a function; it is a homomorphism of K-algebras, as we now verify.

4.12 PROPOSITION F^* is a homomorphism

If $X \subseteq \mathbb{A}^m$ and $Y \subseteq \mathbb{A}^n$ are affine varieties and $F : X \to Y$ is a polynomial map, then

$$F^* : K[Y] \to K[X]$$

is a homomorphism of K-algebras.

PROOF To show that F^* is a K-algebra homomorphism, we must check that it preserves addition, multiplication, and scalar multiplication. We formally prove the first of these and leave the other two to Exercise 4.2.6.

Those with background in advanced linear algebra may recognize the pullback homomorphism as a generalization of the dual of a linear map; see Exercise 4.2.12.

To see that F^* respects addition, let $G_1, G_2 \in K[Y]$ be two polynomial functions. Evaluating $F^*(G_1 + G_2)$ at any value $a \in X$, we obtain

$$\begin{aligned}
\left(F^*(G_1 + G_2)\right)(a) &= \left((G_1 + G_2) \circ F\right)(a) & \text{(definition of pullback)} \\
&= (G_1 + G_2)(F(a)) & \text{(definition of composition)} \\
&= G_1(F(a)) + G_2(F(a)) & \text{(definition of $+$ in $K[Y]$)} \\
&= (G_1 \circ F)(a) + (G_2 \circ F)(a) & \text{(definition of composition)} \\
&= \left((G_1 \circ F) + (G_2 \circ F)\right)(a) & \text{(definition of $+$ in $K[X]$)} \\
&= (F^*G_1 + F^*G_2)(a) & \text{(definition of pullback).}
\end{aligned}$$

Thus, $F^*(G_1 + G_2) = F^*G_1 + F^*G_2$, verifying that F^* preserves addition. □

Recall that our motivation for introducing the pullback homomorphism was to equip ourselves with algebraic tools for determining whether or not two affine varieties are isomorphic. Since the definition of "isomorphism" (in any category) requires checking that the composition of two morphisms is the identity, a preliminary result toward this objective is to establish that pullbacks behave well with respect to compositions and the identity function.

4.13 PROPOSITION *Pullbacks preserve compositions and the identity*

Let $X \subseteq \mathbb{A}^\ell$, $Y \subseteq \mathbb{A}^m$, and $Z \subseteq \mathbb{A}^n$ be affine varieties.

1. If $F : X \to Y$ and $G : Y \to Z$ are polynomial maps, then

$$(G \circ F)^* = F^* \circ G^*.$$

2. The pullback of the identity function is the identity function:

$$(\mathrm{id}_X)^* = \mathrm{id}_{K[X]}.$$

PROOF To prove the first statement, suppose that $H \in K[Z]$. Using associativity of compositions, we then compute

In the language of category theory, these properties of the pullback go by the name "functoriality."

$$(G \circ F)^*(H) = H \circ (G \circ F) = (H \circ G) \circ F = F^*(H \circ G) = F^*(G^*(H)),$$

which shows that $(G \circ F)^* = F^* \circ G^*$. For the second statement, let $H \in K[X]$. Then

$$(\mathrm{id}_X)^*(H) = H \circ \mathrm{id}_X = H,$$

so $(\mathrm{id}_X)^*$ is indeed the identity function on $K[X]$. □

Our aim of using coordinate rings to detect whether affine varieties are isomorphic can now be accomplished.

4.2. PULLBACK HOMOMORPHISMS

> **4.14 COROLLARY** *Pullbacks of isomorphisms are isomorphisms*
>
> Let $X \subseteq \mathbb{A}^m$ and $Y \subseteq \mathbb{A}^n$ be affine varieties. If $F : X \to Y$ is an isomorphism, then $F^* : K[Y] \to K[X]$ is an isomorphism.

PROOF Let $F : X \to Y$ be an isomorphism with inverse $F^{-1} : Y \to X$. To prove that F^* is an isomorphism, it suffices to prove that F^* and $(F^{-1})^*$ are inverse to each other. We verify this using Proposition 4.13:
$$F^* \circ (F^{-1})^* = (F^{-1} \circ F)^* = (\mathrm{id}_X)^* = \mathrm{id}_{K[X]}$$
and
$$(F^{-1})^* \circ F^* = (F \circ F^{-1})^* = (\mathrm{id}_Y)^* = \mathrm{id}_{K[Y]}. \qquad \square$$

The converse of Corollary 4.14 is also true, but we require additional tools in order to prove it, so we defer this discussion until the next section. In the meantime, let us take a look at a few example applications of the previous result.

4.15 EXAMPLE $\mathcal{V}(xy) \not\cong \mathbb{A}^1$

Consider the affine variety $X = \mathcal{V}(xy) \subseteq \mathbb{A}^2$. Let us prove that X is not isomorphic to $\mathbb{A}^1$. Intuitively, this should be somewhat clear from the depiction of the two varieties below. In particular, X consists of two affine lines meeting at a point, which certainly looks quite different than a single affine line.

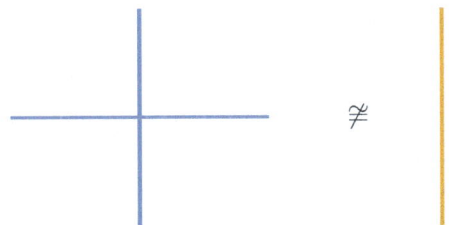

To make this intuition precise, notice that $X = \mathcal{V}(x) \cup \mathcal{V}(y)$ is a reducible affine variety. Thus, $\mathcal{I}(X)$ is not a prime ideal, implying that
$$K[X] = \frac{K[x,y]}{\mathcal{I}(X)}$$
is not an integral domain. Since $K[\mathbb{A}^1] = K[z]$ is a single-variable polynomial ring, it is an integral domain. Thus, given that the property of being an integral domain is preserved under isomorphism, we see that $K[X] \not\cong K[\mathbb{A}^1]$, and we conclude from the contrapositive of Corollary 4.14 that $X \not\cong \mathbb{A}^1$.

The previous example quickly generalizes: if one affine variety is irreducible and another is not, then they cannot be isomorphic, because one of their coordinate rings is an integral domain and the other is not. The next example, on the other hand, illustrates an instance of proving that two *irreducible* affine varieties are not isomorphic, which can be trickier. This example also concludes the discussion that we began in Example 4.8, that bijective polynomial maps between affine varieties are not necessarily isomorphisms.

4.16 EXAMPLE $\mathcal{V}(x^2 - y^3) \not\cong \mathbb{A}^1$

As in Example 4.8, let $X = \mathcal{V}(x^2 - y^3) \subseteq \mathbb{A}^2$. In order to show that the polynomial bijection $F : \mathbb{A}^1 \to X$ defined by $f(a) = (a^3, a^2)$ does not have a polynomial inverse, we prove that $X \not\cong \mathbb{A}^1$. To do so, we analyze the coordinate rings. Since $x^2 - y^3$ is irreducible, the vanishing ideal of X is

$$\mathcal{I}(X) = \langle x^2 - y^3 \rangle.$$

Thus,

$$K[X] = \frac{K[x,y]}{\langle x^2 - y^3 \rangle}.$$

In order to prove that $X \not\cong \mathbb{A}^1$, it suffices to find one ring-theoretic property of the single-variable polynomial ring $K[\mathbb{A}^1] = K[z]$ that is not satisfied by $K[X]$. An example of such a property is that $K[X]$ is not a UFD. Indeed, it can be shown (Exercise 0.3.15) that $[x]$ and $[y]$ are distinct irreducible elements of $K[X]$, so the equality

$$[x]^2 = [y]^3$$

expresses the same element in two inequivalent ways as a product of irreducibles.

In this section, we have discussed a way to associate a K-algebra homomorphism to every polynomial map. In particular, we now have an association from the category of affine varieties to the category of K-algebras that is defined on objects and morphisms by

$$X \longmapsto K[X]$$
$$(F : X \to Y) \longmapsto (F^* : K[Y] \to K[X]).$$

Along with the conditions in Proposition 4.13, such an association is called a *functor* between these categories. In the next section, we show that the association of morphisms is invertible: every K-algebra homomorphism between coordinate rings arises from a unique polynomial map between the corresponding affine varieties. This is a key step in proving the converse to Corollary 4.14, which we accomplish in Corollary 4.21.

Exercises for Section 4.2

4.2.1 Let $F : \mathbb{A}^2 \to \mathbb{A}^2$ be the polynomial map defined by

$$F(a_1, a_2) = (a_1^2 + 2a_1 a_2, a_1 - a_2).$$

Calculate $F^*(G)$, where $G \in K[\mathbb{A}^2]$ is the function defined by

$$G(b_1, b_2) = b_1 - b_2^2.$$

4.2.2 Let $Y = \mathcal{V}(y_3 - y_1^3) \subseteq \mathbb{A}^3$, and let $F : \mathbb{A}^1 \to Y$ be the polynomial map defined by

$$F(a) = (a, a^2, a^3).$$

Calculate $F^*([g]) \in K[\mathbb{A}^1]$, where $g = y_1 y_2 + y_3^2 \in K[y_1, y_2, y_3]$.

4.2. PULLBACK HOMOMORPHISMS

4.2.3 Let X and Y be affine varieties, and let $F : X \to Y$ be a function. Prove the following: if F is a polynomial map, then $G \circ F \in K[X]$ for any $G \in K[Y]$.

4.2.4 Let X and Y be affine varieties, and let $F : X \to Y$ be a function. Prove the following: if $F \circ G \in K[X]$ for all $G \in K[Y]$, then F is a polynomial map. (This is the converse of the previous problem.)

4.2.5 Let $X, Y \subseteq \mathbb{A}^n$ be affine varieties with $X \subseteq Y$, and let $F : X \to Y$ be the inclusion. Describe $F^* : K[Y] \to K[X]$.

4.2.6 Let X and Y be affine varieties, and let $F : X \to Y$ be a polynomial map. Prove that the pullback $F^* : K[Y] \to K[X]$ preserves ring multiplication and scalar multiplication.

4.2.7 For each of the following pairs of affine varieties X and Y, decide whether $X \cong Y$ and prove your answer.
(a) $X = \mathbb{A}^1$ and $Y = \mathbb{A}^2$
(b) $X = \mathcal{V}(x_1, x_2) \subseteq \mathbb{A}^2$ and $Y = \mathcal{V}(y_1 - y_2, y_1^2 - y_2) \subseteq \mathbb{A}^2$
(c) $X = \mathcal{V}(x_3 - x_1^2 + x_1 x_2) \subseteq \mathbb{A}^3$ and $Y = \mathbb{A}^2$
(d) $X = \mathcal{V}(x_1 x_2) \subseteq \mathbb{A}^2$ and $Y = \mathcal{V}(y_1^2 - y_2^2) \subseteq \mathbb{A}^2$

4.2.8 Prove that $\mathcal{V}(xy - 1) \subseteq \mathbb{A}^2$ is not isomorphic to $\mathbb{A}^1$.

4.2.9 Let $X \subseteq \mathbb{A}^m$ and $Y \subseteq \mathbb{A}^n$ be affine varieties, and let $F : X \to Y$ be a polynomial map.
(a) Prove that if F is surjective, then F^* is injective.
(b) Based on part (a), we might hope that if F is injective, then F^* is surjective. Prove that this is false by showing that the polynomial map
$$F : \mathcal{V}(xy - 1) \to \mathbb{A}^1$$
$$F(a, b) = a$$
is injective but F^* is not surjective.

4.2.10 (a) Let $Y \subseteq \mathbb{A}^n$ be an affine variety and let $X \subseteq Y$ be a subset. We say that X is *dense* in Y if there does not exist an affine variety $Z \subseteq \mathbb{A}^n$ such that $X \subseteq Z \subsetneq Y$. Prove that X is dense in Y if and only if the only polynomial function $G \in K[Y]$ that vanishes on X is the zero function.
(b) Let $X \subseteq \mathbb{A}^m$ and $Y \subseteq \mathbb{A}^n$ be affine varieties. We say that a polynomial map $F : X \to Y$ is *dominant* if $F(X)$ is dense in Y. Prove that a polynomial map is dominant if and only if its pullback is injective.

4.2.11 Let $X \subseteq \mathbb{A}^m$ and $Y \subseteq \mathbb{A}^n$ be affine varieties. We say that a polynomial map $F : X \to Y$ is a *closed embedding* if there exists an affine variety $Z \subseteq \mathbb{A}^n$ such that $F(X) = Z$ and the induced map $F : X \to Z$ is an isomorphism. Prove that a polynomial map is a closed embedding if and only if its pullback is surjective.

4.2.12 (For students with some knowledge of advanced linear algebra) Let V and W be K-vector spaces, and let $F : V \to W$ be a linear map. Choose bases for V and W in order to fix isomorphisms $V \cong K^m$ and $W \cong K^n$; via these isomorphisms, we can identify V and W with affine spaces $\mathbb{A}_K^m$ and $\mathbb{A}_K^n$, and hence as affine varieties. Denoting by $V^\vee$ and $W^\vee$ the dual vector spaces, explain why
$$V^\vee \subseteq K[V] \quad \text{and} \quad W^\vee \subseteq K[W],$$
and verify that $F^*|_{W^\vee}$ coincides with the dual map $F^\vee : W^\vee \to V^\vee$.

Section 4.3 Pulling back is a bijection

If $X \subseteq \mathbb{A}^m$ and $Y \subseteq \mathbb{A}^n$ are affine varieties, then the association that takes a polynomial map to its pullback provides a function between two sets of morphisms:

$$\{\text{polynomial maps } X \to Y\} \to \{K\text{-algebra homomorphisms } K[Y] \to K[X]\}$$
$$F \mapsto F^*.$$

The main result of this section is that this function is a bijection.

4.17 PROPOSITION *Pulling back polynomial functions is a bijection*

Let $X \subseteq \mathbb{A}^m$ and $Y \subseteq \mathbb{A}^n$ be affine varieties. The correspondence $F \mapsto F^*$ is a bijection between the set of polynomial maps $X \to Y$ and the set of K-algebra homomorphisms $K[Y] \to K[X]$.

> In the language of category theory, the bijection of Proposition 4.17 says that the functor taking X to $K[X]$ and F to F^* is "fully faithful."

To prove Proposition 4.17, it suffices to produce an inverse to the procedure that takes F to F^*. That is, given any K-algebra homomorphism $\varphi : K[Y] \to K[X]$, it suffices to show that there is a unique polynomial map $F : X \to Y$ such that $F^* = \varphi$. The proof of this statement can be difficult to follow notationally, so we begin with a concrete example, and we encourage the reader to refer to this example while studying the proof.

4.18 EXAMPLE Inverting the pullback

Consider the affine varieties of Example 4.3:

$$X = \mathcal{V}(x_2 - x_1^2) \subseteq \mathbb{A}^2 \quad \text{and} \quad Y = \mathcal{V}(y_1 y_2 - y_3) \subseteq \mathbb{A}^3.$$

Using the fact that $x_2 - x_1^2$ and $y_1 y_2 - y_3$ are both irreducible, we compute

$$K[X] = \frac{K[x_1, x_2]}{\langle x_2 - x_1^2 \rangle} \quad \text{and} \quad K[Y] = \frac{K[y_1, y_2, y_3]}{\langle y_1 y_2 - y_3 \rangle}.$$

Let us consider a K-algebra homomorphism $\varphi : K[Y] \to K[X]$. Such a homomorphism is determined by sending each generator $[y_i]$ to $[f_i]$ for some polynomial $f_i \in K[x_1, x_2]$; for example, consider the homomorphism

$$\varphi : K[Y] \to K[X]$$
$$\varphi([y_1]) = [x_1 + x_2]$$
$$\varphi([y_2]) = [x_1]$$
$$\varphi([y_3]) = [x_1 x_2 + x_2].$$

The image of any element $[g] \in K[Y]$ is determined by the fact that φ is a K-algebra homomorphism; for example,

$$\varphi([y_1^2 + y_2 - y_2 y_3]) = [(x_1 + x_2)^2 + x_1 - x_1(x_1 x_2 + x_2)].$$

More generally, for any $g \in K[y_1, y_2, y_3]$, we have

$$\varphi([g]) = [g(x_1 + x_2, x_1, x_1 x_2 + x_2)].$$

Not just any choice of three polynomials f_1, f_2, f_3 would have given a well-defined K-algebra homomorphism; it must be the case that $[0] = [y_1 y_2 - y_3]$ is sent to $[0] \in K[X]$, or in other words that $f_1 f_2 - f_3$ lies in $\langle x_2 - x_1^2 \rangle$. This is indeed the case for our particular choice of f_1, f_2, f_3:

(4.19) $$(x_1 + x_2)x_1 - (x_1 x_2 + x_2) = x_1^2 - x_2 \in \langle x_2 - x_1^2 \rangle.$$

Having described φ, can we find a polynomial map $F : X \to Y$ for which $F^* = \varphi$? Such a map sends elements of $X \subseteq \mathbb{A}^2$ to elements of $Y \subseteq \mathbb{A}^3$, so it is defined by three polynomials in two variables. Can you think of any candidates for three such polynomials? There is a natural choice: the three polynomials f_1, f_2, f_3 that were used to define φ. In other words, consider the function

$$F : X \to Y$$
$$F(a_1, a_2) = (a_1 + a_2, \, a_1, \, a_1 a_2 + a_2).$$

While F is manifestly a polynomial map, we should confirm that it indeed sends elements of X to elements of Y, or in other words that $F(a_1, a_2)$ satisfies the defining equation of Y. Explicitly, we verify that

(4.20) $$(a_1 + a_2)a_1 - (a_1 a_2 + a_2) = a_1^2 - a_2 = 0,$$

where the last equality is because $(a_1, a_2) \in X = \mathcal{V}(x_2 - x_1^2)$. Note the similarity in equations (4.19) and (4.20): what was needed in order to verify that φ was well-defined was precisely what was needed in order to verify that F mapped X to Y.

Finally, to confirm that $F^* = \varphi$, it suffices to check that these two homomorphisms agree on the generators $[y_1], [y_2], [y_3]$ of $K[Y]$. Viewing $[y_i]$ as a function $Y \to K$, it is simply the coordinate function

$$[y_i] : Y \to K$$
$$[y_i](b_1, b_2, b_3) = b_i.$$

Thus, by definition of the pullback,

$$F^*([y_1])(a_1, a_2) = ([y_1] \circ F)(a_1, a_2) = [y_1](a_1 + a_2, a_1, a_1 a_2 + a_2) = a_1 + a_2.$$

In other words,

$$F^*([y_1]) = [x_1 + x_2],$$

implying that $F^*([y_1]) = \varphi([y_1])$. Similarly, F^* agrees with φ on the other two generators, so we have successfully constructed a polynomial map F for which $F^* = \varphi$. Moreover, the construction essentially illustrates the uniqueness of F: the requirement that $F^*([y_i]) = \varphi([y_i])$ determines the ith component function of F, and these component functions uniquely determine F.

Generalizing the ideas of this example, we are ready to prove Proposition 4.17.

4.3. PULLING BACK IS A BIJECTION

PROOF OF PROPOSITION 4.17   Let $X \subseteq \mathbb{A}^m$ and $Y \subseteq \mathbb{A}^n$ be affine varieties with

$$K[X] = \frac{K[x_1, \ldots, x_m]}{\mathcal{I}(X)} \quad \text{and} \quad K[Y] = \frac{K[y_1, \ldots, y_n]}{\mathcal{I}(Y)}.$$

In order to prove the proposition, we show that, for every K-algebra homomorphism $\varphi : K[Y] \to K[X]$, there exists a unique polynomial map $F : X \to Y$ with $F^* = \varphi$.

(**Existence**) Suppose $\varphi : K[Y] \to K[X]$ is a K-algebra homomorphism. Following the procedure in the example, let $F_i = \varphi([y_i]) \in K[X]$, and consider the polynomial map

$$F = (F_1, \ldots, F_n) : X \to \mathbb{A}^n.$$

As in the example, we must verify that the image of F lies in Y and therefore gives rise to a polynomial map $F : X \to Y$. To prove this, suppose that $a \in X$; we must show that

$$(F_1(a), \ldots, F_n(a)) \in Y.$$

Since $Y = \mathcal{V}(\mathcal{I}(Y))$, it suffices to check that, for all $h \in \mathcal{I}(Y)$,

$$h(F_1(a), \ldots, F_n(a)) = 0.$$

Let $h \in \mathcal{I}(Y)$. Then $[h] = 0 \in K[Y]$, so $\varphi([h]) = 0 \in K[X]$, since φ is a homomorphism. But this implies that

$$\varphi([h(y_1, \ldots, y_n)]) = h(\varphi([y_1]), \ldots, \varphi([y_n])) = h(F_1, \ldots, F_n)$$

is the zero function on X, implying that $h(F_1(a), \ldots, F_n(a)) = 0$. Thus, F is indeed a polynomial map from X to Y.

It remains to show that $F^* = \varphi$. As in the example, for every $i = 1, \ldots, n$,

$$F^*([y_i]) = [y_i] \circ (F_1, \ldots, F_n) = F_i = \varphi([y_i]).$$

Thus, F^* and φ agree on the generators $[y_i]$ of $K[Y]$, implying that $F^* = \varphi$.

(**Uniqueness**) Suppose $F, G : X \to Y$ are polynomial maps with $F^* = G^*$. We must show that $F = G$. By definition,

$$F = (F_1, \ldots, F_n) \quad \text{and} \quad G = (G_1, \ldots, G_n)$$

where $F_i, G_i \in K[X]$ are polynomial functions on X. Evaluating F^* and G^* on $[y_i]$, we have

$$F^*([y_i]) = [y_i] \circ (F_1, \ldots, F_n) = F_i \in K[X],$$

and, similarly, $G^*([y_i]) = G_i \in K[X]$. Since $F^* = G^*$, it follows that

$$F_i = F^*([y_i]) = G^*([y_i]) = G_i$$

for all $i = 1, \ldots, n$, so $F = G$. □

The payoff for the work undertaken to prove Proposition 4.17 is that we can now precisely detect whether two affine varieties are isomorphic by studying their coordinate rings.

4.21 COROLLARY *Coordinate rings detect isomorphisms*

Let $X \subseteq \mathbb{A}^m$ and $Y \subseteq \mathbb{A}^n$ be affine varieties. Then $F : X \to Y$ is an isomorphism if and only if $F^* : K[Y] \to K[X]$ is an isomorphism, and

$$X \cong Y \iff K[X] \cong K[Y].$$

PROOF   The forward implication in the first assertion is the content of Corollary 4.14. To prove the converse, suppose that $F^* : K[Y] \to K[X]$ is an isomorphism with inverse $(F^*)^{-1} : K[X] \to K[Y]$. By the surjectivity in Theorem 4.17, there exists a polynomial map $G : Y \to X$ such that $G^* = (F^*)^{-1}$. We claim that $G = F^{-1}$. Indeed, by Proposition 4.13,

$$(F \circ G)^* = G^* \circ F^* = (F^*)^{-1} \circ F^* = \mathrm{id}_{K[Y]} = (\mathrm{id}_Y)^*.$$

Since $F \circ G$ and id_Y have the same pullback, the injectivity in Theorem 4.17 implies that $F \circ G = \mathrm{id}_Y$. Similarly, $G \circ F = \mathrm{id}_X$, from which it follows that $G = F^{-1}$, implying that F is an isomorphism.

Now consider the second assertion in the statement of the result. The forward implication is a consequence of the first assertion: if $X \cong Y$, then there is an isomorphism $F : X \to Y$ and its pullback gives an isomorphism $F^* : K[Y] \to K[X]$, showing that $K[X] \cong K[Y]$. The backward implication also follows from the first assertion after using the surjectivity in Theorem 4.17 to observe that any isomorphism between $K[X]$ and $K[Y]$ must be the pullback of a polynomial map between X and Y. $\square$

Exercises for Section 4.3

4.3.1 Suppose that $m \leq n$ and consider the natural injection

$$\varphi : K[x_1, \ldots, x_m] \to K[x_1, \ldots, x_n].$$

Describe the corresponding polynomial map $F : \mathbb{A}^n \to \mathbb{A}^m$.

4.3.2 Suppose that $m \leq n$ and consider the surjection

$$\varphi : K[x_1, \ldots, x_n] \to K[x_1, \ldots, x_m]$$
$$f(x_1, \ldots, x_n) \mapsto f(x_1, \ldots, x_m, 0, \ldots, 0).$$

Describe the corresponding polynomial map $F : \mathbb{A}^m \to \mathbb{A}^n$.

4.3.3 Consider the homomorphism of K-algebras

$$\varphi : K[x, y] \to K[t]$$

defined by

$$\varphi(f) = f(t + 1, t^2 + t).$$

For which affine varieties X and Y and which polynomial map $F : X \to Y$ do we have $\varphi = F^*$?

4.3. PULLING BACK IS A BIJECTION

4.3.4 Let $X = \mathcal{V}(x^2 + y^2 + z^2 - 1) \subseteq \mathbb{A}^3_{\mathbb{C}}$, for which
$$\mathbb{C}[X] = \frac{\mathbb{C}[x,y,z]}{\langle x^2 + y^2 + z^2 - 1 \rangle}.$$
Consider the homomorphism of $\mathbb{C}$-algebras
$$\varphi : \mathbb{C}[u,v] \to \mathbb{C}[X]$$
$$\varphi(f) = [f(x+y+z, xyz)].$$
For which polynomial map $F : X \to \mathbb{A}^2$ do we have $\varphi = F^*$?

4.3.5 Let $f_1, f_2, f_3 \in K[w]$ and let $X = \mathcal{V}(x+y-z) \subseteq \mathbb{A}^3$.
(a) Under what conditions on $f_1, f_2,$ and f_3 does
$$F : \mathbb{A}^1 \to X$$
$$F(a) = (f_1(a), f_2(a), f_3(a))$$
give a well-defined polynomial map to X? Give an explicit example of $f_1, f_2, f_3 \in K[w]$ for which this is the case, and an explicit example for which it is not the case.
(b) Under what conditions on $f_1, f_2,$ and f_3 does there exist a well-defined K-algebra homomorphism
$$\varphi : \frac{K[x,y,z]}{\langle x+y-z \rangle} \to K[w]$$
defined on the generators by
$$\varphi([x]) = f_1(w)$$
$$\varphi([y]) = f_2(w)$$
$$\varphi([z]) = f_3(w)?$$
Give an explicit example of $f_1, f_2, f_3 \in K[w]$ for which this is the case, and an explicit example for which it is not the case.
(c) Using parts (a) and (b), explicitly describe the bijection between the set of polynomial maps $\mathbb{A}^1 \to X$ and the set of K-algebra homomorphisms $K[X] \to K[\mathbb{A}^1]$.

4.3.6 Let $X \subseteq \mathbb{A}^m$ be an affine variety, and let $Y \subseteq \mathbb{A}^n$ be a single point. There is only one possible polynomial map $X \to Y$, so by Theorem 4.17, there is only one possible K-algebra homomorphism $K[Y] \to K[X]$. What is $K[Y]$, and what is the one K-algebra homomorphism $K[Y] \to K[X]$?

4.3.7 Suppose that $X \subseteq \mathbb{A}^m$ is an irreducible affine variety and that $Y \subseteq \mathbb{A}^n$ consists of two distinct points. Prove that there are exactly two polynomial maps $F : X \to Y$. Describe the two maps and their pullbacks explicitly.

4.3.8 If $X \subseteq \mathbb{A}^n$ is any affine variety, then the set of polynomial maps $X \to \mathbb{A}^1$ is precisely $K[X]$, so Theorem 4.17 implies there is a bijection between $K[X]$ and the set of K-algebra homomorphisms $K[\mathbb{A}^1] \to K[X]$. Describe this bijection explicitly.

Section 4.4 The equivalence of algebra and geometry

Combining the results of this chapter and the previous one, we now prove that the passage from X to $K[X]$ truly is a dictionary between affine varieties and finitely-generated reduced K-algebras, where we view objects on both sides as "the same" if they are isomorphic. In the terminology of *isomorphism classes*—equivalence classes under the equivalence relation of being isomorphic (which is an equivalence relation in the setting of K-algebras, of affine varieties, or more generally, in any category)—the results we have proven lead to the following theorem.

4.22 THEOREM *Equivalence of algebra and geometry*

The association $X \mapsto K[X]$ induces a bijection

$$\left\{ \begin{array}{c} \text{isomorphism classes of} \\ \text{affine varieties} \end{array} \right\} \longrightarrow \left\{ \begin{array}{c} \text{isomorphism classes of} \\ \text{finitely-generated reduced } K\text{-algebras} \end{array} \right\}.$$

PROOF   Recalling that each coordinate ring is a finitely-generated reduced K-algebra (Proposition 3.33), we can view the association $X \to K[X]$ as a function

$$\{\text{affine varieties}\} \longrightarrow \left\{ \begin{array}{c} \text{isomorphism classes of} \\ \text{finitely-generated reduced } K\text{-algebras} \end{array} \right\}.$$

To see that this function is well-defined on isomorphism classes of affine varieties, we notice that $K[X] \cong K[Y]$ whenever $X \cong Y$; this is one direction of Corollary 4.21. To see that this function is injective on isomorphism classes, we notice that $X \cong Y$ whenever $K[X] \cong K[Y]$; this is the other direction of Corollary 4.21. Finally, to justify surjectivity, we notice that every finitely-generated reduced K-algebra is the coordinate ring of some affine variety (Proposition 3.33).   □

> *In category-theoretic language, the bijection of Theorem 4.22 reflects an "equivalence of categories" between affine varieties and finitely-generated reduced K-algebras.*

Put more loosely, Theorem 4.22 asserts that all of the geometric information about the affine variety X is encoded in the K-algebra $K[X]$. But perhaps we should be a bit more careful: the particular affine space $\mathbb{A}^n$ in which X lives is "geometric information" about X, and yet this information cannot be recovered from the isomorphism class of $K[X]$. For instance, the coordinate rings of $\mathcal{V}(y - x^2) \subseteq \mathbb{A}^2$ and $\mathcal{V}(y - x^2, z) \subseteq \mathbb{A}^3$ are members of the same isomorphism class despite arising from affine varieties in different ambient affine spaces.

The reason the ambient affine space of X cannot be recovered from the isomorphism class of $K[X]$ is that isomorphic affine varieties can live in different affine spaces. A property of affine varieties—or indeed, of any mathematical objects—that is not preserved under isomorphism can be thought of as a "coincidental" property, one that depends on some extraneous choice. By contrast, a property preserved by isomorphisms is one that pertains to the object's "essence." To make these ideas precise, it is useful to have the following definition.

4.4. THE EQUIVALENCE OF ALGEBRA AND GEOMETRY

4.23 DEFINITION *Intrinsic/extrinsic property*

Let $\mathcal{C}$ be a set of mathematical objects with an equivalence relation called isomorphism. A property $\mathcal{P}$ of objects in $\mathcal{C}$ is said to be *intrinsic* if, whenever two objects are isomorphic, one of them has property $\mathcal{P}$ if and only if the other has property $\mathcal{P}$. A property that is not intrinsic is said to be *extrinsic*.

The property of being a vanishing set in $\mathbb{A}^2$, for instance, is an extrinsic property on the set of all affine varieties, since $\mathcal{V}(y - x^2) \subseteq \mathbb{A}^2$ has this property but the isomorphic variety $\mathcal{V}(y - x^2, z) \subseteq \mathbb{A}^3$ does not. Here are some further examples of intrinsic and extrinsic properties in the context of both algebra and geometry.

4.24 EXAMPLE Intrinsic properties of rings

Being reduced is an intrinsic property of rings. To prove this, let $\varphi : R \to S$ be a ring isomorphism; we must show that R is reduced if and only if S is reduced. Assume that R is not reduced. Then there exists $a \in R$ such that $a \neq 0$ and $a^m = 0$ for some $m \geq 1$. Using standard properties of ring isomorphisms, we see that $\varphi(a) \neq 0$ and

$$\varphi(a)^m = \varphi(a^m) = \varphi(0) = 0.$$

Thus, $\varphi(a)$ is a nonzero nilpotent, showing that S is not reduced. The same proof applied to $\varphi^{-1} : S \to R$ shows that, if S is not reduced, then R is not reduced.

Similar arguments show that being an integral domain, a UFD, a PID, or a field are all intrinsic properties of rings (Exercise 4.4.1).

4.25 EXAMPLE Number of generators is extrinsic

Every finitely-generated K-algebra is isomorphic to a quotient $K[x_1, \ldots, x_n]/I$, but the number n of generators is extrinsic. For example, the K-algebras

$$\frac{K[x, y]}{\langle y - x^2 \rangle} \quad \text{and} \quad \frac{K[x, y, z]}{\langle y - x^2, z \rangle}$$

are isomorphic, even though the first has two generators and the second has three. On the other hand, the *minimal* number of generators is an intrinsic property.

4.26 EXAMPLE Irreducibility is intrinsic

The property of being an irreducible affine variety is intrinsic. This can be proved directly using the definition of polynomial maps and irreducibility (Exercise 4.4.2), but we can also prove it using our dictionary between geometry and algebra. To do so, suppose $X \subseteq \mathbb{A}^m$ and $Y \subseteq \mathbb{A}^n$ are affine varieties and $X \cong Y$. Then

$$X \text{ is irreducible} \iff \mathcal{I}(X) \text{ is prime} \quad \text{(Proposition 2.25)}$$
$$\iff K[X] \text{ is an integral domain} \quad \text{(Propositions 3.22 and 0.38)}.$$

By assumption, $X \cong Y$, and therefore $K[X] \cong K[Y]$ (Corollary 4.21). Since being an integral domain is an intrinsic property of K-algebras, we conclude that X is irreducible if and only if Y is irreducible. Thus, irreducibility is an intrinsic property.

An intrinsic property $\mathcal{P}$ of mathematical objects of class $\mathcal{C}$ can be viewed as a subset of the set of isomorphism classes, consisting of those isomorphism classes in which one (and hence every) representative has property $\mathcal{P}$. In particular, an intrinsic property of affine varieties—such as irreducibility—can be viewed as a subset of the set of isomorphism classes of affine varieties. Via the bijection of Theorem 4.22, this can then be identified with a subset of the set of isomorphism classes of finitely-generated reduced K-algebras, which can then be viewed as an intrinsic property of finitely-generated reduced K-algebras. Which algebraic property is it? In the case of irreducibility, the answer is that irreducibility of affine varieties corresponds to the property of a finitely-generated reduced K-algebra being an integral domain.

More generally, we can now ask a very broad question: given an intrinsic geometric property or construction applicable to affine varieties, what is its manifestation in the category of K-algebras? Or, conversely, given an intrinsic algebraic property or construction applicable to K-algebras, what is its manifestation in the category of affine varieties? Both algebra and geometry are illuminated by these questions, and specific examples of such phenomena form the backbone of the algebraic geometry to come.

Exercises for Section 4.4

4.4.1 Prove that the following are intrinsic properties of rings:
   (a) Being an integral domain;
   (b) Being a field;
   (c) Being a principal ideal domain;
   (d) Being a unique factorization domain.

4.4.2 Let $X \subseteq \mathbb{A}^m$ and $Y \subseteq \mathbb{A}^n$ be affine varieties, and let $F : X \to Y$ be an isomorphism.
   (a) Prove that, if a subset $X_1 \subseteq X$ is an affine variety in $\mathbb{A}^m$, then $F(X_1)$ is an affine variety in $\mathbb{A}^n$.
   (b) Prove directly from Definition 2.22 that irreducibility is intrinsic.

4.4.3 Prove that the number of irreducible components of an affine variety is an intrinsic property. Describe the corresponding property of finitely-generated reduced K-algebras.

4.4.4 Prove that being a finite set is an intrinsic property of affine varieties. Describe the corresponding property of finitely-generated reduced K-algebras.

4.4.5 We say that two morphisms $F_1 : A_1 \to B_1$ and $F_2 : A_2 \to B_2$ are *isomorphic* if there exist isomorphisms $G_1 : A_1 \to A_2$ and $G_2 : B_1 \to B_2$ such that
$$F_1 = G_2^{-1} \circ F_2 \circ G_1.$$
   (a) Prove that isomorphism is an equivalence relation on the set of morphisms.
   (b) Prove that there is a bijection between isomorphism classes of polynomial maps between affine varieties and isomorphism classes of K-algebra homomorphisms between finitely-generated reduced K-algebras.

Chapter 5

Proof of the Nullstellensatz

> LEARNING OBJECTIVES FOR CHAPTER 5
> - Define and give examples of modules and algebras over rings.
> - Describe the differences between finitely-generated algebras and finitely-generated modules.
> - Determine, via the concept of integrality, when a finitely-generated algebra is, in fact, a finitely-generated module.
> - Describe the general structure of finitely-generated K-algebras via Noether normalization.
> - Use Noether normalization to prove the Nullstellensatz.

Now that we have collected, in the form of the equivalence of algebra and geometry, some evidence of the power of the Nullstellensatz, the time has come to prove it. The journey to a proof of the Nullstellensatz necessitates a rather long interlude into purely algebraic material, including a tour of R-modules and R-algebras, culminating with the Noether Normalization Theorem in Section 5.4.

In addition to being a key step in the proof of the Nullstellensatz, the Noether Normalization Theorem is a powerful result of independent interest about the structure of finitely-generated K-algebras. It says that, while such an algebra certainly need not be finitely generated as a K-vector space (for example, the polynomial ring $K[x]$ is not), it can always be expressed as a finitely-generated "vector space" with scalars in a subalgebra that is isomorphic to a polynomial ring. We put the word "vector space" in quotes because the scalars here do not form a field, and hence, in order to make sense of Noether normalization, we must generalize the definition of a vector space to allow scalars from a general ring (which, as usual, we assume to be commutative with unity). These generalized vector spaces, which we introduce in Section 5.1, are referred to as *modules*.

While the basic definition of a module over a commutative ring with unity is no different from that of a vector space over a field, the theory in this more general setting leads to a number of new ideas, the most important of which is the notion of integrality and its relationship to finite generation. The first four sections of this chapter are devoted to the development of the theory of R-modules and R-algebras, culminating in Section 5.4 with a proof of Noether normalization. In Section 5.5, we receive the payoff for this work: the proof of the Nullstellensatz, and thus, a complete justification of the equivalence of algebra and geometry.

Section 5.1 Modules

To motivate the concept of modules, we begin with a discussion of the algebraic structure of a few familiar coordinate rings. Consider

$$K[x,y] \quad \text{and} \quad K[x,y]/\langle x^2+y^2-1\rangle,$$

which are the coordinate rings of the affine plane and the unit circle, respectively. The affine plane certainly does not feel like it should be isomorphic to the unit circle, suggesting that there must be some algebraic property that we can use to distinguish between these two K-algebras. Our aim is to describe such a property using ideas from linear algebra.

To start, consider these coordinate rings as vector spaces. Notice that every element of $K[x,y]$ can be written uniquely as a K-linear combination of elements of

$$\mathcal{B} = \{x^i y^j \mid i, j \in \mathbb{N}\},$$

which tells us that $K[x,y]$ is an infinite-dimensional vector space over K with basis $\mathcal{B}$. If we consider the ring $K[x,y]/\langle x^2+y^2-1\rangle$, on the other hand, then we can repeatedly use the relation $[y^2] = [1-x^2]$ to write every element uniquely as

$$[f(x) + g(x)y]$$

for some $f, g \in K[x]$. In other words, $K[x,y]/\langle x^2+y^2-1\rangle$ is also an infinite-dimensional vector space over K, but it has a basis given by the smaller set

$$\mathcal{B}' = \{[x^i y^j] \mid i \in \mathbb{N},\ j \in \{0,1\}\}.$$

Even though $\mathcal{B}'$ can be viewed as a proper subset of $\mathcal{B}$, both $\mathcal{B}$ and $\mathcal{B}'$ are countably infinite, which implies that these two vector spaces are, in fact, isomorphic. Since the two coordinate rings are isomorphic as vector spaces, we see that the theory of vector spaces alone is not enough to distinguish between them.

However, if we allow ourselves to enlarge our "scalars," replacing K with the ring $R = K[x]$, then we notice that every element of $K[x,y]$ can be written uniquely as an R-linear combination of elements of the infinite set

$$\mathcal{S} = \{1, y, y^2, y^3, \dots\},$$

whereas, for the ring $K[x,y]/\langle x^2+y^2-1\rangle$, in order to write every element as an R-linear combination, we only require the two-element set

$$\mathcal{S}' = \{[1], [y]\}.$$

In other words, if we pretend for a moment that $R = K[x]$ is a field (it's not!), then we have observed that $K[x,y]$ is an infinite-dimensional "vector space" over R while $K[x,y]/\langle x^2+y^2-1\rangle$ is finite-dimensional. Thus, we seem to have found a property that differentiates these two coordinate rings.

To make this hypothetical argument a reality, we require an extension of the notion of vector spaces to the setting where the scalars are allowed to be a ring, but not necessarily a field—a setting that is captured by the important algebraic concept of modules. We begin our discussion of modules in this section with the definition and some foundational notions. As always, R denotes a ring, and all rings are assumed to be commutative with unity.

5.1. MODULES

> **5.1 DEFINITION** *R-module*
>
> An *R-module* is an abelian group M (with operation denoted $+$) together with a *scalar multiplication* function
>
> $$R \times M \to M$$
> $$(r, a) \mapsto r \cdot a$$
>
> satisfying the following axioms.
> 1. $r \cdot (a + b) = r \cdot a + r \cdot b$ for all $r \in R$ and all $a, b \in M$.
> 2. $(r + s) \cdot a = r \cdot a + s \cdot a$ for all $r, s \in R$ and all $a \in M$.
> 3. $(rs) \cdot a = r \cdot (s \cdot a)$ for all $r, s \in R$ and all $a \in M$.
> 4. $1 \cdot a = a$ for all $a \in M$.

When $R = K$ is a field, Definition 5.1 is nothing more than the definition of a vector space over K. Many, but not all, of the notions of vector spaces naturally generalize to the module setting. Let us begin our discussion of modules with several examples that will be helpful to keep in mind.

5.2 EXAMPLE R^n is an R-module

The standard example of a vector space is K^n, and this generalizes to the R-module setting. More specifically, consider the Cartesian product

$$R^n = \{(a_1, \ldots, a_n) \mid a_i \in R \text{ for each } i\}.$$

The set R^n is naturally an R-module, with addition and scalar multiplication defined exactly as in the vector-space setting:

$$(a_1, \ldots, a_n) + (b_1, \ldots, b_n) = (a_1 + b_1, \ldots, a_n + b_n),$$
$$r \cdot (a_1, \ldots, a_n) = (ra_1, \ldots, ra_n).$$

5.3 EXAMPLE Polynomial rings are modules

The polynomial ring $R[x_1, \ldots, x_n]$ is an R-module, with addition and scalar multiplication defined in the usual way:

$$\left(\sum_\alpha b_\alpha x^\alpha\right) + \left(\sum_\alpha c_\alpha x^\alpha\right) = \sum_\alpha (b_\alpha + c_\alpha) x^\alpha \quad \text{and} \quad r \cdot \left(\sum_\alpha b_\alpha x^\alpha\right) = \sum_\alpha (rb_\alpha) x^\alpha.$$

The module axioms are a straightforward consequence of the ring axioms.

5.4 EXAMPLE Extension rings

Generalizing the previous example, if $R \subseteq S$ is a subring, then S can be viewed as an R-module, where for $r \in R$ and $s \in S$, we define $r \cdot s$ by the ring multiplication inside S. The module axioms, again, are a consequence of the ring axioms.

As a special case that arose in the discussion at the beginning of this section, we can consider $M = K[x, y]$ as a module over the subring $R = K[x]$. That is, elements in $K[x, y]$ can be added as usual and any element in $K[x, y]$ can be multiplied by a "scalar" in $K[x]$.

5.5 EXAMPLE Abelian groups are $\mathbb{Z}$-modules

Let M be any abelian group. Then M can be viewed as a $\mathbb{Z}$-module, where for $n \in \mathbb{Z}$ and $a \in M$ the scalar multiplication is defined by

$$n \cdot a = \begin{cases} \underbrace{a + \cdots + a}_{n \text{ times}} & \text{if } n > 0, \\ 0 & \text{if } n = 0, \\ \underbrace{(-a) + \cdots + (-a)}_{-n \text{ times}} & \text{if } n < 0. \end{cases}$$

In fact, one can check from the module axioms that this is the only definition of scalar multiplication for which M is a $\mathbb{Z}$-module; see Exercise 5.1.11. It follows that every abelian group is a $\mathbb{Z}$-module in a *canonical* way.

Conversely, every $\mathbb{Z}$-module is an abelian group by forgetting scalar multiplication. Thus, $\mathbb{Z}$-modules and abelian groups are really two different names for the same thing.

Of course, our discussion of R-modules is not complete without introducing the relevant notion of morphisms between them. Given that a module is an abelian group with the additional structure of scalar multiplication, it is natural to define a module homomorphism as a group homomorphism that preserves scalar multiplication.

5.6 DEFINITION *Homomorphisms of R-modules*

Let M and N be R-modules. An *R-module homomorphism* $\varphi : M \to N$ is a group homomorphism for which

$$\varphi(r \cdot a) = r \cdot \varphi(a)$$

for all $r \in R$ and $a \in M$. We say that φ is an *isomorphism of R-modules* and write $M \cong N$ if φ has an inverse that is also an R-module homomorphism.

In other words, an R-module homomorphism is simply a function that preserves both the operations of scalar multiplication and addition:

$$\varphi(r \cdot a) = r \cdot \varphi(a) \quad \text{and} \quad \varphi(a + b) = \varphi(a) + \varphi(b)$$

for all $a, b \in M$ and $r \in R$. In particular, if $R = K$ is a field, so that M and N are K-vector spaces, then a K-module homomorphism is precisely the same as a *linear map* of vector spaces over K.

Another important module-theoretic notion is that of a submodule.

5.7 DEFINITION *Submodule*

Let M be an R-module. A *submodule* $N \subseteq M$ is a subgroup for which $r \cdot a \in N$ for all $r \in R$ and $a \in N$.

5.1. MODULES

In other words, a submodule is a subgroup that is also closed under scalar multiplication, thereby forming a module in its own right. Since a nonempty subset of an additive group is a subgroup if and only if it contains differences of its elements, it follows that a nonempty subset $N \subseteq M$ is a submodule if and only if

$$a - b \in N \quad \text{and} \quad r \cdot a \in N$$

for all $a, b \in N$ and $r \in R$. The notion of submodules is a natural generalization of the notion of linear subspaces from the study of vector spaces.

Given an R-module M and a submodule $N \subseteq M$, we can form the group quotient M/N, and this group quotient naturally inherits the structure of an R-module, with scalar multiplication defined by

$$r \cdot [a] = [r \cdot a].$$

The reader is encouraged to check that scalar multiplication is well-defined and that the quotient M/N satisfies the R-module axioms (Exercise 5.1.8). Just like for groups, rings, and K-algebras, there is a version of the First Isomorphism Theorem for R-modules; this is the content of Exercise 5.1.9.

In the definition of an R-module, we started with an additive abelian group. However, in many cases relevant to us, such as the setting of polynomial rings and their quotients, the additive abelian group will also have a multiplicative structure that endows it with the structure of a ring. In this case, we call the resulting structure an R-algebra, made precise in the following definition.

5.8 DEFINITION R-algebra

An R-algebra is a ring A together with a scalar multiplication function

$$R \times A \to A$$
$$(r, a) \mapsto r \cdot a$$

satisfying the four axioms of an R-module as well as

$$r \cdot (ab) = (r \cdot a)b = a(r \cdot b)$$

for all $r \in R$ and all $a, b \in A$.

The reader should notice that the definition of an R-algebra is not new: after replacing R with K, it is identical to the definition of a K-algebra from Section 3.2. In fact, most of the concepts we discussed concerning K-algebras—such as homomorphisms, subalgebras, ideals, quotients, the First Isomorphism Theorem, and generators—carry over verbatim to the R-algebra setting, and we do not restate them here. As was the case for K-algebras, the prototypical R-algebra is the polynomial ring $R[x_1, \ldots, x_n]$.

In contrast to the setting of K-algebras, however, we gain some flexibility in our perspective now that we do not require our scalars to form a field. For example, even if our motivation is to study polynomials over a field, we now have the ability to view one of the variables as a "scalar" and write

$$K[x_1, \ldots, x_n] = R[x_1, \ldots, x_{n-1}] \quad \text{where} \quad R = K[x_n].$$

This opens up the possibility of proving assertions concerning K-algebras by using induction arguments in the more general R-algebra setting. It is essentially for this reason that the setting of R-algebras is the correct level of algebraic generality that we require for our development of algebraic geometry.

Given an R-algebra, there is a unique underlying R-module obtained by forgetting the multiplicative structure. On the other hand, if you start with an R-module, then there are typically many ways to put a multiplicative structure on it to endow it with the structure of an R-algebra. We illustrate this in the next example.

5.9 EXAMPLE Different R-algebras with the same underlying R-module

Consider the R-module $M = R^2$. A natural way to make M into an R-algebra is to define multiplication componentwise:

$$(a,b) \cdot (c,d) = (ac, bd).$$

However, this is not the only way that we can make M into an R-algebra; another way is given by defining multiplication as follows:

$$(a,b) \cdot (c,d) = (ac, ad + bc).$$

While this second multiplication might feel a bit strange at first glance, the resulting R-algebra is actually isomorphic to the familiar quotient $R[x]/\langle x^2 \rangle$, as the reader is encouraged to verify in Exercise 5.1.10.

In the next section, we turn to a discussion of module generators, which allows us to generalize the important notion of finite-dimensionality from linear algebra to the module setting.

Exercises for Section 5.1

5.1.1 Let $M = K[x,y]$. Find at least three different rings R for which M can be viewed naturally as an R-module.

5.1.2 Let R be a ring and M an R-module. Use the module axioms to prove that

$$0 \cdot a = 0 \quad \text{for all} \quad a \in M$$

and

$$r \cdot 0 = 0 \quad \text{for all} \quad r \in R.$$

5.1.3 Give an example of a ring R, an R-module M, and a subgroup $N \subseteq M$ that is not an R-module.

5.1.4 Let M be an R-module and $N \subseteq M$ a nonempty subset. Prove that N is a submodule if and only if it is closed under addition and scalar multiplication:

$$a + b \in N \quad \text{and} \quad r \cdot a \in N \quad \text{for all} \quad a, b \in N \quad \text{and} \quad r \in R.$$

5.1.5 Consider R as a submodule over itself, where scalar multiplication is usual multiplication. Prove that $I \subseteq R$ is an ideal if and only if it is a submodule.

5.1. MODULES

5.1.6 Prove that an R-module homomorphism is an isomorphism if and only if it is bijective.

5.1.7 Prove that R is an $R[x]$-module under the scalar multiplication defined by
$$(r_0 + r_1 x + r_2 x^2 + \cdots + r_n x^n) \cdot a = r_0 a.$$

5.1.8 Let M be an R-module, $N \subseteq M$ a submodule, and M/N the group quotient.
  (a) Suppose that $[a_1] = [a_2] \in M/N$. Prove that
  $$[r \cdot a_1] = [r \cdot a_2].$$
  Conclude that scalar multiplication is well-defined in M/N.
  (b) Prove that M/N satisfies the R-module axioms.

5.1.9 Let $\varphi : M \to N$ be a homomorphism of R-modules.
  (a) Prove that $\ker(\varphi)$ is a submodule of M.
  (b) Prove that $\operatorname{im}(\varphi)$ is a submodule of N.
  (c) Prove that the function
  $$[\varphi] : M/\ker(\varphi) \to \operatorname{im}(\varphi)$$
  $$[a] \mapsto \varphi(a)$$
  is a well-defined isomorphism of R-modules.

5.1.10 Let A be the R-algebra defined by endowing R^2 with the multiplication
$$(a,b) \cdot (c,d) = (ac, ad + bc).$$
Prove that $A \cong R[x]/\langle x^2 \rangle$ as R-algebras.

5.1.11 Let M be an abelian group. Prove that the only definition of scalar multiplication that makes M into a $\mathbb{Z}$-module is the one given in Example 5.5.

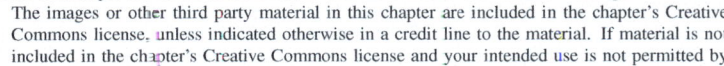

Section 5.2 Module generators

As we learned in the previous section, a module is an algebraic structure that generalizes vector spaces to the setting where the scalars form a ring but not necessarily a field. In this section, we generalize the important vector-space concept of finite-dimensionality to the module setting. The key notions we require for this generalization are those of linear combinations and generators.

5.10 DEFINITION *Linear combination, generators*

Let M be an R-module and let $S \subseteq M$ be a subset. A *linear combination of S* is an element of M of the form

$$r_1 a_1 + \cdots + r_n a_n$$

for some $n \in \mathbb{N}$, $r_i \in R$, and $a_i \in S$. The set of all linear combinations of S is called the *submodule of M generated by S*, and it is denoted RS.

It is a worthwhile exercise to verify that RS is, in fact, a submodule of M, and that it is the smallest submodule of M that contains the set S (Exercise 5.2.1). Let us consider a few examples.

5.11 EXAMPLE Submodules of $R[x]$

Consider $R[x]$ as an R-module. Then the submodule generated by $\{x^2, x^3\}$ is

$$R\{x^2, x^3\} = \{ax^2 + bx^3 \mid a, b \in R\} \subseteq R[x].$$

In other words, it consists of polynomials whose only potentially nonzero coefficients occur in the x^2 and x^3 terms. Similarly,

$$R\{1, x^2, x^4, x^6, \ldots\} = \left\{ \sum_{i=0}^{n} a_i x^{2i} \mid n \in \mathbb{N},\ a_i \in R \right\} \subseteq R[x]$$

consists of polynomials whose nonzero coefficients occur with even powers of x.

5.12 EXAMPLE Submodules of $\mathbb{Z}$

Consider $\mathbb{Z}$ as a $\mathbb{Z}$-module. Then the submodule generated by $\{4, 6\}$ is

$$\mathbb{Z}\{4, 6\} = \{a \cdot 4 + b \cdot 6 \mid a, b \in \mathbb{Z}\}.$$

Noting that $2 = (-1) \cdot 4 + 1 \cdot 6 \in \mathbb{Z}\{4, 6\}$, it is not too hard to see that every even integer can be obtained as a linear combination of 4 and 6, which proves that $2\mathbb{Z} \subseteq \mathbb{Z}\{4, 6\}$. On the other hand, every linear combination of 4 and 6 is even, so $\mathbb{Z}\{4, 6\} \subseteq 2\mathbb{Z}$. Taking these together, we have proved that

$$\mathbb{Z}\{4, 6\} = 2\mathbb{Z}.$$

If, instead, we consider the submodule generated by 2 and 3, we see that

$$1 = (-1) \cdot 2 + 1 \cdot 3 \in \mathbb{Z}\{2, 3\},$$

which implies that $\mathbb{Z}\{2, 3\} = \mathbb{Z}$.

5.2. MODULE GENERATORS

5.13 EXAMPLE *The coordinate ring of the unit circle as a $K[x]$-module*

Consider the coordinate ring of the unit circle $X = V(x^2 + y^2 - 1) \subseteq \mathbb{A}^2$:

$$K[X] = \frac{K[x,y]}{\langle x^2 + y^2 - 1 \rangle}.$$

We can view $K[X]$ as a $K[x]$-module in a natural way by defining

$$f \cdot [g] = [fg]$$

for any $f \in K[x]$ and $[g] \in K[X]$. By repeated use of the equation $[y^2] = [1 - x^2]$, every element of $K[X]$ can be written in the form

$$[f_1(x) + f_2(x)y] = f_1(x) \cdot [1] + f_2(x) \cdot [y],$$

which shows that $K[X] = K[x]\{[1], [y]\}$.

We are especially interested in whether a module can be generated by a finite set, generalizing the concept of finite-dimensionality from the study of vector spaces.

5.14 DEFINITION *Finitely-generated module*

We say that M is a *finitely-generated R-module* if there exist $a_1, \ldots, a_n \in M$ such that

$$M = R\{a_1, \ldots, a_n\}.$$

Example 5.13 shows that the coordinate ring of the unit circle is a finitely-generated $K[x]$-module, generated by $[1]$ and $[y]$. The next example illustrates a familiar module that is not finitely generated.

5.15 EXAMPLE $R[x]$ *is not a finitely-generated R-module*

Consider the polynomial ring $R[x]$. To prove that $R[x]$ is not finitely generated, suppose $f_1, \ldots, f_n \in R[x]$ is any finite set of polynomials and consider the submodule

$$R\{f_1, \ldots, f_n\} \subseteq R[x].$$

We must prove that this submodule is not all of $R[x]$. To do so, let d be the maximum degree of the polynomials $f_1, \ldots, f_n$. Then any linear combination of these polynomials must have degree bounded above by d. In particular, $x^{d+1} \notin R\{f_1, \ldots, f_n\}$.

In many ways, modules behave like vector spaces, but it is important to note their key differences. The reader might recall a standard result in linear algebra that says that every finite-dimensional

> *Further differences between modules and vector spaces are discussed in Exercise 5.2.7.*

vector space over K is isomorphic to K^n for some n. In the module setting, this is not the case; for example, given any nontrivial finite group M, we may view it as a $\mathbb{Z}$-module (Example 5.5), and it is finitely generated because it is generated by all of its elements. However, it is not the case that $M \cong \mathbb{Z}^n$ for any n because $1 < |M| < \infty$, but $\mathbb{Z}^n$ is either infinite (if $n > 0$) or has a single element (if $n = 0$).

In practice, most of the modules in this book will arise naturally with a multiplicative operation, giving them the structure of an algebra. Given an R-algebra, we can talk about its module properties, which pertain to just addition and scalar multiplication (in other words, *linear* algebra), or we can talk about its algebra properties, which also include the multiplication operation (in other words, *polynomial* algebra). We contrast these two perspectives in the next example.

5.16 EXAMPLE Submodule versus subalgebra generated by a set

In Example 5.11, we saw that the submodule of $R[x]$ generated by x^2 and x^3 is

$$R\{x^2, x^3\} = \{ax^2 + bx^3 \mid a, b \in R\}.$$

To contrast this with the algebra setting, consider the subalgebra generated by these same two elements. As defined in Section 3.3, the subalgebra $R[x^2, x^3]$ consists of all polynomial combinations of x^2 and x^3, so, in addition to containing the linear combinations as above, it contains additional elements, such as

$$(x^2)^2 = x^4, \quad x^2 \cdot x^3 = x^5 \quad \text{and} \quad (x^3)^2 = x^6.$$

In fact, one can show (Exercise 5.2.2) that $R[x^2, x^3]$ consists of all polynomials in $R[x]$ in which the linear coefficient is zero:

$$R[x^2, x^3] = R\{1, x^2, x^3, x^4, \dots\}.$$

Thus, we see that the subalgebra generated by x^2 and x^3 is much larger than the submodule; it is not even finitely generated as a module.

If A is an R-algebra and $\mathcal{S} \subseteq A$ is a subset, then

$$R\mathcal{S} \subseteq R[\mathcal{S}],$$

simply because every linear combination is a special type of polynomial combination. It follows that, if A is finitely generated as a module, then it must be finitely generated as an algebra. On the other hand, given a finitely-generated algebra, it is usually not the case that it is finitely generated as a module; the polynomial ring $R[x_1, \dots, x_n]$, for example, is finitely generated as an R-algebra by $x_1, \dots, x_n$, but not finitely generated as an R-module. Thus, being finitely generated in the module sense is much more restrictive than being finitely generated in the algebra sense.

In light of this, it can be interesting and useful to ask whether a given finitely-generated algebra is finitely generated as a module. For example, if we consider $\mathbb{R}$ as a $\mathbb{Z}$-module and choose a real number $a \in \mathbb{R}$, then the subalgebra $\mathbb{Z}[a]$ is finitely generated as a $\mathbb{Z}$-algebra. Is it finitely generated as a module? In the next two examples, we investigate this question for two different values of a.

5.17 EXAMPLE $\mathbb{Z}[\sqrt{2}]$ is a finitely-generated $\mathbb{Z}$-module

Consider the real numbers $\mathbb{R}$ as a $\mathbb{Z}$-algebra, and let us investigate the finitely-generated subalgebra

$$\mathbb{Z}[\sqrt{2}] \subseteq \mathbb{R}.$$

5.2. MODULE GENERATORS

Elements of $\mathbb{Z}[\sqrt{2}]$ are those real numbers that can be obtained as a polynomial combination $f(\sqrt{2})$ for some $f \in \mathbb{Z}[x]$. Consider, for example, the polynomial

$$f = 3 + 5x + 4x^2 - x^3.$$

Then, by definition, $f(\sqrt{2}) \in \mathbb{Z}[\sqrt{2}]$. Simplifying, we see that

$$f(\sqrt{2}) = 3 + 5\sqrt{2} + 4(\sqrt{2})^2 - (\sqrt{2})^3$$
$$= 3 + 5\sqrt{2} + 4 \cdot 2 - 2\sqrt{2}$$
$$= 11 + 3\sqrt{2}.$$

Thus, the polynomial combination $f(\sqrt{2})$ is the same as the linear combination $11 + 3\sqrt{2} \in \mathbb{Z}\{1, \sqrt{2}\}$. Generalizing this same trick to any polynomial, it can be shown that $\mathbb{Z}[\sqrt{2}]$ is a finitely-generated $\mathbb{Z}$-module (Exercise 5.2.4):

$$\mathbb{Z}[\sqrt{2}] = \mathbb{Z}\{1, \sqrt{2}\}.$$

5.18 EXAMPLE $\mathbb{Z}[1/2]$ is not a finitely-generated $\mathbb{Z}$-module

Consider again the real numbers $\mathbb{R}$ as a $\mathbb{Z}$-algebra and let us investigate the finitely-generated subalgebra

$$\mathbb{Z}[1/2] \subseteq \mathbb{R}.$$

This subalgebra consists of polynomial combinations of $1/2$, so taking, for example, $f = 3 + 5x + 4x^2 - x^3$, we see that

$$f(1/2) = 3 + 5(1/2) + 4(1/4) - (1/8) = 51/8 \in \mathbb{Z}[1/2].$$

Notice that, for a polynomial f of degree d, the largest power of 2 that will appear in a denominator of one of the terms in $f(1/2)$ is 2^d. This implies that, upon combining the terms and writing the rational number $f(1/2)$ as a reduced fraction, the denominator will not be divisible by 2^{d+1}. We now use this observation to prove that $\mathbb{Z}[1/2]$ is not finitely generated as a $\mathbb{Z}$-module.

Consider any finitely-generated submodule $\mathbb{Z}\{a_1, \ldots, a_n\} \subseteq \mathbb{Z}[1/2]$. Since each a_i is an element of $\mathbb{Z}[1/2]$, we know that $a_i = f_i(1/2)$ for some polynomial f_i in $\mathbb{Z}[x]$. Let d be the maximum degree of the f_i. Then, upon writing each a_i as a reduced fraction, none of the denominators is divisible by 2^{d+1}. Since taking $\mathbb{Z}$-linear combinations will never introduce additional powers of 2 in the denominators, after reducing, this proves that

$$\frac{1}{2^{d+1}} \notin \mathbb{Z}\{a_1, \ldots, a_n\}.$$

However, since $1/2^{d+1} = f(1/2)$ for $f = x^{d+1} \in \mathbb{Z}[x]$, it follows that

$$\frac{1}{2^{d+1}} \in \mathbb{Z}[1/2],$$

and we conclude that

$$\mathbb{Z}\{a_1, \ldots, a_n\} \neq \mathbb{Z}[1/2].$$

It follows that $\mathbb{Z}[1/2]$ is not a finitely-generated $\mathbb{Z}$-module, as claimed.

Let us pause to ponder the previous two examples. In both examples, we considered a $\mathbb{Z}$-algebra generated by a single real number. We might expect these two algebras to be very similar, but one of them turned out to be a finitely-generated module while the other did not. What, then, is the distinction between the numbers $\sqrt{2}$ and $1/2$ that led to this very different behavior? In the next section, we answer this question by giving a general criterion for determining whether a finitely-generated algebra is actually finitely generated as a module.

Exercises for Section 5.2

5.2.1 Let M be an R-module and $\mathcal{S} \subseteq M$ a subset. Prove the following.
   (a) The set $R\mathcal{S}$ is a submodule of M.
   (b) If $N \subseteq M$ is any submodule containing $\mathcal{S}$, then $R\mathcal{S} \subseteq N$.

5.2.2 Prove that
$$R[x^2, x^3] = R\{1, x^2, x^3, x^4, \ldots\}.$$

5.2.3 Consider $\mathbb{Z}$ as a $\mathbb{Z}$-module and let $a, b \in \mathbb{Z}$. Prove that
$$\mathbb{Z}\{a, b\} = \gcd(a, b)\mathbb{Z}.$$

5.2.4 Prove that $\mathbb{Z}[\sqrt{2}] = \mathbb{Z}\{1, \sqrt{2}\}$.

5.2.5 Prove that $\mathbb{Z}[\pi]$ is not finitely generated as a $\mathbb{Z}$-module.

5.2.6 Let $R \subseteq S \subseteq T$ be rings. Prove that if S is a finitely-generated R-module and T is a finitely-generated S-module, then T is a finitely-generated R-module.

5.2.7 Recall from Example 5.12 that $\mathbb{Z} = \mathbb{Z}\{2, 3\}$.
   (a) Prove that $\{2, 3\}$ is a minimal generating set, in the sense that no proper subset of $\{2, 3\}$ generates $\mathbb{Z}$ as a $\mathbb{Z}$-module.
   (b) Prove that, although every element of $\mathbb{Z}$ can be expressed as
$$r_1 \cdot 2 + r_2 \cdot 3$$
   for some $r_1, r_2 \in \mathbb{Z}$, this expression is not unique.
   (c) Prove that $\{1\}$ also generates $\mathbb{Z}$ as a $\mathbb{Z}$-module, and that it is a minimal generating set.

   This highlights two differences between R-modules and vector spaces. First, if V is a vector space, then a minimal generating set for V is necessarily a basis. By parts (a) and (b), this is not the case for R-modules. Second, if V is a vector space, then every minimal generating set has the same size. Parts (a) and (c) show that this is not the case for R-modules.

Section 5.3 Integrality

Consider a ring inclusion $R \subseteq S$, and let us view S as an R-module. Under what conditions is S a finitely-generated R-module? At the end of the last section, we studied two examples of this setup:

1. $R = \mathbb{Z}$ and $S = \mathbb{Z}[\sqrt{2}]$, and
2. $R = \mathbb{Z}$ and $S = \mathbb{Z}[1/2]$.

In the first case, we observed that $\mathbb{Z}[\sqrt{2}]$ is, in fact, a finitely-generated $\mathbb{Z}$-module, whereas in the second case, we argued that $\mathbb{Z}[1/2]$ is not. Looking back at those examples, one major difference we see between $\sqrt{2}$ and $1/2$ is that taking powers of $\sqrt{2}$ eventually brings us to an element of $\mathbb{Z}$, while taking powers of $1/2$ never brings us back to $\mathbb{Z}$. Indeed, the fact that $(\sqrt{2})^2 = 2$ is what allowed us to reduce all polynomial expressions in $\sqrt{2}$ to linear polynomials.

The goal of this section is to formalize the above observation for general rings. The key new concept for this discussion is the notion of *integrality*. For the following definition, recall that a *monic polynomial* is a nonzero polynomial whose leading coefficient is one; that is, a monic polynomial in $R[x]$ has the form

$$x^n + a_{n-1}x^{n-1} + \cdots + a_1 x + a_0$$

for some $n \geq 0$ and $a_0, \ldots, a_{n-1} \in R$.

5.19 DEFINITION *Algebraic and integral elements*

Let $R \subseteq S$ be rings, and let $a \in S$. We say that a is *algebraic over R* if there exists a nonzero polynomial $f \in R[x]$ such that $f(a) = 0 \in S$. If, moreover, there exists a monic polynomial $f \in R[x]$ such that $f(a) = 0$, then we say that a is *integral over R*.

If R is a field, then an element is algebraic if and only if it is integral, since we can simply divide any polynomial by its leading coefficient to obtain a monic polynomial. In more general rings, where division may not make sense, being integral is stronger than being algebraic.

Let us consider several examples in the setting where $R = \mathbb{Z}$ and $S = \mathbb{R}$.

5.20 EXAMPLE $\sqrt{2}$ is integral over $\mathbb{Z}$

Since $a = \sqrt{2}$ is a root of the monic polynomial

$$x^2 - 2 \in \mathbb{Z}[x],$$

we see that $\sqrt{2}$ is integral over $\mathbb{Z}$.

5.21 EXAMPLE π is not algebraic over $\mathbb{Z}$

At some point in your mathematical journey, you may have learned that π is a *transcendental number*, which simply means that it does not satisfy any polynomial equations over $\mathbb{Z}$. Thus, π is not algebraic over $\mathbb{Z}$.

5.22 EXAMPLE   1/2 is algebraic but not integral over $\mathbb{Z}$

The element $a = 1/2$ is algebraic over $\mathbb{Z}$, since it is a root of the polynomial

$$g(x) = 2x - 1 \in \mathbb{Z}[x],$$

but it is not integral. This is not immediately obvious; although g is not monic, one might still hope that a monic polynomial with $1/2$ as a root exists. But if

$$h(x) = x^n + a_{n-1}x^{n-1} + \cdots + a_1 x + a_0 \in \mathbb{Z}[x]$$

were such a polynomial, then multiplying both sides of the equation $h(1/2) = 0$ by 2^{n-1} would yield

$$\frac{1}{2} + a_{n-1} + a_{n-2} \cdot 2 + a_{n-3} \cdot 2^2 + \cdots + a_1 \cdot 2^{n-2} + a_0 \cdot 2^{n-1} = 0.$$

Moving everything but the first term to the right-hand side, we have expressed $1/2$ as sum of integers, which is impossible, proving that $1/2$ is not integral.

> More generally, a rational number $a \in \mathbb{Q}$ is integral over $\mathbb{Z}$ if and only if $a \in \mathbb{Z}$ (Exercise 5.3.2).

It may happen that every element of S is algebraic, or even integral, over R. When this occurs, we use the following terminology.

5.23 DEFINITION   Algebraic and integral extensions

Let $R \subseteq S$ be rings. We say that S is *algebraic over R* (respectively, *integral over R*) if every element of S is algebraic (respectively, integral) over R.

5.24 EXAMPLE $\mathbb{Z}[\sqrt{2}]$ is integral over $\mathbb{Z}$

In order to prove that $\mathbb{Z}[\sqrt{2}]$ is integral over $\mathbb{Z}$, we must show that *every* element of $\mathbb{Z}[\sqrt{2}]$ is integral over $\mathbb{Z}$. To do this, first recall (Example 5.17) that

$$\mathbb{Z}[\sqrt{2}] = \{r + s\sqrt{2} \mid r, s \in \mathbb{Z}\}.$$

Thus, given an element of $\mathbb{Z}[\sqrt{2}]$ we can write it as $r + s\sqrt{2}$ for some $r, s \in \mathbb{Z}$. Squaring, we obtain the equation

$$(r + s\sqrt{2})^2 = (r^2 + 2s^2) + (2rs)\sqrt{2}.$$

Rearranging and squaring again yields the equation

$$\left((r + s\sqrt{2})^2 - (r^2 + 2s^2)\right)^2 = (2rs)^2 \cdot 2.$$

This last equation implies that $r + s\sqrt{2}$ is a root of the degree-four monic polynomial

$$f(x) = \left(x^2 - (r^2 + 2s^2)\right)^2 - (2rs)^2 \cdot 2 \in \mathbb{Z}[x].$$

Thus, every element of $\mathbb{Z}[\sqrt{2}]$ is integral over $\mathbb{Z}$.

5.3. INTEGRALITY

5.25 EXAMPLE $\mathbb{R}$ is not algebraic over $\mathbb{Z}$

Since the real numbers contain transcendental numbers, such as π, we conclude that $\mathbb{R}$ is not algebraic over $\mathbb{Z}$.

5.26 EXAMPLE $\mathbb{Q}$ is algebraic but not integral over $\mathbb{Z}$

The ring $\mathbb{Q}$ is algebraic over $\mathbb{Z}$ because $a = p/q \in \mathbb{Q}$ is a root of the polynomial $f(x) = qx - p \in \mathbb{Z}[x]$. However, $\mathbb{Q}$ is not integral over $\mathbb{Z}$, by Example 5.22.

We are now ready to return to our goal of determining when a finitely-generated R-algebra is finitely generated as an R-module, which is closely related to the question of integrality. In particular, the next result tells us that a finitely-generated R-algebra is finitely generated as an R-module if and only if it is integral over R. Moreover, in order to check integrality, it suffices to check that the algebra generators are integral over R.

5.27 THEOREM *Finite generation and integrality*

Let $R \subseteq S$ be rings with $S = R[a_1, \ldots, a_n]$. The following are equivalent:

(i) S is a finitely-generated R-module;

(ii) S is integral over R;

(iii) a_i is integral over R for each $i = 1, \ldots, n$.

Before we begin the proof, we mention that the arguments involve certain manipulations of matrices whose entries come from the ring R, and the reader may not have previously worked with matrices in this generality. All such manipulations (matrix-vector products, for example, or determinants of matrices) are defined by the same formulas that define them in the more familiar setting where the entries come from $\mathbb{R}$ or some other field. These definitions make sense with entries in any ring because they involve only sums and products of elements.

PROOF OF THEOREM 5.27 We prove (i) $\Rightarrow$ (ii) $\Rightarrow$ (iii) $\Rightarrow$ (i).

(i) $\Rightarrow$ (ii): Suppose that S is a finitely-generated R-module, which means that there exist $v_1, \ldots, v_m \in S$ such that $S = R\{v_1, \ldots, v_m\}$. We must prove that an arbitrary element $b \in S$ is integral over R. To do so, first multiply b by each of the module generators and express the product as a linear combination of these generators:

$$bv_i = c_{i1}v_1 + c_{i2}v_2 + \cdots + c_{im}v_m,$$

where $c_{ij} \in R$ for $i, j = 1, \ldots, m$. Moving all the terms of each of these equations to the left-hand side, we obtain a system of linear equations

$$(b - c_{11})v_1 + (-c_{12})v_2 + \cdots + (-c_{1m})v_m = 0$$
$$(-c_{21})v_1 + (b - c_{22})v_2 + \cdots + (-c_{2m})v_m = 0$$
$$\vdots$$
$$(-c_{m1})v_1 + (-c_{m2})v_2 + \cdots + (b - c_{mm})v_m = 0.$$

Such a system is more conveniently expressed in matrix-vector form: if C is the matrix whose (i, j) entry is c_{ij}, then we have

$$(5.28) \qquad (bI - C) \cdot \begin{pmatrix} v_1 \\ \vdots \\ v_m \end{pmatrix} = \begin{pmatrix} 0 \\ \vdots \\ 0 \end{pmatrix},$$

where I is the $m \times m$ identity matrix.

At this point, we appeal to *Cramer's Rule*, a result about matrices that the reader may have seen when studying linear algebra. We state it here and direct the reader to Exercise 5.3.6 for a proof.

5.29 LEMMA *Cramer's Rule*

Let A be an $m \times m$ matrix with entries in a ring R. Let $v, w \in R^m$, which we view as column vectors, and suppose that

$$Av = w.$$

Then, for all $i = 1, \ldots, m$, we have

$$\det(A) \cdot v_i = \det(A_i),$$

where A_i is the matrix obtained from A by replacing its ith column by w.

Equipped with this tool, we apply it to the matrix equation (5.28), obtaining

$$(5.30) \qquad \det(bI - C) \cdot v_i = 0$$

for all i; notice, here, that the right-hand side is zero because it is the determinant of a matrix with a column of zeros. Recalling that v_i are generators for S as an R-module, we can express the element $1 \in S$ as a linear combination of $v_1, \ldots, v_m$:

$$1 = d_1 v_1 + \cdots + d_m v_m.$$

Multiplying both sides by $\det(bI - C)$ and applying (5.30) yields

$$\det(bI - C) = 0.$$

Therefore, b is a root of the polynomial

$$f(x) = \det(xI - C) \in R[x].$$

Properties of determinants imply that the leading term of $f(x) = \det(xI - C)$, as a polynomial in x, is the leading term of the product of the diagonal entries:

$$f(x) = \prod_{i=1}^{m} (x - c_{ii}) + \text{terms of degree at most } m - 1 \text{ in } x.$$

From this, we see that $f(x)$ is monic, proving that b is integral over R.

5.3. INTEGRALITY

(ii) ⇒ (iii): If S is integral over R, then every element of S is integral over R, so in particular, each a_i is integral.

(iii) ⇒ (i): Suppose that each a_i is integral over R, so there exist monic polynomials $f_1, \ldots, f_n \in R[x]$ such that $f_i(a_i) = 0$ for each i. Let $d_i > 0$ denote the degree of f_i. Our aim is to prove that S is generated as an R-module by the finite set

$$\mathcal{T} = \{a_1^{k_1} \cdots a_n^{k_n} \mid 0 \leq k_i < d_i\}.$$

The first step in proving that $S = R\mathcal{T}$ is to prove that, for every $i = 1, \ldots, n$,

(5.31) $$R[a_i] = R\{a_i^{k_i} \mid 0 \leq k_i < d_i\}.$$

This step follows from an induction argument, using the relation $f_i(a_i) = 0$ to reduce the degree of polynomial expressions in a_i (Exercise 5.3.7), similarly to how we argued that $\mathbb{Z}[\sqrt{2}] = \mathbb{Z}\{1, \sqrt{2}\}$ in Example 5.17. Once (5.31) is established, it then follows that, for any $\ell_i \geq 0$, we can write $a_i^{\ell_i} \in R[a_i]$ as an R-linear combination of $\{a_i^{k_i} \mid 0 \leq k_i < d_i\}$. Multiplying these linear combinations together, and expanding, we then see that, for any $\ell_1, \ldots, \ell_n \geq 0$, the element

(5.32) $$a_1^{\ell_1} \cdots a_n^{\ell_n} \in S$$

can be written as an R-linear combination of elements in $\mathcal{T}$.

Since every element of $S = R[a_1, \ldots, a_n]$ can be written as an R-linear combination of expressions of the form (5.32) with $\ell_1, \ldots, \ell_n \geq 0$, and since each expression of the form (5.32) with $\ell_1, \ldots, \ell_n \geq 0$ can be written as an R-linear combination of elements in $\mathcal{T}$, we conclude that every element of S can be written as an R-linear combination of elements in $\mathcal{T}$, proving that $S = R\mathcal{T}$, as desired. □

In general, given a ring extension $R \subseteq S$, the ring S need not be finitely generated over the ring R, either as an R-module or as an R-algebra, nor does it need to be integral over R. However, Theorem 5.27 tells us that those extensions that are both finitely generated as R-algebras and integral over R are exactly the same as those that are finitely generated as R-modules. We illustrate this result in the following Venn diagram.

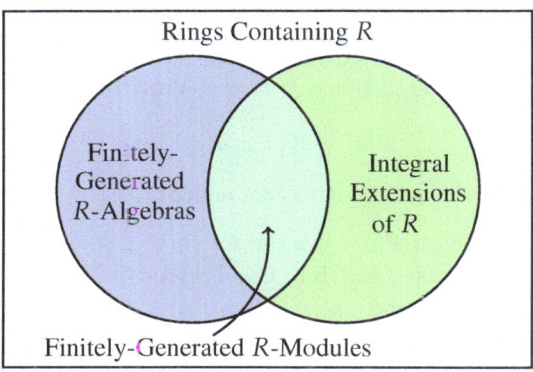

Exercises for Section 5.3

5.3.1 Is the element $\frac{\sqrt{2}}{2} \in \mathbb{R}$ algebraic over $\mathbb{Z}$? Is it integral over $\mathbb{Z}$?

5.3.2 Prove that $a \in \mathbb{Q}$ is integral over $\mathbb{Z}$ if and only if $a \in \mathbb{Z}$.
This motivates the terminology "integral."

5.3.3 Prove that $\mathbb{C}$ is integral (or equivalently, algebraic) over $\mathbb{R}$. (Hint: Mimic the argument in Example 5.24.)

5.3.4 For each of the following rings, let $a = [x]$ be the coset of x, and determine whether the ring is integral over the subring $R[a]$.
(a) $R[x,y]$
(b) $R[x,y]/\langle y^2 - x^2 \rangle$
(c) $R[x,y]/\langle xy \rangle$
(d) $R[x,y,z]/\langle y^2 - x^3, z^3 - x^4 \rangle$

5.3.5 Since, by Example 5.26, $\mathbb{Q}$ is not integral over $\mathbb{Z}$, it follows from Theorem 5.27 that $\mathbb{Q}$ is not a finitely-generated $\mathbb{Z}$-module. Verify this directly by showing that, for any $a_1, \ldots, a_n \in \mathbb{Q}$, there is a strict containment
$$\mathbb{Z}\{a_1, \ldots, a_n\} \subsetneq \mathbb{Q}.$$

5.3.6 This exercise proves Cramer's Rule, assuming some comfort with matrix multiplication and determinants. Let A be an $m \times m$ matrix with entries in a ring R and let $v, w \in R^m$ be vectors such that $Av = w$.
(a) Let I_i denote the matrix obtained from the $m \times m$ identity matrix I by replacing its ith column by v. Prove that
$$AI_i = A_i,$$
where A_i is obtained from A by replacing its ith column by w.
(b) Use your favorite method for computing determinants to prove that
$$\det(I_i) = v_i,$$
then use the multiplicativity of determinants to conclude that
$$\det(A) \cdot v_i = \det(A_i).$$

5.3.7 Let $R \subseteq S$ be rings and let $a \in S$. Suppose that there is a degree-d monic polynomial $f \in R[x]$ such that $f(a) = 0$. Prove that
$$R[a] = R\{1, a, a^2, \ldots, a^{d-1}\}.$$
Where does your proof fail if f is not monic?

5.3.8 Let $R \subseteq S$ be rings. Prove that the set $\{s \in S \mid s \text{ is integral over } R\}$ is a subring of S containing R. (Hint: Use Theorem 5.27.)
This subring is called the *integral closure* of R in S.

5.3.9 Let $R = \mathbb{Z}$. Describe examples of rings S that lie in every region of the Venn diagram from this section.

Section 5.4 Noether normalization

In the last three sections, we developed a number of module-theoretic notions, and we now bring those developments to bear on the particular type of algebraic objects that are at the center of algebraic geometry: finitely-generated K-algebras.

The Noether Normalization Theorem is a structural result about all finitely-generated K-algebras. To help frame the statement of this theorem, we recall that the prototype of a finitely-generated K-algebra is the polynomial ring

$$K[x_1, \ldots, x_d].$$

Not every finitely-generated K-algebra is isomorphic to a polynomial ring, and the Noether Normalization Theorem seeks to answer the question: How closely can we "approximate" finitely-generated K-algebras with polynomial rings? As we will see, the answer is that, given a finitely-generated K-algebra A, we can always find a subalgebra $B \subseteq A$ such that

1. $B \cong K[x_1, \ldots, x_d]$ for some d, and
2. A is a finitely-generated B-module or, equivalently, A is integral over B.

In other words, the Noether Normalization Theorem ensures that every finitely-generated K-algebra is finitely generated as a module (the smallest type of ring extension) over a polynomial ring (the simplest type of K-algebra).

Before stating and proving the Noether Normalization Theorem, we pause to introduce the notion of algebraic independence, which generalizes linear independence to the setting of polynomial algebra and will help us discuss a criterion for when a subalgebra is isomorphic to a polynomial ring.

5.33 **DEFINITION** *Algebraic (in)dependence*

Let A be an R-algebra. Elements $a_1, \ldots, a_d \in A$ are said to be *algebraically dependent over R* if there exists a nonzero polynomial $f \in R[x_1, \ldots, x_d]$ such that

$$f(a_1, \ldots, a_d) = 0.$$

If no such polynomial exists, then $a_1, \ldots, a_d$ are said to be *algebraically independent over R*.

A set $S \subseteq A$ is said to be *algebraically dependent over R* if S contains distinct elements $a_1, \ldots, a_d$ that are algebraically dependent over R. If no such elements exist, then S is said to be *algebraically independent over R*.

We mention one small subtlety about the terminology. Because lists of elements allow for repetition while sets do not account for repetition, it is possible for the elements $a_1, \ldots, a_d$ to be algebraically dependent even though the set $\{a_1, \ldots, a_d\}$ is not. This happens precisely when there is repetition among the elements. For example, if $a_1 = a_2 = \pi$, then the elements a_1 and a_2 are algebraically dependent over $\mathbb{Z}$—because the nonzero polynomial $f = x_1 - x_2 \in \mathbb{Z}[x_1, x_2]$ vanishes when evaluated at (a_1, a_2)—but the set $\{a_1, a_2\} = \{\pi\}$ is algebraically independent over $\mathbb{Z}$. Whenevever it is necessary to do so, we will emphasize that $a_1, \ldots, a_d$ are distinct in order to avoid this subtlety.

Finite sets of algebraically independent elements generate subalgebras that are isomorphic to polynomial rings, as described in the next result, which is an application of the First Isomorphism Theorem (Exercise 5.4.1).

5.34 PROPOSITION *Algebraically independent generators*

Let A be an R-algebra. Then $a_1, \ldots, a_d \in A$ are algebraically independent over R if and only if the evaluation map

$$R[x_1, \ldots, x_d] \to R[a_1, \ldots, a_d] \subseteq A$$
$$f(x_1, \ldots, x_d) \mapsto f(a_1, \ldots, a_d)$$

is an R-algebra isomorphism.

In other words, Proposition 5.34 asserts that, given an R-algebra A, finding a subalgebra B that is isomorphic to a polynomial ring is equivalent to finding a finite set of algebraically independent elements. This idea connects our discussion to the following statement of the Noether Normalization Theorem, valid for any field K.

5.35 THEOREM *Noether Normalization Theorem*

If A is a finitely-generated K-algebra, then there exists a finite subset $\mathcal{S} \subseteq A$ such that

(i) $\mathcal{S}$ is algebraically independent over K, and

(ii) A is integral over $K[\mathcal{S}]$.

In particular, if we write the distinct elements of the set $\mathcal{S}$ in the theorem as $a_1, \ldots, a_d$, then $K[a_1, \ldots, a_d]$ is isomorphic to a polynomial ring in d variables and A is integral over $K[a_1, \ldots, a_d]$. Although it may not be possible to find a finite algebraically independent set that generates all of A as a K-algebra (since A may not be isomorphic to a polynomial ring), the fact that A is integral over $K[a_1, \ldots, a_d]$, and thus a finitely-generated $K[a_1, \ldots, a_d]$-module, says that the elements $a_1, \ldots, a_d$ generate a polynomial ring that is "as close as possible to A."

The Noether Normalization Theorem motivates the following definition.

5.36 DEFINITION *Noether basis*

Let A be a finitely-generated K-algebra. A *Noether basis* of A over K is a finite subset $\mathcal{S} \subseteq A$ such that

(i) $\mathcal{S}$ is algebraically independent over K, and

(ii) A is integral over $K[\mathcal{S}]$.

In the language of Noether bases, the Noether Normalization Theorem is simply the assertion that every finitely-generated K-algebra admits a Noether basis. Before delving into the proof of the Noether Normalization Theorem, let us consider a few concrete examples of Noether bases.

5.4. NOETHER NORMALIZATION

5.37 EXAMPLE A Noether basis for $K[x,y]/\langle x^2 + y^2 - 1 \rangle$

Consider the K-algebra
$$A = \frac{K[x,y]}{\langle x^2 + y^2 - 1 \rangle}.$$

Notice that A is generated by $[x]$ and $[y]$, but these two elements are not algebraically independent because they satisfy a nonzero polynomial relation:
$$[x]^2 + [y]^2 - 1 = [x^2 + y^2 - 1] = 0.$$

Let $a = [x] \in A$. We prove that $\{a\}$ is a Noether basis of A.

First, we observe that a is algebraically independent over K. To prove this, we must show that $f(a) \neq 0$ for any nonzero single-variable polynomial $f \in K[z]$. Notice that, for any such f, we have $f(x) \notin \langle x^2 + y^2 - 1 \rangle$. Therefore,
$$f(a) = [f(x)] \neq 0 \in \frac{K[x,y]}{\langle x^2 + y^2 - 1 \rangle}.$$

Next, notice that $K[a] \neq A$, because we have no way of writing $[y] \in A$ as a polynomial expression in $a = [x]$. However, we have already seen in Example 5.13 that A is generated by $[1]$ and $[y]$ as a $K[a]$-module:
$$A = K[a]\{[1], [y]\}.$$

This proves that A is a finitely-generated $K[a]$-module, and thus it is integral over $K[a]$. Given that a is algebraically independent over K and A is integral over $K[a]$, we conclude that $\{a\}$ is a Noether basis.

This example readily generalizes to the coordinate ring of the unit n-sphere
$$A = \frac{K[x_1, \ldots, x_n]}{\langle x_1^2 + \cdots + x_n^2 - 1 \rangle}$$
to show that $\{a_1 = [x_1], \ldots, a_{n-1} = [x_{n-1}]\}$ is a Noether basis (Exercise 5.4.2).

5.38 EXAMPLE A Noether basis for $K[x,y]/\langle xy - 1 \rangle$

Consider the K-algebra
$$A = \frac{K[x,y]}{\langle xy - 1 \rangle}.$$

As in Example 5.37, A is generated by $[x]$ and $[y]$, but these are not algebraically independent, so they do not form a Noether basis. One might naturally guess, then, that either of the single elements $[x]$ or $[y]$ would form a Noether basis. However, this is not the case. For example, while $[x]$ is certainly algebraically independent over K, it can be checked that A is *not* integral over $K[x]$ (Exercise 5.4.3).

Even though neither $[x]$ nor $[y]$ alone forms a Noether basis for A, the Noether Normalization Theorem guarantees that a Noether basis must exist. In this case, a Noether basis is given by the element $a = [x + y]$. A proof of this assertion is outlined in Exercise 5.4.4.

Now that we have seen a few examples of Noether bases, let us turn toward a proof of the Noether Normalization Theorem, which uses a clever induction argument that crucially relies on Theorem 5.27.

PROOF OF THEOREM 5.35 Any finitely-generated K-algebra can, by definition, be expressed as $A = K[b_1, \ldots, b_n]$ for some $b_1, \ldots, b_n \in A$, and we prove the theorem by induction on the number n of generators.

(Base Case) If $n = 0$, then $A = K$ and the empty set is a Noether basis.

(Induction Step) Suppose that Noether bases exist for all K-algebras with fewer than n generators, and let $A = K[b_1, \ldots, b_n]$. If $b_1, \ldots, b_n$ are algebraically independent over K, then $\{b_1, \ldots, b_n\}$ is a Noether basis of A over K, and we are done. Assume, then, that $b_1, \ldots, b_n$ are not algebraically independent over K, meaning that there exists a nonzero polynomial $f \in K[x_1, \ldots x_n]$ such that

$$f(b_1, \ldots, b_n) = 0.$$

We will manipulate this relation to obtain a relation that is monic in b_1.

Choose N to be an integer greater than the maximum exponent appearing in any monomial with nonzero coefficient in f, and define new elements $c_2, \ldots, c_n \in A$ by

$$c_2 = b_2 - b_1^N, \ c_3 = b_3 - b_1^{N^2}, \ldots, \ c_n = b_n - b_1^{N^{n-1}},$$

so that

$$0 = f(b_1, \ldots, b_n) = f\left(b_1,\ c_2 + b_1^N,\ c_3 + b_1^{N^2}, \ldots,\ c_n + b_1^{N^{n-1}}\right).$$

Let us consider the following single-variable polynomial

$$g(z) = f\left(z,\ c_2 + z^N,\ c_3 + z^{N^2}, \ldots,\ c_n + z^{N^{n-1}}\right) \in K[c_2, \ldots, c_n][z].$$

Notice that $g(b_1) = 0$, and we now argue that the leading coefficient of g is a unit, implying that b_1 is integral over $K[c_2, \ldots, c_n]$.

Evaluating any monomial $x_1^{\alpha_1} \cdots x_n^{\alpha_n}$ of f at $\left(z, c_2 + z^N, \ldots, c_n + z^{N^{n-1}}\right)$ gives

(5.39) $$z^{\alpha_1} \left(c_2 + z^N\right)^{\alpha_2} \left(c_3 + z^{N^2}\right)^{\alpha_3} \cdots \left(c_n + z^{N^{n-1}}\right)^{\alpha_n}.$$

Collecting terms with the same power of z, we can rearrange (5.39) to

$$z^{\alpha_1 + \alpha_2 N + \alpha_3 N^2 + \cdots + \alpha_n N^{n-1}} + \text{lower-degree terms in } z.$$

Because N was chosen to be larger than $\alpha_1, \ldots, \alpha_n$, the exponent

$$\alpha_1 + \alpha_2 N + \alpha_3 N^2 + \cdots + \alpha_n N^{n-1}$$

uniquely determines the numbers $\alpha_1, \ldots, \alpha_n$. (This assertion is the "uniqueness of base-N expansions"; for example, if $N = 10$, it is the assertion that the digits of a number are uniquely determined by the number itself—see Exercise 5.4.7.) In particular, it follows that different monomials

$$x_1^{\alpha_1} \cdots x_n^{\alpha_n} \quad \text{and} \quad x_1^{\alpha'_1} \cdots x_n^{\alpha'_n}$$

will have different degrees in z after evaluating them at $\left(z, c_2 + z^N, \ldots, c_n + z^{N^{n-1}}\right)$.

5.4. NOETHER NORMALIZATION

Therefore, there is a unique monomial with nonzero coefficient in f for which the quantity $\alpha_1 + \alpha_2 N + \alpha_3 N^2 + \cdots + \alpha_n N^{n-1}$ is maximal, and if $c \in K$ is the nonzero coefficient of this monomial in f, then c is also the leading coefficient of g. Thus, $c^{-1}g \in K[c_2, \ldots, c_n][z]$ is a monic polynomial that vanishes at b_1, so b_1 is integral over $K[c_2, \ldots, c_n]$. It then follows from Theorem 5.27 that $K[c_2, \ldots, c_n][b_1]$ is integral over $K[c_2, \ldots, c_n]$. But since we can freely convert between polynomial combinations in $b_1, b_2, \ldots, b_n$ and $b_1, c_2, \ldots, c_n$, we have

$$K[c_2, \ldots, c_n][b_1] = K[b_1, c_2, \ldots, c_n] = K[b_1, b_2, \ldots, b_n] = A.$$

Thus, we have proven that $A = K[b_1, \ldots, b_n]$ is integral over $K[c_2, \ldots, c_n]$.

By the induction hypothesis, there exists a Noether basis $\mathcal{S} \subseteq K[c_2, \ldots, c_n]$. We claim that $\mathcal{S}$ is also a Noether basis of A; to prove this, we must argue that $\mathcal{S}$ is algebraically independent over K and that A is integral over $K[\mathcal{S}]$. That $\mathcal{S}$ is algebraically independent over K simply follows from the fact that $\mathcal{S}$ is a Noether basis of $K[c_1, \ldots, c_n]$. To prove that A is integral over $K[\mathcal{S}]$, notice that each of the following rings is integral over the one that precedes it:

$$K[\mathcal{S}] \subseteq K[c_2, \ldots, c_n] \subseteq A,$$

where the integrality of the first inclusion follows from the assumption that $\mathcal{S}$ is a Noether basis of $K[c_1, \ldots, c_n]$ and the integrality of the second inclusion was the heart of our argument above. By Theorem 5.27, this means that each of these rings is finitely generated as a module over the one that precedes it. Thus, by Exercise 5.2.6, we conclude that A is finitely generated as a module over $K[\mathcal{S}]$, and therefore—by Theorem 5.27 again—we conclude that A is integral over $K[\mathcal{S}]$. Hence, $\mathcal{S}$ is a Noether basis of A, concluding the induction argument. □

The Noether Normalization Theorem is the culmination of the last four sections of algebraic developments, and the payoffs for all of this hard work will be plentiful. Most immediately, as we will see in the next section, Noether normalization can be used to prove the Nullstellensatz, and therefore, the equivalence of algebra and geometry. However, the payoff does not end there; Noether normalization is also closely connected to the notion of dimension, a concept we will introduce in the next chapter. As we will see, if X is an affine variety, the number of elements in any Noether basis of the coordinate ring $K[X]$ is equal to the dimension of X, a fact that will allow us to use Noether bases to prove fundamental properties about dimension.

Exercises for Section 5.4

5.4.1 Prove Proposition 5.34.

5.4.2 Consider the K-algebra

$$A = \frac{K[x_1, \ldots, x_n]}{\langle x_1^2 + \cdots + x_n^2 - 1 \rangle}.$$

and let $a_i = [x_i]$ for all $i = 1, \ldots, n-1$.
 (a) Prove that $a_1, \ldots, a_{n-1}$ are algebraically independent.
 (b) Prove that A is integral over $K[a_1, \ldots, a_{n-1}]$.

5.4.3 Let $A = K[x,y]/\langle xy - 1 \rangle$ and let $a = [x]$.
  (a) Prove that a is algebraically independent over K.
  (b) Prove that A is *not* integral over $K[a]$.

5.4.4 Let $A = K[x,y]/\langle xy - 1 \rangle$ and let $a = [x+y]$.
  (a) Prove that a is algebraically independent over K.
  (b) Prove that A is integral over $K[a]$. (Hint: Argue that both $[x]$ and $[y]$ are zeros of monic quadratic polynomials over $K[a]$.)

5.4.5 Let A be a K-algebra. Prove that every Noether basis of A over K is a maximal algebraically independent set. (In other words, any set properly containing a Noether basis must be algebraically dependent.)

5.4.6 Show by example that the converse of Exercise 5.4.5 is false: that is, if A is a finitely-generated K-algebra and $\mathcal{S}$ is a maximal algebraically independent set, it need not be the case that A is integral over $K[\mathcal{S}]$.

5.4.7 Let N be a positive integer and set $[N] = \{0, 1, \ldots, N-1\}$. For $k \geq 0$, consider the function

$$\psi_n : [N]^{k+1} \to \mathbb{N}$$
$$(\alpha_0, \ldots, \alpha_k) \mapsto \alpha_0 + \alpha_1 N + \alpha_2 N^2 + \cdots + \alpha_k N^k.$$

  (a) If $N = 10$, how would you describe the number $\psi_k(\alpha_0, \ldots, \alpha_k)$?
  (b) Prove that ψ_k is injective for all $k \geq 0$. (In other words, the number $\alpha_0 + \alpha_1 N + \alpha_2 N^2 + \cdots + \alpha_k N^k$ uniquely determines $\alpha_0, \ldots, \alpha_k$.)

Section 5.5 Proof of the Nullstellensatz

We have now built up the necessary algebraic tools to discuss a proof of the Nullstellensatz and thus a complete justification of the equivalence of algebra and geometry that was developed in Chapters 1 – 4. We begin with two useful lemmas that will allow us to prove a seemingly weaker version of the Nullstellensatz. Then, we prove that—surprisingly—this weak version implies the full Nullstellensatz.

The first lemma we require simply tells us that algebraically closed fields do not have any nontrivial algebraic field extensions.

5.40 LEMMA *Algebraic over algebraically closed fields*

Let $K \subseteq L$ be fields such that L is algebraic over K. If K is algebraically closed, then $L = K$.

PROOF Suppose, toward a contradiction, that $a \in L \setminus K$. Then, since a is algebraic over K, there exists a polynomial $f \in K[x]$ such that $f(a) = 0$. There may be many such polynomials, but let f be one of minimum possible degree. Because K is algebraically closed, there exists $b \in K$ such that $f(b) = 0$, so Corollary 0.48 tells us that we can factor $f(x)$ as

> *The name "algebraically closed" reflects the fact that any attempt to adjoin an element to K that is algebraic over K produces only elements that are already there, so K is "closed under adjoining algebraic elements."*

$$f(x) = (x - b)g(x)$$

for some $g \in K[x]$ with $\deg(g) < \deg(f)$. Evaluating both sides at $x = a$ yields

$$0 = (a - b)g(a).$$

We have assumed that $a \notin K$ and $b \in K$, so $a \neq b$. Hence, the above is only possible if $g(a) = 0$, which contradicts the assumption that f is a minimum-degree polynomial in $K[x]$ that vanishes at a. □

The next result is the primary consequence of the Noether Normalization Theorem that we require in this section. It was first proved by Zariski, though by a different method than what we present here; it is valid for all fields K.

5.41 LEMMA *Zariski's Lemma*

Let $K \subseteq L$ be fields. If L is a finitely-generated K-algebra, then L is algebraic over K.

PROOF Let $K \subseteq L$ be fields such that L is a finitely-generated K-algebra. Applying the Noether Normalization Theorem to L, we see that there exists a $d \geq 0$ and elements $a_1, \ldots, a_d \in L$ that are algebraically independent over K and such that L is integral—and thus algebraic—over $K[a_1, \ldots, a_d]$. To show that L is algebraic over K, it suffices to prove that $d = 0$.

Toward a contradiction, suppose that $d > 0$. Since L is a field, there exists $a_1^{-1} \in L$. Because L is integral over $K[a_1, \ldots, a_d]$, the element a_1^{-1} is a solution to a monic polynomial with coefficients in $K[a_1, \ldots, a_d]$:

$$(a_1^{-1})^m + c_1(a_1^{-1})^{m-1} + \cdots + c_{m-1}(a_1^{-1}) + c_m = 0,$$

where $c_1, \ldots, c_m \in K[a_1, \ldots, a_d]$. Multiplying both sides by a_1^m, we obtain

$$1 + c_1 a_1 + \cdots + c_{m-1} a_1^{m-1} + c_m a_1^m = 0.$$

This is a polynomial relation among $a_1, \ldots, a_d$, contradicting the algebraic independence of these elements. The contradiction implies that $d = 0$, as desired. □

The following result, which we will prove from the previous two lemmas, is commonly referred to as the "weak Nullstellensatz," because if we had already proven the Nullstellensatz, it would follow quite immediately (Exercise 5.5.1).

5.42 PROPOSITION *Weak Nullstellensatz*

Let K be algebraically closed. For any proper ideal $I \subsetneq K[x_1, \ldots, x_n]$, we have $\mathcal{V}(I) \neq \emptyset$.

PROOF Suppose that K is algebraically closed, and let $I \subsetneq K[x_1, \ldots, x_n]$ be a proper ideal. By Exercise 5.5.3, there exists a maximal ideal J of $K[x_1, \ldots, x_n]$ containing I. Since $I \subseteq J$ implies $\mathcal{V}(J) \subseteq \mathcal{V}(I)$, it suffices to prove that $\mathcal{V}(J) \neq \emptyset$.

Since J is maximal, the quotient ring

$$L = K[x_1, \ldots, x_n]/J$$

is a field. Moreover, L is a finitely-generated K-algebra (generated by $[x_1], \ldots, [x_n]$) and hence Zariski's Lemma implies that L is algebraic over K. But K is algebraically closed, so Lemma 5.40 implies that $L = K$.

As $L = K$, there is a K-algebra isomorphism $\varphi : L \to K$. Define $a_i = \varphi([x_i])$. Using the fact that φ is a K-algebra homomorphism, we see that

$$\varphi([x_i - a_i]) = \varphi([x_i]) - \varphi([a_i]) = a_i - a_i = 0.$$

Since φ is injective, this implies that $[x_i - a_i] = 0$ for all i. In other words,

$$\langle x_1 - a_1, \ldots, x_n - a_n \rangle \subseteq J.$$

Since $\langle x_1 - a_1, \ldots, x_n - a_n \rangle$ is a maximal ideal and J is a proper ideal, we conclude that $J = \langle x_1 - a_1, \ldots, x_n - a_n \rangle$, and it follows that

$$\mathcal{V}(J) = \{(a_1, \ldots, a_n)\} \neq \emptyset,$$

as required. □

The weak Nullstellensatz is a blunt tool that simply ensures $\mathcal{V}(I)$ is nonempty. The (strong) Nullstellensatz, on the other hand, tells us exactly which polynomials vanish on $\mathcal{V}(I)$. We now prove that this strengthening is actually implied by the weak Nullstellensatz. Since we have proven the weak version without using the strong version, with this the proof of the Nullstellensatz will at last be complete.

5.5. PROOF OF THE NULLSTELLENSATZ

5.43 THEOREM *Nullstellensatz recalled*

Let K be an algebraically closed field. Then, for any ideal $I \subseteq K[x_1,\ldots,x_n]$, we have
$$\mathcal{I}(\mathcal{V}(I)) = \sqrt{I}.$$

PROOF The inclusion $\mathcal{I}(\mathcal{V}(I)) \supseteq \sqrt{I}$ holds over any field and follows directly from the definitions; see Exercise 1.5.3. Therefore, it remains to prove the inclusion $\mathcal{I}(\mathcal{V}(I)) \subseteq \sqrt{I}$. Since $K[x_1,\ldots,x_n]$ is Noetherian, write $I = \langle f_1,\ldots,f_m\rangle$.

Let $g \in \mathcal{I}(\mathcal{V}(I))$. Introduce a new variable x_{n+1} and consider the ideal
$$J = \langle f_1,\ldots,f_m,h\rangle \subseteq K[x_1,\ldots,x_n,x_{n+1}],$$
where
$$h(x_1,\ldots,x_{n+1}) = 1 - x_{n+1}g(x_1,\ldots,x_n).$$

We claim that $\mathcal{V}(J) = \emptyset \subseteq \mathbb{A}^{n+1}$. To see this, suppose toward a contradiction that $a = (a_1,\ldots,a_n,a_{n+1}) \in \mathcal{V}(J)$. Then $f_i(a) = 0$ for all i, but since f_i only involves the variables $x_1,\ldots,x_n$, this is equivalent to the statement that $f_i(a_1,\ldots,a_n) = 0$ for all i. That is, $(a_1,\ldots,a_n) \in \mathcal{V}(I)$. Since $g \in \mathcal{I}(\mathcal{V}(I))$, it follows that $g(a_1,\ldots,a_n) = 0$, and hence

> The trick of adding an extra variable is a clever device for bringing the weak Nullstellensatz to bear on the (strong) Nullstellensatz.

$$h(a) = 1 - a_{n+1}g(a_1,\ldots,a_n) = 1 \neq 0.$$

Since $h \in J$, this contradicts the fact that $a \in \mathcal{V}(J)$.

We have thus shown that $\mathcal{V}(J) = \emptyset$, so the weak Nullstellensatz implies that $J = K[x_1,\ldots,x_{n+1}]$. In particular, the constant polynomial 1 can be expressed in terms of the generators of J:

(5.44) $$1 = q_1 f_1 + \cdots + q_m f_m + r(1 - x_{n+1}g),$$

where $q_1,\ldots,q_m,r \in K[x_1,\ldots,x_{n+1}]$. We would like to isolate g in this equation, but doing so necessitates division by x_{n+1}. To make sense of this, consider the larger ring $K(x_{n+1})[x_1,\ldots,x_n]$, where we allow rational functions in x_{n+1} (we were introduced to this ring in Section 0.6). The equation (5.44) still holds in this larger ring, and now we can divide both sides by a sufficiently high power of x_{n+1} so that no positive powers of x_{n+1} appear:

(5.45) $$x_{n+1}^{-k} = \tilde{q}_1 f_1 + \cdots + \tilde{q}_m f_m + \tilde{r}(x_{n+1}^{-1} - g),$$

where $\tilde{q}_1,\ldots,\tilde{q}_m,\tilde{r} \in K[x_1,\ldots,x_n,x_{n+1}^{-1}] \subseteq K(x_{n+1})[x_1,\ldots,x_n]$.

Now that we have an equation in the polynomial ring $K[x_1,\ldots,x_n,x_{n+1}^{-1}]$, we can make the substitution $x_{n+1}^{-1} = g \in K[x_1,\ldots,x_n]$, which yields the equation
$$g^k = \hat{q}_1 f_1 + \cdots + \hat{q}_m f_m,$$
where $\hat{q}_i \in K[x_1,\ldots,x_n]$ is obtained from $\tilde{q}_i$ by setting $x_{n+1}^{-1} = g$. This last equation shows that $g^k \in I$, so $g \in \sqrt{I}$, completing the proof. □

The last step of the proof, where we set $x_{n+1}^{-1} = g$, often strikes readers as suspicious on a first pass; why can we simply choose a value for x_{n+1}^{-1}? We stress that the reason this is valid is that x_{n+1}^{-1} is a variable like any other in the ring $K[x_1, \ldots, x_n, x_{n+1}^{-1}]$. Let us carry out this procedure explicitly in a small example.

5.46 EXAMPLE An illustration of the proof of the Nullstellensatz

Let $J = \langle x^2, y \rangle \subseteq K[x, y]$, and let $g = x + y$. Denoting the variable x_{n+1} by z in this case, equation (5.44) reads

$$1 = q_1(x, y, z) \cdot x^2 + q_2(x, y, z) \cdot y + r(x, y, z) \cdot (1 - z(x + y)),$$

and it is straightforward to check that this equation holds for the following choice of q_1, q_2, and r:

$$1 = (z^2) \cdot x^2 + (2xz^2 + yz^2) \cdot y + (1 + xz + yz) \cdot (1 - z(x + y)).$$

Note that this equation takes place in $K[x, y, z]$. Working within the larger ring $K(z)[x, y]$, we can divide both sides by z^2, eliminating all positive powers of z and yielding the equation

$$z^{-2} = (1) \cdot x^2 + (2x + y) \cdot y + (z^{-1} + x + y) \cdot (z^{-1} - (x + y)),$$

and this new equation is valid in the subring $K[x, y, z^{-1}] \subseteq K(z)[x, y]$. In this subring, which is just a polynomial ring, we may substitute one variable for an expression in the others. Choosing the subsitution $z^{-1} = g = x + y$ results in the equation

$$(x + y)^2 = (1) \cdot x^2 + (2x + y) \cdot y,$$

which is manifestly true in $K[x, y]$ and illustrates that $g^2 \in J$.

Exercises for Section 5.5

5.5.1 Prove that the Nullstellensatz implies the weak Nullstellensatz.

5.5.2 Give an example to show that the weak Nullstellensatz can fail if K is not algebraically closed. Discuss why your example is consistent with Zariski's Lemma even though it is inconsistent with the weak Nullstellensatz.

5.5.3 Let $I \subseteq K[x_1, \ldots, x_n]$ be an ideal with $I \neq K[x_1, \ldots, x_n]$. Prove that there exists a maximal ideal containing I. (Hint: We know that $K[x_1, \ldots, x_n]$, being Noetherian, satisfies the ascending chain condition.)

5.5.4 Let K be algebraically closed, and let $I \subsetneq K[x_1, \ldots, x_n]$ be a proper ideal.
(a) Interpret the statement "J is a maximal ideal containing I" in terms of the affine varieties $\mathcal{V}(I)$ and $\mathcal{V}(J)$.
(b) Using this interpretation in terms of affine varieties, find two different maximal ideals containing $I = \langle y^2 - x^3 - x^2 \rangle \subseteq K[x, y]$.

5.5. PROOF OF THE NULLSTELLENSATZ

5.5.5 Let $I = \langle x^2 + y^2 - 1, y - 1 \rangle \subseteq K[x,y]$ and let $g = x \in K[x,y]$.
 (a) Calculate $\mathcal{V}(I) \subseteq \mathbb{A}^2$, and confirm (without citing the Nullstellensatz) that $g \in \mathcal{I}(\mathcal{V}(I))$ but $g \notin I$.
 (b) Let $f_1 = x^2 + y^2 - 1$ and $f_2 = y - 1$ be the generators of I, and let $J = \langle f_1, f_2, 1 - zx \rangle$, as in the proof of the Nullstellensatz. Find $q_1, q_2, r \in K[x,y,z]$ such that equation (5.44) holds.
 (c) Set $z = 1/g$ in the equation from part (b) and clear denominators to deduce an equation that exhibits $g \in \sqrt{I}$.
 (d) You have just verified that the ideal $I = \langle x^2 + y^2 - 1, y - 1 \rangle$ is not radical. On the other hand,

$$I = \langle x^2 + y^2 - 1 \rangle + \langle y - 1 \rangle,$$

which implies that

$$\mathcal{V}(I) = \mathcal{V}(x^2 + y^2 - 1) \cap \mathcal{V}(y - 1).$$

Draw a picture of $\mathcal{V}(x^2 + y^2 - 1)$ and $\mathcal{V}(y - 1)$ over the real numbers. Do you have a guess about what geometric feature of these varieties is responsible for $\langle x^2 + y^2 - 1 \rangle + \langle y - 1 \rangle$ not being radical?

5.5.6 Prove that a field K is algebraically closed if and only if every maximal ideal $I \subseteq K[x_1, \ldots, x_n]$ has the form

$$I = \langle x_1 - a_1, \ldots, x_n - a_n \rangle$$

for some $a_1, \ldots, a_n \in K$.

Chapter 6

Dimension

> LEARNING OBJECTIVES FOR CHAPTER 6
> - Describe the dimension of an affine variety, both intuitively and in terms of algebraic independence in coordinate rings.
> - Define the function field of an affine variety and compute it in examples.
> - Use transcendence bases to measure field extensions.
> - Carefully compute the dimensions of affine varieties in examples.
> - Prove the Fundamental Theorem of Dimension Theory as an application of the Noether Normalization Theorem.

Given a set of polynomial equations, arguably the most fundamental geometric question one could ask about its solution set is how "big" it is. This is the question on which we aim to shed light in this chapter by introducing the important concept of *dimension* of affine varieties.

Intuitively, dimension measures the freedom to be able to move within a set. So how do we measure the freedom to move within an affine variety? As we will see, the key to answering this question lies within the coordinate ring. One of the motivating ideas of this chapter is that *the dimension of an affine variety is the maximum number of algebraically independent elements in its coordinate ring.*

This motivating idea might remind the reader of their knowledge of dimension from linear algebra, where the dimension of a (finite-dimensional) vector space can be defined as the maximum number of linearly independent elements. Indeed, there are many analogies between the ideas developed in this chapter and the ideas concerning dimension in linear algebra. In particular, the definition and properties of *transcendence bases*, introduced in Section 6.3, will closely parallel ideas concerning bases of vector space. In order to obtain a robust theory of transcendence bases, one requires working with fields, instead of rings, which is why Section 6.2 is devoted to the notion of *function fields* of irreducible affine varieties.

Once the groundwork regarding function fields and transcendence bases has been laid, Section 6.5 defines and describes some basic properties of dimension. For example, we see that $\dim(\mathbb{A}^r) = n$ and, if $X \subseteq Y$, then $\dim(X) \leq \dim(Y)$. The ideas of this chapter then culminate in Section 6.6 with what we call the Fundamental Theorem of Dimension Theory. This result, which is central in algebraic geometry, essentially says that each polynomial equation in a set decreases the dimension of the solution set by at most one.

Section 6.1 Motivating ideas

Dimension is a notion with which we are all intimately familiar; for example, at a young age, most of us probably came to grips with the understanding that we live in a three-dimensional world. The way in which we understand those three dimensions is that we have three basic directions in which we can move: forward-backward, left-right, and up-down. Of course, there are more than just these three basic directions, but every other direction is a combination of just these three.

For many of us, linear algebra is the first mathematical subject in which we learn a precise definition of dimension, and it naturally captures the notions of "basic" and "combination" alluded to in the previous paragraph. Recall that a vector space V over a field K is said to have *dimension n* if it has a basis of size n. Importantly, one must check that any two bases of V have the same size in order to make sure that this definition is well-defined. One way to

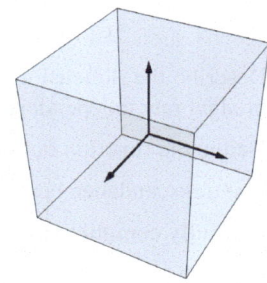

view basis vectors is as the basic directions of movement inside V. In the image above, we have depicted the three standard basis vectors in $\mathbb{R}^3$: any direction of movement in $\mathbb{R}^3$ can be uniquely expressed as a linear combination of these.

> Note the important role the field K plays in vector-space dimension: the usual plane is two-dimensional as an $\mathbb{R}$-vector space but only one-dimensional as a $\mathbb{C}$-vector space.

The idea that dimension measures our freedom to move within a set is the intuition that will guide our definition of dimension in algebraic geometry. We further motivate our development with the following familiar nonlinear example.

6.1 EXAMPLE Dimension of the sphere

Consider the sphere $X = \mathcal{V}(x^2 + y^2 + z^2 - 1) \subseteq \mathbb{A}^3$. Over the real numbers, this is simply the familiar unit sphere, and we can even imagine the unit sphere as a model for the surface of the Earth.

We naturally view the surface of our planet as having two dimensions because, at any location, we have two basic directions in which we can move; when we are not at one of the poles, we call these directions north-south and east-west. But how do we make our real-world intuition of these two dimensions of freedom algebraically precise?

The Blue Marble, photo by NASA

One way to argue is the following: if we choose two of the coordinates of a point in X, say $x = a$ and $y = b$, for some $a, b \in K$, then the defining equation for X tells us that the third coordinate is constrained by these choices: we must have

$$z^2 = 1 - a^2 - b^2.$$

In particular, there are at most two possible solutions for the third coordinate.

6.1. MOTIVATING IDEAS

In other words, in order to describe a point on X, we have two variables of freedom—we can choose the x- and y-coordinates without constraint—whereupon the third coordinate is then determined up to finitely many values. This reflects the algebraic fact that the elements $[x], [y] \in K[X]$ are algebraically independent over K—fixing the value of one does not constrain the other—whereas the elements $[x], [y], [z]$ are algebraically dependent over K because they satisfy a polynomial relation:
$$[x]^2 + [y]^2 + [z]^2 - 1 = [x^2 + y^2 + z^2 - 1] = 0 \in K[X].$$

The previous example suggests that our "freedom to move" within X is measured by algebraically independent functions: anytime we have algebraically independent coordinate functions in $K[X]$, then we can choose the values of those coordinates freely, without constraint. Since dimension should measure our maximum freedom of movement, we arrive at the following motivating idea.

6.2 KEY IDEA *Dimension of an affine variety*

The *dimension* of an affine variety X should be the maximum number of elements in $K[X]$ that are algebraically independent over K.

In fact, one could take this key idea as the definition of dimension; the reason we do not is because it is not particularly easy to work with. For starters, it is not even clear from this description whether or not dimension is finite: given an affine variety X, how do we know that it is not possible to find larger and larger sets of algebraically independent functions in $K[X]$? In order to argue that this cannot happen, we need to lay some groundwork first. Developing this groundwork carefully will require some effort on our part, undertaken in Sections 6.2 – 6.4, before we finally present a more robust definition of dimension in Section 6.5 and show that it is equivalent to the description above (Corollary 6.37).

In the meantime, we can keep Key Idea 6.2 in the back of our minds as motivation and turn to a discussion of our aspirations for dimension. What are the essential properties that we should expect in a notion of dimension for affine varieties? We begin by listing those properties, which we call the *axioms* for dimension.

6.3 DEFINITION *Axioms for dimension of affine varieties*

We say that a function
$$D: \left\{ \begin{array}{c} \text{isomorphism classes of} \\ \text{nonempty affine varieties} \end{array} \right\} \longrightarrow \mathbb{N}$$
is a *dimension function* if it satisfies the following properties:

1. $D(X) = 0$ if X consists of a single point;
2. $D(X_1 \cup \cdots \cup X_n) = \max\{D(X_1), \ldots, D(X_n)\}$;
3. If $X \subseteq \mathbb{A}^n$ is irreducible and $f \in K[x_1, \ldots, x_n]$ is such that $X \cap \mathcal{V}(f)$ is neither empty nor all of X, then $D(X \cap \mathcal{V}(f)) = D(X) - 1$.

Let us briefly interpret the axioms for dimension. First of all, since two isomorphic affine varieties share the same essential properties, we should expect their dimensions to be the same; this is the reason that we view dimension as a function from the set of isomorphism classes of affine varieties. It should also be somewhat clear from our intuition why we require Axiom 1: there is no freedom to move within a single point, so we should expect the dimension of a point to be zero.

Regarding Axiom 2, consider as an example the union of a line L and a plane P in $\mathbb{A}^3$. If the line is contained in the plane, then their union is simply equal to the plane, so $\dim(L \cup P) = \dim(P)$, the maximum of the two individual dimensions. If the line is not contained in the plane, however, then our intuition for "freedom to move" breaks down: the number of directions in which you can move within $L \cup P$ depends on the point at which you stand. To resolve this ambiguity, we simply declare that the larger dimension trumps the smaller. More generally, we will find that dimension is only readily definable for irreducible varieties, and the dimension of a reducible variety will be defined as the maximum dimension of its irreducible components.

Finally, for Axiom 3, let us parse the statement further by writing X as the vanishing set $\mathcal{V}(f_1, \ldots, f_m)$. Then

$$X \cap \mathcal{V}(f) = \mathcal{V}(f_1, \ldots, f_m, f).$$

In other words, Axiom 3 is essentially saying that if we impose one additional equation, then the dimension should go down by exactly one. It is hard to overstate how important this property of dimension is in algebraic geometry, which is why, in this text, we refer to this property as the *Fundamental Theorem of Dimension Theory*.

> *In the vector-space setting, Axiom 3 is a consequence of the Rank-Nullity Theorem (see Exercise 6.1.6).*

We note that the assumptions in the hypothesis of Axiom 3 are all necessary. We assume that X is irreducible because, if not, then it could consist of two irreducible components of the same dimension, and intersecting with $\mathcal{V}(f)$ might just pick out one of these components. For example, in $\mathbb{A}^2$, the one-dimensional variety $X = \mathcal{V}(xy)$ is the union of the two axes, and intersecting with $\mathcal{V}(x)$ simply picks out the one-dimensional y-axis, violating the conclusion of Axiom 3. We assume that $X \cap \mathcal{V}(f)$ is not empty because the dimension of $\varnothing$ is undefined, and we assume that $X \cap \mathcal{V}(f) \neq X$ because the conclusion of Axiom 3 would be violated if this were so.

One might naturally ask why we do not include more axioms for dimension. For example, it might be natural for us to assume that $\mathbb{A}^n$ has dimension n and, more generally, that the dimension of an affine variety defined by linear equations is equal to its dimension as a vector space. The reason we do not assume more is because the short list of axioms listed in Definition 6.3 already uniquely determines the dimension of all affine varieties, as stated in the next result.

6.4 PROPOSITION *There exists at most one dimension function*

If D_1 and D_2 are both dimension functions, then $D_1(X) = D_2(X)$ for all affine varieties X.

PROOF   A proof is outlined in Exercise 6.1.8.   □

6.1. MOTIVATING IDEAS

In light of Proposition 6.4, the path ahead is clear: our goal is to define a function

$$D : \left\{ \begin{array}{c} \text{isomorphism classes of} \\ \text{nonempty affine varieties} \end{array} \right\} \longrightarrow \mathbb{N}$$

and to prove that it satisfies Axioms 1 – 3 of Definition 6.3. The next three sections are devoted to developing the algebraic tools we require in order to define dimension—our primary focus will be on further understanding the notions of algebraic (in)dependence and developing the key concept of transcendence degree. Upon rigorously defining dimension in Section 6.5 in terms of transcendence degree, Axioms 1 and 2 will be rather straightforward to prove; in fact, they can already be proved using Key Idea 6.2 (see Exercises 6.1.1, 6.1.3, and 6.1.4). Proving Axiom 3, on the other hand, is quite involved, with a key step coming from the Noether Normalization Theorem; we undertake this in Section 6.6.

Exercises for Section 6.1

6.1.1 Assuming Key Idea 6.2, prove that the dimension of a single point is zero.

6.1.2 Assuming Key Idea 6.2, prove that $\dim(\mathbb{A}^n) \geq n$.

6.1.3 Let $X, Y \subseteq \mathbb{A}^n$ be affine varieties with $X \subseteq Y$.
(a) Let $f_1, \ldots, f_m \in K[x_1, \ldots, x_n]$ be polynomials such that
$$f_1|_X, \ldots, f_m|_X \in K[X]$$
are algebraically independent over K. Prove that
$$f_1|_Y, \ldots, f_m|_Y \in K[Y]$$
are algebraically independent over K.
(b) Assume that, using Key Idea 6.2, $\dim(X) = d_1$ and $\dim(Y) = d_2$. Prove that $d_1 \leq d_2$.

6.1.4 Let $X_1, \ldots, X_m \subseteq \mathbb{A}^n$ be affine varieties and let $X = X_1 \cup \cdots \cup X_m$.
(a) Let $f_1, \ldots, f_\ell \in K[x_1, \ldots, x_n]$ be polynomials such that
$$f_1|_X, \ldots, f_\ell|_X \in K[X]$$
are algebraically independent over K. Prove that
$$f_1|_{X_i}, \ldots, f_\ell|_{X_i} \in K[X_i]$$
are algebraically independent over K for some i.
(b) Assume that, using Key Idea 6.2, $\dim(X_i) = d_i$ for $i = 1, \ldots, m$. Prove that
$$\dim(X) = \max\{d_1, \ldots, d_m\}.$$

(Hint: One inequality uses (a) and the other uses the previous exercise.)

6.1.5 Let D be a dimension function. Prove that $D(\mathbb{A}^n) = n$.

6.1.6 Let $\ell_1, \ldots, \ell_m$ be linear homogeneous polynomials in n variables and consider the linear subspace $V \subseteq K^n$ defined by their vanishing:
$$V = \mathcal{V}(\ell_1, \ldots, \ell_m) \subseteq \mathbb{A}^n = K^n.$$

(a) Interpret V as the kernel of a specific matrix M, and use the Rank-Nullity Theorem to write the dimension of V in terms of n and $\mathrm{rk}(M)$.

(b) Let ℓ be another homogeneous linear equation and consider the subspace
$$W = V \cap \mathcal{V}(\ell).$$
Interpret W as the kernel of a specific matrix M'. How are M and M' related?

(c) Using the relationship between M and M', describe how the dimensions of V and W are related? How does this compare to Axiom 3 in Definition 6.3?

(Hint: For (c), there are two possible cases to consider; what are they?)

6.1.7 Let $X \subseteq \mathbb{A}^n$ be an irreducible affine variety that is not a single point. Prove that there exists $f \in K[x_1, \ldots, x_n]$ such that
$$\emptyset \subsetneq X \cap \mathcal{V}(f) \subsetneq X.$$

(This exercise is useful for the next one.)

6.1.8 Let D_1 and D_2 be dimension functions.

(a) Prove that $D_1(X) = 0$ if and only if X is a finite union of points. Conclude that $D_1(X) = D_2(X)$ whenever $D_1(X) = 0$.

(b) Assume that $m \geq 1$ and that $D_1(X) = D_2(X)$ when $D_1(X) < m$. Prove that $D_1(X) = D_2(X)$ for all X with $D_1(X) = m$.

(c) Combine (a) and (b) into an inductive proof of Proposition 6.4.

6.1.9 This exercise proves that a dimension function does not necessarily exist when K is not algebraically closed. Let $K = \mathbb{R}$ and, toward a contradiction, assume that a dimension function D exists.

(a) Prove that $D(\mathbb{A}^2) = 2$ and $D(\{(0,0)\}) = 0$.

(b) Prove that $\{(0,0)\} = \mathcal{V}(f) \cap \mathbb{A}^2$ for some f.

(c) Argue that parts (a) and (b) contradict Axiom 3 in Definition 6.3.

Open Access This chapter is licensed under the terms of the Creative Commons Attribution-NonCommercial 4.0 International License (http://creativecommons.org/licenses/by-nc/4.0/), which permits any noncommercial use, sharing, adaptation, distribution and reproduction in any medium or format, as long as you give appropriate credit to the original author(s) and the source, provide a link to the Creative Commons license and indicate if changes were made.

The images or other third party material in this chapter are included in the chapter's Creative Commons license, unless indicated otherwise in a credit line to the material. If material is not included in the chapter's Creative Commons license and your intended use is not permitted by statutory regulation or exceeds the permitted use, you will need to obtain permission directly from the copyright holder.

Section 6.2 Function fields

As we learned in the last section, the dimension of an affine variety should be a measure of the maximum number of algebraically independent elements in its coordinate ring. However, working with algebraic (in)dependence in K-algebras can be rather difficult, and it is much easier to work with these notions in the context of field extensions of K. Therefore, in this section, we study a way of passing from an irreducible affine variety X to a corresponding field $K(X)$, its *function field*.

6.5 DEFINITION *Function field*

Let X be an irreducible affine variety. The *function field* of X, denoted $K(X)$, is the fraction field of its coordinate ring:

$$K(X) = \mathrm{Frac}(K[X]).$$

The elements of $K(X)$ are called *rational functions on X*.

Recall that fraction fields were defined in Section 0.6, where we learned that elements of $K(X)$ are ratios of the form f/g with $f, g \in K[X]$ and $g \neq 0$, and fractions f_1/g_1 and f_2/g_2 are equal if and only if $f_1 g_2 = f_2 g_1 \in K[X]$. The fraction field is, indeed, a field, with the operations of addition and multiplication defined by

$$\frac{f_1}{g_1} + \frac{f_2}{g_2} = \frac{f_1 g_2 + f_2 g_1}{g_1 g_2} \quad \text{and} \quad \frac{f_1}{g_1} \cdot \frac{f_2}{g_2} = \frac{f_1 f_2}{g_1 g_2}.$$

As fraction fields are defined only for integral domains (Exercise 0.6.4), function fields are defined only for irreducible affine varieties. Let us turn to a few examples.

6.6 EXAMPLE *Function field of affine space*

Since the coordinate ring of $\mathbb{A}^n$ is $K[x_1, \ldots, x_n]$, the function field is the usual field of rational functions

$$K(\mathbb{A}^n) = K(x_1, \ldots, x_n),$$

which we already encountered in Section 0.6.

6.7 EXAMPLE *Function field of $\mathcal{V}(x^2 - y^3)$*

Let $X = \mathcal{V}(x^2 - y^3) \subseteq \mathbb{A}^2$. Computing the coordinate ring, we have

$$K[X] = K[x,y]/\langle x^2 - y^3 \rangle \quad \Longrightarrow \quad K(X) = \mathrm{Frac}(K[x,y]/\langle x^2 - y^3 \rangle).$$

This looks like a rather complicated field; however, we can identify it with a more familiar one. Consider the K-algebra homomorphism

$$\varphi : K[X] \to K[t]$$
$$[f(x,y)] \mapsto f(t^3, t^2).$$

This is the pullback of the polynomial map defined in Example 4.8. Notice that φ is an injection, but not an isomorphism, because the polynomial t is not in the image.

If we pass to function fields, then we can extend φ to a field homomorphism

$$\overline{\varphi} : K(X) \to K(t)$$
$$\frac{[f(x,y)]}{[g(x,y)]} \mapsto \frac{f(t^3, t^2)}{g(t^3, t^2)}.$$

Now that we are allowed to divide, we see that t actually lies in the image of $\overline{\varphi}$:

$$t = \frac{\varphi([x])}{\varphi([y])} = \overline{\varphi}\left(\frac{[x]}{[y]}\right).$$

This allows us to invert $\overline{\varphi}$:

$$\overline{\varphi}^{-1}\left(\frac{f(t)}{g(t)}\right) = \frac{f([x]/[y])}{g([x]/[y])} = \frac{[y^{\max(\deg(f),\deg(g))}]f(x/y)]}{[y^{\max(\deg(f),\deg(g))}]g(x/y)]},$$

where the second equality simply serves to clear denominators in f and g so that the image is explicitly expressed as an element of $K(X)$. It follows that $K(X) \cong K(t)$. We encourage the reader to check the details of this argument in Exercise 6.2.3.

What we can conclude from this example is that, even though X and $\mathbb{A}^1$ are not isomorphic, their function fields are, as they are both isomorphic to $K(t)$. Thus, even though the coordinate ring knows everything about the isomorphism class of an affine variety, some of that information is lost upon passing to the function field.

Although affine varieties with isomorphic function fields may not be isomorphic, in Chapter 12 we will learn that they are "birational."

In the previous two examples, both of the function fields we considered were isomorphic to the usual field of rational functions $K(x_1, \ldots, x_d)$ for some d. If an irreducible variety has a function field isomorphic to the usual field of rational functions, we say that the variety is *rational*. It turns out that rational varieties are actually quite special—it is a general fact beyond the scope of this text that "most" varieties are not rational. In particular, for a "random" polynomial $f \in K[x_1, \ldots, x_n]$ of sufficiently large degree, $\mathcal{V}(f)$ will not be rational. The next example illustrates perhaps the simplest example of a non-rational variety.

6.8 EXAMPLE $\mathcal{V}(x^3 + y^3 + 1)$ is not rational

Set $K = \mathbb{C}$ and let $X = \mathcal{V}(x^3 + y^3 + 1) \subseteq \mathbb{A}^2$. The real points of X are depicted in the image to the right. We argue that X is not rational. This argument hinges on the fact that, if $f_1, f_2, f_3 \in K[x_1, \ldots, x_n]$ are nonzero such that (i) no two of them share a common irreducible factor and (ii) $f_1^3 + f_2^3 + f_3^3 = 0$, then it must be the case that $f_1, f_2,$ and f_3 are all constant (see Exercise 6.2.4 for a proof).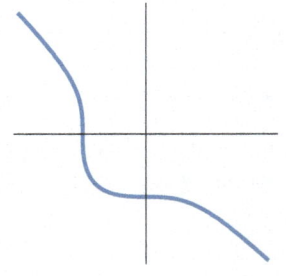

To argue that X is not rational, first notice that $x^3 + y^3 + 1$ is irreducible, so

$$K[X] = K[x,y]/\langle x^3 + y^3 + 1\rangle \implies K(X) = \operatorname{Frac}(K[x,y]/\langle x^3 + y^3 + 1\rangle).$$

6.2. FUNCTION FIELDS

Suppose that
$$\varphi : K(X) \to K(x_1, \ldots, x_d)$$
is a field homomorphism for some $d > 0$; we prove that $\text{im}(\varphi) \subseteq K$, so φ cannot be an isomorphism. Notice that $\varphi([x]) = f/g$ and $\varphi([y]) = h/k$ for some polynomials $f, g, h, k \in K[x_1, \ldots, x_d]$ with $g, k \neq 0$. Since $K[x_1, \ldots, x_n]$ is a UFD, we may factor each polynomial into irreducibles and reduce the quotients; in other words, we can assume that neither f and g nor h and k share common irreducible factors. As $[x^3 + y^3 + 1] = 0 \in K(X)$, we obtain the relation

$$0 = \varphi([x^3 + y^3 + 1]) = \left(\frac{f}{g}\right)^3 + \left(\frac{h}{k}\right)^3 + 1.$$

Clearing denominators gives

(6.9) $$(fk)^3 + (gh)^3 + (gk)^3 = 0.$$

Equation (6.9) tells us that $g^3 \mid (fk)^3$, but since f and g are assumed not to have any common irreducible factors, it follows that $g^3 \mid k^3$. A parallel argument tells us that $k^3 \mid g^3$. Thus, since $g^3 \mid k^3$ and $k^3 \mid g^3$, it follows that $g^3 = ak^3$ for some constant a. Substituting this into equation (6.9) and canceling k^3, we arrive at the relation

(6.10) $$f^3 + ah^3 + g^3 = 0.$$

Notice that the irreducible factors of g are the same as those of k, and thus distinct from both f and h, by assumption. Furthermore, f and h cannot have an irreducible factor in common, or else, by equation (6.10), this would also be an irreducible factor of g, contradicting our assumptions regarding f and g. Therefore, no two of the terms in equation (6.10) share a common irreducible factor, and we may apply Exercise 6.2.4 to conclude that f, g, h, and k are all constant, as claimed.

The argument above shows that $K(X)$ is not isomorphic to $K(x_1, \ldots, x_d)$ for any $d > 0$, but what if $d = 0$? Given that X has more than one point, we know that $K \subsetneq K[X]$, implying that $K \subsetneq K(X)$ and hence $K(X)$ is not isomorphic to K. Taking this together with the previous argument, we conclude that X is not rational.

○───○

We conclude this section with a brief comment on terminology: even though elements of $K(X)$ are called *rational functions*, they are not actually functions on X. In particular, given an element $f/g \in K(X)$ with $g \neq 0$, it is possible that $g(a) = 0$ for some (but not all) values $a \in X$. Thus, f/g only defines a function on the subset of X where $g \neq 0$. It is common to use dashed arrows for rational functions to remind the reader that they are not defined on the entire domain:

$$\frac{f}{g} : X \dashrightarrow K.$$

If we restrict the domain to the complement of $\mathcal{V}(g)$, then every rational function gives rise to an actual function that assigns a value to each element of the domain:

$$\frac{f}{g} : X \setminus \mathcal{V}(g) \longrightarrow K.$$

This concept and notation will be vastly generalized in Section 12.1.

Exercises for Section 6.2

6.2.1 Let $X = \mathcal{V}(xy - 1) \subseteq \mathbb{A}^2$. Prove that $K(X) \cong K(t)$.

6.2.2 Let R and S be integral domains and $\varphi : R \to S$ an injective ring homomorphism. Prove that there exists a well-defined field homomorphism
$$\overline{\varphi} : \operatorname{Frac}(R) \to \operatorname{Frac}(S)$$
$$\frac{a}{b} \mapsto \frac{\varphi(a)}{\varphi(b)}.$$

6.2.3 Let φ and $\overline{\varphi}$ be the homomorphisms of Example 6.7.
  (a) Prove that φ is injective, and use this to prove that $\overline{\varphi}$ is well-defined.
  (b) Prove that $\overline{\varphi}$ and $\overline{\varphi}^{-1}$ are, in fact, inverse field homomorphisms.

6.2.4 Assume that $\operatorname{char}(K) \neq 3$, and let $f_1, f_2, f_3 \in K[x_1, \ldots, x_n]$ be nonzero polynomials such that (i) no two of them share a common irreducible factor and (ii) $f_1^3 + f_2^3 + f_3^3 = 0$. This exercise outlines a proof of the fact that f_1, f_2, and f_3 must all be constant.
  (a) Toward a contradiction, assume that f_1 is nonconstant in x_1 and let f_i' denote the derivative of f_i with respect to x_1. Prove that
  $$f_1^2(f_1'f_3 - f_1f_3') = f_2^2(f_2f_3' - f_2'f_3).$$
  (Hint: Start by differentiating $f_1^3 + f_2^3 + f_3^3 = 0$ using the chain rule.)
  (b) Prove that $f_1'f_3 - f_1f_3' \neq 0$ and conclude that $f_2f_3' - f_2'f_3 \neq 0$.
  (Hint: Apply the quotient rule to differentiate f_1/f_3.)
  (c) Prove that $f_1^2 \mid (f_2f_3' - f_2'f_3)$ and conclude that
  $$2\deg_{x_1}(f_1) \leq \deg_{x_1}(f_2) + \deg_{x_1}(f_3) - 1.$$
  (d) Prove that the procedure in (a) – (c) can be repeated two more times with f_1 replaced by f_2 and f_3. What inequalities do you obtain?
  (e) Add the three inequalities to find a contradiction.

6.2.5 Assume $\operatorname{char}(K) \neq 2$ and let $X = \mathcal{V}(x^2 + y^2 - 1) \subseteq \mathbb{A}^2$. Prove that the map
$$\varphi : K(X) \to K(t)$$
defined by
$$\varphi\left(\frac{[f(x,y)]}{[g(x,y)]}\right) = \frac{f\left(\frac{1-t^2}{1+t^2}, \frac{2t}{1+t^2}\right)}{g\left(\frac{1-t^2}{1+t^2}, \frac{2t}{1+t^2}\right)}$$
is a well-defined field isomorphism.

Section 6.3 Transcendence bases

As we discussed in Section 6.1, the dimension of an affine variety X should be equal to the maximum number of algebraically independent elements in $K[X]$. As it turns out, studying algebraic (in)dependence is more straightforward in the field setting, which is why, in Section 6.2, we introduced the function field $K(X)$ associated to an irreducible affine variety. Given an irreducible affine variety, there is a natural set of inclusions

$$K \subseteq K[X] \subseteq K(X),$$

and algebraically independent elements in $K[X]$ remain algebraically independent in $K(X)$. With this motivation, we now turn to studying algebraic (in)dependence in the setting of field extensions of the form $K \subseteq K(X)$.

To help us discuss field extensions, we introduce the notion of field generators.

6.11 DEFINITION *Field generators*

Let $K \subseteq L$ be fields and let $\mathcal{S} \subseteq L$ be a subset. The *field extension of K generated by $\mathcal{S}$* is

$$K(\mathcal{S}) = \{ab^{-1} : a, b \in K[\mathcal{S}],\ b \neq 0\} \subseteq L.$$

We say that L is a *finitely-generated field extension* of K if $L = K(\mathcal{S})$ for a finite set $\mathcal{S} \subseteq L$.

Notice that $K(\mathcal{S})$ is the smallest subfield of L that contains both K and $\mathcal{S}$ (Exercise 6.3.1), and it is canonically isomorphic to $\mathrm{Frac}(K[\mathcal{S}])$, where $K[\mathcal{S}]$ is the K-algebra generated by $\mathcal{S}$. In addition, we have (Exercise 6.3.2)

$$K(\mathcal{S}_1)(\mathcal{S}_2) = K(\mathcal{S}_1 \cup \mathcal{S}_2).$$

When $\mathcal{S} = \{a_1, \ldots, a_n\}$, we omit the set brackets and write $K(\mathcal{S}) = K(a_1, \ldots, a_n)$.

6.12 EXAMPLE Fields of rational functions

For our purposes, the prototype of a field extension of K is the field of rational functions, which was introduced in Section 0.6:

$$K \subseteq K(x_1, \ldots, x_n),$$

As the notation suggests, $K(x_1, \ldots, x_n)$ is finitely generated over K by $x_1, \ldots, x_n$.

6.13 EXAMPLE Function fields

Let $X \subseteq \mathbb{A}^n$ be an irreducible affine variety with coordinate ring $K[X]$ and function field $K(X)$. The coordinate functions $[x_1], \ldots, [x_n] \in K[X]$ can naturally be viewed as elements of $K(X)$, and every element of $K(X)$ can be written as

$$\frac{[f(x_1, \ldots, x_n)]}{[g(x_1, \ldots, x_n)]} = \frac{f([x_1], \ldots, [x_n])}{g([x_1], \ldots, [x_n])}$$

for some $f, g \in K[x_1, \ldots, x_n]$ with $[g] \neq 0$. By definition, the final expression lies in $K([x_1], \ldots, [x_n])$, implying that $K(X)$ is finitely generated by $[x_1], \ldots, [x_n]$.

We now aim to define the key notion of transcendence bases, which is an analogue in the field-extension setting of vector-space bases. Recall that a basis of a vector space must satisfy two conditions: (i) it must be linearly independent and (ii) it must span the vector space. The first condition places an upper bound on the size of a basis—if you have too many elements in a subset of a vector space, then they will be linearly dependent—and the second condition places a lower bound on the size of a basis—if you have too few elements, then they will not span the entire vector space. In the same way, transcendence bases are required to satisfy two properties and, as we will see in the next section, the first places an upper bound on the size of transcendence bases and the second places a lower bound.

6.14 **DEFINITION** *Transcendence basis*

Let $K \subseteq L$ be fields. A subset $S \subseteq L$ is a *transcendence basis* of L over K if it satisfies the following two conditions:

(i) S is algebraically independent over K, and

(ii) L is algebraic over $K(S)$.

By clearing denominators, (ii) is equivalent to the seemingly stronger condition that L is algebraic over the K-algebra $K[S]$ (Exercise 6.3.3).

Before discussing concrete examples of transcendence bases, it will be helpful to become familiar with a few properties regarding algebraic field extensions like the one appearing in condition (ii) of Definition 6.14. The next result, which is the field analogue of Theorem 5.27, is the key tool that we require.

6.15 **PROPOSITION** *Extending fields by algebraic elements*

Let $K \subseteq L$ be fields and let $a_1, \ldots, a_n \in L$. The following are equivalent:

(i) a_i is algebraic over K for each $i = 1, \ldots, n$;

(ii) $K[a_1, \ldots, a_n]$ is a finite-dimensional vector space over K;

(iii) $K[a_1, \ldots, a_n]$ is algebraic over K;

(iv) $K[a_1, \ldots, a_n] = K(a_1, \ldots, a_n)$.

PROOF The equivalence of conditions (i) – (iii) is a special case of Theorem 5.27 when $R = K$ is a field. In this setting, algebraicity is equivalent to integrality, and finitely-generated modules are the same as finite-dimensional vector spaces.

To prove that (iv) is equivalent to the other three, it suffices to prove that (iv) is equivalent to (iii). That (iv) implies (iii) is essentially the statement of Zariski's Lemma (Lemma 5.41). More specifically, if $K[a_1, \ldots, a_n] = K(a_1, \ldots, a_n)$, then $K[a_1, \ldots, a_n]$ is a field containing K that is finitely generated as a K-algebra, from which Zariski's Lemma tells us that $K[a_1, \ldots, a_n]$ is algebraic over K.

It remains to prove that (iii) implies (iv), so assume that $K[a_1, \ldots, a_n]$ is algebraic over K; we aim to prove that $K[a_1, \ldots, a_n] = K(a_1, \ldots, a_n)$. The inclusion $K[a_1, \ldots, a_n] \subseteq K(a_1, \ldots, a_n)$ follows directly from the definitions, so let us turn toward proving the other inclusion.

6.3. TRANSCENDENCE BASES

Since $K(a_1,\ldots,a_n)$ is the smallest subfield of L containing K and $a_1,\ldots,a_n$, we can prove the inclusion $K(a_1,\ldots,a_n) \subseteq K[a_1,\ldots,a_n]$ by arguing that the ring $K[a_1,\ldots,a_n]$ is actually a field. To do so, suppose $a \in K[a_1,\ldots,a_n]$ is nonzero; we must prove that a has a multiplicative inverse. Using the assumption that $K[a_1,\ldots,a_n]$ is algebraic over K, we know that there is a nontrivial algebraic relation of the form

$$c_d a^d + \cdots + c_1 a + c_0 = 0 \implies c_d a^d + \cdots + c_{i+1} a^{i+1} + c_i a^i = 0,$$

where $c_0,\ldots,c_d \in K$ and c_i is the first nonzero coefficient. Solving for a^i, we obtain

$$a^i = -c_i^{-1}(c_d a^d + \cdots + c_{i+1} a^{i+1}).$$

Canceling the factor of a^i from both sides—which we can do because this computation is happening within the field L—and factoring out a on the right, we conclude that a has an inverse:

$$1 = \left[-c_i^{-1}(c_d a^{d-i-1} + \cdots + c_{i+1})\right] a.$$

This shows that $K[a_1,\ldots,a_n]$ is a field, as desired. $\square$

The one condition in Proposition 6.15 that is not a special case of Theorem 5.27 is condition (iv). We illustrate this condition in the following example.

6.16 EXAMPLE $\mathbb{Q}(\sqrt{2}) = \mathbb{Q}[\sqrt{2}]$

Consider $\sqrt{2} \in \mathbb{R}$. Since $\sqrt{2}$ is a root of the polynomial $x^2 - 2 \in \mathbb{Q}[x]$, we see that $\sqrt{2}$ is algebraic over $\mathbb{Q}$. Thus, Proposition 6.15 asserts that $\mathbb{Q}(\sqrt{2}) = \mathbb{Q}[\sqrt{2}]$, which we now verify.

Using the argument of Example 5.17, we compute that $\mathbb{Q}[\sqrt{2}] = \mathbb{Q}\{1,\sqrt{2}\}$. It then follows from the definition of field generators that every element of $\mathbb{Q}(\sqrt{2})$ has the form

$$\frac{a + b\sqrt{2}}{c + d\sqrt{2}},$$

where $a,b,c,d \in \mathbb{Q}$ with c and d not both zero. We can "rationalize" the quotient:

$$\frac{a + b\sqrt{2}}{c + d\sqrt{2}} = \frac{(a + b\sqrt{2})(c - d\sqrt{2})}{(c + d\sqrt{2})(c - d\sqrt{2})} = \frac{(ac - 2bd) + (bc - ad)\sqrt{2}}{c^2 - 2d^2} = r + q\sqrt{2},$$

where

$$r = \frac{ac - 2bd}{c^2 - 2d^2} \in \mathbb{Q} \quad \text{and} \quad q = \frac{bc - ad}{c^2 - 2d^2} \in \mathbb{Q}.$$

This shows that every element of $\mathbb{Q}(\sqrt{2})$ is actually an element of $\mathbb{Q}[\sqrt{2}]$. Since the other inclusion follows from the definitions, we conclude that $\mathbb{Q}(\sqrt{2}) = \mathbb{Q}[\sqrt{2}]$.

One of the most important consequences of Proposition 6.15 is the next result, the failure of which in the ring setting—see Example 6.18 below—is the essential reason why studying algebraic (in)dependence is so much better behaved in the field setting than it is in the ring setting.

6.17 COROLLARY *Algebraic over algebraic is algebraic*

Let $K \subseteq L \subseteq M$ be finitely-generated field extensions. If L is algebraic over K and M is algebraic over L, then M is algebraic over K.

PROOF By Proposition 6.15, the claim is equivalent to proving that, if L is a finite-dimensional vector space over K, and M is a finite-dimensional vector space over L, then M is a finite-dimensional vector space over K. This follows from the observation (see Exercise 5.2.6) that, if $L = K\{a_1, \ldots, a_m\}$ and $M = L\{b_1, \ldots, b_n\}$, then
$$M = K\{a_i b_j \mid 1 \leq i \leq m,\ 1 \leq j \leq n\}. \qquad \square$$

The next example illustrates the failure of Corollary 6.17 in the ring setting.

6.18 EXAMPLE *Corollary 6.17 fails for ring extensions*

Consider the ring extensions
$$\mathbb{C} \subseteq \frac{\mathbb{C}[x]}{\langle x^2 - 1 \rangle} \subseteq \frac{\mathbb{C}[x,y]}{\langle x^2 - 1, xy - y \rangle}.$$

We show that the second ring is algebraic over the first, and that the third ring is algebraic over the second, but that the third ring is not algebraic over the first.

Beginning with the first extension, notice that, by using the relation $[x^2] = 1$, any element of $\mathbb{C}[x]/\langle x^2 - 1\rangle$ can be written as $[ax + b]$ for some $a, b \in \mathbb{C}$. A direct computation shows that such an element $[ax + b]$ is a root of the polynomial
$$f(z) = z^2 - 2bz + b^2 - a^2 \in \mathbb{C}[z],$$
proving that the first extension is algebraic.

To show that the second extension is algebraic, notice that, by using the relations $[x^2] = 1$ and $[xy] = [y]$, any element of $\mathbb{C}[x,y]/\langle x^2 - 1, xy - y\rangle$ can be written as $[ax + b + yg(y)]$ for some $a, b \in \mathbb{C}$ and $g(y) \in \mathbb{C}[y]$. A direct computation shows that such an element $[ax + b + yg(y)]$ is a root of the polynomial
$$f(z) = [x - 1]z + [ax - bx + b - a] \in \left(\mathbb{C}[x]/\langle x^2 - 1\rangle\right)[z],$$
proving that the second extension is also algebraic.

However, even though both extensions are algebraic, the extension
$$\mathbb{C} \subseteq \frac{\mathbb{C}[x,y]}{\langle x^2 - 1, xy - y \rangle}$$
is not algebraic. To convince ourselves of this, it is enough to argue that $[y]$ is not algebraic over $\mathbb{C}$. Indeed, every nonzero element of $\langle x^2 - 1, xy - y\rangle \subseteq \mathbb{C}[x,y]$ is necessarily a multiple of $x - 1$, and therefore, given any nonzero polynomial $f(z) \in \mathbb{C}[z]$, it will never be the case that
$$f([y]) = [f(y)] = 0 \in \frac{\mathbb{C}[x,y]}{\langle x^2 - 1, xy - y \rangle}.$$

6.3. TRANSCENDENCE BASES

While Corollary 6.17 fails over rings in general, by passing to fraction fields, it still holds in the special setting of integral domains (Exercise 6.3.4).

We now close this section with several concrete examples to illustrate the notion of transcendence bases. The examples range from the empty example to a long-standing open problem concerning transcendental numbers in $\mathbb{R}$.

6.19 EXAMPLE Transcendence bases of algebraic extensions

A field extension $K \subseteq L$ has an empty transcendence basis if and only if L is algebraic over K. To verify this, first notice that $\emptyset \subseteq L$ is algebraically independent for vacuous reasons: since $\emptyset$ does not contain any elements, there cannot be an algebraic relation among them. Thus, by definition, $\emptyset \subseteq L$ is a transcendence basis if and only if L is algebraic over $K(\emptyset) = K$.

6.20 EXAMPLE Transcendence basis for the unit sphere

Let $K = \mathbb{C}$ and let $X = \mathcal{V}(x^2 + y^2 + z^2 - 1) \subseteq \mathbb{A}^3$. We verify that $\{[x], [y]\}$ is a transcendence basis of the field extension $\mathbb{C} \subseteq \mathbb{C}(X)$. To do this, we have two conditions to check.

First, we verify that $\{[x], [y]\}$ is algebraically independent over $\mathbb{C}$. Suppose that there is a polynomial relation $0 = f([x], [y])$; we must prove that $f = 0$. Since $0 = f([x], [y]) = [f(x, y)] \in \mathbb{C}[X]$, we have

$$f(x, y) \in \mathcal{I}(X) = \langle x^2 + y^2 + z^2 - 1 \rangle.$$

However, since every nonzero element of $\langle x^2 + y^2 + z^2 - 1 \rangle$ has positive degree in z and $f(x, y)$ does not, this implies that $f(x, y)$ must be the zero polynomial, and we conclude that $\{[x], [y]\}$ is algebraically independent over $\mathbb{C}$.

In order to prove that $\mathbb{C}(X)$ is algebraic over $\mathbb{C}([x], [y])$, it suffices—by Proposition 6.15 and the observation that $\mathbb{C}(X) = \mathbb{C}([x], [y], [z])$—to prove that $[z]$ is algebraic over $\mathbb{C}([x], [y])$. This follows from the observation that $[z]$ is a zero of the polynomial

$$g(w) = w^2 + ([x]^2 + [y]^2 - 1) \in \mathbb{C}([x], [y])[w].$$

6.21 EXAMPLE Transcendence basis of $\mathcal{V}(f)$

Generalizing the previous example, if $X = \mathcal{V}(f)$ for some irreducible polynomial $f \in K[x_1, \ldots, x_n]$ that is not contained in $K[x_1, \ldots, x_{n-1}]$, a similar argument as that used in the previous example shows (Exercise 6.3.5) that $\{[x_1], \ldots, [x_{n-1}]\}$ is a transcendence basis of the field extension

$$K \subseteq K(X) = \mathrm{Frac}(K[x_1, \ldots, x_n]/\langle f \rangle)$$

6.22 EXAMPLE π and e

It is a long-standing open problem to determine whether π and e are algebraically independent over $\mathbb{Q}$. If they are algebraically independent, then $\{\pi, e\}$ is a transcendence basis of $\mathbb{Q}(\pi, e)$ over $\mathbb{Q}$; otherwise, each of the single-element sets $\{\pi\}$ and $\{e\}$ would be a transcendence basis of $\mathbb{Q}(\pi, e)$ over $\mathbb{Q}$.

Now that we have seen a few examples of transcendence bases, some natural questions arise. Do transcendence bases exist for all field extensions? If so, what can be said about the size of transcendence bases? In the next section, we prove that, if $K \subseteq L$ is a finitely-generated field extension, then finite transcendence bases exist, and they all have the same size. This will allow us to define the transcendence degree of such a field extension as the size of any transcendence basis, similarly to the way one can define the dimension of a vector space. The notion of transcendence degree will be the key to making a rigorous definition of dimension of affine varieties in Section 6.5.

Exercises for Section 6.3

6.3.1 Let $K \subseteq L$ be a field extension and let $\mathcal{S} \subseteq L$ be a subset.
   (a) Prove that $K(\mathcal{S})$ is a subfield of L.
   (b) If M is a subfield of L that contains K and $\mathcal{S}$, prove that $K(\mathcal{S}) \subseteq M$.

6.3.2 Let $K \subseteq L$ be a field extension and let $\mathcal{S}_1, \mathcal{S}_2 \subseteq L$ be subsets. Prove that
$$K(\mathcal{S}_1)(\mathcal{S}_2) = K(\mathcal{S}_1 \cup \mathcal{S}_2).$$

6.3.3 Let $K \subseteq L$ be a field extension and let $\mathcal{S} \subseteq L$ be a set. Prove that L is algebraic over $K(\mathcal{S})$ if and only if L is algebraic over $K[\mathcal{S}]$.

6.3.4 Let K be a field and suppose that $A \subseteq B$ are finitely-generated K-algebras such that A is algebraic over K and B is algebraic over A. Assuming that A and B are both integral domains, prove that B is algebraic over K.
   (Hint: Use fraction fields.)

6.3.5 Let $f \in K[x_1, \ldots, x_n]$ be an irreducible polynomial that is not an element of $K[x_1, \ldots, x_{n-1}]$ and consider the irreducible affine variety $X = \mathcal{V}(f)$. Prove that $\{[x_1], \ldots, [x_{n-1}]\}$ is a transcendence basis of $K(X)$ over K.

6.3.6 Let A be a finitely-generated K-algebra that is also an integral domain. Prove that every Noether basis of A over K is also a transcendence basis of $\mathrm{Frac}(A)$ over K.

Section 6.4 Transcendence degree

In the last section, we familiarized ourselves with transcendence bases. In many ways, transcendence bases of field extensions are similar to bases of vector spaces. Importantly, bases of vector spaces provide a way of measuring how big a vector space is: all bases have the same size and the dimension of a vector space is the size of any basis. In this section, we develop the analogous notions for transcendence bases. In particular, we show that every finitely-generated field extension has a finite transcendence basis and that all such bases have the same size. This allows us to define the *transcendence degree* of a field extension $K \subseteq L$, which is an algebraic (as opposed to merely linear-algebraic) measure of the size of L relative to K.

We begin by proving the existence of finite transcendence bases.

6.23 PROPOSITION *Existence of finite transcendence bases*

Let $K \subseteq L$ be fields and $a_1, \ldots, a_n \in L$. If L is algebraic over $K(a_1, \ldots, a_n)$, then $\{a_1, \ldots, a_n\}$ contains a transcendence basis of L over K. In particular, finite transcendence bases exist for all finitely-generated field extensions.

PROOF We prove the result by induction on n. If $n = 0$, then L is algebraic over K and $\emptyset$ is a transcendence basis (Example 6.19), proving the base case.

Suppose the result holds for fewer than n elements, and let $a_1, \ldots, a_n \in L$ be such that L is algebraic over $K(a_1, \ldots, a_n)$. If $a_1, \ldots, a_n$ are algebraically independent, then they form a transcendence basis of L over K, and we're done. Otherwise, there exists a nonzero polynomial $f \in K[x_1, \ldots, x_n]$ such that $f(a_1, \ldots, a_n) = 0$. Among all such f, choose one that has positive degree in the minimal number of variables. Since f must have positive degree in at least one x_i, assume without loss of generality that it has positive degree in x_n. If we write $f = \sum g_i x_n^i$ with $g_i \in K[x_1, \ldots, x_{n-1}]$, then at least one g_i must be nonzero, and furthermore, any nonzero g_i cannot vanish at $(a_1, \ldots, a_{n-1})$, as this would violate the minimality in the choice of f. It follows that the single-variable polynomial

$$f(a_1, \ldots, a_{n-1}, x_n) \in K[a_1, \ldots, a_{n-1}][x_n]$$

is nonzero and vanishes at a_n. Therefore, a_n is algebraic over $K[a_1, \ldots, a_{n-1}]$, and thus algebraic over $K(a_1, \ldots, a_{n-1})$. By Proposition 6.15, we then see that

$$K(a_1, \ldots, a_{n-1}) \subseteq K(a_1, \ldots, a_n).$$

is an algebraic field extension. We now have a sequence of algebraic extensions

$$K(a_1, \ldots, a_{n-1}) \subseteq K(a_1, \ldots, a_n) \subseteq L,$$

and Corollary 6.17 implies that $K(a_1, \ldots, a_{n-1}) \subseteq L$ is algebraic. By the induction hypothesis, $\{a_1, \ldots, a_{n-1}\}$ contains a transcendence basis of L over K, showing that $\{a_1, \ldots, a_n\}$ contains a transcendence basis of L over K, and completing the induction argument.

To justify the final assertion in the lemma, suppose that L is finitely generated over K. Then L is equal to—and thus algebraic over—$K(a_1, \ldots, a_n)$ for some $a_1, \ldots, a_n \in L$, so $\{a_1, \ldots, a_n\}$ contains a finite transcendence basis. □

> With a little more work and the Axiom of Choice, one can prove that (possibly infinite) transcendence bases exist for all field extensions.

Our next task is to prove that every transcendence basis has the same size. The next lemma, named after the strategy used in the proof, is the key result that we require.

6.24 LEMMA *The exchange lemma*

Let $K \subseteq L$ be fields and let $a_1, \ldots, a_n \in L$ be distinct elements of L such that L is algebraic over $K(a_1, \ldots, a_n)$. If a set $\mathcal{S} \subseteq L$ is algebraically independent over K, then $\mathcal{S}$ is finite and $|\mathcal{S}| \leq n$.

PROOF Suppose that $a_1, \ldots, a_n \in L$ are distinct elements of L with L algebraic over $K(a_1, \ldots, a_n)$, and let $\mathcal{S} \subseteq L$ be algebraically independent over K. Toward a contradiction, suppose that $|\mathcal{S}| > n$ (this includes the possibility that $\mathcal{S}$ is infinite). To derive a contradiction, we recursively "exchange" elements in $\{a_1, \ldots, a_n\}$ for elements of $\mathcal{S}$. More specifically, at the completion of the kth step in this process, we will have chosen distinct elements $b_1, \ldots, b_k \in \mathcal{S}$ such that—after possibly reordering $a_1, \ldots, a_n$ if necessary—L is algebraic over $K(b_1, \ldots, b_k, a_{k+1}, \ldots, a_n)$. As we will see, the contradiction arises after the nth step of this process. We now describe the kth step of this recursive exchange process precisely.

Suppose that, for some $k \in \{0, \ldots, n\}$, we have already found distinct elements $b_1, \ldots, b_{k-1} \in \mathcal{S}$ such that—after possible reordering $a_1, \ldots, a_n$ if necessary—L is algebraic over $K(b_1, \ldots, b_{k-1}, a_k, \ldots, a_n)$. Since $|\mathcal{S}| > n \geq k$, we may choose another element $b_k \in \mathcal{S} \setminus \{b_1, \ldots, b_{k-1}\}$. Since L is algebraic over the field $K(b_1, \ldots, b_{k-1}, a_k, \ldots, a_n)$, it follows that L is also algebraic over the algebra $K[b_1, \ldots, b_{k-1}, a_k, \ldots, a_n]$ (Exercise 6.3.3). Since $b_k \in L$, this implies that there exists a nonzero polynomial $f \in K[x_1, \ldots, x_k, y_k, \ldots, y_n]$ such that

$$f(b_1, \ldots, b_k, a_k, \ldots, a_n) = 0.$$

Among all such f, choose one that has positive degree in the minimal number of variables $y_k, \ldots, y_n$. Since $b_1, \ldots, b_k$ are distinct elements of the algebraically independent set $\mathcal{S}$, it follows that f must have positive degree in at least one of the variables $y_k, \ldots, y_n$; after possibly relabeling indices, suppose that f has positive degree in y_k. Viewing f as a single-variable polynomial in y_k with coefficients in $K[x_1, \ldots, x_k, y_{k+1}, \ldots, y_n]$, notice that none of its nonzero coefficients can vanish when evaluated at $(b_1, \ldots, b_k, a_{k+1}, \ldots, a_n)$, as this would contradict the minimality in the choice of f. It then follows that

$$f(b_1, \ldots, b_k, y_k, a_{k+1}, \ldots, a_n) \in K[b_1, \ldots, b_k, a_{k+1}, \ldots, a_n][y_k]$$

is a nonzero polynomial that vanishes at a_k, implying that a_k is algebraic over $K[b_1, \ldots, b_k, a_{k+1}, \ldots, a_n]$, and thus also algebraic over $K(b_1, \ldots, b_k, a_{k+1}, \ldots, a_n)$. Therefore, each field in the following chain is algebraic over the prior one:

$$K(b_1, \ldots, b_k, a_{k+1}, \ldots, a_n) \subseteq K(b_1, \ldots, b_k, a_k, \ldots, a_n) \subseteq L.$$

Corollary 6.17 then implies that L is algebraic over $K(b_1, \ldots, b_k, a_{k+1}, \ldots, a_n)$, completing the kth step of the exchange process.

6.4. TRANSCENDENCE DEGREE

After completing the nth step in the recursion described above, we obtain distinct elements $b_1, \ldots, b_n \in \mathcal{S}$ such that L is algebraic over $K(b_1, \ldots, b_n)$. By assumption, $|\mathcal{S}| > n$, so there exists $b \in \mathcal{S} \setminus \{b_1, \ldots, b_n\} \subseteq L$, so b is algebraic over $K(b_1, \ldots, b_n)$, and thus also algebraic over $K[b_1, \ldots, b_n]$. But then there exists a nontrivial polynomial relation $f(b_1, \ldots, b_n, b) = 0$, contradicting the algebraic independence of $\mathcal{S}$. The contradiction implies that $\mathcal{S}$ is finite and $|\mathcal{S}| \leq n$. $\square$

The exchange lemma allows us to deduce the following important result.

6.25 PROPOSITION *Transcendence bases all have the same size*

If L is a finitely-generated field extension of K, then every transcendence basis of L over K has the same finite size.

PROOF   Suppose that L is a finitely-generated field extension of K. By Proposition 6.23, L contains a finite transcendence basis, so let $\{a_1, \ldots, a_n\}$ be a transcendence basis of size n. Let $\mathcal{S}$ be any other transcendence basis. Since L is algebraic over $K(a_1, \ldots, a_n)$ and $\mathcal{S}$ is algebraically independent over K, the exchange lemma implies that $\mathcal{S}$ is finite and $|\mathcal{S}| \leq n$. Write $\mathcal{S} = \{b_1, \ldots, b_m\}$ where $b_1, \ldots, b_m \in L$ are distinct and $m \leq n$. Since L is algebraic over $K(b_1, \ldots, b_m)$ and $a_1, \ldots, a_n$ are algebraically independent over K, the exchange lemma implies that $m \geq n$. Thus, we conclude that $|\mathcal{S}| = m = n$. $\square$

We now come to the following central definition, which will be the key to defining and studying the dimension of affine varieties in the remainder of this chapter.

6.26 DEFINITION *Transcendence degree*

Let L be a finitely-generated field extension of K. The *transcendence degree* of L over K, denoted $\mathrm{trdeg}_K(L)$, is the size of any transcendence basis of L over K.

We revisit the examples from the previous section.

6.27 EXAMPLE   Transcendence degree of algebraic extensions

Recall from Example 6.19 that $K \subseteq L$ is algebraic if and only if $\emptyset$ is a transcendence basis. In other words, we see that $K \subseteq L$ is algebraic if and only if $\mathrm{trdeg}_K(L) = 0$. In this sense, algebraic field extensions are the smallest type of field extensions.

6.28 EXAMPLE   Transcendence degree and the unit sphere

Let $K = \mathbb{C}$ and let $X = \mathcal{V}(x^2 + y^2 + z^2 - 1) \subseteq \mathbb{A}^3$. As we saw in Example 6.20, the two-element set $\{[x], [y]\} \subseteq K(X)$ is a transcendence basis over K. Thus, $\mathrm{trdeg}_K K(X) = 2$.

6.29 EXAMPLE   Transcendence degree of $\mathbb{Q}(\pi, e)$

To rephrase the open problem of Example 6.22: as of this writing, it is unknown whether $\mathrm{trdeg}_\mathbb{Q} \mathbb{Q}(\pi, e)$ is 1 or 2.

To conclude this section, we list a few useful properties for bounding transcendence degree. Notice that Proposition 6.23 provides a tool for computing transcendence bases and transcendence degree "from above": every set $\mathcal{S} \subseteq L$ such that L is algebraic over $K(\mathcal{S})$ can be trimmed down to a transcendence basis of L, analogously to the way in which a spanning set of a vector space can be trimmed down to a basis. The next result provides a tool for computing transcendence bases and transcendence degree "from below": if $\mathcal{S} \subseteq L$ is an algebraically independent set, then it can be built up to a transcendence basis of L over K.

6.30 PROPOSITION *Independent sets are contained in bases*

Let $K \subseteq L$ be a finitely-generated field extension. If a subset $\mathcal{S} \subseteq L$ is algebraically independent over K, then $\mathcal{S}$ is finite and contained in a transcendence basis of L over K.

PROOF   Exercise 6.4.1.   □

As a consequence of Propositions 6.23 and 6.30, we obtain the following result that provides a tool for finding upper and lower bounds on transcendence degree.

6.31 COROLLARY *Bounding transcendence degree*

Let $K \subseteq L$ be a finitely-generated field extension and let $\mathcal{S} \subseteq L$ be a subset.
1. If L is algebraic over $K(\mathcal{S})$, then $\mathrm{trdeg}_K(L) \leq |\mathcal{S}|$.
2. If $\mathcal{S}$ is algebraically independent over K, then $\mathrm{trdeg}_K(L) \geq |\mathcal{S}|$.

PROOF   Since transcendence degree is finite, the first assertion is true if $\mathcal{S}$ is infinite, so we can assume that $\mathcal{S}$ is finite. If L is algebraic over $K(\mathcal{S})$, then Proposition 6.23 implies that $\mathcal{S}$ contains a transcendence basis, and it then follows from the definition of transcendence degree that $\mathrm{trdeg}_K(L) \leq |\mathcal{S}|$. Similarly, if $\mathcal{S}$ is algebraically independent over K, then Proposition 6.30 implies that $\mathcal{S}$ is finite and contained in a transcendence basis, and it then follows that $\mathrm{trdeg}_K(L) \geq |\mathcal{S}|$.   □

Exercises for Section 6.4

6.4.1 Prove Proposition 6.30.

6.4.2 Suppose that $L \subseteq M$ are fields that are finitely generated over K. Prove that $\mathrm{trdeg}_K(L) \leq \mathrm{trdeg}_K(M)$.

6.4.3 Compute the transcendence degree of $\mathbb{Q}(\pi, i, \sqrt{2})$ over $\mathbb{Q}$.

6.4.4 Let $K \subseteq L$ be a finitely-generated field extension with $\mathrm{trdeg}_K(L) = d$, and let $a_1, \ldots, a_d \in L$ be distinct elements.
   (a) Prove that $\{a_1, \ldots, a_d\} \subseteq L$ is a transcendence basis of L over K if and only if $a_1, \ldots, a_d$ are algebraically independent over K.
   (b) Prove that $\{a_1, \ldots, a_d\} \subseteq L$ is a transcendence basis of L over K if and only if L is algebraic over $K(a_1, \ldots, a_d)$.

6.4. TRANSCENDENCE DEGREE

6.4.5 Suppose that $K \subseteq L$ is a finitely-generated field extension.
   (a) Prove that $\mathrm{trdeg}_K(L)$ is the maximum number of algebraically independent elements in L.
   (b) Prove that $\mathrm{trdeg}_K(L)$ is the minimum number of elements of L that generate a field over which L is algebraic.

6.4.6 Using the notions of vector-space dimension, span, and linear independence, state the linear-algebra analogues of all of the results in this section, as well as Exercises 6.4.4 and 6.4.5.

Section 6.5 Dimension: definition and first properties

Now that we have familiarized ourselves with the concepts of function fields and transcendence degree, we are ready to state the formal definition of dimension.

> **6.32 DEFINITION** *Dimension of an affine variety*
>
> The *dimension* of an irreducible affine variety $X \subseteq \mathbb{A}^n$ is the transcendence degree of its function field over K:
>
> $$\dim(X) = \mathrm{trdeg}_K(K(X)).$$
>
> The *dimension* of a nonempty reducible affine variety is the maximum of the dimensions of its irreducible components.

There are many terms regarding dimension that are commonly used. For example, varieties of dimension one and of dimension two are commonly referred to as *curves* and *surfaces*, respectively, while varieties of dimension n are often called *n-folds*. The *codimension* of an affine variety $X \subseteq \mathbb{A}^n$ is defined by

$$\mathrm{codim}(X) = n - \dim(X),$$

and a codimension-one variety is commonly referred to as a *hypersurface*. More generally, if $X \subseteq Y \subseteq \mathbb{A}^n$ are affine varieties, then the *codimension of X in Y* is defined by

$$\mathrm{codim}_Y(X) = \dim(Y) - \dim(X).$$

Let us compute dimension in a few concrete examples.

6.33 EXAMPLE Affine space $\mathbb{A}^n$ has dimension n

As we saw in Example 6.6, the function field of $\mathbb{A}^n$ is the field of rational functions:

$$K(\mathbb{A}^n) = K(x_1, \ldots, x_n).$$

Since $x_1, \ldots, x_n \in K(\mathbb{A}^n)$ are algebraically independent and generate $K(\mathbb{A}^n)$, they form a transcendence basis. Thus,

$$\dim(\mathbb{A}^n) = \mathrm{trdeg}_K(K(\mathbb{A}^n)) = n.$$

6.34 EXAMPLE The unit sphere has dimension 2

Let $K = \mathbb{C}$ and let $X = \mathcal{V}(x^2 + y^2 + z^2 - 1) \subseteq \mathbb{A}^3$. As we saw in Example 6.20, the two elements

$$[x], [y] \in \mathbb{C}(X)$$

form a transcendence basis. Thus,

$$\dim(X) = \mathrm{trdeg}_\mathbb{C}(\mathbb{C}(X)) = 2.$$

More generally, it can be shown that an affine variety $X \subseteq \mathbb{A}^n$ can be written as $X = \mathcal{V}(f)$ for some nonzero polynomial $f \in K[x_1, \ldots, x_n]$ if and only if every irreducible component of X has dimension $n - 1$ (Exercise 6.5.4).

6.5. DIMENSION: DEFINITION AND FIRST PROPERTIES

We required the function field in order to ensure that transcendence degree was well-defined; however, now that we have laid the groundwork, we can discuss dimension purely in terms of coordinate rings. The next result describes how we can use the coordinate rings of (possibly reducible) varieties to bound their dimension.

6.35 PROPOSITION *Polynomial functions and dimension*

Let $X \subseteq \mathbb{A}^n$ be a nonempty affine variety and let $\mathcal{S} \subseteq K[X]$.
1. If $\mathcal{S}$ is algebraically independent over K, then $|\mathcal{S}| \leq \dim(X)$.
2. If $K[X]$ is integral over $K[\mathcal{S}]$, then $\dim(X) \leq |\mathcal{S}|$.

PROOF   We prove the first statement and leave the second as Exercise 6.5.5. Let $\mathcal{S} \subseteq K[X]$ be algebraically independent over K. Consider the irreducible decomposition $X = X_1 \cup \cdots \cup X_m$. Restricting the domain of any function $F \in K[X]$ to a component X_j, we obtain a function $F|_{X_j} \in K[X_j]$. Given distinct functions $F_1, \ldots, F_d \in \mathcal{S}$, we claim that the restrictions

$$F_1|_{X_j}, \ldots, F_d|_{X_j} \in K[X_j]$$

are algebraically independent over K for at least one j. To see why, assume this is not the case. Then, for each $j = 1, \ldots, m$, there exists a nonzero polynomial $g_j \in K[x_1, \ldots, x_d]$ such that

$$g_j(F_1, \ldots, F_d)|_{X_j} = g_j(F_1|_{X_j}, \ldots, F_d|_{X_j}) = 0.$$

But given that X is the union of $X_1, \ldots, X_m$, the product $g_1 \cdots g_m \in K[x_1, \ldots, x_d]$ then gives a nontrivial polynomial relation on X:

$$g_1(F_1, \ldots, F_d) \cdots g_m(F_1, \ldots, F_d) = 0 \in K[X],$$

contradicting the assumption that $\mathcal{S}$ is algebraically independent over K.

Thus, given distinct elements $F_1, \ldots, F_d \in \mathcal{S}$, the functions $F_1|_{X_j}, \ldots, F_d|_{X_j}$ are algebraically independent over K for at least one j, and for such a j, it follows that

$$\dim(X) \geq \dim(X_j) = \operatorname{trdeg}_K(K(X_j)) \geq d,$$

where the first inequality and the equality follow from the definition of dimension, while the second inequality follows from Corollary 6.31, Part 2. As this holds for any d distinct elements of $\mathcal{S}$, we conclude that $|\mathcal{S}| \leq \dim(X)$, as claimed. □

As a consequence of Proposition 6.35, we have the following result, which says that every Noether basis of a coordinate ring has the same size, and that this size is equal to the dimension of the affine variety.

6.36 COROLLARY *Noether bases and dimension*

If $X \subseteq \mathbb{A}^n$ is a nonempty affine variety and $\{F_1, \ldots, F_d\} \subseteq K[X]$ is a Noether basis of size d, then $d = \dim(X)$.

PROOF By definition of Noether bases, $F_1, \ldots, F_d$ are algebraically independent over K and $K[X]$ is integral over $K[F_1, \ldots, F_d]$. Thus, the two parts of Proposition 6.35 imply that $d \leq \dim(X)$ and $\dim(X) \leq d$, from which we conclude that $d = \dim(X)$. □

We can also tie our formal discussion of dimension back to the characterization of dimension given in Key Idea 6.2, which is the first part of the next result.

6.37 COROLLARY *Characterizations of dimension*

Let $X \subseteq \mathbb{A}^n$ be a nonempty affine variety.

1. The dimension of X is the maximum number of algebraically independent functions in $K[X]$.
2. The dimension of X is the minimum number of $F_1, \ldots, F_d \in K[X]$ such that $K[X]$ is integral over $K[F_1, \ldots, F_d]$.

PROOF We prove the first part and leave the second to Exercise 6.5.6.

If $\mathcal{S} \subseteq K[X]$ is algebraically independent, then the first part of Proposition 6.35 tells us that $|\mathcal{S}| \leq \dim(X)$. Thus, the maximum number of algebraically independent functions in $K[X]$ is bounded above by $\dim(X)$. On the other hand, Noether normalization tells us that $K[X]$ contains a Noether basis, which, by Corollary 6.36, has size $\dim(X)$. This implies that $K[X]$ contains at least one algebraically independent set of size $\dim(X)$. Taking these together, we conclude that the maximum number of algebraically independent elements in $K[X]$ is equal to $\dim(X)$. □

We now have a number of ways to think about the dimension of an affine variety, and we can use these ideas to prove that dimension satisfies certain natural properties. The first property regards inclusions of affine varieties. Since dimension measures the size of an affine variety, we should certainly expect that $X \subseteq Y$ implies $\dim(X) \leq \dim(Y)$. Less obvious, though, is the fact that this implication becomes strict when we restrict to irreducible affine varieties. In other words, if $Y \subseteq \mathbb{A}^n$ is an irreducible affine variety, then the only affine varieties that are strictly contained in Y must have strictly smaller dimension. This is the content of the next result.

6.38 PROPOSITION *Dimension and inclusions*

If $X \subseteq Y$ are nonempty affine varieties in $\mathbb{A}^n$, then $\dim(X) \leq \dim(Y)$. If, furthermore, Y is irreducible and $X \subsetneq Y$, then $\dim(X) < \dim(Y)$.

PROOF Let $X, Y \subseteq \mathbb{A}^n$ be nonempty affine varieties with $X \subseteq Y$, and define $d = \dim(X)$. By Corollary 6.37, there exists an algebraically independent subset of $K[X]$ of size d. Let $f_1, \ldots, f_d \in K[x_1, \ldots, x_n]$ be polynomials such that $\{f_1|_X, \ldots, f_d|_X\}$ is an algebraically independent subset of $K[X]$. Given that the functions $f_1, \ldots, f_d$ do not satisfy a polynomial relation when restricted to X, they certainly do not satisfy a polynomial relation when restricted to the larger set Y (the reader is encouraged to pause and convince themselves of this assertion). Thus, $\{f_1|_Y, \ldots, f_d|_Y\}$ is an algebraically independent set in $K[Y]$, and by Corollary 6.37, we conclude that $\dim(Y) \geq d = \dim(X)$.

6.5. DIMENSION: DEFINITION AND FIRST PROPERTIES

To prove the second statement, suppose, in addition, that Y is irreducible and that $X \subsetneq Y$. Our aim is to prove that $\dim(X) < \dim(Y)$. Toward a contradiction, assume that $\dim(Y) = \dim(X) = d$. Using the assumption that $X \subsetneq Y$, there exists $f \in K[x_1, \ldots, x_n]$ such that $f \in \mathcal{I}(X) \setminus \mathcal{I}(Y)$, or in other words, such that $f|_Y \neq 0 \in K[Y]$ but $f|_X = 0 \in K[X]$. Since the maximum number of algebraically independent elements in $K[Y]$ is $\dim(Y) = d$, the set $\{f_1|_Y, \ldots, f_d|_Y, f|_Y\}$ cannot be algebraically independent (here, $f_1, \ldots, f_d$ are the polynomials that were chosen in the previous paragraph). Thus, we can choose a nonzero polynomial

$$g(z) \in K[f_1|_Y, \ldots, f_d|_Y][z]$$

such that $g(f|_Y) = 0$. Without loss of generality, we may assume that g is such a polynomial of smallest possible degree. Writing the relation term by term, we have

$$0 = g(f|_Y) = g_k(f_1|_Y, \ldots, f_d|_Y)(f|_Y)^k + \cdots$$
$$\cdots + g_1(f_1|_Y, \ldots, f_d|_Y)f|_Y + g_0(f_1|_Y, \ldots, f_d|_Y)$$

for some $g_0, \ldots, g_k \in K[z_1, \ldots, z_d]$. Notice that g_0 cannot be the zero polynomial, as otherwise we could cancel one factor of $f|_Y$—using our assumption that Y is irreducible, so $K[Y]$ is an integral domain—producing a polynomial of smaller degree that vanishes when evaluated at $f|_Y$, thereby contradicting the minimality in the choice of g. Upon restricting to X, the fact that $f|_X$ is the zero function then implies that g_0 gives a polynomial relation among $f_1|_X, \ldots, f_d|_X$, contradicting the assumption that these functions are algebraically independent elements of $K[X]$. The contradiction implies that $\dim(X) < \dim(Y)$, as claimed. □

Note that the second statement of Proposition 6.38 fails without the assumption that Y is irreducible. For example, the x-axis $\mathcal{V}(y) \subseteq \mathbb{A}^2$ is strictly contained in the union of the axes $\mathcal{V}(xy) \subseteq \mathbb{A}^2$, but both varieties have dimension one.

We now have a rigorous definition of dimension, and it is not too difficult to prove that it satisfies the first two axioms of Definition 6.3 (see Exercises 6.5.1, 6.5.2, and 6.5.3). Therefore, it remains to prove Axiom 3, which is the content of the remaining section in this chapter.

Exercises for Section 6.5

6.5.1 Using the results of this section, prove that two isomorphic (possibly reducible) affine varieties have the same dimension.

6.5.2 Let X be an affine variety. Prove that $\dim(X) = 0$ if and only if X is finite.

6.5.3 Let $X_1, \ldots, X_m \subseteq \mathbb{A}^n$ be (possibly reducible) affine varieties. Prove that

$$\dim(X_1 \cup \cdots \cup X_m) = \max\{\dim(X_1), \ldots, \dim(X_m)\}.$$

6.5.4 Let $X \subseteq \mathbb{A}^n$ be an affine variety. Prove that $X = \mathcal{V}(f)$ for some nonzero polynomial $f \in K[x_1, \ldots, x_n]$ if and only if every irreducible component of X has dimension $n - 1$.

6.5.5 Prove Proposition 6.35, Part 2.

6.5.6 Prove Corollary 6.37, Part 2.

Section 6.6 The Fundamental Theorem

We now arrive at the technical heart of dimension theory: the Fundamental Theorem of Dimension Theory. This result says that, if $X \subseteq \mathbb{A}^n$ is an irreducible affine variety and $f \in K[x_1, \ldots, x_n]$ is a polynomial, then either $X \cap \mathcal{V}(f)$ is a trivial intersection (empty or all of X) or $\dim(X \cap \mathcal{V}(f)) = \dim(X) - 1$.

The proof of this result is rather formidable, requiring some fortitude on the part of the reader. It draws on many of the ideas introduced in prior chapters, with a key step coming from the Noether Normalization Theorem. In addition, it requires several new ideas regarding minimal polynomials and their relation to determinants of certain linear maps of algebraic field extensions. We begin with a discussion of the new ideas, which are captured in Lemmas 6.39, 6.41, and 6.42, all of which will then be used in the proof of the Fundamental Theorem (Theorem 6.45).

6.39 LEMMA/DEFINITION *Minimal polynomial*

Let $L \subseteq M$ be a field extension. For any $a \in M$ that is algebraic over L, there exists a unique irreducible monic polynomial $\mu_a \in L[x]$ such that $\mu_a(a) = 0 \in M$. We call μ_a the *minimal polynomial of a over L*.

PROOF Let $L \subseteq M$ be a field extension and let $a \in M$ algebraic over L. By algebraicity, there exists a nonzero polynomial in $L[x]$ that vanishes at a. We may find such a polynomial of minimal possible degree, and by dividing by the leading coefficient, we can even find one that is monic. Let μ_a be a monic polynomial of minimal degree that vanishes at a. By the minimality of its degree, μ_a is irreducible.

To prove the lemma, it remains to prove that every monic irreducible polynomial that vanishes at a is equal to μ_a. Suppose that $g \in L[x]$ is a monic irreducible polynomial that vanishes at a. By the division algorithm,

$$g = q\mu_a + r$$

where $r = 0$ or $\deg(r) < \deg(\mu_a)$. Evaluating at a and using $g(a) = \mu_a(a) = 0$, we see that $r(a) = 0$. Thus, the minimality of the degree of μ_a implies that $r = 0$. Therefore, $g = q\mu_a$, from which the irreducibility of g implies that q is a unit. Since both g and μ_a are monic, we must have $q = 1$, and we conclude that $g = \mu_a$. □

6.40 EXAMPLE Minimal polynomial of $\sqrt{2}$ over $\mathbb{Q}$

The minimal polynomial of $\sqrt{2}$ over $\mathbb{Q}$ is $x^2 - 2 \in \mathbb{Q}[x]$. To prove this, we simply observe that $x^2 - 2$ is monic, irreducible, and vanishes at $\sqrt{2}$, from which Lemma 6.39 implies that it is the minimal polynomial of $\sqrt{2}$ over $\mathbb{Q}$.

Minimal polynomials play a key role in the proof of Theorem 6.45. In that setting, we will have a Noether basis $\{F_1, \ldots, F_d\}$ of a coordinate ring $K[X]$ (where X is irreducible), and we will be considering the field extension

$$K(F_1, \ldots, F_d) \subseteq K(X).$$

The key result we require, which follows from the next lemma, is that, for any $F \in K[X] \subseteq K(X)$, the coefficients of μ_F lie in $K[F_1, \ldots, F_d] \subseteq K(F_1, \ldots, F_d)$.

6.6. THE FUNDAMENTAL THEOREM

> **6.41 LEMMA** *Minimal polynomials and Noether bases*
>
> Let A be a finitely-generated K-algebra that is also an integral domain, and let $\{a_1, \ldots, a_d\}$ be a Noether basis of A over K. Then the following hold.
> 1. The field extension $K(a_1, \ldots, a_d) \subseteq \text{Frac}(A)$ is algebraic.
> 2. For any element $a \in A \subseteq \text{Frac}(A)$, the minimal polynomial μ_a of a over $K(a_1, \ldots, a_d)$ satisfies
> $$\mu_a \in K[a_1, \ldots, a_d][x].$$

By definition, the minimal polynomial μ_a is an element of $K(a_1, \ldots, a_d)[x]$, meaning that the coefficients are *rational functions* in the elements of the Noether basis. The assertion of the second part of the lemma is that the coefficients of the minimal polynomial are actually *polynomials* in the elements of the Noether basis.

PROOF OF LEMMA 6.41 To prove that $K(a_1, \ldots, a_d) \subseteq \text{Frac}(A)$ is algebraic, choose algebra generators $b_1, \ldots, b_n$ of A, so that $A = K[b_1, \ldots, b_n]$. These algebra generators become field generators upon passing to the fraction field:
$$\text{Frac}(A) = K(b_1, \ldots, b_n).$$
Since A is integral over $K[a_1, \ldots, a_d]$, each b_i is integral over $K[a_1, \ldots, a_d]$. In particular, each b_i is algebraic over $K(a_1, \ldots, a_d) \subseteq \text{Frac}(A)$. Thus, Proposition 6.15 implies that the field extension
$$K(a_1, \ldots, a_d) \subseteq \text{Frac}(A)$$
is algebraic.

To prove Part 2, let $a \in A$. Using that A is integral over $K[a_1, \ldots, a_d]$, there exists a monic polynomial $f \in K[a_1, \ldots, a_d][x]$ such that $f(a) = 0$. Choose f to be such a polynomial that has minimal degree among all monic polynomials that vanish at a, in which case f must also be irreducible. Assuming without loss of generality that $a_1, \ldots, a_d$ are distinct, the definition of Noether bases implies that $a_1, \ldots, a_d$ are algebraically independent, so f is an irreducible element of the multivariable polynomial ring
$$f \in K[a_1, \ldots, a_d][x] = K[a_1, \ldots, a_d, x].$$
By repeated use of Proposition 0.59, the monic polynomial f remains irreducible in the larger ring $K(a_1, \ldots, a_d)[x]$, and since it vanishes at a, it must be equal to the minimal polynomial of a. Therefore, $\mu_a = f \in K[a_1, \ldots, a_d][x]$. □

To set up the final lemma required for the proof of Theorem 6.45, let $L \subseteq M$ be a finitely-generated algebraic field extension, which, by Proposition 6.15, implies that M is a finite-dimensional vector space over L. For any element $a \in M$, define
$$T_a : M \to M$$
$$b \mapsto ab.$$
Notice that this function is a linear map of M as a vector space over L. More precisely, for any $b_1, b_2 \in M$ and any $c \in L$, we check that
$$T_a(b_1 + cb_2) = a(b_1 + cb_2) = ab_1 + cab_2 = T_a(b_1) + cT_a(b_2).$$

As $T_a : M \to M$ is L-linear, it has a well-defined determinant $\det(T_a) \in L$, which can be computed by picking a basis of M over L, writing the linear map as a matrix, and computing the determinant of the matrix using any of the usual formulas for determinants. Importantly, the determinant is independent of the basis. The next lemma relates this determinant to the minimal polynomial of a.

6.42 LEMMA *Determinants of multiplication transformations*

Let $L \subseteq M$ be a finitely-generated algebraic field extension. Then, for any $a \in M$ with minimal polynomial $\mu_a(x) \in L[x]$, we have

$$\det(T_a) = \pm \mu_a(0)^\ell$$

for some positive integer ℓ.

PROOF Since $L \subseteq M$ is a finitely-generated algebraic field extension, M is a finite-dimensional vector space over L. Given any nonzero $a \in M$ with minimal polynomial

$$\mu_a(x) = x^d + \mu_{a,d-1} x^{d-1} + \cdots + \mu_{a,1} x + \mu_{a,0},$$

Exercise 6.6.1 shows that we can find a basis for M (as a vector space over L) of the following form for some $b_1, \ldots, b_\ell \in M$:

$$\{b_1, b_1 a, \ldots, b_1 a^{d-1}, b_2, b_2 a, \ldots, b_2 a^{d-1}, \ldots, b_\ell, b_\ell a, \ldots, b_\ell a^{d-1}\}.$$

Writing the linear map T_a as a matrix in terms of this basis, T_a is a block diagonal matrix

$$T_a = \begin{pmatrix} T'_a & \cdots & 0 \\ \vdots & \ddots & \vdots \\ 0 & \cdots & T'_a \end{pmatrix},$$

where there are ℓ blocks, each of the form

$$T'_a = \begin{pmatrix} 0 & 0 & \cdots & 0 & -\mu_{a,0} \\ 1 & 0 & \cdots & 0 & -\mu_{a,1} \\ 0 & 1 & \cdots & 0 & -\mu_{a,2} \\ \vdots & \vdots & \ddots & \vdots & \vdots \\ 0 & 0 & \cdots & 1 & -\mu_{a,d-1} \end{pmatrix}.$$

Expanding the determinant of T'_a along the top row, we have

$$\det(T_a) = \det(T'_a)^\ell = (\pm \mu_{a,0})^\ell = \pm \mu_a(0)^\ell. \qquad \square$$

6.43 EXAMPLE $\det(T_a)$ for $a \in L$

Let $L \subseteq M$ be a finitely-generated algebraic field extension and suppose that $a \in L$. Then the minimal polynomial of a is $\mu_a = x - a$. Therefore, Lemma 6.42 implies that $\det(T_a) = \pm a^\ell$ for some positive integer ℓ. This conclusion can also be argued directly using the fact that, in any basis of M as a vector space over L, we have $T_a = aI$ where I is the identity matrix. Thus, $\det(T_a) = \det(aI) = a^\ell$, where ℓ is the dimension of M as an L-vector space.

6.6. THE FUNDAMENTAL THEOREM

6.44 EXAMPLE $\det(T_a)$ with $a = \sqrt{2} \in \mathbb{Q}(\sqrt{2}, \sqrt{3})$

Consider $a = \sqrt{2} \in \mathbb{Q}(\sqrt{2}, \sqrt{3})$. The minimal polynomial of a is $\mu_a = x^2 - 2$, and a basis for $\mathbb{Q}(\sqrt{2}, \sqrt{3})$ as a vector space over $\mathbb{Q}$ is given by

$$\{1, \sqrt{2}, \sqrt{3}, \sqrt{2}\sqrt{3}\}.$$

In terms of this basis, we can write

$$T_a = \begin{pmatrix} 0 & 2 & 0 & 0 \\ 1 & 0 & 0 & 0 \\ 0 & 0 & 0 & 2 \\ 0 & 0 & 1 & 0 \end{pmatrix}.$$

From this matrix expression, we compute $\det(T_a) = (-2)^2 = \mu_a(0)^2$.

We are now prepared to prove the main result of this section. The following result is a restatement of Axiom 3 in Definition 6.3 and thus completes the proof that the notion of dimension developed in this chapter is the unique dimension function on the set of isomorphism classes of affine varieties.

6.45 THEOREM *Fundamental Theorem of Dimension Theory*

If $X \subseteq \mathbb{A}^n$ is an irreducible affine variety and $f \in K[x_1, \ldots, x_n]$ is a polynomial such that $X \cap \mathcal{V}(f)$ is neither empty nor all of X, then

$$\dim(X \cap \mathcal{V}(f)) = \dim(X) - 1.$$

PROOF Let $X \subseteq \mathbb{A}^n$ be an irreducible affine variety and $f \in K[x_1, \ldots, x_n]$ such that $X \cap \mathcal{V}(f)$ is neither empty nor all of X. For convenience, let us define $Y = X \cap \mathcal{V}(f)$ and set $F = f|_X \in K[X]$. We aim to prove that

$$\dim(Y) = \dim(X) - 1.$$

Since Y is an affine variety strictly contained in the irreducible affine variety X, Proposition 6.38 tells us that $\dim(Y) \leq \dim(X) - 1$. Thus, it remains to prove

$$\dim(Y) \geq \dim(X) - 1.$$

Choose a Noether basis $\{F_1, \ldots, F_d\} \subseteq K[X]$ over K, and assume that $F_1, \ldots, F_d$ are distinct. Corollary 6.36 implies that $\dim(X) = d$. Let μ_F be the minimal polynomial of $F \in K[X] \subseteq K(X)$ over $K(F_1, \ldots, F_d)$. By Lemma 6.41,

$$\mu_F \in K[F_1, \ldots, F_d][x].$$

Write

$$\mu_F(x) = x^k + \mu_{F,k-1}x^{k-1} + \cdots + \mu_{F,1}x + \mu_{F,0}.$$

Since $\mu_F(F) = 0$ and $F|_Y = 0$, we see that $\mu_{F,0}|_Y = 0$. Notice that $\mu_{F,0}$ is neither zero nor a unit; if it were zero, then $\mu_F(x)$ would be a reducible polynomial, and if it were a unit, then $\mu_{F,0}|_Y \neq 0$.

The following technical claim is central to the proof.
Claim: If $G \in K[F_1, \ldots, F_d] \subseteq K[X]$, then

$$G|_Y = 0 \quad \text{if and only if} \quad G \in \sqrt{\langle \mu_{F,0} \rangle} \subseteq K[F_1, \ldots, F_d].$$

($\Rightarrow$) Suppose that $G \in K[F_1, \ldots, F_d]$ and $G|_Y = 0$. Since $Y = X \cap V(f)$, it follows from the Nullstellensatz (Exercise 6.6.2) that

$$G \in \sqrt{\langle f|_X \rangle} = \sqrt{\langle F \rangle} \subseteq K[X].$$

Thus, $G^m = HF$ for some $H \in K[X]$. Considering the algebraic field extension $K(F_1, \ldots, F_d) \subseteq K(X)$, we see that there exist positive integers ℓ_1, ℓ_2, ℓ_3 such that

$$\begin{aligned}
G^{m\ell_1} &= \det(T_{G^m}) \\
&= \det(T_{HF}) \\
&= \det(T_H)\det(T_F) \\
&= \pm \mu_{H,0}^{\ell_2} \mu_{F,0}^{\ell_3}.
\end{aligned}$$

The first equality follows from $G^m \in K[F_1, \ldots, F_d]$ (as in Example 6.43), the second from $G^m = HF$, the third from multiplicativity of determinants and the fact that $T_{HF} = T_H T_F$, and the fourth from Lemma 6.42. Since $H \in K[X]$, Lemma 6.41 implies that $\mu_{H,0} \in K[F_1, \ldots, F_d]$. Thus, the equation

$$G^{m\ell_1} = \pm \mu_{H,0}^{\ell_2} \mu_{F,0}^{\ell_3}$$

lives in the polynomial ring $K[F_1, \ldots, F_d]$, so $G \in \sqrt{\langle \mu_{F,0} \rangle} \subseteq K[F_1, \ldots, F_d]$.

($\Leftarrow$) If $G^m = H\mu_{F,0}$ for some positive integer m and some $H \in K[F_1, \ldots, F_d]$, then $\mu_{F,0}|_Y = 0$ implies that $G^m|_Y = 0$, from which it follows that $G|_Y = 0$.

Having proved the claim, we now prove the theorem. Consider the restriction

$$\begin{aligned}
\varphi : K[F_1, \ldots, F_d] &\to K[Y] \\
G &\mapsto G|_Y.
\end{aligned}$$

The claim implies that $\ker(\varphi) = \sqrt{\langle \mu_{F,0} \rangle}$. Since $F_1, \ldots, F_d$ are algebraically independent, $K[F_1, \ldots, F_d]$ is a polynomial ring, so Proposition 1.31 implies that $\sqrt{\langle \mu_{F,0} \rangle} = \langle Q \rangle$ where Q is the product of the distinct irreducible factors of $\mu_{F,0}$. Thus, by the First Isomorphism Theorem, we obtain an injection

$$\begin{aligned}
[\varphi] : \frac{K[F_1, \ldots, F_d]}{\ker(\varphi) = \langle Q \rangle} &\to K[Y] \\
[G] &\mapsto G|_Y.
\end{aligned}$$

Since $\mu_{F,0}$ is not a unit, then neither is Q, so it must depend on at least one of the generators; without loss of generality, assume that it depends on F_d. Then $[F_1], \ldots, [F_{d-1}]$ are algebraically independent in the domain of $[\varphi]$, and by injectivity, it follows that $F_1|_Y, \ldots, F_{d-1}|_Y$ are algebraically independent in $K[Y]$. Thus, Proposition 6.35 implies that $\dim(Y) \geq d - 1$, as desired. $\square$

6.6. THE FUNDAMENTAL THEOREM

While the version of the Fundamental Theorem of Dimension Theory proved in Theorem 6.45 is sufficient to uniquely determine the dimension of all affine varieties, we can actually prove a stronger form of the result. Theorem 6.45 asserts that *at least one* of the irreducible components of $X \cap \mathcal{V}(f)$ has dimension $\dim(X) - 1$. However, it actually turns out that *every* irreducible component of $X \cap \mathcal{V}(f)$ has dimension $\dim(X) - 1$, as we now verify.

6.46 THEOREM *Strong Fundamental Theorem of Dimension Theory*

If $X \subseteq \mathbb{A}^n$ is an irreducible affine variety and $f \in K[x_1, \ldots, x_n]$ is a polynomial such that $X \cap \mathcal{V}(f)$ is neither empty nor all of X, then every irreducible component of $X \cap \mathcal{V}(f)$ has dimension $\dim(X) - 1$.

PROOF As in the proof of Theorem 6.45, set $Y = X \cap \mathcal{V}(f)$. Let Y' be any irreducible component of Y and let Y'' be the union of the other irreducible components. Given that

$$Y'' \subsetneq Y' \cup Y'',$$

> In the context of quasiprojective varieties, the procedure in this proof can be interpreted as the restriction to an "affine open subset" that omits all but one irreducible component.

we may choose a polynomial $g \in K[x_1, \ldots, x_n]$ such that $g|_{Y''} = 0$ but $g|_{Y'} \neq 0$. Choose defining equations $X = \mathcal{V}(f_1, \ldots, f_m)$, and define two new affine varieties in $\mathbb{A}^{n+1}$ by

$$\widetilde{X} = \mathcal{V}(f_1, \ldots, f_m, x_{n+1}g - 1) \quad \text{and} \quad \widetilde{Y} = \mathcal{V}(f_1, \ldots, f_m, f, x_{n+1}g - 1).$$

Notice that $\widetilde{Y} = \widetilde{X} \cap \mathcal{V}(f)$, and furthermore, $\widetilde{Y} \neq \emptyset$ because it contains all points of the form $(a_1, \ldots, a_n, g(a_1, \ldots, a_n)^{-1})$, where $(a_1, \ldots, a_n) \in Y' \setminus \mathcal{V}(g)$, and this latter set is nonempty by our choice of g.

It can be shown (see Exercise 6.6.3) that $\widetilde{X}$ and $\widetilde{Y}$ are irreducible with

$$K(\widetilde{X}) \cong K(X) \quad \text{and} \quad K(\widetilde{Y}) \cong K(Y').$$

Since dimension is defined only in terms of the function field, it follows that

$$\dim(\widetilde{X}) = \dim(X) \quad \text{and} \quad \dim(\widetilde{Y}) = \dim(Y').$$

Since Y' it not all of X, we know that $\dim(Y') < \dim(X)$, which then implies that $\dim(\widetilde{Y}) < \dim(\widetilde{X})$, so $\widetilde{Y}$ is not all of $\widetilde{X}$. Thus, $\widetilde{Y} = \widetilde{X} \cap \mathcal{V}(f)$ is nonempty and also not all of $\widetilde{X}$, so the hypotheses of Theorem 6.45 are all met with respect to $\widetilde{X}$ and $\widetilde{Y}$, and we conclude that

$$\dim(Y') = \dim(\widetilde{Y}) = \dim(\widetilde{X}) - 1 = \dim(X) - 1.$$

Since Y' was chosen arbitrarily among the irreducible components of $X \cap \mathcal{V}(f)$, we conclude that every irreducible component has dimension $\dim(X) - 1$. □

By repeatedly applying Theorem 6.46, one obtains the following concrete application, which says that the codimension of an affine variety is bounded above by the number of defining equations.

> **6.47 COROLLARY** *Defining equations and dimension*
>
> If $X = \mathcal{V}(f_1, \ldots, f_k) \subseteq \mathbb{A}^n$ is nonempty, then $\dim(X) \geq n - k$.

PROOF Exercise 6.6.4. □

In proving the previous corollary, the reader will recognize why the strong form of the Fundamental Theorem of Dimension Theory is so much more useful than the weak. Theorem 6.46 also implies the following alternative characterization of dimension, which, in many algebraic geometry textbooks, is taken as the definition.

> **6.48 COROLLARY** *Chains of inclusions and dimension*
>
> If $X \subseteq \mathbb{A}^n$ is an affine variety, then $\dim(X)$ is equal to the maximum d such that there exist irreducible affine varieties $X_0, \ldots, X_d \subseteq \mathbb{A}^n$ with
>
> $$X_0 \subsetneq X_1 \subsetneq \cdots \subsetneq X_d \subseteq X.$$

PROOF Exercise 6.6.6. □

In concluding this chapter, we mention that it is totally reasonable to study dimension of algebraic varieties over fields that are not algebraically closed, such as the real numbers $\mathbb{R}$. In this setting, the Fundamental Theorem of Dimension Theory fails wildly (see Exercise 6.6.7, for example). However, the results discussed in Section 6.5 continue to hold, because they did not require the Nullstellensatz. Surprisingly, even though our proof of Corollary 6.48 required the Nullstellensatz, one can actually circumvent the Nullstellensatz and prove that this characterization is equal to the transcendence degree definition over general fields. For this, and more general results on dimension theory, we direct the reader to a more advanced text on commutative algebra.

Exercises for Section 6.6

6.6.1 Let $L \subseteq M$ be an algebraic extension and let a be an element of M whose minimal polynomial has degree d.
   (a) Prove that $\{1, a, \ldots, a^{d-1}\}$ is a basis of $L(a)$ as a vector space over L.
   (b) If $b_1, \ldots, b_\ell \subseteq M$ is a basis for M as a vector space over $L(a)$, prove that $\{b_i a^j \mid 1 \leq i \leq \ell,\ 0 \leq j \leq d-1\}$ is a basis of M as a vector space over L.
   (c) Given a basis of the form in Part (b), prove that the matrix associated to the linear map $T_a : M \to M$ is block diagonal of the form given in the proof of Lemma 6.42.

6.6.2 Let $X \subseteq \mathbb{A}^n$ be an affine variety and $f \in K[x_1, \ldots, x_n]$. Use the Nullstellensatz to prove that $F \in K[X]$ vanishes on $X \cap \mathcal{V}(f)$ if and only if

$$F \in \sqrt{\langle f|_X \rangle}.$$

6.6. THE FUNDAMENTAL THEOREM

6.6.3 Let $X = \mathcal{V}(f_1,\ldots,f_m) \subseteq \mathbb{A}^n$ and let X' be any irreducible component of X. Let $g \in K[x_1,\ldots,x_n]$ be a polynomial that vanishes on every irreducible component of X except X', and define
$$\widetilde{X} = \mathcal{V}(f_1,\ldots,f_m,x_{n+1}g-1) \subseteq \mathbb{A}^{n+1}.$$
This exercise proves that $\widetilde{X}$ is irreducible and that $K(\widetilde{X}) \cong K(X')$.

(a) Consider the homomorphism
$$\varphi : K[x_1,\ldots,x_n,x_{n+1}] \to K(X')$$
$$h(x_1,\ldots,x_n,x_{n+1}) \mapsto [h(x_1,\ldots,x_n,g^{-1})].$$
Prove that $\ker(\varphi) = \mathcal{I}(\widetilde{X})$.

(b) Use (a) to explain why $\widetilde{X}$ is irreducible.

(c) Use $[\varphi]$ to prove that $K(\widetilde{X}) \cong K(X')$.

6.6.4 Prove Corollary 6.47.

6.6.5 Suppose that $f,g \in K[x,y,z]$ have at least one common zero. Prove that they have infinitely many common zeros.

6.6.6 Prove Corollary 6.48.

6.6.7 Prove that every affine variety over $\mathbb{R}$ can be realized as the vanishing set of a single polynomial. In particular, over $\mathbb{R}$, the dimension of a variety has nothing to do with the number of defining equations.

6.6.8 Let X be an affine variety. Prove that the dimension of X is equal to the maximum d such that there exist prime ideals $P_0, \ldots, P_d \subseteq K[X]$ with
$$P_0 \subsetneq P_1 \subsetneq \cdots \subsetneq P_d \subsetneq K[X].$$

The supremum of lengths of chains of prime ideals in a ring is called the *Krull dimension* of the ring, and this exercise proves that the dimension of an affine variety is equal to the Krull dimension of its coordinate ring.

Chapter 7

Smoothness

> LEARNING OBJECTIVES FOR CHAPTER 7
> - Define tangent spaces of affine varieties intuitively, via linearizations.
> - Calculate, in examples, the linearization of an affine variety at a point.
> - Define and work with the dual of a vector space.
> - Characterize tangent spaces intrinsically in terms of coordinate rings.
> - Describe the relationship between the dimension of an affine variety and the dimensions of its tangent spaces
> - Determine the smooth and singular points of affine varieties.

There is a very good chance that the pictures of affine varieties that we have depicted in this book have taken you back to your days of calculus. You probably recall computing derivatives and tangent spaces in your calculus class, and you may also remember that derivatives and tangent spaces provide a way to characterize where graphs have "singularities." For example, the fact that the derivative of the absolute value function $f(x) = |x|$ is undefined at $x = 0$ corresponds to the geometric observation that the graph has a "corner" over $x = 0$. Our goal in this chapter is to introduce the "calculus" of algebraic geometry; in particular, we aim to develop the notions of linearizations and tangent spaces and to use them to give a precise meaning of singular points of affine varieties.

We begin this chapter by defining tangent spaces in Section 7.1. Given an affine variety $X \subseteq \mathbb{A}^n$ and a point $a \in X$, the tangent space $T_a X$ is a vector subspace of K^n whose elements can geometrically be viewed as vectors that are tangent to X at a. This perspective is extrinsic, however: it relies on viewing X as sitting inside of a particular affine space. We rectify this situation in Sections 7.2 and 7.3, introducing key ideas of vector-space duality, which we then use to establish an intrinsic interpretation of tangent spaces in terms of the coordinate ring $K[X]$. In particular, this implies that isomorphic affine varieties have isomorphic tangent spaces at corresponding points.

The dimension of the tangent space $T_a X$ as a vector space is always bounded below by the dimension of X, as we prove in Section 7.4. The points of a variety where the tangent space has the same dimension as the variety are the *smooth* points of the variety, while the special points at which the dimension of the tangent space becomes larger than the dimension of the variety are the *singular* points. We close this chapter by discussing properties of smooth and singular points in Section 7.5.

Section 7.1 Linearizations and tangent spaces

Tangent lines and tangent planes play a starring role in most single- and multivariable calculus classes. In these classes, one typically starts with a definition of the derivative in terms of limits, then uses the limit definition to derive the standard rules for differentiation. For example, one of the standard differentiation rules states that the derivative of x^k is kx^{k-1} for any $k \in \mathbb{N}$. Since limits do not make sense over a general ground field K, the starting point for our discussion of tangency in algebraic geometry is not with limits, but with the differentiation rules for polynomials.

We assume that the reader is familiar with the standard rules for differentiating multivariable polynomials, and we extend these formulas to any ring R. In other words, if
$$f = a_m x^m + a_{m-1} x^{m-1} + \cdots + a_1 x + a_0 \in R[x],$$
then the *derivative* of f with respect to x is defined by
$$\frac{\partial f}{\partial x} = m a_m x^{m-1} + (m-1) a_{m-1} x^{m-2} + \cdots + a_1 \in R[x],$$
where $m a_m$ refers to adding a_m to itself m times. If $f \in R[x_1, \ldots, x_n]$ is a multivariable polynomial, then we can define the *partial derivative* $\partial f / \partial x_i$ for each variable x_i by differentiating f as an element of $R'[x_i]$, where
$$R' = R[x_1, \ldots, x_{i-1}, x_{i+1}, \ldots, x_n].$$
In terms of these partial derivatives, the key definition required for our development of tangent spaces is the following.

7.1 DEFINITION *Linearization of a polynomial at a point*

If $f \in K[x_1, \ldots, x_n]$ and $a = (a_1, \ldots, a_n) \in \mathbb{A}^n$, then the *linearization of f at a* is defined by
$$L_a f = f(a) + \sum_{i=1}^{n} \left(\frac{\partial f}{\partial x_i}(a) \cdot (x_i - a_i) \right) \in K[x_1, \ldots, x_n].$$

We note that $\frac{\partial f}{\partial x_i} \in K[x_1, \ldots, x_n]$ is a polynomial and $\frac{\partial f}{\partial x_i}(a) \in K$ denotes the evaluation of that polynomial at a, so $L_a f \in K[x_1, \ldots, x_n]$ is a linear polynomial. In multivariable calculus (in other words, when $K = \mathbb{R}$), the linearization of f at a is introduced as the linear function that most closely approximates f near a, and it is sometimes called the *linear approximation of f at a*.

The linearization of an affine variety at a point is defined as the vanishing set of the linearizations of all polynomials in the vanishing ideal.

7.2 DEFINITION *Linearization of an affine variety at a point*

Let $X \subseteq \mathbb{A}^n$ be an affine variety and $a \in X$. The *linearization of X at a* is
$$L_a X = \mathcal{V}(\{L_a f \mid f \in \mathcal{I}(X)\}) \subseteq \mathbb{A}^n.$$

7.1. LINEARIZATIONS AND TANGENT SPACES

The reader is encouraged to check (Exercise 7.1.1) that, if $\mathcal{I}(X) = \langle f_1, \ldots, f_m \rangle$, then
$$L_a X = \mathcal{V}(L_a f_1, \ldots, L_a f_n).$$
Since every vanishing ideal is finitely generated, this implies that the linearization can always be defined by a finite set of linear polynomials. We also note that the initial term $f_i(a)$ in the definition of each $L_a f_i$ vanishes, since $f_i \in \mathcal{I}(X)$ and $a \in X$.

Let us consider a few examples of linearizations.

7.3 EXAMPLE Linearizations of a parabola

Consider the parabola $X = \mathcal{V}(y - x^2) \subseteq \mathbb{A}^2$, for which $\mathcal{I}(X) = \langle y - x^2 \rangle$. To compute the linearization at the origin, we must compute the linearization of the generator $f = y - x^2$ at $a = (0,0)$. From the definition, we have
$$L_a f = 0(x - 0) + 1(y - 0) = y.$$
Thus, the linearization of the parabola at the origin is the x-axis: $L_a X = \mathcal{V}(y)$.

If we consider the point $b = (1,1)$, on the other hand, we compute

$$L_b f = -2(x - 1) + 1(y - 1) = -2x + y + 1.$$

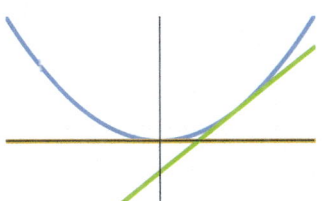

From this, we see that the linearization of the parabola at $(1,1)$ is $L_b X = \mathcal{V}(-2x + y + 1)$. We have depicted both of these linearizations in the image to the right, which is consistent with our intuition from calculus.

7.4 EXAMPLE Linearizations of a sphere

Consider the sphere $Y = \mathcal{V}(x^2 + y^2 + z^2 - 1) \subseteq \mathbb{A}^3_\mathbb{C}$, for which $\mathcal{I}(Y)$ is generated by $g = x^2 + y^2 + z^2 - 1$. To compute the linearization at $a = (0, 0, -1) \in Y$, we start by computing the linearization of the generator:
$$L_a g = 0(x - 0) + 0(y - 0) + -2(z + 1) = -2(z + 1).$$
From this, we see that $L_a Y = \mathcal{V}(-2(z + 1)) = \mathcal{V}(z + 1)$, which is the plane containing a that is parallel to the xy-plane.

If, on the other hand, we consider the point $b = (1/\sqrt{3}, 1/\sqrt{3}, 1/\sqrt{3}) \in Y$, then the linearization of the generator is

$$L_b g = \frac{2}{\sqrt{3}}(x + y + z - \sqrt{3}).$$

Thus, we have

$$L_b Y = \mathcal{V}(x + y + z - \sqrt{3}).$$

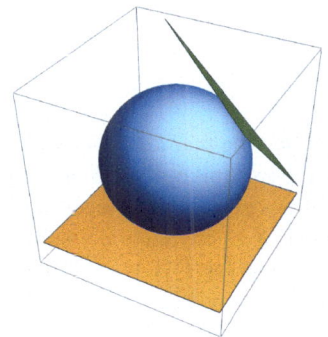

These linearizations are depicted over the real numbers in the image, and are again consistent with our intuition from calculus.

7.5 EXAMPLE Linearizations of a cusp

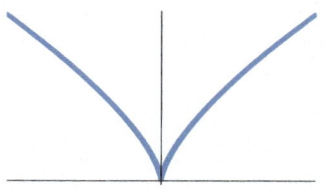

Consider the variety $X = \mathcal{V}(x^2 - y^3) \subseteq \mathbb{A}^2$, pictured to the right over $\mathbb{R}$. As you can see, this curve has a "cusp" at the origin, whereas it looks "smooth" at all other points. This geometric observation is reflected in the linearizations: as the reader is encouraged to check in Exercise 7.1.3, the linearization of X at the origin is two-dimensional (all of $\mathbb{A}^2$), whereas the linearization at any other point is one-dimensional. We will see in Section 7.5 that the dimension of the linearization is a way of detecting the "singular" points in a variety, such as the cusp in X.

We now use the notion of linearizations to define tangent spaces.

7.6 DEFINITION *Tangent vector and tangent space at a point*

Let $X \subseteq \mathbb{A}^n$ be an affine variety and $a = (a_1, \ldots, a_n) \in X$. For any $b = (b_1, \ldots, b_n) \in L_a X$, the *tangent vector associated to b* is defined by

$$\vec{ab} = (b_1 - a_1, \ldots, b_n - a_n) \in K^n.$$

The *tangent space of X at a* is the collection of tangent vectors:

$$T_a X = \{\vec{ab} \mid b \in L_a X\} \subseteq K^n.$$

It is common to view the tangent vector $\vec{ab}$ geometrically as an arrow from a to b within the linearization $L_a X$. Returning to the unit sphere from Example 7.4 and taking $a = (0, 0, -1)$, we may picture some examples of tangent vectors as drawn in the image to the right.

At this point the reader may (rightfully) be confused about the distinction between $L_a X$ and $T_a X$—these two objects feel very similar. In fact, the natural function

$$L_a X \to T_a X$$
$$b \mapsto \vec{ab}$$

is a bijection that identifies $L_a X$ with $T_a X$ as sets. Why, then, do we choose to give these two similar objects different names and notation? The reason is that we view $L_a X$ as a geometric object—an affine variety consisting of points in $\mathbb{A}^n$—while we view $T_a X$ as an algebraic object, consisting of vectors in K^n. In fact, as we will see below, the set $T_a X$ is a vector subspace of K^n, and this key fact allows us to use standard tools from linear algebra to study tangent spaces.

Before discussing why the tangent space is a vector subspace of K^n, we first discuss a useful reinterpretation of $T_a X$ using gradient vectors.

7.1. LINEARIZATIONS AND TANGENT SPACES

7.7 DEFINITION *Gradient vector*

For any $a \in \mathbb{A}^n$ and $f \in K[x_1, \ldots, x_n]$ the *gradient of f at a* is the vector

$$\nabla f(a) = \left(\frac{\partial f}{\partial x_1}(a), \ldots, \frac{\partial f}{\partial x_n}(a)\right) \in K^n.$$

The next result characterizes tangent spaces in terms of gradient vectors, similarly to how tangent planes are usually computed in multivariable calculus.

7.8 PROPOSITION *Gradient characterization of tangent spaces*

Let $X \subseteq \mathbb{A}^n$ be an affine variety and $a \in X$. Then

$$T_a X = \{v \in K^n \mid \nabla f(a) \cdot v = 0 \text{ for all } f \in \mathcal{I}(X)\}.$$

> The "$\cdot$" appearing in Proposition 7.8 is the standard dot product.

PROOF Let $v = (v_1, \ldots, v_n) \in K^n$. By the definition of $T_a X$, we see that $v \in T_a X$ if and only if $v = \vec{ab}$ for some $b \in L_a X$. Unwinding this, we find it to be equivalent to the requirement that $b = (v_1 + a_1, \ldots, v_n + a_n) \in L_a X$. By the definition of $L_a X$, we have that $b \in L_a X$ if and only if, for all $f \in \mathcal{I}(X)$,

$$0 = L_a f(b) = \sum_{i=1}^{n} \left[\frac{\partial f}{\partial x_i}(a)\right](b_i - a_i) = \nabla f(a) \cdot v.$$

Thus, $v \in T_a X$ if and only if $\nabla f(a) \cdot v = 0$ for all $f \in \mathcal{I}(X)$. $\square$

As in the case of $L_a X$, the reader is encouraged to verify that, if we have a finite set of generators $\mathcal{I}(X) = \langle f_1, \ldots, f_m \rangle$, then the vanishing of Proposition 7.8 need only be checked on the generators:

$$T_a X = \{v \in K^n \mid \nabla f_i(a) \cdot v = 0 \text{ for all } i = 1, \ldots, m\}.$$

As a consequence of Proposition 7.8, we now prove that $T_a X$ is a vector space.

7.9 COROLLARY *The tangent space is a vector space*

Let $X \subseteq \mathbb{A}^n$ be an affine variety and $a \in X$. The tangent space $T_a X$ is a vector subspace of K^n.

PROOF First, note that $0 \in T_a X$, so checking that $T_a X$ is a subspace then amounts to verifying that it is closed under the operations. Suppose that $v, w \in T_a X$ and $r \in K$; we must check that $v - rw \in T_a X$. For any $f \in \mathcal{I}(X)$, we have

$$\nabla f(a) \cdot (v - rw) = \nabla f(a) \cdot v - r \nabla f(a) \cdot w = 0,$$

where the first equality follows from the linearity of the dot product and the second equality follows from Proposition 7.8 and the assumption that $v, w \in T_a X$. Therefore, by Proposition 7.8, we see that $v + rw \in T_a X$, concluding the proof. $\square$

Starting with an affine variety $X \subseteq \mathbb{A}^n$ and a point $a \in X$, we have now defined a vector space $T_a X \subseteq K^n$. However, the definition of $T_a X$ we have given is extrinsic: it depends heavily on the inclusion $X \subseteq \mathbb{A}^n$. In Section 7.3, our primary aim is to give an intrinsic characterization of $T_a X$ that depends only on the K-algebra $K[X]$ and the maximal ideal $I_a \subseteq K[X]$ comprised of polynomial functions on X that vanish at a. This characterization will be given in terms of the linear-algebraic notion of dual vector spaces, so we must first take a small detour in the next section to discuss the notion of vector-space duality.

Exercises for Section 7.1

7.1.1 Let $X \subseteq \mathbb{A}^n$ be an affine variety with $a \in X$ and $\mathcal{I}(X) = \langle f_1, \ldots, f_m \rangle$.
   (a) Prove that
$$L_a X = \mathcal{V}(L_a f_1, \ldots, L_a f_m).$$
   (b) Prove that
$$T_a X = \{v \in K^n \mid \nabla f_i(a) \cdot v = 0 \text{ for all } i = 1, \ldots, m\}.$$

7.1.2 Give an example of an affine variety $X = \mathcal{V}(f) \subseteq \mathbb{A}^n$ such that
$$L_a X \neq \mathcal{V}(L_a f).$$

In other words, when computing linearizations, it does not suffice to use defining polynomials; rather, one requires generators for the vanishing ideal.

7.1.3 Let $X = \mathcal{V}(x^2 - y^3) \subseteq \mathbb{A}^2$. Prove that the linearization at $(0,0)$ is two-dimensional and that the linearization at any other point is one-dimensional.

7.1.4 Let $X = \mathcal{V}(x^2 + y^2 - z) \subseteq \mathbb{A}^3$.
   (a) Draw a picture of X over $\mathbb{R}$.
   (b) Prove that the linearization at $(0,0,0)$ is three-dimensional and that the linearization at any other point of X is two-dimensional.

7.1.5 For any affine variety $X \subseteq \mathbb{A}^n$ and $a \in X$, prove that
$$\dim(T_a X) = \dim(L_a X),$$
where the left-hand side is the dimension as a vector space and the right-hand side is the dimension as an affine variety.

Open Access This chapter is licensed under the terms of the Creative Commons Attribution-NonCommercial 4.0 International License (http://creativecommons.org/licenses/by-nc/4.0/), which permits any noncommercial use, sharing, adaptation, distribution and reproduction in any medium or format, as long as you give appropriate credit to the original author(s) and the source, provide a link to the Creative Commons license and indicate if changes were made.

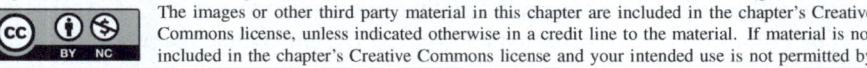
The images or other third party material in this chapter are included in the chapter's Creative Commons license, unless indicated otherwise in a credit line to the material. If material is not included in the chapter's Creative Commons license and your intended use is not permitted by statutory regulation or exceeds the permitted use, you will need to obtain permission directly from the copyright holder.

Section 7.2 Duality of vector spaces

This section presents the key notions regarding duality of vector spaces that we will need in order to study tangent spaces intrinsically, with the culmination of our discussion being the property that every finite-dimensional vector space is canonically isomorphic to its double dual.

We begin with the definition of the dual of a vector space, which simply collects all linear maps from the vector space to the ground field.

7.10 DEFINITION *Dual of a vector space*

Let V be a vector space. The *dual* of V, denoted $V^\vee$, is the set of all linear functions $\varphi : V \to K$.

We note that $V^\vee$ is, itself, a vector space with respect to the usual addition and scalar multiplication of functions to the ground field K (Exercise 7.2.1). Moreover, given a basis of V, there is a natural "dual" basis on $V^\vee$. Its definition, given below, relies on the fact that any linear function $V \to K$ is uniquely determined by its values on the elements of a basis of V.

7.11 DEFINITION *Dual of a basis*

If V is a finite-dimensional vector space with basis $\{v_1, \ldots, v_n\}$ of size n, then the *dual* of $\{v_1, \ldots, v_n\}$ is the set of functions $\{v_1^\vee, \ldots, v_n^\vee\} \subseteq V^\vee$ determined by setting

$$v_i^\vee(v_j) = \begin{cases} 1 & i = j, \\ 0 & i \neq j. \end{cases}$$

Let us consider a concrete example.

7.12 EXAMPLE *Dual of a basis of K^2*

Let $V = K^2$, and consider the basis consisting of the two vectors

$$v_1 = (1, 1), \quad v_2 = (1, -1).$$

Notice that $v_1^\vee : V \to K$ sends $(1, 1)$ to 1 and $(1, -1)$ to 0, while $v_2^\vee$ sends $(1, 1)$ to 0 and $(1, -1)$ to 1. Thus, given any $(a, b) \in K^2$, writing

$$(a, b) = \frac{a+b}{2}(1, 1) + \frac{a-b}{2}(1, -1)$$

shows that

$$v_1^\vee(a, b) = \frac{a+b}{2} \quad \text{and} \quad v_2^\vee(a, b) = \frac{a-b}{2}.$$

As one might expect, the dual of a basis of V is a basis of $V^\vee$, as we now verify.

7.13 PROPOSITION Dual of a basis is a basis

If V is a vector space with basis $\{v_1, \ldots, v_n\}$ of size n, then $\{v_1^\vee, \ldots, v_n^\vee\}$ is a basis of $V^\vee$.

PROOF We prove that $\{v_1^\vee, \ldots, v_n^\vee\}$ is linearly independent and spans $V^\vee$.

To prove linear independence, suppose that $a_1 v_1^\vee + \cdots + a_n v_n^\vee = 0$ for some $a_1, \ldots, a_n \in K$. Evaluating at v_i, we have

$$0 = (a_1 v_1^\vee + \cdots + a_n v_n^\vee)(v_i) = a_i.$$

Thus, $a_1 = \cdots = a_n = 0$, showing that $\{v_1^\vee, \ldots, v_n^\vee\}$ is linearly independent.

To prove that $\{v_1^\vee, \ldots, v_n^\vee\}$ spans $V^\vee$, suppose $\varphi \in V^\vee$, so that $\varphi : V \to K$ is a linear function. Consider the linear combination $\varphi(v_1) v_1^\vee + \cdots + \varphi(v_n) v_n^\vee$, and notice that, for every $j = 1, \ldots, n$, the definition of $v_1^\vee, \ldots, v_n^\vee$ implies that

$$(\varphi(v_1) v_1^\vee + \cdots + \varphi(v_n) v_n^\vee)(v_j) = \varphi(v_j).$$

As $\{v_1, \ldots, v_n\}$ is a basis of V and the linear functions $\varphi(v_1) v_1^\vee + \cdots + \varphi(v_n) v_n^\vee$ and φ agree on each basis element, it follows that $\varphi = \varphi(v_1) v_1^\vee + \cdots + \varphi(v_n) v_n^\vee$. As φ was arbitrary, this shows that $\{v_1^\vee, \ldots, v_n^\vee\}$ spans $V^\vee$. $\square$

The previous result implies that the dual of a finite-dimensional vector space is, itself, a finite-dimensional vector space of the same dimension. In particular, given a basis of V, there is an isomorphism $V \cong V^\vee$ that takes each element of the basis to the corresponding element of the dual basis. This raises an important question: is this isomorphism independent of the basis? The answer is "no," as the next example illustrates.

> Note that Proposition 7.13 fails for infinite-dimensional vector spaces; see Exercise 7.2.3.

7.14 EXAMPLE Different bases define different isomorphisms $V \cong V^\vee$

Let $V = K^2$, and consider the two bases given by the following pairs of elements:

$$v_1 = (1,0),\ v_2 = (0,1) \quad \text{and} \quad w_1 = (1,0),\ w_2 = (1,1).$$

One checks that the dual basis of $\{v_1, v_2\}$ is given by

$$v_1^\vee(a,b) = a \quad \text{and} \quad v_2^\vee(a,b) = b,$$

while the dual basis of $\{w_1, w_2\}$ is given by

$$w_1^\vee(a,b) = a - b \quad \text{and} \quad w_2^\vee(a,b) = b.$$

In particular, since $v_1 = w_1$ but $v_1^\vee \neq w_1^\vee$, we conclude that the isomorphism between V and $V^\vee$ mapping v_1, v_2 to $v_1^\vee, v_2^\vee$ is different from the isomorphism mapping w_1, w_2 to $w_1^\vee, w_2^\vee$.

7.2. DUALITY OF VECTOR SPACES

While we have now seen how to construct isomorphisms between V and $V^\vee$ from the choice of a basis, the previous example demonstrates that these isomorphisms are generally different for different choices of bases. In fact, try as one might, it is impossible to construct an isomorphism between a nonzero vector space and its dual without making some extraneous choice, like the choice of a basis. It may, then, come as a bit of a surprise that, upon taking a second dual, we find that there exists a *canonical* isomorphism between V and $(V^\vee)^\vee$.

> We typically refer to $(V^\vee)^\vee$ as the "double dual" of V.

7.15 THEOREM $V = (V^\vee)^\vee$

If V is a finite-dimensional vector space, then there is a canonical isomorphism between V and $(V^\vee)^\vee$.

PROOF Define $\omega : V \to (V^\vee)^\vee$ to be the function that takes a vector v in V to the corresponding "evaluation function" on $V^\vee$ defined by

$$e_v : V^\vee \to K$$
$$\varphi \mapsto \varphi(v).$$

Some reflection should convince the reader that $e_{av+w} = ae_v + e_w$ for every $a \in K$ and $v, w \in V$; in other words, ω is a linear map. Notice that the linear map ω is, indeed, canonically defined, as no extraneous choices (such as the choice of a basis) were required in the definition of ω.

It remains to prove that ω is an isomorphism, and we do this by choosing a basis. Suppose that V has dimension n and let $\{v_1, \ldots, v_n\} \subseteq V$ be a basis of V. The evaluation function e_{v_i} evaluated on the dual basis element $v_j^\vee \in V^\vee$ gives

$$e_{v_i}(v_j^\vee) = v_j^\vee(v_i) = \begin{cases} 1 & i = j, \text{ and} \\ 0 & i \neq j. \end{cases}$$

In other words, this shows that $\{e_{v_1}, \ldots, e_{v_n}\}$ is the dual basis of the dual basis $\{v_1^\vee, \ldots, v_n^\vee\}$, and is thus a basis by Lemma 7.13. Since ω is a linear map that sends a basis of V to a basis of $(V^\vee)^\vee$, standard linear-algebra considerations then imply that ω is an isomorphism. □

> We note that $\omega : V \to (V^\vee)^\vee$ is injective for any vector space V, but surjective only if $\dim(V) < \infty$.

Having now familiarized ourselves with duals (and double duals) of vector spaces, we return to the topic of tangent spaces in the next section, where we develop an intrinsic description of tangent spaces that does not depend on how an affine varieties sit within affine space.

Exercises for Section 7.2

7.2.1 Let V be a vector space. Prove that $V^\vee$ is also a vector space.

7.2.2 Give an example of distinct bases $\{v_1, v_2\}$ and $\{w_1, w_2\}$ of K^2 such that the isomorphisms $V \to V^\vee$ given by $v_i \mapsto v_i^\vee$ and $w_i \mapsto w_i^\vee$ are the same.

7.2.3 Consider the infinite-dimensional vector space $K[x]$, along with the basis $\{1, x, x^2, x^3, \dots\}$. Prove that the linear map

$$K[x] \to K$$
$$f \mapsto f(1)$$

is not in the span of $1^\vee, x^\vee, (x^2)^\vee, (x^3)^\vee, \dots$.

7.2.4 Let V be a vector space and let W be the set of linear maps $K \to V$. Prove that V and W are canonically isomorphic vector spaces.

7.2.5 Let V and W be vector spaces such that W has finite dimension n, and let $\{w_1, \dots, w_n\}$ be a basis of W. Construct an isomorphism between $(V^\vee)^n$ and the vector space of linear maps $V \to W$.

7.2.6 Let V and W be vector spaces. Given a linear map $\varphi : V \to W$, define the *dual* of φ by

$$\varphi^\vee : W^\vee \to V^\vee$$
$$\psi \mapsto \psi \circ \varphi.$$

(a) Explain why $\psi \circ \varphi \in V^\vee$ for any linear map $\varphi : V \to W$ and $\psi \in W^\vee$.
(b) Prove that $\varphi^\vee$ is a linear map.
(c) Suppose that V and W are finite-dimensional. Prove that $(\varphi^\vee)^\vee = \varphi$.

Section 7.3 Tangent spaces from coordinate rings

In this section, our aim is to reinterpret the tangent space T_aX purely in terms of the coordinate ring $K[X]$. One useful consequence of this intrinsic description of the tangent space is that it will imply that tangent spaces are preserved under isomorphism. In other words, if $F : X \to Y$ is an isomorphism and $a \in X$, then we can conclude that T_aX and $T_{F(a)}Y$ are isomorphic vector spaces.

In order to start our intrinsic description of the tangent space, we first require an intrinsic way of thinking about a point $a \in X$. Given $a \in X$, define

$$I_a = \{F \in K[X] \mid F(a) = 0\} \subseteq K[X].$$

From the definition, we see that I_a is a subset of $K[X]$. However, more is true: I_a is a maximal ideal in $K[X]$, as the reader is encouraged to verify in Exercise 7.3.1. Notice that I_a is defined intrinsically, without reference to an ambient affine space in which X sits.

In order to get a more concrete description of I_a, let us consider an (extrinsic) quotient description of $K[X]$. More specifically, suppose that $X \subseteq \mathbb{A}^n$, so that we can write $K[X] = K[x_1, \ldots, x_n]/\mathcal{I}(X)$ and $a = (a_1, \ldots, a_n) \in X$, and for any $F \in K[X]$, we can write $F = [f]$ for some $f \in K[x_1, \ldots, x_n]$. Then $F \in I_a$ if and only if $f(a) = 0$, and since every polynomial that vanishes at a can be written as

$$f = (x_1 - a_1)f_1 + \cdots + (x_n - a_n)f_n$$

for some polynomials $f_1, \ldots, f_n \in K[x_1, \ldots, x_n]$, it follows that

(7.16) $$I_a = \langle [x_1 - a_1], \ldots, [x_n - a_n] \rangle \subseteq \frac{K[x_1, \ldots, x_n]}{\mathcal{I}(X)}.$$

Observe that I_a can naturally be viewed as a vector space: adding two functions that vanish at a results in a function that vanishes at a, and multiplying a function that vanishes at a by a scalar results in a function that vanishes at a.

Consider now the ideal $I_a^2 = I_a \cdot I_a$, which is the ideal product of I_a with itself. Using the generators for I_a in equation (7.16), it follows that

(7.17) $$I_a^2 = \langle [(x_i - a_i)(x_j - a_j)] \mid 1 \leq i, j \leq n \rangle \subseteq I_a.$$

Notice that I_a^2 is a vector subspace of I_a, so we can consider the vector-space quotient I_a/I_a^2. In terms of the generators in equation (7.16), the quotient is finite-dimensional, spanned by the linear functions $[x_i - a_i]$ (Exercise 7.3.2):

$$I_a/I_a^2 = K\{[x_1 - a_1], \ldots, [x_n - a_n]\}.$$

Our primary result in this section is the following, which interprets the finite-dimensional vector space I_a/I_a^2 as the dual of the tangent space T_aX.

7.18 THEOREM $I_a/I_a^2 = (T_aX)^\vee$

Let $X \subseteq \mathbb{A}^n$ be an affine variety and let $a \in X$. Then there is a canonical vector-space isomorphism

$$I_a/I_a^2 = (T_aX)^\vee.$$

Before presenting a proof, let us explore a few familiar examples of tangent spaces to help parse the new ideas arising in Theorem 7.18.

7.19 EXAMPLE Tangent space of a parabola

Consider the parabola $X = \mathcal{V}(y - x^2) \subseteq \mathbb{A}^2$, and let $a = (0,0)$. In Example 7.3, we computed that X has a one-dimensional linearization $L_a X$, and thus a one-dimensional tangent space $T_a X$. By Theorem 7.18, it then follows that I_a / I_a^2 should be a one-dimensional vector space. To confirm this explicitly, notice that

$$I_a = \langle [x], [y] \rangle = K\{[x], [y], [x^2], [xy], [y^2], [x^3], [x^2 y], [xy^2], [y^3], \ldots\} \subseteq K[X]$$

and

$$I_a^2 = \langle [x^2], [xy], [y^2] \rangle = K\{[x^2], [xy], [y^2], [x^3], [x^2 y], [xy^2], [y^3], \ldots\} \subseteq K[X].$$

Observing that $[y] = [x^2]$ in $K[X]$, we then see that, with the exception of $[x]$, every generator of I_a also appears in I_a^2. Thus, $I_a / I_a^2 = K\{[x]\}$, which is a one-dimensional vector space over K, as expected. Moreover, notice that the function $[x] \in K[X]$ gives a linear function on the linearization $L_a X = \mathcal{V}(y) \subseteq \mathbb{A}^2$, and upon identifying $L_a X = \{(a,0) \mid a \in K\}$ with $T_a X = K$, we may naturally identify the vector space $I_a / I_a^2 = K\{[x]\}$ with the dual space $(T_a X)^\vee$.

7.20 EXAMPLE Tangent space of a cusp

As in Example 7.5, consider the cusp $X = \mathcal{V}(x^2 - y^3) \subseteq \mathbb{A}^2$, and let $a = (0,0)$. In Exercise 7.1.3, the reader was encouraged to show that X has a two-dimensional linearization $L_a X = \mathbb{A}^2$, and thus a two-dimensional tangent space $T_a X = K^2$. By Theorem 7.18, it then follows that I_a / I_a^2 should be a two-dimensional vector space.

To confirm this expectation explicitly, we compute I_a and I_a^2 exactly as in the previous example, but observe that, unlike in the previous example, $[x]$ and $[y]$ remain linearly independent in the quotient I_a / I_a^2. Thus, $I_a / I_a^2 = K\{[x], [y]\}$ is a two-dimensional vector space, and upon identifying $L_a X = \mathbb{A}^2$ with $T_a X = K^2$, we can naturally view the vector space $I_a / I_a^2 = K\{[x], [y]\}$ as the dual space $(T_a X)^\vee$.

PROOF OF THEOREM 7.18 We define a canonical surjective linear map

$$\varphi : I_a \to (T_a X)^\vee$$

and prove that $\ker(\varphi) = I_a^2$. To define φ, suppose that $F \in I_a$ and write $F = [f]$ for some $f \in K[x_1, \ldots, x_n]$. We define $\varphi(F)$ to be the linear map

$$\varphi(F) : T_a X \to K$$
$$v \mapsto \nabla f(a) \cdot v.$$

Of course, we should be worried that this definition depends on the choice of representative f. However, due to Proposition 7.8, it follows (Exercise 7.3.3) that

$$[f] = [g] \implies \nabla f(a) \cdot v = \nabla g(a) \cdot v \quad \text{for all} \quad v \in T_a X,$$

so $\varphi(F) \in (T_a X)^\vee$ is independent of the choice of representative f. To check that φ is linear, we require that $\varphi(F + rG) = \varphi(F) + r\varphi(G)$ for any $F, G \in K[X]$ and $r \in K$; this follows from linearity of derivatives (Exercise 7.3.4). Thus, it remains to prove that φ is surjective and that $\ker(\varphi) = I_a^2$.

7.3. TANGENT SPACES FROM COORDINATE RINGS

To prove that φ is surjective, let $\rho : T_a X \to K$ be a linear map, and extend it to a linear map $\bar{\rho} : K^n \to K$. Define a linear polynomial

$$f = \bar{\rho}(e_1)(x_1 - a_1) + \cdots + \bar{\rho}(e_n)(x_n - a_n) \in K[x_1, \ldots, x_n],$$

where $e_1, \ldots, e_n$ are the standard basis vectors of K^n, and notice that $[f] \in I_a$. We claim that $\varphi([f]) = \rho$. Unraveling the definitions, we see that $\nabla f(a) \cdot e_i = \bar{\rho}(e_i)$ for all $i = 1, \ldots, n$. Since a linear map on $T_a X \subseteq K^n$ is uniquely determined by the values of any extension on the standard basis vectors, we have $\nabla f(a) \cdot v = \rho(v)$ for all $v \in T_a X$, and we conclude that $\varphi([f]) = \rho$. Thus, φ is surjective.

To prove that $\ker(\varphi) = I_a^2$, first suppose that $F \in I_a^2$. By equation (7.17), it follows that $F = [f]$ where f has the form

$$f = \sum_{i,j=1}^{n} (x_i - a_i)(x_j - a_j) f_{i,j} \quad \text{for some} \quad f_{i,j} \in K[x_1, \ldots, x_n].$$

Using the product rule to compute derivatives of f, one calculates that $\nabla f(a) = 0$, from which it follows that $F \in \ker(\varphi)$. Thus, $I_a^2 \subseteq \ker(\varphi)$.

To prove the other inclusion, suppose that $F \in \ker(\varphi)$ and write $F = [f]$ for some $f \in K[x_1, \ldots, x_n]$. The assumption that $F \in \ker(\varphi)$ means that

$$\nabla f(a) \cdot v = 0 \quad \text{for all} \quad v \in T_a X.$$

With reference to a choice of generators $\mathcal{I}(X) = \langle g_1, \ldots, g_m \rangle$, Proposition 7.8 and linear-algebra considerations (Exercise 7.3.5) then imply that

$$\nabla f(a) = \sum_{i=1}^{m} a_i \nabla g_i(a)$$

for some values $a_1, \ldots, a_m \in K$. Since gradients act linearly on polynomials and since $\mathcal{I}(X)$ is closed under taking linear combinations, we then see that

$$\nabla f(a) = \nabla g(a) \quad \text{for} \quad g = \sum_{i=1}^{m} a_i g_i \in \mathcal{I}(X).$$

Using that $F = [f]$ is in the domain of φ and thus $[f] \in I_a$, we can write

$$f = \sum_{i=1}^{n} b_i (x_i - a_i) + \sum_{i,j=1}^{n} f_{i,j}(x_i - a_i)(x_j - a_j).$$

Writing a similar expression for g, the equality $\nabla f(a) = (b_1, \ldots, b_n) = \nabla g(a)$ implies that

$$f - g = \sum_{i,j=1}^{n} (f_{i,j} - g_{i,j})(x_i - a_i)(x_j - a_j).$$

Since $g \in \mathcal{I}(X)$, we have $[f] = [f - g]$, and we conclude from the expression above that

$$F = [f - g] \in \langle [(x_i - a_i)(x_j - a_j)] \mid 1 \le i, j \le n \rangle = I_a^2,$$

finishing the proof. $\square$

Taking duals of both sides in Theorem 7.18 and applying Theorem 7.15, we arrive at the following intrinsic characterization of tangent spaces.

7.21 COROLLARY $T_a X = (I_a / I_a^2)^\vee$

Let $X \subseteq \mathbb{A}^n$ be an affine variety and let $a \in X$. Then there is a canonical vector-space isomorphism

$$T_a X = (I_a / I_a^2)^\vee.$$

The primary upshot of Theorem 7.18 and Corollary 7.21 is that we can now study tangent spaces of affine varieties without ever needing to reference the affine space in which an affine variety lives. More specifically, we have seen that any point $a \in X$ corresponds to a maximal ideal I_a in the coordinate ring $K[X]$, and we can now simply study the tangent space $T_a X$ abstractly as the vector space $(I_a / I_a^2)^\vee$. As you might guess, then, tangent spaces are indeed intrinsic, meaning that they are preserved by isomorphisms. We now close this section by explicitly spelling out the intrinsic nature of tangent spaces.

7.22 COROLLARY *Tangent spaces are intrinsic*

Let X and Y be affine varieties, $F : X \to Y$ an isomorphism, and $a \in X$. Then F induces a vector-space isomorphism

$$T_a X \cong T_{F(a)} Y.$$

PROOF As the reader is encouraged to verify in Exercise 7.3.6, the pullback isomorphism $F^* : K[Y] \to K[X]$ identifies the maximal ideals $I_{F(a)}$ and I_a:

$$F^*(I_{F(a)}) = I_a.$$

It then follows that

$$F^*(I_{F(a)}^2) = I_a^2.$$

Therefore, F^* induces a vector-space isomorphism $I_{F(a)} / I_{F(a)}^2 \cong I_a / I_a^2$, and taking duals gives an isomorphism $T_a X \cong T_{F(a)} Y$. □

Exercises for Section 7.3

7.3.1 Let X be an affine variety and $a \in X$ a point. Prove that

$$I_a = \{F \in K[X] \mid F(a) = 0\}$$

is a maximal ideal in $K[X]$.

7.3.2 Let $X \subseteq \mathbb{A}^n$ be an affine variety and $a \in X$. Prove that

$$I_a / I_a^2 = K\{[x_1 - a_1], \ldots, [x_n - a_n]\}.$$

7.3. TANGENT SPACES FROM COORDINATE RINGS

7.3.3 Let $X \subseteq \mathbb{A}^n$ be an affine variety and $a \in X$. Prove that if $[f] = [g] \in K[X]$, then $\nabla f(a) \cdot v = \nabla g(a) \cdot v$ for all $v \in T_a X$.

7.3.4 Let φ be defined as in the proof of Theorem 7.18. For any $F, G \in K[X]$ and $r \in K$, use linearity of derivatives to prove that
$$\varphi(F + rG) = \varphi(F) + r\varphi(G).$$

7.3.5 Let $v_1, \ldots, v_m \in K^n$ and let $V \subseteq K^n$ be the linear subspace defined by
$$V = \{v \in K^n \mid v_i \cdot v = 0 \text{ for all } i = 1, \ldots, m\}.$$
Suppose that $w \in K^n$ satisfies $w \cdot v = 0$ for all $v \in V$. Prove that w is in the span of $v_1, \ldots, v_m$. (Hint: Consider the matrix M with rows $v_1, \ldots, v_m$ and the matrix M' obtained from M by appending the row w. Use a rank-nullity argument to prove that $\text{rk}(M) = \text{rk}(M')$.)

7.3.6 Let $X \subseteq \mathbb{A}^m$ and $Y \subseteq \mathbb{A}^n$ be affine varieties, $F : X \to Y$ an isomorphism, and $a \in X$. Prove that $F^* : K[Y] \to K[X]$ identifies the maximal ideals $I_{F(a)}$ and I_a:
$$F^*(I_{F(a)}) = I_a.$$

Section 7.4 Tangent spaces and dimension

In Section 7.1, we saw several examples of linearizations and tangent spaces. One thing you may have observed is that the dimensions of the tangent spaces were sometimes bigger than, but never smaller than, the dimension of the variety itself. That the dimension of an irreducible variety gives a lower bound for the dimension of the tangent space at any point on that variety is the main result of this section.

> **7.23 PROPOSITION** *Lower bound on tangent-space dimension*
> Let $X \subseteq \mathbb{A}^n$ be an irreducible affine variety and $a \in X$. Then
> $$\dim(T_a X) \geq \dim(X).$$

We note that the dimension appearing in the left-hand side of the inequality is the dimension of $T_a X \subseteq K^n$ as a vector space, while the dimension in the right-hand side is the dimension of $X \subseteq \mathbb{A}^n$ as an affine variety. If one prefers, they may interpret both sides of the inequality in Proposition 7.23 as dimensions of affine varieties in $\mathbb{A}^n$ by noting (Exercise 7.1.5) that $\dim(T_a X) = \dim(L_a X)$.

Before proving Proposition 7.23, we first prove a stronger biconditional result in the zero-dimensional setting.

> **7.24 LEMMA** *Zero-dimensional tangent spaces*
> If $X \subseteq \mathbb{A}^n$ is an irreducible affine variety and $a \in X$, then $\dim(X) = 0$ if and only if $\dim(T_a X) = 0$.

PROOF   If $\dim(X) = 0$, then X consists of the single point a, and it can be checked from the definitions that $T_a X = \{0\}$ (Exercise 7.4.1).

To prove the other direction, suppose that $\dim(T_a X) = 0$ and consider the maximal ideal $I_a \subseteq K[X]$ comprised of polynomial functions that vanish at a. Our aim is to show that I_a is the zero ideal. It then follows that the zero ideal is maximal in $K[X]$, so $K[X]$ is a field, and the Nullstellensatz implies that X is a single point.

Since a finite-dimensional vector space and its dual have the same dimension, Theorem 7.18 and the assumption that $\dim(T_a X) = 0$ imply that $I_a = I_a^2$. Choose generators
$$I_a = \langle F_1, \ldots, F_m \rangle.$$
Since $I_a = I_a^2$, we can view each F_i as an element of I_a^2, allowing us to find equations of the form
$$F_i = G_{i,1} F_1 + \cdots + G_{i,m} F_m,$$
where $G_{i,j} \in I_a$ for each $i, j = 1, \ldots, m$. Thus, we obtain a system of linear equations
$$(I - G) \cdot \begin{pmatrix} F_1 \\ \vdots \\ F_m \end{pmatrix} = \begin{pmatrix} 0 \\ \vdots \\ 0 \end{pmatrix},$$
where I is the $m \times m$ identity matrix and G is the matrix with entries $G_{i,j}$.

7.4. TANGENT SPACES AND DIMENSION

An application of Cramer's Rule then implies that

$$\det(I - G)F_i = 0 \in K[X]$$

for all $i = 1, \ldots, m$. Notice that $\det(I - G)$ is a polynomial function on X, and because $G_{i,j}(a) = 0$ for all i, j, we see that $\det(I - G)$ takes the value 1 at a. Thus, $\det(I - G)$ is not the zero function on X, and because $K[X]$ is an integral domain (here, we use the assumption that X is irreducible), we conclude that $F_i = 0$ for all i. Thus, I_a is the zero ideal, concluding the proof. □

We now prove Proposition 7.23. The proof uses induction on $\dim(X)$, where the induction step (which requires Lemma 7.24) is accomplished by slicing both X and $L_a X$ with a hyperplane—a variety defined by a single linear equation—and applying the Fundamental Theorem of Dimension Theory.

PROOF OF PROPOSITION 7.23   Given an irreducible affine variety $X \subseteq \mathbb{A}^n$ and a point $a \in X$, we prove that $\dim(T_a X) \geq \dim(X)$ by induction on $\dim(X)$.

(**Base case**) If $\dim(X) = 0$, then the fact that vector-space dimensions are nonnegative implies $\dim(T_a X) \geq \dim(X)$. (In fact, by Lemma 7.24, we know more: $\dim(T_a X) = \dim(X)$ in this case.)

(**Induction step**) Let $X \subseteq \mathbb{A}^n$ be an irreducible affine variety of positive dimension, and suppose that the inequality of the proposition holds for all irreducible affine varieties of dimension $\dim(X) - 1$. Let $a = (a_1, \ldots, a_n) \in X$.

Since we have assumed that $\dim(X) > 0$, the "if" direction of Lemma 7.24 implies that $T_a X \supsetneq \{0\}$, which is equivalent to $L_a X \supsetneq \{a\}$. Thus, we can choose a point $b \in L_a X \setminus \{a\}$. The two points $a, b \in \mathbb{A}^n$ must differ in at least one coordinate; without loss of generality, assume that they differ in the first coordinate and define the hyperplane

$$H = \mathcal{V}(x_1 - a_1) \subseteq \mathbb{A}^n.$$

Set $Y = X \cap H$. Notice first that Y cannot be all of X. Indeed, if $Y = X$, then $X \subseteq H$, which would imply that $L_a X \subseteq H$ (Exercises 7.4.2 and 7.4.3). But H was chosen specifically so that $b \in L_a X$ but $b \notin H$, so $L_a X \not\subseteq H$. Thus, $Y \neq X$. Moreover, since $a \in Y$, we conclude that

(7.25) $$\emptyset \subsetneq Y \subsetneq X,$$

from which the strong form of the Fundamental Theorem of Dimension Theory implies that every irreducible component of Y has dimension $\dim(X) - 1$.

Let Z be an irreducible component of Y that contains a. Since $Z \subseteq X$ and $Z \subseteq H$, we have that $L_a Z \subseteq L_a X$ (Exercise 7.4.2) and $L_a Z \subseteq H$ (Exercise 7.4.3). This implies that

(7.26) $$L_a Z \subseteq L_a X \cap H \subsetneq L_a X,$$

from which it follows that $T_a Z \subsetneq T_a X$, so $\dim(T_a Z) < \dim(T_a X)$.

Putting everything together and using the induction hypothesis on Z, we have that

$$\dim(X) - 1 = \dim(Z) \leq \dim(T_a Z) \leq \dim(T_a X) - 1,$$

from which it follows that $\dim(X) \leq \dim(T_a X)$, completing the proof. □

While Proposition 7.23 gives a lower bound for the dimension of $T_a X$, there does not exist an upper bound in terms of $\dim(X)$. To illustrate this point, the next example describes a method for constructing affine curves containing a point at which the tangent space has arbitrarily high dimension.

7.27 EXAMPLE Small varieties with big tangent spaces

Fix $n \geq 1$ and consider the affine variety
$$X_n = \mathcal{V}(\{x_i^{n+j-1} - x_j^{n+i-1} \mid 1 \leq i < j \leq n\}) \subseteq \mathbb{A}^n.$$
Note that X_2 is the cusp of Example 7.5. It can be shown (Exercise 7.4.4) that X_n is a one-dimensional irreducible affine variety and that its points take the form
$$X_n = \{(b^n, b^{n+1}, \ldots, b^{2n-1}) \mid b \in \mathbb{A}^1\}.$$
Let $a = (0, \ldots, 0) \in X$; we argue that $L_a X_n = \mathbb{A}^n$, implying that $T_a X_n = K^n$. Recall that $L_a X_n = \mathcal{V}(\{L_a f \mid f \in \mathcal{I}(X)\})$. Consider a polynomial
$$f \in \mathcal{I}(X_n) \subseteq K[x_1, \ldots, x_n].$$
Since f vanishes at $a = (0, \ldots, 0) \in X_n$, it has a vanishing constant term and we can write
$$f = \sum_{i=1}^n c_i x_i + g,$$
where $c_i \in K$ and the constant and linear terms in g are all zero. Evaluating at a point $(b^n, \ldots, b^{2n-1}) \in X_n$, we have
$$0 = f(b^n, \ldots, b^{2n-1}) = \sum_{i=1}^n c_i b^{n+i-1} + g(b^n, \ldots, b^{2n-1}).$$
Since this relation holds for all $b \in \mathbb{A}^1$, it follows that the associated polynomial is the zero polynomial:
$$0 = \sum_{i=1}^n c_i y^{n+i-1} + g(y^n, \ldots, y^{2n-1}) \in K[y].$$
Since the terms in $\sum_{i=1}^n c_i y^{n+i-1}$ all have distinct powers of y that are less than $2n$ whereas the terms in $g(y^n, \ldots, y^{2n-1})$ all have degree at least $2n$, it follows that $c_i = 0$ for all i. Therefore, the linear terms in f vanish, so $L_a f = 0$. Since this holds for all $f \in \mathcal{I}(X_n)$, we conclude that $L_a X_n = \mathcal{V}(0) = \mathbb{A}^n$ and $T_a X_n = K^n$.

The previous example has an interesting consequence about embedding affine varieties that one might contrast with the Whitney Embedding Theorem in the study of smooth manifolds. Since the tangent space of X_n at the origin is

> The Whitney Embedding Theorem asserts that any smooth manifold of dimension n can be viewed as a submanifold of $\mathbb{R}^{2n}$.

n-dimensional, then any affine variety isomorphic to X_n will also have a tangent space that is n-dimensional. In particular, the curve X_n cannot be isomorphic to any affine variety within $\mathbb{A}^m$ for any $m < n$. Thus, for any $m > 0$, this example shows that there exist one-dimensional affine varieties that cannot be embedded in $\mathbb{A}^m$.

Exercises for Section 7.4

7.4.1 If $X = \{a\} \subseteq \mathbb{A}^n$ is a single point, prove that $T_a X = \{0\} \subseteq K^n$.

7.4.2 Let $X \subseteq Y \subseteq \mathbb{A}^n$ be affine varieties. For any $a \in X$, prove that

$$L_a X \subseteq L_a Y.$$

Conclude that $T_a X \subseteq T_a Y$.

7.4.3 Suppose that $X = \mathcal{V}(\ell_1, \ldots, \ell_m) \subseteq \mathbb{A}^n$, where $\ell_1, \ldots, \ell_m \in K[x_1, \ldots, x_n]$ are all linear polynomials. Prove that $L_a X = X$ for any $a \in X$.

7.4.4 Let X_n be the affine variety defined in Example 7.27.
  (a) Prove that
$$X_n = \{(b^n, b^{n+1}, \ldots, b^{2n-1}) \mid b \in \mathbb{A}^1\}.$$
  (Hint: Take b to be the quotient of the first two coordinates.)
  (b) Prove that $\dim(X_n) = 1$. (Hint: Prove that $\{[x_1]\}$ is a Noether basis.)
  (c) Prove that X_n is irreducible. (Hint: Part (a) may be helpful.)

7.4.5 This exercise illustrates some strange behavior of tangent spaces that occurs over the real numbers. Consider the irreducible real affine variety

$$X = \mathcal{V}(y^2 - x(x+1)^2) \subseteq \mathbb{A}_\mathbb{R}^2.$$

  (a) Prove that $(-1, 0)$ is an isolated point of X. In other words, prove that $(-1, 0) \in X$ and you can find a circular disk D of some positive radius centered at $(-1, 0)$ such that $D \cap X = \{(-1, 0)\}$.
  (b) Given that $(-1, 0)$ is an isolated point of X, you might expect that $\dim(T_{(-1,0)} X) = 0$. To the contrary, show that $\dim(T_{(-1,0)} X) = 2$.
  (c) As we have seen in previous examples, the real solutions of polynomial equations do not typically see the whole picture; often one needs to look at the complex solutions. If we consider the complex variety

$$X_\mathbb{C} = \mathcal{V}(y^2 - x(x+1)^2) \subseteq \mathbb{A}_\mathbb{C}^2,$$

  what do you think is happening at the point $(-1, 0) \in X_\mathbb{C}$ that helps explain your answer to part (b)?

Section 7.5 Smooth and singular points

One of the most important aspects of tangent spaces is that they detect when affine varieties have "kinks" or "cusps." For instance, we saw in Example 7.5 that the "cusp" point in the variety $\mathcal{V}(x^2 - y^3) \subseteq \mathbb{A}^2$ is special in that the tangent space at this point is two-dimensional, while the tangent spaces at all other points are one-dimensional. Such special points in a variety where the dimension of the tangent space jumps are called singular points, as we make precise in the next definition.

> **7.28 DEFINITION** *Smooth and singular points*
>
> Let X be an irreducible affine variety and let $a \in X$. We say that X is *smooth at a* if $\dim(T_a X) = \dim(X)$; otherwise, we say that X is *singular at a*. We say that X is *smooth* if it is smooth at every point $a \in X$; otherwise, we say that X is *singular*.

> By Proposition 7.23, singular points are points $a \in X$ for which
> $$\dim(T_a X) > \dim(X).$$

One can also define smoothness of reducible varieties, but because different components can have different dimensions, the dimension of X should be replaced with its *local dimension at a*, which is the maximum dimension of all irreducible components that contain a. For simplicity, we restrict our focus to irreducible varieties throughout our discussion of smoothness.

7.29 EXAMPLE Affine space is smooth

Since $\mathcal{I}(\mathbb{A}^n) = \{0\} \subseteq K[x_1, \ldots, x_n]$, it follows that, for any $a \in \mathbb{A}^n$, we have
$$L_a \mathbb{A}^n = \mathcal{V}(L_a 0) = \mathcal{V}(0) = \mathbb{A}^n,$$
which implies that $T_a \mathbb{A}^n = K^n$. Therefore,
$$\dim(T_a \mathbb{A}^n) = \dim(K^n) = n = \dim(\mathbb{A}^n),$$
showing that $\mathbb{A}^n$ is smooth at all points.

7.30 EXAMPLE The cusp is singular at the origin

Consider the variety $X = \mathcal{V}(x^2 - y^3) \subseteq \mathbb{A}^2$ of Examples 7.5 and 7.20. The tangent space $T_a X$ is two-dimensional at $a = (0,0)$ and one-dimensional at all other points of X. Since X is itself a one-dimensional variety, this shows that X is singular at the origin but smooth at all other points.

7.31 EXAMPLE The cone is singular

Let $X = \mathcal{V}(x^2 + y^2 - z^2) \subseteq \mathbb{A}^3_\mathbb{C}$, which is the cone whose real points are depicted to the right. It can be shown that X is singular at the origin, where the surface is pinched down to a point, but smooth at all other points (Exercise 7.5.5).

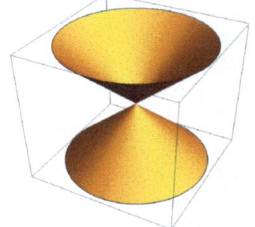

7.5. SMOOTH AND SINGULAR POINTS

Given that the tangent spaces of an affine variety $X \subseteq \mathbb{A}^n$ are vector subspaces of K^n, one might rightfully expect that tools from linear algebra can be used to determine the singular points of a variety. The key object we require to import tools of linear algebra into studying tangent spaces and singularities is the Jacobian matrix, which simply organizes the partial derivatives of a finite collection of polynomials.

7.32 DEFINITION *Jacobian matrix*

Given polynomials $f_1, \ldots, f_m \in K[x_1, \ldots, x_n]$, the *Jacobian matrix* of $f_1, \ldots, f_m$ is the $m \times n$ matrix

$$\mathrm{Jac}_{f_1,\ldots,f_m} = \begin{pmatrix} \frac{\partial f_1}{\partial x_1} & \cdots & \frac{\partial f_1}{\partial x_n} \\ \vdots & \ddots & \vdots \\ \frac{\partial f_m}{\partial x_1} & \cdots & \frac{\partial f_m}{\partial x_n} \end{pmatrix}.$$

Notice that the entries in the matrix $\mathrm{Jac}_{f_1,\ldots,f_m}$ are all elements of $K[x_1, \ldots, x_n]$, so it makes sense to evaluate them at any point $a \in \mathbb{A}^n$, obtaining an $m \times n$ matrix

$$\mathrm{Jac}_{f_1,\ldots,f_m}(a) \in K^{mn}.$$

The next result interprets the tangent space at a as the kernel of a Jacobian matrix, which leads to a determinantal characterization of the singular points of a variety.

7.33 PROPOSITION *Jacobian criterion for smoothness*

If $X \subseteq \mathbb{A}^n$ is an irreducible affine variety with $\mathcal{I}(X) = \langle f_1, \ldots, f_m \rangle$, then

$$T_a X = \ker(\mathrm{Jac}_{f_1,\ldots,f_m}(a)).$$

Consequently, X is singular at $a \in X$ if and only if

$$\mathrm{rk}(\mathrm{Jac}_{f_1,\ldots,f_m}(a)) < \mathrm{codim}(X).$$

PROOF Notice that the rows of $\mathrm{Jac}_{f_1,\ldots,f_m}(a)$ are the gradients of the functions $f_1, \ldots, f_m$:

$$\mathrm{Jac}_{f_1,\ldots,f_m}(a) = \begin{pmatrix} \nabla f_1(a) \\ \vdots \\ \nabla f_m(a) \end{pmatrix}.$$

By definition of matrix multiplication, we have that $v \in \ker(\mathrm{Jac}_{f_1,\ldots,f_m}(a))$ if and only if $\nabla f_i(a) \cdot v = 0$ for all $i = 1, \ldots, m$. Therefore, Proposition 7.8 and the comments immediately following its proof imply that

$$T_a X = \ker(\mathrm{Jac}_{f_1,\ldots,f_m}(a)).$$

The Rank-Nullity Theorem then implies that

$$\dim(T_a X) = n - \mathrm{rk}(\mathrm{Jac}_{f_1,\ldots,f_m}(a)).$$

By Proposition 7.23, it follows that

$$\text{rk}(\text{Jac}_{f_1,\ldots,f_m}(a)) \leq \text{codim}(X),$$

and the definition of smooth and singular points then implies that X is singular at a if and only if equality fails. □

The Jacobian criterion is especially simple in the case of hypersurfaces.

7.34 EXAMPLE Jacobian criterion for hypersurfaces

Let $X = \mathcal{V}(f)$ where $f \in K[x_1,\ldots,x_n]$ is irreducible, so that $\mathcal{I}(X) = \langle f \rangle$. Then $\text{codim}(X) = 1$ and the Jacobian criterion says that X is singular at a if and only if the $1 \times n$ matrix

$$\left(\frac{\partial f}{\partial x_1}(a) \quad \cdots \quad \frac{\partial f}{\partial x_n}(a) \right)$$

has rank zero. Thus, X is singular at a if and only if

$$\frac{\partial f}{\partial x_1}(a) = \cdots = \frac{\partial f}{\partial x_n}(a) = 0.$$

By Proposition 7.33, the singular points of a variety are characterized as the points where the Jacobian drops rank. A convenient attribute of this characterization is that the points at which the Jacobian drops rank are, themselves, the solutions of a system of polynomial equations. This leads to the following important consequence.

7.35 COROLLARY Singular points are closed

If $X \subseteq \mathbb{A}^n$ is an irreducible affine variety and $\text{Sing}(X) \subseteq X$ is the set of singular points of X, then $\text{Sing}(X)$ is also an affine variety.

Before presenting the proof, we first recall that an $r \times r$ *minor* of a matrix M is the determinant of an $r \times r$ matrix obtained by removing some subset of rows and columns from M, and an important result in linear algebra states that $\text{rk}(M) < r$ if and only if all $r \times r$ minors of M vanish.

PROOF OF COROLLARY 7.35 Choose a generating set $\mathcal{I}(X) = \langle f_1,\ldots,f_m \rangle$ and, for simplicity, set $r = \text{codim}(X)$. By Proposition 7.33, a point $a \in \mathbb{A}^n$ is a singular point of X if and only if $f_i(a) = 0$ for all $i = 1,\ldots,m$ and

$$\text{rk}(\text{Jac}_{f_1,\ldots,f_m}(a)) < r.$$

Thus, $\text{Sing}(X)$ is the vanishing set of the polynomials $f_1,\ldots,f_m \in K[x_1,\ldots,x_n]$ together with the $r \times r$ minors of $\text{Jac}_{f_1,\ldots,f_m}$, each of which is a polynomial in $K[x_1,\ldots,x_n]$. □

Knowing that $\text{Sing}(X)$ is an affine variety allows us to study $\text{Sing}(X)$ using all of the tools in our affine-variety toolkit. For instance, we can compute the dimension of $\text{Sing}(X)$, and the next example argues that "almost all points" (in a dimension-theoretic sense) of an irreducible hypersurface are smooth.

7.5. SMOOTH AND SINGULAR POINTS

7.36 EXAMPLE Generic smoothness of irreducible hypersurfaces

Let $X = \mathcal{V}(f) \subseteq \mathbb{A}^n$ where $f \in K[x_1, \ldots, x_n]$ is irreducible. We claim that X has at least one smooth point. To justify this, suppose to the contrary that every point of X is singular. By the Jacobian criterion, this implies that

$$\frac{\partial f}{\partial x_1}, \ldots, \frac{\partial f}{\partial x_n} \in K[x_1, \ldots, x_n]$$

all vanish on X, so they are elements of the vanishing ideal $\mathcal{I}(X) = \langle f \rangle$. Since f is irreducible, it cannot be constant; without loss of generality, suppose that f has positive degree in x_1. Then $\partial f/\partial x_1$ is not the zero polynomial and has strictly smaller degree in x_1 than f, contradicting that it is an element of $\langle f \rangle$.

Thus, X has at least one smooth point, from which it follows that $\mathrm{Sing}(X) \subsetneq X$. Knowing that both $\mathrm{Sing}(X)$ and X are affine varieties, and that X is irreducible, it then follows from Proposition 6.38 that $\dim(\mathrm{Sing}(X)) < \dim(X)$. In other words, since the dimension of $\mathrm{Sing}(X)$ is strictly smaller than X, this says that the singular points form a very special subset of points of X: almost all points of X are smooth, a property that is often referred to as *generic smoothness*.

We note that generic smoothness holds for general affine varieties, following from the fact that every affine variety has at least one smooth point. A proof of this general fact is slightly beyond the scope of our current discussion.

Exercises for Section 7.5

7.5.1 Prove that smoothness is an intrinsic property of affine varieties.

7.5.2 Consider the affine variety $X = \mathcal{V}(y^2 - x^3 - x^2) \subseteq \mathbb{A}^2_\mathbb{C}$.
   (a) Draw a picture of the real points of X and make a conjecture about where X might be singular.
   (b) Use the Jacobian criterion to determine all points where X is singular.

7.5.3 Let $f \in K[x_1, \ldots, x_n]$ be a square-free polynomial and consider the affine variety $X = \mathcal{V}(x_{n+1}^2 - f) \in \mathbb{A}^{n+1}$.
   (a) Prove that X is irreducible.
   (b) Prove that X is smooth.

7.5.4 Prove that the complex unit sphere $X = \mathcal{V}(x_1^2 + \cdots + x_n^2 - 1) \subseteq \mathbb{A}^n_\mathbb{C}$ is smooth.

7.5.5 Prove that the complex cone $\mathcal{V}(x^2 + y^2 - z^2) \subseteq \mathbb{A}^3_\mathbb{C}$ is singular at the origin and smooth at all other points.

7.5.6 Find all singular points of the *Whitney umbrella*: $\mathcal{V}(x^2 + y^2 z) \subseteq \mathbb{A}^3$. (See Example 1.6 for an image of the real points of the Whitney umbrella.)

7.5.7 Consider the affine twisted cubic curve $X = \mathcal{V}(y - x^2, z - x^3) \subseteq \mathbb{A}^3$. We know that $X \cong \mathbb{A}^1$, and it follows that X is smooth. Give a different proof of smoothness using the Jacobian criterion.

Chapter 8

Products

> LEARNING OBJECTIVES FOR CHAPTER 8
> - Prove that products of affine varieties are affine varieties.
> - Determine the vanishing ideal $\mathcal{I}(X \times Y)$ from $\mathcal{I}(X)$ and $\mathcal{I}(Y)$.
> - Describe how irreducibility, dimension, and smoothness of a product of affine varieties relate to its factors.
> - Describe elements of tensor products of R-modules, and use the tensor product rules to detect when two elements are equal.
> - Identify, in concrete examples, tensor products of R-modules and R-algebras with more familiar modules and algebras.
> - Compute the coordinate ring of $X \times Y$ as a tensor product.

A common way in which to build new mathematical objects from old is to take Cartesian products. The product of two additive groups, for example, is a group when equipped with componentwise addition, and the product of two topological spaces is a topological space when equipped with the product topology. In this chapter, we seek to understand the Cartesian product of two affine varieties.

It takes a small bit of work to convince oneself that the product of affine varieties is itself an affine variety, and after this is carried out, we can study the geometric attributes of products, such as irreducibility, dimension, and smoothness; the first two sections of this chapter are devoted to this development. The culmination of these efforts is a proof of the fact that the product of two smooth affine varieties X and Y is a smooth affine variety of dimension $\dim(X) + \dim(Y)$.

After familarizing ourselves with the geometric attributes of products of affine varieties, the equivalence of algebra and geometry invites a natural question: what is the algebraic operation on coordinate rings that corresponds to the geometric operation of Cartesian product? In other words, how is the K-algebra $K[X \times Y]$ related to $K[X]$ and $K[Y]$?

The answer to this question is that $K[X \times Y]$ is the *tensor product* of $K[X]$ and $K[Y]$. In the special case where $X = Y = \mathbb{A}^1$, this is the statement that $K[x, y]$ is the tensor product of $K[x]$ and $K[y]$. Because tensor products are likely to be unfamiliar—or if not unfamiliar, perhaps intimidating—to many readers, we do not assume any prior knowledge of them. Instead, we start from the goal of defining an algebraic operation that combines $K[x]$ and $K[y]$ to produce $K[x, y]$, and we build up the definition slowly to suit that goal. The chapter culminates with a proof that $K[X \times Y] = K[X] \otimes_K K[Y]$ for any affine varieties X and Y.

Section 8.1 Products of affine varieties

Given affine varieties $X \subseteq \mathbb{A}^m$ and $Y \subseteq \mathbb{A}^n$, the Cartesian product $X \times Y$ is defined set-theoretically simply as the collection of ordered pairs:

$$X \times Y = \{(a,b) \mid a \in X, b \in Y\} \subseteq \mathbb{A}^m \times \mathbb{A}^n.$$

On the other hand, $\mathbb{A}^m \times \mathbb{A}^n$ can be canonically identified with $\mathbb{A}^{m+n}$, so we can readily view $X \times Y$ as a subset of an affine space:

In what follows, we use different variables for $\mathbb{A}^m$ and $\mathbb{A}^n$ to avoid confusing the two affine spaces.

$$X \times Y \subseteq \mathbb{A}^{m+n}.$$

Our first goal is to prove that, within $\mathbb{A}^{m+n}$, the set $X \times Y$ is itself an affine variety.

8.1 PROPOSITION *Products of affine varieties are affine varieties*

If $X = \mathcal{V}(f_1, \ldots, f_r) \subseteq \mathbb{A}^m$ and $Y = \mathcal{V}(g_1, \ldots, g_s) \subseteq \mathbb{A}^n$, where $f_1, \ldots, f_r \in K[x_1, \ldots, x_m]$ and $g_1, \ldots, g_s \in K[y_1, \ldots, y_n]$, then

$$X \times Y = \mathcal{V}(f_1, \ldots, f_r, g_1, \ldots, g_s) \subseteq \mathbb{A}^{m+n},$$

where we view the polynomials $f_1, \ldots, f_r$ and $g_1, \ldots, g_s$ as elements of the larger polynomial ring $K[x_1, \ldots, x_m, y_1, \ldots, y_n]$.

While we have stated Proposition 8.1 for products of two affine varieties, it naturally generalizes to finite Cartesian products (Exercise 8.1.4).

PROOF By definition, a point $(a,b) = (a_1, \ldots, a_m, b_1, \ldots, b_n)$ of $\mathbb{A}^{m+n}$ lies in $X \times Y$ if and only if $a \in X$ and $b \in Y$. By definition of X and Y, this is true if and only if

$$f_1(a) = \cdots = f_r(a) = 0 \quad \text{and} \quad g_1(b) = \cdots = g_s(b) = 0.$$

If we now view $f_1, \ldots, f_r$ as elements of $K[x_1, \ldots, x_m, y_1, \ldots, y_n]$ that happen to involve only the first m variables, then $f_i(a,b) = f_i(a)$. Similarly, $g_j(a,b) = g_j(b)$. Thus, $(a,b) \in X \times Y$ if and only if

$$f_1(a,b) = \cdots = f_r(a,b) = g_1(a,b) = \cdots = g_s(a,b) = 0,$$

which says precisely that $X \times Y = \mathcal{V}(f_1, \ldots, f_r, g_1, \ldots, g_s)$. □

8.2 EXAMPLE Parallel parabolas in $\mathbb{A}^3$

Let $X = \mathcal{V}(y - x^2) \subseteq \mathbb{A}^2$ and $Y = \mathcal{V}(z^2 - 1) \subseteq \mathbb{A}^1$, so that X is a parabola in $\mathbb{A}^2$ and Y is the two points $\{\pm 1\} \subseteq \mathbb{A}^1$. Then

$$X \times Y = \mathcal{V}(y - x^2, z^2 - 1) \subseteq \mathbb{A}^3,$$

which can be visualized as two parallel parabolas that are contained in the planes that are parallel to the xy-plane at a height of ± 1.

8.1. PRODUCTS OF AFFINE VARIETIES

8.3 EXAMPLE A cylinder

Let $X = \mathcal{V}(x^2 + y^2 - 1) \subseteq \mathbb{A}^2$ and $Y = \mathbb{A}^1$. Then
$$X \times Y = \mathcal{V}(x^2 + y^2 - 1) \subseteq \mathbb{A}^3,$$
which, over the real numbers, is the circular cylinder depicted below.

Knowing that products of affine varieties are, themselves, affine varieties is an important first step in studying products, but in order to study coordinate rings, dimension, and tangent spaces, we require an understanding of their vanishing ideals.

8.4 PROPOSITION *Vanishing ideal of a product*

Let $X \subseteq \mathbb{A}^m$ and $Y \subseteq \mathbb{A}^n$ be affine varieties. Then
$$\mathcal{I}(X \times Y) = \langle \mathcal{I}(X) \rangle + \langle \mathcal{I}(Y) \rangle,$$
where we view $\mathcal{I}(X) \subseteq K[x_1, \ldots, x_m]$ and $\mathcal{I}(Y) \subseteq K[y_1, \ldots, y_n]$ as subsets of the larger polynomial ring $K[x_1, \ldots, x_m, y_1, \ldots, y_n]$.

We note that $\mathcal{I}(X)$ and $\mathcal{I}(Y)$ are not ideals when viewed as subsets of $K[x_1, \ldots, x_m, y_1, \ldots, y_n]$, which is why we first consider the ideals generated by them before taking the ideal sum. Given generators $\mathcal{I}(X) = \langle f_1, \ldots, f_k \rangle$ and $\mathcal{I}(Y) = \langle g_1, \ldots, g_\ell \rangle$, Proposition 8.4 implies (Exercise 8.1.5) that

> Recall that the ideal sum is defined by $I + J = \{a + b \mid a \in I, b \in J\}$.

$$\mathcal{I}(X \times Y) = \langle f_1, \ldots, f_k, g_1, \ldots g_\ell \rangle \subseteq K[x_1, \ldots, x_m, y_1, \ldots, y_n].$$

PROOF OF PROPOSITION 8.4 Note that a general element of $\langle \mathcal{I}(X) \rangle$ is of the form
$$h_1 f_1 + \cdots + h_k f_k$$
for some $h_1, \ldots, h_k \in K[x_1, \ldots, x_m, y_1, \ldots, y_n]$ and $f_1, \ldots, f_k \in \mathcal{I}(X)$. Evaluating any such polynomial at $(a, b) \in X \times Y$, we obtain
$$h_1(a,b) f_1(a) + \cdots + h_k(a,b) f_k(a) = 0,$$
because $f_1(a) = \cdots = f_k(a) = 0$ for all $a \in X$. A similar argument shows that any polynomial in $\langle \mathcal{I}(Y) \rangle$ vanishes at every point of $X \times Y$, and it then follows that every polynomial of $\langle \mathcal{I}(X) \rangle + \langle \mathcal{I}(Y) \rangle$ vanishes on $X \times Y$, showing that
$$\langle \mathcal{I}(X) \rangle + \langle \mathcal{I}(Y) \rangle \subseteq \mathcal{I}(X \times Y).$$

To prove the other inclusion, suppose that $h \in \mathcal{I}(X \times Y)$. It is always possible to find an expression of the form

(8.5) $$h = f_1 g_1 + \cdots + f_k g_k$$

with $f_1, \ldots, f_k \in K[x_1, \ldots, x_m]$ and $g_1, \ldots, g_k \in K[y_1, \ldots, y_n]$ (for example, write h as a sum of monomials and then separate the x-variables from the y-variables in each monomial). In any expression of the form (8.5), we may assume after rearranging that $g_1, \ldots, g_j \notin \mathcal{I}(Y)$ and $g_{j+1}, \ldots, g_k \in \mathcal{I}(Y)$; fix an expression of this form for which j is minimal. For this particular expression with j minimal, we claim that $f_1, \ldots, f_j \in \mathcal{I}(X)$, from which it follows that

$$h = f_1 g_1 + \cdots + f_j g_j + f_{j+1} g_{j+1} + \cdots + f_k g_k \in \langle \mathcal{I}(X) \rangle + \langle \mathcal{I}(Y) \rangle,$$

and the result follows.

To prove the claim, suppose toward a contradiction that one of $f_1, \ldots, f_j$ does not lie in $\mathcal{I}(X)$. Without loss of generality, suppose that $f_1(a) \neq 0$ for some $a \in X$. Define

$$g = h(a, y) = f_1(a) g_1 + f_2(a) g_2 + \cdots + f_k(a) g_k \in K[y_1, \ldots, y_n].$$

Since h vanishes on $X \times Y$, it follows that $g \in \mathcal{I}(Y)$. Solving for g_1, we have

$$g_1 = f_1(a)^{-1}(g - f_2(a) g_2 - \cdots - f_k(a) g_k).$$

Replacing g_1 by this expression in (8.5), we arrive at an expression for h of the form

$$h = (f_2 - f_1(a)^{-1} f_2(a) f_1) g_2 + \cdots + (f_k - f_1(a)^{-1} f_k(a) f_1) g_k + f_1(a)^{-1} f_1 g.$$

In this new expression, $g_2, \ldots, g_j \notin \mathcal{I}(Y)$ while $g_{j+1}, \ldots, g_k, g \in \mathcal{I}(Y)$, contradicting the minimality of j and completing the proof. $\square$

Proposition 8.4 allows us to compute coordinate rings of products:

$$K[X \times Y] = \frac{K[x_1, \ldots, x_m, y_1, \ldots, y_n]}{\langle \mathcal{I}(X) \rangle + \langle \mathcal{I}(Y) \rangle}.$$

However, there is an unsatisfactory nature to this description, as it requires viewing both X and Y extrinsically as subsets of affine spaces, not merely as isomorphism classes of affine varieties. Thus, we are met with a natural question: can we compute the coordinate ring $K[X \times Y]$ intrinsically in terms of the two coordinate rings $K[X]$ and $K[Y]$?

The best possible answer to this question would be to specify an algebraic operation $\star$ such that

$$K[X] \star K[Y] = K[X \times Y]$$

for all affine varieties X and Y. Whatever $\star$ might be, a special case would necessarily be that

$$K[x] \star K[y] = K[x, y],$$

which comes from taking $X = Y = \mathbb{A}^1$ and therefore $X \times Y = \mathbb{A}^2$.

8.1. PRODUCTS OF AFFINE VARIETIES

The reader might search her mind at this moment for ways that she currently knows to combine two K-algebras A and B to produce a new K-algebra $A \star B$. The first such operation that might come to mind is the direct sum $A \oplus B$. This is not what we seek, though; the natural ring structure on $A \oplus B$ is componentwise:

$$(a_1, b_1) + (a_2, b_2) = (a_1 + a_2, b_1 + b_2),$$
$$(a_1, b_1) \cdot (a_2, b_2) = (a_1 a_2, b_1 b_2),$$

and the reader can verify (Exercise 8.1.7) that this never produces an integral domain. Thus, $K[x, y]$ cannot be the same as $K[x] \oplus K[y]$.

As we will soon see, the correct choice for the operation $\star$ turns out to be the *tensor product* of K-algebras. In Sections 8.3 and 8.4, we develop the algebraic foundations of tensor products, culminating with a proof in Section 8.5 of the fact that the tensor product indeed captures the coordinate ring of a product of affine varieties. Before diving into tensor products, however, we first devote Section 8.2 to studying the key geometric attributes of products, answering natural questions regarding irreducibility, dimension, and smoothness.

Exercises for Section 8.1

8.1.1 Let $X = \mathcal{V}(x^2 - 1) \subseteq \mathbb{A}^1$ and let $Y = \mathcal{V}(y - z) \subseteq \mathbb{A}^2$. Draw a picture over $\mathbb{R}$ of $X \times Y \subseteq \mathbb{A}^3$.

8.1.2 Draw a picture over $\mathbb{R}$ of the affine variety

$$Z = \mathcal{V}(x + z^2 - z, x) \subseteq \mathbb{A}^3 = \mathbb{A}^2 \times \mathbb{A}^1.$$

Is $Z = X \times Y$ for affine varieties $X \subseteq \mathbb{A}^2$ and $Y \subseteq \mathbb{A}^1$? If so, give an example of such X and Y. If not, how can you tell?

8.1.3 Draw a picture over $\mathbb{R}$ of the affine variety

$$Z = \mathcal{V}(xyz) \subseteq \mathbb{A}^3 = \mathbb{A}^2 \times \mathbb{A}^1.$$

Is $Z = X \times Y$ for affine varieties $X \subseteq \mathbb{A}^2$ and $Y \subseteq \mathbb{A}^1$? If so, give an example of such X and Y. If not, how can you tell?

8.1.4 Generalize and prove Proposition 8.1 for a finite product of affine varieties.

8.1.5 Let $X \subseteq \mathbb{A}^m$ and $Y \subseteq \mathbb{A}^n$ be affine varieties. If $\mathcal{I}(X) = \langle f_1, \ldots, f_k \rangle$ and $\mathcal{I}(Y) = \langle g_1, \ldots, g_\ell \rangle$, prove that

$$\mathcal{I}(X \times Y) = \langle f_1, \ldots, f_k, g_1, \ldots g_\ell \rangle \subseteq K[x_1, \ldots, x_m, y_1, \ldots, y_n].$$

8.1.6 Let X, X', Y, Y' be affine varieties with $X \cong X'$ and $Y \cong Y'$. Prove that

$$X \times Y \cong X' \times Y'.$$

In other words, products of affine varieties are intrinsic.

8.1.7 Prove that, if A and B are nonzero rings, then the ring $A \oplus B$ with componentwise addition and multiplication is not an integral domain.

Section 8.2 Attributes of products

Let us begin our study of the geometric attributes of products with the following central result regarding irreducibility, which will be useful in the study of both dimension and smoothness of products.

> **8.6 PROPOSITION** *Products of irreducible varieties are irreducible*
>
> If X and Y are irreducible affine varieties, then $X \times Y$ is irreducible.

Before embarking on the proof, let us pause to consider a few concrete examples to illustrate how irreducibility interacts with products.

8.7 EXAMPLE An irreducible product

If $X = \mathcal{V}(y - x^2) \subseteq \mathbb{A}^2$ and $Y = \mathcal{V}(z - 1) \subseteq \mathbb{A}^1$, then both X and Y are irreducible, and the product

$$X \times Y = \mathcal{V}(y - x^2, z - 1) \subseteq \mathbb{A}^3$$

is a parabola in $\mathbb{A}^3$. The irreducibility of $X \times Y$ can be seen from the isomorphism $X \times Y \cong X$, which is simply a restriction of the vertical projection $\mathbb{A}^3 \to \mathbb{A}^2$.

8.8 EXAMPLE A reducible product

Let $X = \mathcal{V}(xy) \subseteq \mathbb{A}^2$ and let $Y = \mathbb{A}^1$. Then $X = \mathcal{V}(x) \cup \mathcal{V}(y)$ is reducible, as is the product:

$$X \times Y = \mathcal{V}(x) \cup \mathcal{V}(y) \subseteq \mathbb{A}^3,$$

depicted in the image below.

While this example is rather simplified by the fact that Y is an entire affine space, it contains the kernel of the proof of Proposition 8.6: if Y is irreducible, then a decomposition of $X \times Y$ as a union of two affine varieties can only come from a similar decomposition of X.

PROOF OF PROPOSITION 8.6 Suppose that $X \subseteq \mathbb{A}^m$ and $Y \subseteq \mathbb{A}^n$ are irreducible affine varieties. To prove that $X \times Y$ is irreducible, we must show that one can only have

(8.9) $$X \times Y = Z_1 \cup Z_2$$

for affine varieties $Z_1, Z_2 \subseteq X \times Y \subseteq \mathbb{A}^{m+n}$ if either $Z_1 = X \times Y$ or $Z_2 = X \times Y$.

8.2. ATTRIBUTES OF PRODUCTS

Suppose that (8.9) holds, and for each point $a \in X$, consider the set

$$\{a\} \times Y \subseteq X \times Y.$$

For example, if $X \times Y$ is the cylinder of Example 8.3 or the intersecting planes of Example 8.8, then the sets $\{a\} \times Y$ are the vertical lines in the product. Each set $\{a\} \times Y$ is an affine variety (because both $\{a\}$ and Y are affine varieties), and they are all isomorphic to Y; therefore, since Y is irreducible, $\{a\} \times Y$ is irreducible for all $a \in X$. Intersecting both sides of (8.9) with $\{a\} \times Y$ yields

$$\{a\} \times Y = \Big(\big(\{a\} \times Y\big) \cap Z_1\Big) \cup \Big(\big(\{a\} \times Y\big) \cup Z_2\Big).$$

Given that $\{a\} \times Y$ is irreducible, it must be the case that either

$$(\{a\} \times Y) \cap Z_1 = \{a\} \times Y \quad \text{or} \quad (\{a\} \times Y) \cap Z_2 = \{a\} \times Y.$$

In other words, for any $a \in X$, either $\{a\} \times Y \subseteq Z_1$ or $\{a\} \times Y \subseteq Z_2$. Thus, defining two sets

$$X_1 = \{a \in X \mid \{a\} \times Y \subseteq Z_1\} \quad \text{and} \quad X_2 = \{a \in X \mid \{a\} \times Y \subseteq Z_2\},$$

we have

(8.10) $$X_1 \cup X_2 = X.$$

So far, we only know that X_1 and X_2 are subsets of X, but in fact, they are affine varieties themselves. To see this, recall that Z_1 is an affine variety, so we have

$$Z_1 = \mathcal{V}(f_1, \ldots, f_r)$$

for some polynomials $f_1, \ldots, f_r \in K[x_1, \ldots, x_m, y_1, \ldots, y_n]$. For any $b \in Y$, consider the affine variety

$$(X_1)_b = \mathcal{V}(f_1(x, b), \ldots, f_r(x, b)) \subseteq \mathbb{A}^m,$$

and notice that

$$(X_1)_b = \{a \in \mathbb{A}^m \mid (a, b) \in Z_1\}.$$

Furthermore, we have

$$X_1 = \{a \in X \mid (a, b) \in Z_1 \text{ for all } b \in Y\} = X \cap \bigcap_{b \in Y} (X_1)_b,$$

so X_1 is an intersection of affine varieties and is thus an affine variety. The same argument applies to X_2.

Recalling equation (8.10), we have now expressed X as a union of two affine varieties. Given that X is irreducible, this is only possible if either $X_1 = X$ or $X_2 = X$. The case that $X_1 = X$ means that for all $a \in X$, we have $\{a\} \times Y \subseteq Z_1$, or in other words, $X \times Y \subseteq Z_1$, implying that $X \times Y = Z_1$. Similarly, the case that $X_2 = X$ means that $X \times Y = Z_2$. We have thus shown that $X \times Y = Z_1$ or $X \times Y = Z_2$, and this completes the proof that $X \times Y$ is irreducible. □

Proposition 8.6 leads to the following description of the irreducible decomposition of a product, which the reader is encouraged to prove in Exercise 8.2.3.

8.11 COROLLARY *Irreducible decomposition of a product*

If X and Y are affine varieties with irreducible components $X_1, \ldots, X_k$ and $Y_1, \ldots, Y_\ell$, respectively, then the irreducible components of $X \times Y$ are the affine varieties $X_i \times Y_j$ for all $i = 1, \ldots, k$ and $j = 1, \ldots, \ell$.

We now turn to dimension. Either by analogy to vector-space dimension or by viewing dimension intuitively as the number of independent directions in which a point can move, one might naturally expect that $\dim(X \times Y) = \dim(X) + \dim(Y)$. The next result says that this expectation is indeed satisfied.

8.12 PROPOSITION *Dimension is additive over products*

If X and Y are nonempty affine varieties, then
$$\dim(X \times Y) = \dim(X) + \dim(Y).$$

PROOF   We proceed by induction on $\dim(X) + \dim(Y)$.

(Base case) If $\dim(X) + \dim(Y) = 0$, then $\dim(X) = \dim(Y) = 0$, so X and Y are each a finite set of points. Since Cartesian products of finite sets are finite sets, it follows that $X \times Y$ is a finite set, so $\dim(X \times Y) = 0 = \dim(X) + \dim(Y)$.

(Induction step) Let $d > 0$ and suppose that dimension is additive on products whenever the factors have dimensions adding up to less than d. Let $X \subseteq \mathbb{A}^m$ and $Y \subseteq \mathbb{A}^n$ be nonempty affine varieties with $\dim(X) + \dim(Y) = d$; we must prove that $\dim(X \times Y) = \dim(X) + \dim(Y)$.

Upon decomposing X and Y into their irreducible components, Corollary 8.12 implies that the dimension of $X \times Y$ is the maximum dimension of $X_i \times Y_j$, where X_i and Y_j are irreducible components of X and Y. Thus, if the conclusion of the induction step holds for the irreducible components of X and Y, then it also holds for X and Y. Therefore, it suffices to prove the induction step for irreducible affine varieties, and we henceforth assume that X and Y—and thus $X \times Y$—are irreducible.

Since $\dim(X) + \dim(Y) > 0$, we may assume, without loss of generality, that $\dim(X) > 0$. Thus, X has more than one point and we can find a polynomial $f \in K[x_1, \ldots, x_m]$ such that $\emptyset \subsetneq \mathcal{V}(f) \cap X \subsetneq X$. Viewing this polynomial also as an element of $K[x_1, \ldots, x_m, y_1, \ldots, y_n]$, we have
$$\mathcal{V}(f) \cap (X \times Y) = (\mathcal{V}(f) \cap X) \times Y.$$

Applying the Fundamental Theorem of Dimension Theory to both sides of the above equality, we then obtain the desired conclusion:
$$\begin{aligned}\dim(X \times Y) &= \dim(\mathcal{V}(f) \cap (X \times Y)) + 1 \\ &= \dim((\mathcal{V}(f) \cap X) \times Y) + 1 \\ &= (\dim(X) - 1) + \dim(Y) + 1 = \dim(X) + \dim(Y),\end{aligned}$$
where the third equality uses the induction hypothesis.   □

8.2. ATTRIBUTES OF PRODUCTS

The stage is now set for the final topic of this section: smoothness of products of affine varieties. As smoothness is defined in terms of tangent spaces, and tangent spaces are defined in terms of linearizations, the following result describing linearizations and tangent spaces of products is the key step toward showing that a product of smooth affine varieties is smooth. (The statement uses the direct sum $V \oplus W$ of vector spaces, which, we remind the reader, is a vector space of dimension $\dim(V) + \dim(W)$ consisting of pairs (v, w) with $v \in V$ and $w \in W$.)

8.13 PROPOSITION *Tangent spaces of a product*

If $X \subseteq \mathbb{A}^m$ and $Y \subseteq \mathbb{A}^n$ are affine varieties with $a \in X$ and $b \in Y$, then

$$L_{(a,b)}(X \times Y) = L_a X \times L_b Y \quad \text{and} \quad T_{(a,b)}(X \times Y) = T_a X \oplus T_b Y.$$

PROOF Suppose that $\mathcal{I}(X) = \langle f_1, \ldots, f_k \rangle$ and $\mathcal{I}(Y) = \langle g_1, \ldots, g_\ell \rangle$. By Proposition 8.4 and the discussion immediately thereafter, we have

$$\mathcal{I}(X \times Y) = \langle f_1, \ldots, f_k, g_1, \ldots, g_\ell \rangle.$$

As we saw in Section 7.1, the linearizations can then be computed as

$$L_a X = \mathcal{V}(L_a f_1, \ldots, L_a f_k),$$
$$L_b Y = \mathcal{V}(L_b g_1, \ldots, L_b g_\ell), \quad \text{and}$$
$$L_{(a,b)}(X \times Y) = \mathcal{V}(L_a f_1, \ldots, L_a f_k, L_b g_1, \ldots, L_b g_\ell),$$

where we have used that $L_{(a,b)} f_i = L_a f_i$ because $f_i \in K[x_1, \ldots, x_m]$, and similarly for g_j. Thus, by Proposition 8.1, we conclude that $L_{(a,b)}(X \times Y) = L_a X \times L_b Y$.

To prove the statement regarding tangent spaces, recall from Section 7.1 that

$$T_a X = \{\overrightarrow{aa'} \mid a' \in L_a X\} \subseteq K^m \quad \text{and} \quad T_b Y = \{\overrightarrow{bb'} \mid b' \in L_b Y\} \subseteq K^n.$$

Therefore, the vector space $T_a X \oplus T_b Y \subseteq K^m \oplus K^n$ consists of vectors of the form

$$(\overrightarrow{aa'}, \overrightarrow{bb'})$$

with $a' \in L_a X$ and $b' \in L_b Y$. Under the identification $K^m \oplus K^n = K^{m+n}$, we have

$$(\overrightarrow{aa'}, \overrightarrow{bb'}) = \overrightarrow{(a,b)(a',b')}.$$

Thus, we conclude that

$$T_a X \oplus T_b Y = \left\{ \overrightarrow{(a,b)(a',b')} \mid (a', b') \in L_a X \times L_b Y \right\} = T_{(a,b)}(X \times Y),$$

where the final equality uses that $L_a X \times L_b Y = L_{(a,b)}(X \times Y)$. $\square$

That products of smooth affine varieties are smooth now follows quickly.

8.14 COROLLARY *Products of smooth varieties are smooth*

If X and Y are smooth affine varieties, then so is $X \times Y$.

PROOF   Suppose that X and Y are smooth (irreducible) affine varieties. Given any $(a,b) \in X \times Y$, we have

$$\begin{aligned}\dim(X \times Y) &= \dim(X) + \dim(Y) \\ &= \dim(T_a X) + \dim(T_b Y) \\ &= \dim(T_a X \oplus T_b Y) \\ &= \dim(T_{(a,b)}(X \times Y)),\end{aligned}$$

where the first equality is Proposition 8.12, the second uses the assumption that both X and Y are smooth, the third is the fact that vector-space dimension is additive over direct sums, and the fourth is the second statement in Proposition 8.13. By definition of smoothness, this implies that $X \times Y$ is smooth. □

Having now familiarized ourselves with products of affine varieties and their attributes, the rest of this chapter is devoted to developing an intrinsic construction of the coordinate ring of a product. As we mentioned at the end of the previous section, the key construction that we require is that of tensor products. In the next two sections, we develop the algebraic foundations of tensor products—with an eye toward tensor products of polynomial rings—and in the final section of this chapter, we use our developments to argue that the coordinate ring of a product is indeed the tensor product of the coordinate rings of each factor.

Exercises for Section 8.2

8.2.1 Does the converse of Proposition 8.6 hold? Prove or give a counterexample.

8.2.2 Let $X = \mathcal{V}(xy) \subseteq \mathbb{A}^2$ and let $Y = \mathcal{V}(z^2 - 1) \subseteq \mathbb{A}^1$. Describe $X \times Y$ and calculate its irreducible decomposition.

8.2.3 Prove Corollary 8.11.

8.2.4 Does the converse of Corollary 8.14 hold? More specifically, if X and Y are irreducible and $X \times Y$ is smooth, does this imply that both X and Y are smooth? Prove or give a counterexample.

8.2.5 Choose one of the five results from this section and state a generalization of it that applies to a finite product of affine varieties. Use an induction argument to deduce the general statement from the special case of products with two factors.

Section 8.3 Tensor products of modules and algebras

The goal of this section is to define the *tensor product* $M \otimes_R N$, an R-module that is built from R-modules M and N. Additionally, if M and N happen to be R-algebras, then the tensor product will also inherit the structure of an R-algebra. The model situation is when $M = R[x]$ and $N = R[y]$, in which case we will have

$$R[x] \otimes_R R[y] = R[x, y].$$

In order to understand how to generalize this example to general modules, we focus on one key property of $R[x, y]$, which we have already encountered in the proof of Proposition 8.4: every polynomial $h \in R[x, y]$ can be expressed (non-uniquely) as a sum

$$h = f_1 g_1 + f_2 g_2 + \cdots + f_k g_k$$

with $f_1, \ldots, f_k \in R[x]$ and $g_1, \ldots, g_k \in R[y]$. For instance, $3x^2 y + 2x + 4xy^2$ can be written as a sum of three terms

$$(3x^2) \cdot y + (2x) \cdot 1 + (4x) \cdot y^2.$$

Now suppose that M and N are any R-modules. A "product" $f_i g_i$ in which $f_i \in M$ and $g_i \in N$ can be understood, formally, as an element of $M \times N$, so

> *Recall that, in a ring R (with unity), the number 3 represents $1 + 1 + 1$.*

our first step toward defining $M \otimes_R N$ is to construct an R-module in which it makes sense to add such products. Our starting point for this is with the notion of *formal linear combinations*.

8.15 DEFINITION   *Formal linear combination*

Let S be a set. A *formal R-linear combination* of elements of S is an expression of the form

$$\sum_{s \in S} a_s \cdot s,$$

where $a_s \in R$ for all $s \in S$ and $a_s = 0$ for all but finitely many s. We consider two formal R-linear combinations equal if and only if all of their coefficients agree; that is,

$$\sum_{s \in S} a_s \cdot s = \sum_{s \in S} b_s \cdot s \iff a_s = b_s \text{ for all } s \in S.$$

We stress here that the elements of S are nothing more than symbols that record the information of their coefficients, analogously to the way the variables $x_1, \ldots, x_n$ in a monomial $x_1^{\alpha_1} \cdots x_n^{\alpha_n}$ are nothing more than symbols that record the information of their exponents $\alpha_1, \ldots, \alpha_n$. In particular, the data of a formal linear combination of elements of S is equivalent to the data of a function

$$f : S \to R$$
$$f(s) = a_s$$

for which $f(s) = 0$ for all but finitely many $s \in S$. We choose to express this information as a linear combination only for notational convenience.

8.16 EXAMPLE Formal linear combinations of elements of a finite set

Let $S = \{\diamondsuit, \heartsuit, \clubsuit, \spadesuit\}$. Then

$$3\diamondsuit - 2\heartsuit + 1\clubsuit + 7\spadesuit$$

and

$$0\diamondsuit + 4\heartsuit - 30\clubsuit + 0\spadesuit$$

> We have chosen a strange set S here to emphasize that the elements of S need not have any structure, nor any meaning beyond their role as placeholders for their coefficients.

are examples of formal $\mathbb{Z}$-linear combinations of elements of S.

8.17 EXAMPLE Formal linear combinations of elements of an infinite set

Let $S = \{x_1, x_2, x_3, \ldots\}$. Taking the standard convention of omitting a summand $a_s \cdot s$ when $a_s = 0$, formal $\mathbb{Z}$-linear combinations of elements of S are *finite* sums like

$$7x_2 + 4x_5 - 2x_6 \quad \text{or} \quad 3x_1 + 5x_{17} - 12x_{100} - 9x_{120}.$$

Even though S may not have any structure whatsoever, the set of all formal linear combinations of elements of S forms an R-module in a natural way.

8.18 DEFINITION Free module

The set of all formal R-linear combinations of elements of a set S is called the *free R-module on S* and is denoted RS.

Addition in RS is defined simply by adding corresponding coefficients; for example, the sum of the formal linear combinations in Example 8.16 is

> The term "free" refers to the fact that the elements of S are unrelated.

$$(3\diamondsuit - 2\heartsuit + 1\clubsuit + 7\spadesuit) + (0\diamondsuit + 4\heartsuit - 30\clubsuit + 0\spadesuit) = 3\diamondsuit + 2\heartsuit - 29\clubsuit + 7\spadesuit.$$

Similarly, scalar multiplication works as one might expect:

$$-2(3\diamondsuit - 2\heartsuit + 1\clubsuit + 7\spadesuit) = -6\diamondsuit + 4\heartsuit - 2\clubsuit - 14\spadesuit.$$

In Exercise 8.3.1, the reader is encouraged to verify that the general rules for addition and scalar multiplication in RS satisfy the axioms of an R-module.

Returning to our goal of defining $M \otimes_R N$, let M and N again be R-modules, and consider the free R-module on their Cartesian product:

$$R(M \times N) = \left\{ a_1 \cdot (m_1, n_1) + \cdots + a_k \cdot (m_k, n_k) \;\middle|\; \begin{array}{l} a_1,\ldots,a_k \in R, \\ m_1,\ldots,m_k \in M, \\ n_1,\ldots,n_k \in N \end{array} \right\}.$$

For example, an element of the free R-module on the set $R[x] \times R[y]$ might be

$$3 \cdot (x^2, y) + 2 \cdot (x, 1) + 4 \cdot (x, y^2).$$

8.3. TENSOR PRODUCTS OF MODULES AND ALGEBRAS

Removing commas from the previous expression, we obtain something that looks a lot like the polynomial
$$3x^2y + 2x + 4xy^2 \in R[x,y],$$
so the reader may become hopeful that this free R-module is isomorphic to $R[x,y]$.

Alas, this is not yet the case. While there is an R-module homomorphism from the free R-module on $R[x] \times R[y]$ to $R[x,y]$ defined by
$$a_1 \cdot (f_1, g_1) + \cdots + a_k \cdot (f_k, g_k) \longmapsto a_1 f_1 g_1 + \cdots + a_k f_k g_k,$$
this map is not injective. In particular, the three formal linear combinations
$$3 \cdot (x^2, y) + 2 \cdot (x, 1) + 4 \cdot (x, y^2)$$
$$1 \cdot (3x^2, y) + 1 \cdot (2x, 1) + 1 \cdot (4x, y^2), \text{ and}$$
$$1 \cdot (x^2, 3y) + 1 \cdot (x, 2 + 4y^2)$$
are all distinct elements of the domain that map to $3x^2y + 2x + 4xy^2 \in R[x,y]$.

By taking a quotient of the free R-module by a certain submodule—namely the kernel of the above homomorphism—we can equate elements whose images in $R[x,y]$ are the same. This, at last, gives us the definition of the tensor product. We return to the general context of R-modules to state the precise definition.

8.19 DEFINITION *Tensor product of R-modules*

Let M and N be R-modules. The *tensor product* of M and N is the quotient
$$M \otimes_R N = \frac{R(M \times N)}{H},$$
where H is the submodule generated by elements of the following forms:

$(m, n_1 + n_2) - (m, n_1) - (m, n_2)$, where $m \in M$, $n_1, n_2 \in N$,
$(m_1 + m_2, n) - (m_1, n) - (m_2, n)$, where $m_1, m_2 \in M$, $n \in N$,
$r(m, n) - (rm, n)$, where $r \in R$, $m \in M$, $n \in N$, and
$r(m, n) - (m, rn)$, where $r \in R$, $m \in M$, $n \in N$.

The coset of $a_1(m_1, n_1) + \cdots + a_k(m_k, n_k)$ in $M \otimes_R N$ is denoted

(8.20) $$a_1(m_1 \otimes n_1) + \cdots + a_k(m_k \otimes n_k).$$

Definition 8.19 is a lot to parse, but the main takeaway is that the quotient by H ensures that the following equations hold for all $m, m_1, m_2 \in M$, all $n, n_1, n_2 \in N$, and all $r \in R$:
$$m \otimes (n_1 + n_2) = m \otimes n_1 + m \otimes n_2;$$
$$(m_1 + m_2) \otimes n = m_1 \otimes n + m_2 \otimes n;$$
$$r(m \otimes n) = (rm) \otimes n = m \otimes (rn).$$

We refer to these equations as the "tensor product relations," and we will see in the examples and exercises how they are used to manipulate elements of tensor products.

8.21 EXAMPLE Elements of $R[x] \otimes_R R[y]$

In the tensor product $R[x] \otimes_R R[y]$, an example of an element is

$$3x \otimes y + x \otimes y^2 + 2x^2 \otimes 1.$$

By the tensor product relations, this is the same as the elements

$$x \otimes (3y) + x \otimes y^2 + 2x^2 \otimes 1 \quad \text{and} \quad x \otimes (3y + y^2) + 2x^2 \otimes 1,$$

and it can be expressed in many other equivalent ways. Note that, if $\otimes$ is replaced by multiplication in $R[x,y]$, then the above two-variable polynomials are all equal; this is the motivation for defining the tensor product relations the way we do.

8.22 EXAMPLE Elements of $\mathbb{Z}_2 \otimes_\mathbb{Z} \mathbb{Z}_4$

Consider the two finite abelian groups $\mathbb{Z}_2 = \mathbb{Z}/2\mathbb{Z}$ and $\mathbb{Z}_4 = \mathbb{Z}/4\mathbb{Z}$, which we view as $\mathbb{Z}$-modules. By a slight abuse of notation, we write elements in $\mathbb{Z}_n$ simply as integers, where it is understood that each integer represents an equivalence class modulo n. An example of an element in $\mathbb{Z}_2 \otimes_\mathbb{Z} \mathbb{Z}_4$ might then be written as

(8.23) $$1 \otimes 2 + 1 \otimes 3.$$

This can be rewritten in various ways using the tensor product relations; for example, the first summand equals

$$1 \otimes (2 \cdot 1) = (2 \cdot 1) \otimes 1 = 0 \otimes 1$$

and the second summand equals

$$1 \otimes (3 \cdot 1) = (3 \cdot 1) \otimes 1 = 1 \otimes 1,$$

so the element in (8.23) is the same as

$$0 \otimes 1 + 1 \otimes 1 = (0+1) \otimes 1 = 1 \otimes 1.$$

In Example 8.25 of the next section, we give a complete description of the elements in $\mathbb{Z}_2 \otimes_\mathbb{Z} \mathbb{Z}_4$, but the motivated reader might try to describe them all now.

Notice that an arbitrary element of $M \otimes_R N$ can be expressed as

$$m_1 \otimes n_1 + \cdots + m_k \otimes n_k$$

for some $m_1, \ldots, m_k \in M$ and $n_1, \ldots, n_k \in N$. Indeed, by definition, an element of $M \otimes_R N$ has the form (8.20) for some $a_1, \ldots, a_k \in R$, but the coefficients can be absorbed into $m_1, \ldots, m_k$ by the tensor product relations. We refer to elements of $M \otimes_R N$ of the form $m \otimes n$ as *simple tensors*, so another way of saying the above assertion is that *every element of $M \otimes_R N$ is a sum of simple tensors*. However, the expression as a sum of simple tensors is far from unique, as was seen in the previous examples.

> **Beware!** A common mistake is to think that every element of $M \otimes_R N$ is of the form $m \otimes n$, which is false.

8.3. TENSOR PRODUCTS OF MODULES AND ALGEBRAS

By definition, the tensor product of two R-modules M and N is a quotient of an R-module by a submodule, and thus, an R-module. However, if both M and N happen to be R-algebras—in other words, if they are each endowed with a multiplication that is compatible with their R-module structure—then the tensor product also inherits the structure of an R-algebra in a natural way.

8.24 PROPOSITION *Tensor products of R-algebras are R-algebras*

If M and N are R-algebras, then the tensor product $M \otimes_R N$ is an R-algebra with multiplication defined by setting

$$(m \otimes n)(m' \otimes n') = (mm') \otimes (nn')$$

and extending by linearity.

The stipulation that we "extend by linearity" is nothing more than the standard distributivity of multiplication. For example, in $R[x] \otimes_R R[y]$ we have

$$(x \otimes y^2 + x^3 \otimes y)(1 \otimes y + 2x^2 \otimes y^4) = x \otimes y^3 + 2x^3 \otimes y^6 + x^3 \otimes y^2 + 2x^5 \otimes y^5.$$

Given that elements of $M \otimes_R N$ can be expressed in multiple ways, it is not immediately clear that the multiplication in Proposition 8.24 is well-defined, and that is the main issue that is resolved in the proof.

PROOF OF PROPOSITION 8.24   Assuming that M and N are both R-algebras, we first note that the free R-module $R(M \times N)$ naturally inherits a multiplication defined by setting

$$(m, n)(m', n') = (mm', nn')$$

and extending by linearity. Since elements of $R(M \times N)$ can be written *uniquely* as formal R-linear combinations of pairs (m, n), this multiplication is well-defined, and some reflection should convince the reader that it endows $R(M \times N)$ with the structure of an R-algebra. Since quotients of R-algebras by ideals are R-algebras (see Proposition 3.17), it remains to prove that the submodule H in Definition 8.19 is an ideal. We leave this verification to the reader in Exercise 8.3.6. □

Equipped now with the definition of tensor products, have we achieved our goal? In particular, is it true that $R[x] \otimes_R R[y] = R[x, y]$? We at least have a natural candidate for a canonical isomorphism:

$$\varphi : R[x] \otimes_R R[y] \to R[x, y]$$
$$\varphi(f_1(x) \otimes g_1(y) + \cdots + f_k(x) \otimes g_k(y)) = f_1(x)g_1(y) + \cdots + f_k(x)g_k(y).$$

This matches with our intuition from earlier in the section; for example,

$$\varphi(3x^2 \otimes y + 2x \otimes 1 + 4x \otimes y^2) = 3x^2y + 2x + 4xy^2.$$

But since expressions in the domain are not unique, it is not yet clear whether φ is even well-defined. Thus, we still require a better understanding of how to construct well-defined maps from tensor products, and we accomplish this in the next section through a discussion of bilinearity.

Exercises for Section 8.3

8.3.1 Let S be a set and define addition and scalar multiplication in RS by
$$\sum a_s \cdot s + \sum b_s \cdot s = \sum (a_s + b_s) \cdot s \quad \text{and} \quad r \sum a_s \cdot s = \sum (ra_s) \cdot s.$$
Prove that RS with these operations satisfies the axioms of an R-module.

8.3.2 Prove that the function
$$\varphi : \mathbb{Z}\{\diamondsuit, \heartsuit, \clubsuit, \spadesuit\} \to \mathbb{Z}^4$$
$$\varphi(a\diamondsuit + b\heartsuit + c\clubsuit + d\spadesuit) = (a,b,c,d)$$
is an isomorphism of $\mathbb{Z}$-modules (or equivalently, of abelian groups).

8.3.3 Let S be a set, M an R-module, and $f_0 : S \to M$ a function. Prove that there is a unique R-module homomorphism $f : RS \to M$ such that
$$f(1 \cdot s) = f_0(s) \text{ for all } s \in S.$$

8.3.4 Use the tensor product relations to express the element
$$x^2 \otimes y^3 + x \otimes (2y^3) \in \mathbb{R}[x] \otimes_\mathbb{R} \mathbb{R}[y]$$
in at least three other ways.

8.3.5 Let M and N be R-modules. For any $m \in M$ and $n \in N$, show that
$$m \otimes 0 = 0 = 0 \otimes n.$$
Note that the first 0 is the additive identity in N, the middle 0 is the additive identity in $M \otimes_R N$, and the last 0 is the additive identity in M.

8.3.6 Suppose that M and N are R-algebras and view $R(M \times N)$ as an R-algebra with multiplication defined by setting
$$(m,n)(m',n') = (mm', nn')$$
and extending by linearity. Given any generator of the submodule H in Definition 8.19, prove that its product with (m', n') is also one of the generators. Use this to prove that H is an ideal of $R(M \times N)$.

8.3.7 Let $a, b \geq 2$ be integers. Prove that every element of $\mathbb{Z}_a \otimes_\mathbb{Z} \mathbb{Z}_b$ can be expressed as a simple tensor.

8.3.8 Describe all of the simple tensors in $\mathbb{Z}_2 \otimes_\mathbb{Z} \mathbb{Z}_5$. Which ones are equal to which other ones? Combining with the previous exercise, what can you say about the group $\mathbb{Z}_2 \otimes_\mathbb{Z} \mathbb{Z}_5$? Can you extend this to $\mathbb{Z}_a \otimes_\mathbb{Z} \mathbb{Z}_b$ when a and b are relatively prime?

8.3.9 For any $n \geq 1$, prove that $\mathbb{Z}_n \otimes_\mathbb{Z} \mathbb{Q} = \{0\}$.

8.3.10 Let V and W be vector spaces over a field K. Prove that any element of $V \otimes_K W$ can be expressed as
$$m_1 \otimes n_1 + \cdots + m_\ell \otimes n_\ell$$
in which $m_1, \ldots, m_\ell$ and $n_1, \ldots, n_\ell$ are each linearly independent over K.

Section 8.4 Tensor products and bilinearity

At the end of the last section, we observed that constructing well-defined maps from tensor products is a subtle task, simply because the elements of a tensor product can be expressed in many different ways. In particular, how do we ensure that a proposed map agrees on all of the different possible expressions for each element? In this section, our primary aim is to describe a method for easily constructing well-defined maps from tensor products.

To further motivate the usefulness of constructing well-defined homomorphisms from tensor products, we return to a concrete example from the previous section.

8.25 EXAMPLE How many elements does $\mathbb{Z}_2 \otimes_{\mathbb{Z}} \mathbb{Z}_4$ contain?

An arbitrary element of the tensor product $\mathbb{Z}_2 \otimes_{\mathbb{Z}} \mathbb{Z}_4$ can be expressed as

$$m_1 \otimes n_1 + \cdots + m_k \otimes n_k,$$

where $m_1, \ldots, m_k \in \mathbb{Z}_2$ and $n_1, \ldots, n_k \in \mathbb{Z}_4$. By Exercise 8.3.5, any summands in which $m_i = 0$ vanish, so it suffices to assume that $m_i = 1$ for all i. But

$$1 \otimes n_1 + \cdots + 1 \otimes n_k = 1 \otimes (n_1 + \cdots + n_k),$$

so in fact, any element of $\mathbb{Z}_2 \otimes_{\mathbb{Z}} \mathbb{Z}_4$ is equal to $1 \otimes n$ for some $n \in \mathbb{Z}_4$. This leaves four possibilities:

$$1 \otimes 0, \ 1 \otimes 1, \ 1 \otimes 2, \text{ and } 1 \otimes 3.$$

The first and third of these are in fact equal, since $1 \otimes 0 = 0$ and also

$$1 \otimes 2 = 1 \otimes (2 \cdot 1) = (2 \cdot 1) \otimes 1 = 0 \otimes 1 = 0.$$

Similarly, the elements $1 \otimes 1$ and $1 \otimes 3$ are equal, because

$$1 \otimes 3 = 1 \otimes (3 \cdot 1) = (3 \cdot 1) \otimes 1 = 1 \otimes 1.$$

Thus, we have

$$\mathbb{Z}_2 \otimes_{\mathbb{Z}} \mathbb{Z}_4 = \{0, 1 \otimes 1\}.$$

We now arrive at a conundrum: are the remaining two elements equal, or are they distinct? One approach for resolving this conundrum and showing that the two elements are, in fact, distinct would be to construct a well-defined homomorphism

$$\mathbb{Z}_2 \otimes_{\mathbb{Z}} \mathbb{Z}_4 \to \mathbb{Z}_2$$

that takes $1 \otimes 1$ to 1. If we can do this, then we can definitively conclude that $1 \otimes 1 \neq 0$ because any homomorphism must send 0 to 0. This leads us back to the motivating question of this section: how do we construct such a well-defined map?

The primary tool that will allow us to construct well-defined maps from tensor products is the notion of a bilinear function, which we now introduce.

8.26 DEFINITION *R-bilinear map*

Let M, N, and L be R-modules. A map

$$\varphi : M \times N \to L$$

is *R-bilinear* if the following hold.
- $\varphi(m_1 + m_2, n) = \varphi(m_1, n) + \varphi(m_2, n)$ for all $m_1, m_2 \in M$, $n \in N$.
- $\varphi(m, n_1 + n_2) = \varphi(m, n_1) + \varphi(m, n_2)$ for all $m \in M$, $n_1, n_2 \in N$.
- $r\varphi(m, n) = \varphi(rm, n) = \varphi(m, rn)$ for all $r \in R$, $m \in M$, $n \in N$.

In other words, φ is R-bilinear precisely if it induces an R-module homomorphism on either factor of $M \times N$ when the input to the other factor is held fixed.

8.27 EXAMPLE Multiplication maps are bilinear

Consider the ring R as an R-module, where the scalar multiplication is the usual multiplication. The multiplication map

$$\varphi : R \times R \to R$$
$$\varphi(m, n) = mn$$

is R-bilinear by the distributive, associative, and commutative properties of multiplication in R. More generally, if A is an R-algebra, then the multiplication map

$$\varphi : A \times A \to A$$
$$\varphi(m, n) = mn$$

is R-bilinear by the algebra axioms.

8.28 EXAMPLE Products of linear maps are bilinear

Considering R^2, R^3, and R as R-modules, the map

$$\varphi : R^2 \times R^3 \to R$$
$$\varphi((x, y), (u, v, w)) = (2x + 3y) \cdot (5u - v + w)$$

is R-bilinear. More generally, if M and N are R-modules and A is an R-algebra, and if $\varphi_1 : M \to A$ and $\varphi_2 : N \to A$ are R-module homomorphisms, then

$$\varphi : M \times N \to A$$
$$\varphi(m, n) = \varphi_1(m)\varphi_2(n)$$

is R-bilinear (Exercise 8.4.1).

The definition of R-bilinearity likely reminds the reader of the tensor product relations, so it should come as no surprise that R-bilinearity is the key to constructing well-defined maps from tensor products. The following proposition makes this connection precise.

8.4. TENSOR PRODUCTS AND BILINEARITY

> **8.29 PROPOSITION** *Module homomorphisms from tensor products*
> Let M, N, and L be R-modules. If $\varphi : M \times N \to L$ is R-bilinear, then there exists a unique R-module homomorphism $\widehat{\varphi} : M \otimes_R N \to L$ such that
> $$\widehat{\varphi}(m \otimes n) = \varphi(m, n).$$

Before proving the proposition, let us illustrate its utility in a few familiar examples.

8.30 EXAMPLE $\mathbb{Z}_2 \otimes_{\mathbb{Z}} \mathbb{Z}_4 = \mathbb{Z}_2$

To define a homomorphism from $\mathbb{Z}_2 \otimes_{\mathbb{Z}} \mathbb{Z}_4$ to $\mathbb{Z}_2$, start with the map

$$\varphi : \mathbb{Z}_2 \times \mathbb{Z}_4 \to \mathbb{Z}_2$$
$$\varphi(a, b) = a \cdot \pi(b),$$

where $\pi : \mathbb{Z}_4 \to \mathbb{Z}_2$ is the map that reduces inputs modulo 2. The reader is encouraged to convince themselves directly of the bilinearity of φ, though it also follows from Example 8.28 since π is a $\mathbb{Z}$-module homomorphism. Given this bilinearity, φ induces the homomorphism

$$\widehat{\varphi} : \mathbb{Z}_2 \otimes_{\mathbb{Z}} \mathbb{Z}_4 \to \mathbb{Z}_2$$
$$\widehat{\varphi}(a_1 \otimes b_1 + \cdots + a_k \otimes b_k) = a_1 \pi(b_1) + \cdots + a_k \pi(b_k),$$

which by Proposition 8.29 is a well-defined homomorphism of $\mathbb{Z}$-modules. The fact that $\widehat{\varphi}$ is well-defined is illustrated by the fact that equivalent elements of $\mathbb{Z}_2 \otimes_{\mathbb{Z}} \mathbb{Z}_4$ are mapped to the same element of $\mathbb{Z}_2$; for example, we saw in Example 8.25 that

$$1 \otimes 1 = 1 \otimes 3 \in \mathbb{Z}_2 \otimes_{\mathbb{Z}} \mathbb{Z}_4,$$

and since $\pi(1) = \pi(3) = 1 \in \mathbb{Z}_2$, we now observe from the definition of $\widehat{\varphi}$ that

$$\widehat{\varphi}(1 \otimes 1) = \widehat{\varphi}(1 \otimes 3) = 1 \in \mathbb{Z}_2,$$

as expected. The fact that $\widehat{\varphi}(1 \otimes 1) \neq 0$ implies that $1 \otimes 1 \neq 0$, so we have resolved the issue raised in Example 8.25. In fact, since $\mathbb{Z}_2$ has only two elements, it follows that $\widehat{\varphi}$ is an isomorphism of $\mathbb{Z}$-modules (or, equivalently, of abelian groups).

8.31 EXAMPLE $R[x] \otimes_R R[y] = R[x, y]$

To consider an example that is especially relevant for us, let us define a homomorphism from $R[x] \otimes_R R[y]$ to $R[x, y]$ by starting with the following R-bilinear map:

$$\varphi : R[x] \times R[y] \to R[x, y]$$
$$\varphi(f, g) = f \cdot g.$$

Again, the reader is encouraged to check the R-bilinearity of φ directly, but it is also a special case of Example 8.28. Proposition 8.29 then gives a homomorphism

$$\widehat{\varphi} : R[x] \otimes_R R[y] \to R[x, y].$$

More specifically, the homomorphism $\widehat{\varphi}$ is defined by

$$\widehat{\varphi}(f_1 \otimes g_1 + \cdots + f_k \otimes g_k) = f_1 g_1 + \cdots + f_k g_k.$$

For example, we compute

$$\widehat{\varphi}(x^2 \otimes y + x^2 \otimes 1) = x^2 y + x^2 \quad \text{and} \quad \widehat{\varphi}(x^2 \otimes (y+1)) = x^2(y+1).$$

Observing that $x^2 \otimes y + x^2 \otimes 1 = x^2 \otimes (y+1) \in R[x] \otimes_R R[y]$, we would expect that $\widehat{\varphi}$ sends these elements to the same polynomial in $R[x, y]$, and indeed, we have $x^2 y + x^2 = x^2(y+1) \in R[x, y]$.

Consider, now, the map

$$R[x, y] \to R[x] \otimes_R R[y]$$
$$\sum_{i,j} r_{ij} x^i y^j \mapsto \sum_{i,j} r_{ij} x^i \otimes y^j.$$

In Exercise 8.4.2, the reader is encouraged to prove that this map is inverse to $\widehat{\varphi}$, from which we conclude that $\widehat{\varphi}$ is an isomorphism of R-modules.

○───○

With these examples in mind, let us turn to the proof of the proposition.

PROOF OF PROPOSITION 8.29   First, we note that, if $\widehat{\varphi}$ exists, then it must be unique because its value on any simple tensor along with linearity determines its value on any element of $M \otimes_R N$. Thus, it suffices to prove that $\widehat{\varphi}$ exists.

To prove existence, begin by extending the map $\varphi : M \times N \to L$ to a map $\widetilde{\varphi} : R(M \times N) \to L$ by linearity. That is, define

$$\widetilde{\varphi} : R(M \times N) \to L$$
$$a_1(m_1, n_1) + \cdots + a_k(m_k, n_k) \mapsto a_1 \cdot \varphi(m_1, n_1) + \cdots + a_k \cdot \varphi(m_k, n_k).$$

This map is automatically well-defined (as its domain has no relations), and some reflection should convince the reader that it is an R-module homomorphism.

The claim, now, is that the kernel of $\widetilde{\varphi}$ contains the submodule H of Definition 8.19. If this is the case, then $\widetilde{\varphi}$ induces the well-defined R-module homomorphism

$$\widehat{\varphi} : \frac{R(M \times N)}{H} \to L$$

satisfying the condition of the proposition.

To prove the claim, it suffices to prove that the kernel of $\widetilde{\varphi}$ contains the generators of H, which are listed in Definition 8.19. This is indeed the case; for example,

$$\widetilde{\varphi}((m, n_1 + n_2) - (m, n_1) - (m, n_2)) = \varphi(m, n_1 + n_2) - \varphi(m, n_1) - \varphi(m, n_2),$$

which equals zero by the definition of R-bilinearity, and similar arguments apply to the other three types of generators. It follows that $\widetilde{\varphi}$ sends every generator of H to zero, so $H \subseteq \ker(\widetilde{\varphi})$, completing the proof.   □

So far, we have constructed R-module homomorphisms from tensor products $M \otimes_R N$, but if both M and N happen to be R-algebras, can we construct R-*algebra* homomorphisms from their tensor product? The key to extending module homomorphisms from tensor products to algebra homomorphisms is the following definition.

8.4. TENSOR PRODUCTS AND BILINEARITY

8.32 DEFINITION *Multiplicative map*

Let M, N, and L be R-algebras. A map

$$\varphi : M \times N \to L$$

is *multiplicative* if, for all $m, m' \in M$ and $n, n' \in N$, we have

$$\varphi(mm', nn') = \varphi(m, n)\varphi(m', n').$$

For example, the function $\varphi : R[x] \times R[y] \to R[x, y]$ defined by

$$\varphi(f, g) = fg$$

is multiplicative, as the reader can readily verify. More generally, if M, N, and L are R-algebras and $\varphi_1 : M \to L$ and $\varphi_2 : N \to L$ are R-algebra homomorphisms, then the function $\varphi : M \times N \to L$ defined by

$$\varphi(m, n) = \varphi_1(m)\varphi_2(n)$$

is multiplicative (Exercise 8.4.1).

The next result generalizes Proposition 8.29 to the algebra setting.

8.33 PROPOSITION *Algebra homomorphisms from tensor products*

Let M, N, and L be R-algebras. If $\varphi : M \times N \to L$ is R-bilinear and multiplicative, then there exists a unique R-algebra homomorphism $\widehat{\varphi} : M \otimes_R N \to L$ such that

$$\widehat{\varphi}(m \otimes n) = \varphi(m, n).$$

PROOF We already know that there exists a unique R-module homomorphism $\widehat{\varphi}$, so it remains to prove that $\widehat{\varphi}$ preserves multiplication. Notice that

$$\begin{aligned}
\widehat{\varphi}((m \otimes n)(m' \otimes n')) &= \widehat{\varphi}(mm' \otimes nn') \\
&= \varphi(mm', nn') \\
&= \varphi(m, n)\varphi(m', n') \\
&= \widehat{\varphi}(m \otimes n)\widehat{\varphi}(m' \otimes n'),
\end{aligned}$$

where the first equality uses the definition of multiplication in $M \otimes_R N$, the second and fourth use the definition of $\widehat{\varphi}$, and the third uses the multiplicativity of φ. This argument proves that the R-module homomorphism $\widehat{\varphi}$ preserves multiplication of simple tensors, which we can then extend to multiplication of general elements of $M \otimes_R N$, using the R-linearity of $\widehat{\varphi}$. □

We are finally able to conclude our story of the canonical R-algebra isomorphism $R[x] \otimes_R R[y] \cong R[x, y]$ that has served as motivation for much of our development of tensor products; we accomplish this in the next example.

8.34 EXAMPLE $R[x] \otimes_R R[y] = R[x,y]$ as R-algebras.

Upon observing that the canonical bilinear map $\varphi : R[x] \times R[y] \to R[x,y]$ of Example 8.31 is multiplicative, we deduce from Proposition 8.33 that the R-module homomorphism $\widehat{\varphi}$ is actually an R-algebra homomorphism. Moreover, we saw in Example 8.31 that $\widehat{\varphi}$ has an inverse, and is thus an isomorphism of R-algebras.

In the special case $R = K$, the previous example gives a canonical K-algebra isomorphism
$$K[\mathbb{A}^1] \otimes_K K[\mathbb{A}^1] = K[\mathbb{A}^1 \times \mathbb{A}^1].$$
Thus, it is meaningful to ask whether the natural generalization holds: given affine varieties X and Y, do we have a canonical isomorphism of K-algebras
$$K[X] \otimes_K K[Y] = K[X \times Y]?$$
The goal of the next section is to prove that this is indeed the case.

Exercises for Section 8.4

8.4.1 (a) Let M and N be R-modules, let A be an R-algebra, and suppose that $\varphi_1 : M \to A$ and $\varphi_2 : N \to A$ are R-module homomorphisms. Prove that the map
$$\varphi : M \times N \to A$$
$$\varphi(m,n) = \varphi_1(m)\varphi_2(n)$$
is R-bilinear.
(b) Assume, furthermore, that M and N are R-algebras and φ_1 and φ_2 are R-algebra homomorphisms. Prove that φ is multiplicative.

8.4.2 Let $\widehat{\varphi} : R[x] \otimes_R R[y] \to R[x,y]$ be the R-module map defined in Example 8.31. Prove that the map
$$R[x,y] \to R[x] \otimes_R R[y]$$
$$\sum_{i,j} r_{ij} x^i y^j \mapsto \sum_{i,j} r_{ij} x^i \otimes y^j.$$
is inverse to $\widehat{\varphi}$.

8.4.3 Prove that there is a canonical R-algebra isomorphism
$$R[x_1,\ldots,x_m] \otimes_R R[y_1,\ldots,y_n] = R[x_1,\ldots,x_m,y_1,\ldots,y_n].$$

8.4.4 Prove that, for any natural numbers a and b, there is a canonical $\mathbb{Z}$-algebra isomorphism
$$\mathbb{Z}_a \otimes_\mathbb{Z} \mathbb{Z}_b = \mathbb{Z}_{\gcd(a,b)}.$$
(You may wish to solve Exercise 8.3.7 first.)

8.4. TENSOR PRODUCTS AND BILINEARITY

8.4.5 Let V be a vector space over K with basis $\{e_i\}_{i \in I}$ and let W be a vector space over K with basis $\{f_j\}_{j \in J}$.
  (a) Prove that the set
  $$\{e_i \otimes f_j \mid i \in I, j \in J\}$$
  is a basis of the vector space $V \otimes_K W$.
  (b) Assuming V and W are finite-dimensional, what does part (a) tell you about the relationship between the vector-space dimensions $\dim(V)$, $\dim(W)$, and $\dim(V \otimes_K W)$?

8.4.6 Let M and N be R-modules. Prove that $M \otimes_R N = N \otimes_R M$.

8.4.7 Let M be an R-module. Prove that $R \otimes_R M = M \otimes_R R = M$.

8.4.8 Let M, M', and N be R-modules. Prove that
$$(M \oplus M') \otimes_R N = (M \otimes_R N) \oplus (M' \otimes_R N).$$

8.4.9 (a) Prove that $\mathbb{C} \otimes_\mathbb{R} \mathbb{R}^n = \mathbb{C}^n$.
  (b) Prove that $\mathbb{C} \otimes_\mathbb{R} \mathbb{R}[x] = \mathbb{C}[x]$.

8.4.10 Let R and S be rings with $R \subseteq S$, where we view S as an R-module via the multiplication in S, and let M be an R-module. Prove that $S \otimes_R M$ (which, by construction, is an R-module) in fact has the structure of an S-module, where the scalar multiplication is given by
$$s \cdot (s_1 \otimes m_1 + \cdots + s_k \otimes m_k) = (ss_1) \otimes m_1 + \cdots + (ss_k) \otimes m_k.$$

The passage from M to $S \otimes_R M$ is called *extension of scalars* and is illustrated by the two examples in Exercise 8.4.9.

Section 8.5   The coordinate ring of a product

We proved in Section 8.1 that the product of affine varieties is an affine variety, and we are now ready to compute the coordinate ring of such a product in terms of the coordinate rings of the two factors.

> **8.35 THEOREM**   *The coordinate ring of a product*
>
> For any affine varieties X and Y, there is a canonical K-algebra isomorphism
>
> $$K[X \times Y] = K[X] \otimes_K K[Y].$$

Before we prove the theorem in general, let us refocus on the goal by returning to a concrete example.

8.36 EXAMPLE   The coordinate ring of a parabola in $\mathbb{A}^3$

Let $X = \mathcal{V}(y - x^2) \subseteq \mathbb{A}^2$ and let $Y = \mathcal{V}(z - 1) \subseteq \mathbb{A}^1$, so that

$$X \times Y = \mathcal{V}(y - x^2, z - 1) \subseteq \mathbb{A}^3,$$

a parabola in the $z = 1$ plane of $\mathbb{A}^3$. One can prove (directly, or via the Nullstellensatz) that

$$K[X \times Y] = \frac{K[x, y, z]}{\langle y - x^2, z - 1 \rangle},$$

and upon omitting all occurrences of y and z using the relations $[y] = [x^2]$ and $[z] = 1$, we conclude that

$$K[X \times Y] \cong K[x].$$

On the other hand, considering the coordinate rings of X and Y separately, we have

$$K[X] = \frac{K[x, y]}{\langle y - x^2 \rangle} \cong K[x] \quad \text{and} \quad K[Y] = \frac{K[z]}{\langle z - 1 \rangle} \cong K.$$

Thus, by the result of either Exercise 8.4.3 or Exercise 8.4.7, we have

$$K[X] \otimes_K K[Y] \cong K[x] \otimes_K K = K[x],$$

so indeed, $K[X \times Y]$ and $K[X] \otimes_K K[Y]$ are isomorphic.

PROOF OF THEOREM 8.35   Let X and Y be affine varieties. To define the canonical isomorphism of Theorem 8.35, consider the function

$$\varphi : K[X] \times K[Y] \to K[X \times Y]$$
$$\varphi(F, G) = F \times G,$$

where $F \times G$ is the function on $X \times Y$ given by

$$(F \times G)(a, b) = F(a)G(b).$$

8.5. THE COORDINATE RING OF A PRODUCT

If F and G are polynomial functions, then $F \times G$ is also a polynomial function on $X \times Y$, as the reader is encouraged to verify in Exercise 8.5.1. Furthermore, φ is K-bilinear and multiplicative, so it induces a homomorphism of K-algebras

$$\widehat{\varphi} : K[X] \otimes_K K[Y] \to K[X \times Y].$$

Note that $\widehat{\varphi}$ really is canonical; in particular, it does not depend on choosing representations for X and Y in affine spaces. It remains to prove that $\widehat{\varphi}$ is a bijection, from which it follows that it is an isomorphism.

(**Surjectivity**) Let $H \in K[X \times Y]$. Choosing representations $X \subseteq \mathbb{A}^m$ and $Y \subseteq \mathbb{A}^n$, we can view H as the restriction to $X \times Y$ of a polynomial function $h \in K[x_1, \ldots, x_m, y_1, \ldots, y_r]$. Write

$$h(x,y) = \sum_{i=1}^{\ell} f_i(x) g_i(y)$$

for some polynomials $f_1, \ldots, f_\ell \in K[x_1, \ldots, x_m]$ and $g_1, \ldots, g_\ell \in K[y_1, \ldots, y_n]$. Let $F_1, \ldots, F_\ell \in K[X]$ be the polynomial functions defined by restricting $f_1, \ldots, f_\ell$, and similarly for $G_1, \ldots, G_\ell \in K[Y]$. Then

$$\widehat{\varphi}\left(\sum_{i=1}^{\ell} F_i \otimes G_i\right) = H,$$

as one sees by evaluating both sides on an arbitrary point $(a,b) \in X \times Y$.

(**Injectivity**) Suppose that $H \in K[X] \otimes_K K[Y]$ and $\widehat{\varphi}(H) = 0$; we must argue that $H = 0$. Note that H can be written (in many different ways) as a sum of simple tensors:

$$H = \sum_{i=1}^{\ell} F_i \otimes G_i.$$

Fix one such expression for H such that ℓ is minimal. We will prove that $\ell = 0$, implying that $H = 0$.

Toward a contradiction, suppose that $\ell > 0$. This implies that $F_\ell \neq 0$, as otherwise we could omit the final summand in our expression for H, contradicting the minimality of ℓ. Thus, we can choose an element $a \in X$ such that $F_\ell(a) \neq 0$. Given that $\widehat{\varphi}(H) = 0 \in K[X \times Y]$, it follows that, for any $b \in Y$, we have

$$0 = \widehat{\varphi}(H)(a,b) = \sum_{i=1}^{\ell} F_i(a) G_i(b) \implies 0 = \sum_{i=1}^{\ell} F_i(a) G_i \in K[Y].$$

Since $F_\ell(a) \neq 0$, we can then rewrite $G_\ell \in K[Y]$ as

$$G_\ell = -F_\ell(a)^{-1}(F_1(a) G_1 + \cdots + F_{\ell-1}(a) G_{\ell-1}).$$

Substituting this into our expression for H and simplifying by using the tensor product relations, we obtain

$$H = \sum_{i=1}^{\ell-1} \left(F_i - F_\ell(a)^{-1} F_i(a) F_\ell\right) \otimes G_i,$$

contradicting the minimality of ℓ. This contradiction implies that $\ell = 0$, finishing the proof. □

We close with an example to further illustrate Theorem 8.35.

8.37 EXAMPLE Polynomial functions on a parabola in $\mathbb{A}^3$

Returning to the variety $X \times Y$ of Example 8.36, an example of a polynomial function $H \in K[X \times Y]$ might be

$$H : X \times Y \to K$$
$$H(a,b,c) = ac + b^2,$$

which is the restriction of the polynomial $h(x,y,z) = xz + y^2$. If $F_1, F_2 \in K[X]$ are the restrictions of the polynomials $f_1(x,y) = x$ and $f_2(x,y) = y^2$, respectively, and $G_1, G_2 \in K[Y]$ are the restrictions of the polynomials $g_1(z) = z$ and $g_2(z) = 1$, respectively, then

$$\widehat{\varphi}(F_1 \otimes G_1 + F_2 \otimes G_2) = H.$$

Of course, recalling that $X \times Y = \mathcal{V}(y - x^2, z - 1)$, we could equally well express H as

$$H(a,b,c) = a + a^4,$$

in which case we have

$$\widehat{\varphi}((F_1 + F_1^4) \otimes 1) = H.$$

That these two seemingly different elements in $K[X] \otimes_K K[Y]$ both map to H reflects that they are not actually different:

$$F_1 \otimes G_1 + F_2 \otimes G_2 = (F_1 + F_1^4) \otimes 1 \in K[X] \otimes_K K[Y],$$

as one can verify from the defining equations of X and Y together with the tensor product relations.

Exercises for Section 8.5

8.5.1 Let X and Y be affine varieties and let $F \in K[X]$ and $G \in K[Y]$ be polynomial functions. Prove that the function

$$F \times G : X \times Y \to K$$
$$(a,b) \mapsto F(a)G(b)$$

is a polynomial function on $X \times Y$.

8.5.2 What is the coordinate ring of the cylinder $\mathcal{V}(x^2 + y^2 - 1) \subseteq \mathbb{A}^3$? Express your answer both as a tensor product and as a quotient of $K[x,y,z]$.

8.5.3 Let $X \subseteq \mathbb{A}^m$ be an affine variety and let $Y \subseteq \mathbb{A}^n$ be a set of r points.
   (a) Describe $X \times Y \subseteq \mathbb{A}^m \times \mathbb{A}^n$ geometrically.
   (b) Prove that $K[Y] \cong K^r$, and deduce—using Exercises 8.4.8 and 8.4.7—that $K[X \times Y]$ is isomorphic to the direct sum of r copies of $K[X]$.
   (c) Explain the relationship between the geometric statement in (a) and the algebraic statement in (b).

8.5. THE COORDINATE RING OF A PRODUCT

8.5.4 Let X and Y be affine varieties, and let $P : X \times Y \to X$ be the projection map $P(a,b) = a$. Describe P^* explicitly as a homomorphism

$$K[X] \to K[X] \otimes_K K[Y].$$

8.5.5 Let X, Y, Z be affine varieties, and let $F : Z \to X$ and $G : Z \to Y$ be polynomial maps. Define the function

$$H : Z \to X \times Y$$
$$c \mapsto (F(c), G(c)).$$

(a) Prove that H is a polynomial map.
(b) Describe H^* explicitly as a homomorphism

$$K[X] \otimes_K K[Y] \to K[Z].$$

8.5.6 Prove that the K-algebra $K[x, y, z]/\langle xy, xz, yz \rangle$ cannot be expressed as a tensor product $A \otimes_K B$ where A and B are K-algebras, neither of which is equal to K.

8.5.7 State a generalization of Theorem 8.35 for finite products of affine varieties. Use induction to prove the general statement.

8.5.8 Suppose that K is algebraically closed, and let A and B be finitely-generated reduced K-algebras.

(a) Combine Proposition 8.6 with other results you have learned to show that if A and B are integral domains, then $A \otimes_K B$ is an integral domain.
(b) Prove that the result of part (a) can fail if K is not algebraically closed by arguing that

$$\mathbb{C} \otimes_\mathbb{R} \mathbb{C} \cong \mathbb{C} \oplus \mathbb{C}$$

as $\mathbb{R}$-algebras, and although $\mathbb{C}$ is an integral domain, $\mathbb{C} \oplus \mathbb{C}$ is not.

8.5.9 Write an alternative proof of Proposition 8.12 by arguing that Noether bases for $K[X]$ and $K[Y]$ give rise to a Noether basis for

$$K[X \times Y] = K[X] \otimes_K K[Y].$$

Part II

Projective Algebraic Geometry

Chapter 9

Projective Varieties

> LEARNING OBJECTIVES FOR CHAPTER 9
> - Define and work with projective space from various perspectives.
> - Understand what it means for a polynomial to vanish at a point of projective space, and understand the role of homogeneity in projective vanishing.
> - Define and work with the $\mathcal{V}$- and $\mathcal{I}$-operators in the projective setting, and understand the relationship between them.
> - Calculate affine restrictions of projective varieties and projective closures of affine varieties to pass between the affine and projective settings.
> - State the projective Nullstellensatz, and reduce it to the affine Nullstellensatz via affine cones.

Up until this point in the book, all of the varieties that we have studied have lived inside of affine space $\mathbb{A}^n$. There is a larger ambient space, however, in which the notion of "variety" also makes sense, known as *projective space* and denoted $\mathbb{P}^n$. The goal of this chapter is to define $\mathbb{P}^n$ and the *projective varieties* one obtains as vanishing sets of polynomials inside $\mathbb{P}^n$.

The motivation for this generalization comes from the desire to make uniform statements in settings where a statement about affine varieties has unavoidable exceptions. A key example of this phenomenon is the statement that, in $\mathbb{A}^2$, any pair of lines must intersect—with the exception of parallel lines. Projective space $\mathbb{P}^2$ can be viewed as the result of adding "points at infinity" to $\mathbb{A}^2$ so that each line in $\mathbb{A}^2$ meets a particular point at infinity dictated by the line's slope. With the addition of these extra points, we find in $\mathbb{P}^2$ that every pair of lines intersects, without exception. This is a special case of a beautiful result known as Bézout's Theorem, which states that a pair of curves in $\mathbb{P}^2$, defined by polynomials of degrees r and s, intersect in $r \cdot s$ points when counted appropriately. While the corresponding statement in $\mathbb{A}^2$ is often true, one can easily find exceptions: the parabola $\mathcal{V}(y - x^2)$ and the vertical line $\mathcal{V}(x)$ intersect in only a single point, for example. From the perspective of $\mathbb{P}^2$, this exception again occurs because there is an additional intersection "at infinity" that is hidden when one restricts their attention to $\mathbb{A}^2$.

These observations in plane geometry led algebraic geometers to ultimately understand $\mathbb{P}^n$, and not $\mathbb{A}^n$, as the most natural ambient space in which to study solutions of polynomial systems. While the definition of projective space can be difficult to digest on a first pass, the elegance and uniformity that it will lend to our study of algebraic geometry is certainly worth the effort.

Section 9.1 Projective space

Part of what makes the study of projective varieties challenging on a first encounter—but also what makes it rich and interesting—is the multitude of different ways in which one can define projective space. We will present three different perspectives on projective space, beginning with the one that is the most computationally useful.

> **9.1 DEFINITION** *Projective space, first perspective*
>
> Let $n \in \mathbb{N}$. The n-dimensional *projective space* over K, denoted $\mathbb{P}_K^n$ or simply $\mathbb{P}^n$, is the set
> $$\mathbb{P}^n = \frac{K^{n+1} \setminus \{(0,0,\ldots,0)\}}{\sim},$$
> where $\sim$ is the equivalence relation given by
> $$(a_0, a_1, \ldots, a_n) \sim (b_0, b_1, \ldots, b_n)$$
> $$\iff$$
> $$(\lambda a_0, \lambda a_1, \ldots, \lambda a_n) = (b_0, b_1, \ldots, b_n) \text{ for some } \lambda \in K \setminus \{0\}.$$
> We denote the equivalence class of $(a_0, a_1, \ldots, a_n)$ by $[a_0 : a_1 : \cdots : a_n]$.

> *The reader should pause to convince themselves that $\sim$ is, indeed, an equivalence relation.*

Note that the term "dimension" in this context should be taken, for now, as nothing more than an indication of the number of coordinates; since $\mathbb{P}^n$ is neither a vector space nor an affine variety, it cannot be meaningfully given a dimension in any of the contexts in which that term has been used thus far in this book. Nevertheless, our use of the term "n-dimensional" may make more sense after the following examples.

9.2 EXAMPLE *0-dimensional projective space*

An element of $\mathbb{P}^0$ is an equivalence class $[a]$, where $a \in K \setminus \{0\}$ and $[a] = [b]$ if $\lambda a = b$ for some $\lambda \in K \setminus \{0\}$. In particular, taking $\lambda = 1/a$ shows that $[a] = [1]$ for any $a \in K \setminus \{0\}$, so $\mathbb{P}^0$ has just a single element:
$$\mathbb{P}^0 = \{[1]\}.$$

9.3 EXAMPLE *1-dimensional projective space*

Elements of $\mathbb{P}^1$ are of the form $[a_0 : a_1]$, where $a_0, a_1 \in K$ are not both zero. For instance, $[1 : 2]$ is an element of $\mathbb{P}^1$, and scaling both coordinates by the same $\lambda \in K \setminus \{0\}$ yields different representations of the same element:
$$[1:2] = [2:4] = [3:6] = [-1:-2] = \cdots.$$

It is instructive to divide the elements of $\mathbb{P}^1$ into two types: those whose first coordinate is nonzero and those whose first coordinate is zero. Consider an element of the first type, such as $[3 : 7] \in \mathbb{P}^1$. Scaling both coordinates by $1/3$ shows that
$$[3:7] = [1:7/3] \in \mathbb{P}^1.$$

9.1. PROJECTIVE SPACE

Similarly, any element of $\mathbf{P}^1$ with nonzero first coordinate is equal to $[1:b]$ for some $b \in K$. On the other hand, an element of $\mathbb{P}^1$ whose first coordinate is zero is always equal to $[0:1]$; for instance, scaling both coordinates by $1/4$ shows that $[0:4] = [0:1] \in \mathbb{P}^1$.

The conclusion that we arrive at, then, is that

(9.4) $$\mathbb{P}^1 = \{[1:b] \mid b \in K\} \sqcup \{[0:1]\}.$$

Some reflection should convince the reader that two elements $[1:b]$ and $[1:b']$ with $b \neq b'$ cannot be equal to one another in $\mathbb{P}^1$. As a result, there is a natural bijection between the elements of the form $[1:b] \in \mathbb{P}^1$ and the elements of $\mathbb{A}^1$ given by

> We use the symbol $\sqcup$ for disjoint unions; in other words, $A = B \sqcup C$ means $A = B \cup C$ and $B \cap C = \emptyset$.

$$\{[1:b] \mid b \in K\} \to \mathbb{A}^1$$
$$[1:b] \mapsto b.$$

Under this bijection, the decomposition (9.4) can be viewed as

(9.5) $$\mathbb{P}^1 = \mathbb{A}^1 \sqcup \{[0:1]\}.$$

9.6 EXAMPLE 2-dimensional projective space

As above, the elements of $\mathbb{P}^2$ can be divided into two types, depending on whether their first coordinate is nonzero or zero, and those with nonzero first coordinate can be rescaled to the form $[1:b_1:b_2]$. Thus,

$$\mathbb{P}^2 = \{[1:b_1:b_2] \mid b_1, b_2 \in K\} \sqcup \{[0:b_1:b_2] \mid b_1, b_2 \in K \text{ not both zero}\}.$$

Also analogously to the previous example, elements of the first type are in natural bijection with $\mathbb{A}^2$:

$$\{[1:b_1:b_2] \mid b_1, b_2 \in K\} \to \mathbb{A}^2$$
$$[1:b_1:b_2] \mapsto (b_1, b_2).$$

Now there is not just a single element with first coordinate zero, however, but many; for example, $[0:0:1] \neq [0:1:1]$. In fact, elements of $\mathbb{P}^2$ with first coordinate zero are in natural bijection with a projective space of one dimension lower:

$$\{[0:b_1:b_2] \mid b_1, b_2 \in K \text{ not both zero}\} \to \mathbb{P}^1$$
$$[0:b_1:b_2] \mapsto [b_1:b_2].$$

Under these two bijections, we have shown that

(9.7) $$\mathbb{P}^2 = \mathbb{A}^2 \sqcup \mathbb{P}^1.$$

The decompositions (9.5) and (9.7) can be generalized to any n, and doing so brings us to our second perspective on projective space.

9.8 PROPOSITION *Projective space, second perspective*

For any $n \geq 1$, there is a natural bijection

$$\mathbb{P}^n = \mathbb{A}^n \sqcup \mathbb{P}^{n-1}.$$

The elements of $\mathbb{P}^{n-1}$ inside $\mathbb{P}^n$ are referred to as *points at infinity* in $\mathbb{P}^n$.

The proof of this bijection is the content of Exercise 9.1.3; the key point, as we saw previously in the cases of $\mathbb{P}^1$ and $\mathbb{P}^2$, is that elements of $\mathbb{P}^n$ with nonzero first coordinate correspond to elements of $\mathbb{A}^n$, whereas elements with first coordinate zero correspond to elements of $\mathbb{P}^{n-1}$.

But why the terminology "points at infinity"? To understand this, consider the case of $\mathbb{P}^1_\mathbb{R}$. Under the decomposition

$$\mathbb{P}^1_\mathbb{R} = \mathbb{A}^1_\mathbb{R} \sqcup \{[0:1]\},$$

the elements $1, 2, 3, \ldots \in \mathbb{A}^1_\mathbb{R}$ correspond to the following elements of $\mathbb{P}^1_\mathbb{R}$:

$$[1:1],\ [1:2],\ [1:3], \ldots \in \mathbb{P}^1_\mathbb{R}.$$

By rescaling, though, these can be re-expressed as

$$\left[\tfrac{1}{1}:1\right],\ \left[\tfrac{1}{2}:1\right],\ \left[\tfrac{1}{3}:1\right], \ldots \in \mathbb{P}^1_\mathbb{R}.$$

Thus, as $n \in \mathbb{A}^1_\mathbb{R}$ grows arbitrarily large, the corresponding points $\left[\tfrac{1}{n}:1\right]$ in $\mathbb{P}^1_\mathbb{R}$ tend to $[0:1]$. This explains why we refer to $[0:1]$ as the *point at infinity*, writing

> Our use of "limits" in $\mathbb{P}^n$ is merely intuitive here, since a topology is needed to make limits precise.

$$\mathbb{P}^1_\mathbb{R} = \mathbb{A}^1_\mathbb{R} \sqcup \{\infty\}.$$

Note that the points $-n \in \mathbb{A}^1_\mathbb{R}$—corresponding to $[1:-n] \in \mathbb{P}^1_\mathbb{R}$—also approach $[0:1]$ as $n \to \infty$. Thus, visually, it is illustrative to depict $\mathbb{P}^1_\mathbb{R}$ as a loop: as we go arbitrarily far in either direction of $\mathbb{A}^1_\mathbb{R}$, we tend to the same point $[0:1] \in \mathbb{P}^1_\mathbb{R}$.

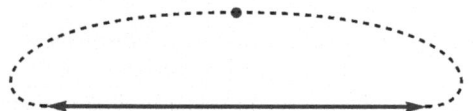

For $n > 1$, it becomes more difficult to give a visual representation of $\mathbb{P}^n$, but the same perspective still holds. For example, we have

$$\mathbb{P}^2_\mathbb{R} = \mathbb{A}^2_\mathbb{R} \sqcup \{\text{points at infinity}\}.$$

In this space, we can tend toward infinity along any line in $\mathbb{A}^2_\mathbb{R}$. To do so, consider the sequence of points $(1, m+b), (2, 2m+b), (3, 3m+b), \ldots$ that radiate outward along the line $y = mx + b$ of slope m and y-intercept b. These correspond in $\mathbb{P}^2_\mathbb{R}$ to

$$[1:1:m+b],\ [1:2:2m+b],\ [1:3:3m+b], \ldots \in \mathbb{P}^2_\mathbb{R}.$$

9.1. PROJECTIVE SPACE

Rescaling, we obtain equivalent points

$$[\tfrac{1}{1}:1:m+\tfrac{b}{1}], [\tfrac{1}{2}:1:m+\tfrac{b}{2}], [\tfrac{1}{3}:1:m+\tfrac{b}{3}], \ldots \in \mathbb{P}^2_{\mathbb{R}},$$

which tend toward $[0:1:m]$. This limit is a "point at infinity" in $\mathbb{P}^2_{\mathbb{R}}$, since it is a point with first coordinate zero. Below, we have depicted $\mathbb{A}^2_{\mathbb{R}}$ along with several points at infinity in $\mathbb{P}^2_{\mathbb{R}}$ that are approached along lines of different slopes.

We see now why $\mathbb{P}^2_{\mathbb{R}}$ has many points at infinity whereas $\mathbb{P}^1_{\mathbb{R}}$ had just one: in $\mathbb{P}^2_{\mathbb{R}}$, the point at infinity that we approach by walking outward along a line depends on the slope of that line. In fact, the idea of "following a line to the point at infinity to which it leads" can be made precise as a bijection between points at infinity in $\mathbb{P}^2_{\mathbb{R}}$ and lines through the origin in $\mathbb{A}^2_{\mathbb{R}}$. This is a special case of a more general phenomenon, which we now state.

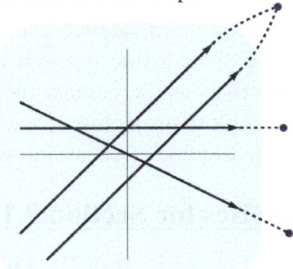

9.9 PROPOSITION *Points at infinity are slopes of lines*

For any $n \in \mathbb{N}$, there is a natural bijection

$$\{\text{points at infinity in } \mathbb{P}^n\} = \{\text{lines through } (0,\ldots,0) \text{ in } \mathbb{A}^n\}.$$

PROOF A line through $(0,\ldots,0) \in \mathbb{A}^n$ is, by definition, a set of points of the form

$$L = \{(\lambda a_1, \lambda a_2, \ldots, \lambda a_n) \mid \lambda \in K\},$$

where $a_i \in K$ are fixed and at least one a_i is nonzero. The desired bijection, then, is given by associating to the line

$$L = \{(\lambda a_1, \lambda a_2, \ldots, \lambda a_n) \mid \lambda \in K\}$$

the point at infinity

$$[0:a_1:a_2:\cdots:a_n] \in \mathbb{P}^n,$$

which (as the reader is encouraged to verify intuitively) can be viewed as the point toward which an outward-radiating sequence of points on L tends. □

Given that the points at infinity in $\mathbb{P}^n$ are also in bijection with $\mathbb{P}^{n-1}$, Proposition 9.9 can be viewed in another light: it gives us our third perspective on projective space, which is often taken as an alternative definition of the space itself.

9.10 COROLLARY *Projective space, third perspective*

For any $n \in \mathbb{N}$, there is a natural bijection

$$\mathbb{P}^n = \{\text{lines through } (0,\ldots,0) \text{ in } \mathbb{A}^{n+1}\}.$$

Tracing through the bijection of Proposition 9.9 explains how to match up the first and third perspectives with one another: an element $[a_0 : a_1 : \cdots : a_n] \in \mathbb{P}^n$ corresponds to the line

$$L = \{(\lambda a_0, \lambda a_1, \ldots, \lambda a_n) \mid \lambda \in K\} \subseteq \mathbb{A}^{n+1},$$

which is the line through the origin passing through $(a_0, a_1, \ldots, a_n) \in \mathbb{A}^{n+1}$.

With these ideas combined, an element of $\mathbb{P}^n$ can be viewed in three different ways: as an equivalence class $[a_0 : a_1 : \cdots : a_n]$, as a point either in $\mathbb{A}^n$ or at infinity, or as a line through the origin in $\mathbb{A}^{n+1}$. Moving fluidly between these perspectives as the context dictates is one of the skills that the reader will develop as we explore projective space and—beginning in the next section—the analogue in projective space of all we know about varieties.

Exercises for Section 9.1

9.1.1 Let $[2:1:3] \in \mathbb{P}^2$. Prove that

$$[2:1:3] = [6:3:9]$$

but that

$$[2:1:3] \neq [6:4:12].$$

In general, which $[a:b:c] \in \mathbb{P}^2$ satisfy $[2:1:3] = [a:b:c]$?

9.1.2 Prove that $[a_0 : \cdots : a_n] = [b_0 : \cdots : b_n] \in \mathbb{P}^n$ if and only if

$$a_i b_j = a_j b_i \quad \text{for all} \quad i, j \in \{0, \ldots, n\}.$$

In this case, we say that the *cross-multiplications* agree.

9.1.3 Prove that there is a natural bijection $\mathbb{P}^n = \mathbb{A}^n \sqcup \mathbb{P}^{n-1}$ in three steps:
   (a) Define

   $$U = \{[a_0 : a_1 : \cdots : a_n] \in \mathbb{P}^n \mid a_0 \neq 0\}$$

   and

   $$V = \{[a_0 : a_1 : \cdots : a_n] \in \mathbb{P}^n \mid a_0 = 0\}.$$

   Prove that $\mathbb{P}^n = U \sqcup V$.
   (b) Prove that there is a natural bijection between U and $\mathbb{A}^n$.
   (c) Prove that there is a natural bijection between V and $\mathbb{P}^{n-1}$.

9.1.4 Prove that there is a natural bijection

$$\mathbb{P}^n = \mathbb{A}^n \sqcup \mathbb{A}^{n-1} \sqcup \mathbb{A}^{n-2} \sqcup \cdots \sqcup \mathbb{A}^1 \sqcup \mathbb{A}^0$$

and describe the points in each component.

9.1.5 Let $[2:1:3] \in \mathbb{P}^2$. Describe this point as
   (a) an element of $\mathbb{A}^2 \sqcup \mathbb{P}^1$, and
   (b) a line through $(0,0,0)$ in $\mathbb{A}^3$.

9.1.6 Repeat Problem 9.1.5 for the point $[0:1:3] \in \mathbb{P}^2$.

9.1.7 Which lines through $(0,0,0) \in \mathbb{A}^3$ correspond to points at infinity in $\mathbb{P}^2$?

Section 9.2 The projective $\mathcal{V}$-operator

Just like an affine variety, a projective variety is defined as the common vanishing set of a collection of polynomials. Taking the perspective on $\mathbb{P}^n$ given in Definition 9.1, the inputs to those polynomials are tuples $(a_0, \ldots, a_n)$. However, since Definition 9.1 involves an equivalence relation, a polynomial might vanish on one representative but not on another, so it is not immediately clear what we mean when we say that a polynomial "vanishes" at a point of projective space.

For instance, suppose we consider points $[a_0 : a_1] \in \mathbb{P}^1$ as inputs to the two-variable polynomial $f = x_0^2 - x_1$. The point $[2:4]$ would seem to be in the vanishing set of f, since

$$f(2,4) = 2^2 - 4 = 0.$$

On the other hand, however, we see that $[2:4] = [4:8]$, and

$$f(4,8) = 4^2 - 8 = 8 \neq 0.$$

Thus, the question of whether f vanishes at the point $[2:4] = [4:8]$ does not seem to have a well-defined answer. The solution to this discrepancy is simply to declare that a polynomial "vanishes" at a point of projective space only if it vanishes when evaluated at *every* representative of that point.

9.11 DEFINITION *Projective vanishing*

Let $f \in K[x_0, \ldots, x_n]$ be a polynomial and let $a \in \mathbb{P}^n$ be a point. We say that f *vanishes at* a and write $f(a) = 0$ if

$$f(a_0, \ldots, a_n) = 0$$

for every representative $a = [a_0 : \cdots : a_n]$.

For example, the polynomial $f(x_0, x_1) = x_0^2 - x_1$ does *not* vanish at the point $a = [2:4] \in \mathbb{P}^1$ because it does not vanish when evaluated at the equivalent representative $a = [4:8]$. A priori, checking that a polynomial vanishes at *every* representative of a point seems to be an arduous task—after all, there are infinitely many representatives for any point. However, this task can be simplified with the introduction of homogeneous polynomials.

9.12 DEFINITION *Homogeneous polynomial*

A polynomial $f \in K[x_0, \ldots, x_n]$ is *homogeneous of degree d* if every nonzero term of f has degree d.

A "term" of f is any summand in the expression $f = \sum_\alpha a_\alpha x^\alpha$. A term $a_\alpha x^\alpha$ is "nonzero" when $a_\alpha \neq 0$.

For example, the two-variable polynomial $f = x_0^2 - x_1$ is inhomogeneous, because it has a nonzero term of degree two and another of degree one, while the polynomial $g = x_0^2 - 2x_0 x_1$ is homogeneous of degree two. The zero polynomial is vacuously homogeneous of every degree, since it does not have any nonzero terms.

In the context of studying vanishing within projective space, the importance of working with homogeneous polynomials is the following result.

9.13 LEMMA *Projective vanishing of homogeneous polynomials*

Let $f \in K[x_0, \ldots, x_n]$ be a homogeneous polynomial and let

$$[a_0 : \cdots : a_n] = [b_0 : \cdots : b_n] \in \mathbb{P}^n.$$

Then

$$f(a_0, \ldots, a_n) = 0 \iff f(b_0, \ldots, b_n) = 0.$$

In other words, when working with homogeneous polynomials, vanishing of a polynomial can be verified by checking vanishing at a single representative. For example, consider the homogeneous polynomial $g = x_0^2 - 2x_0 x_1$. To check that g vanishes at the equivalence class $[2:1] \in \mathbb{P}^1$, it suffices to verify vanishing at one representative:

$$g(2,1) = 2^2 - 2 \cdot 2 \cdot 1 = 0.$$

If we replace $[2:1]$ by the alternative representative $[4:2]$ (or any other representative for this point of $\mathbb{P}^1$), the vanishing persists:

$$g(4,2) = 4^2 - 2 \cdot 4 \cdot 2 = 0.$$

PROOF OF LEMMA 9.13   The key observation we need is that f is homogeneous of degree d if and only if

(9.14) $$f(\lambda a_0, \ldots, \lambda a_n) = \lambda^d f(a_0, \ldots, a_n)$$

for all $\lambda, a_0, \ldots, a_n \in K$ (Exercise 9.2.1). Suppose that f is homogeneous of degree d and that

$$[a_0 : \cdots : a_n] = [b_0 : \cdots : b_n] \in \mathbb{P}^n.$$

By definition of the equivalence relation on $\mathbb{P}^n$, there exists a nonzero $\lambda \in K$ such that

$$(\lambda a_0, \ldots, \lambda a_n) = (b_0, \ldots, b_n).$$

From (9.14) we then see that $f(b_0, \ldots, b_n)$ and $f(a_0, \ldots, a_n)$ differ by the nonzero scalar multiple λ^d, so one vanishes if and only if the other does. □

Lemma 9.13 shows that we can readily determine whether a homogeneous polynomial vanishes at a point of projective space simply by checking a single representative, but what about the vanishing of an inhomogeneous polynomial? To address this question, we introduce the homogeneous components of a polynomial.

9.15 DEFINITION *Homogeneous components*

Given a polynomial $f \in K[x_0, \ldots, x_n]$, the *dth homogeneous component of f*, denoted f_d, is the sum of all nonzero terms of f of degree d. If f does not have any nonzero terms of degree d, then $f_d = 0$.

9.2. THE PROJECTIVE $\mathcal{V}$-OPERATOR

For example, the nonzero homogeneous components of the polynomial

$$f = x^4 z + x^2 y + xyz + x + y + 5$$

are

$$f_5 = x^4 z, \quad f_3 = x^2 y + xyz, \quad f_1 = x - y, \quad f_0 = 5.$$

The next result describes the vanishing of a polynomial at a point of projective space in terms of the vanishing of its homogeneous components.

9.16 LEMMA *Projective vanishing and homogeneous components*

Let $f \in K[x_0, \ldots, x_n]$ be a polynomial and $a \in \mathbb{P}^n$ a point. Then f vanishes at a if and only if every homogeneous component of f vanishes at a.

In other words, in order to determine whether a general polynomial vanishes at (every representative of) a point of $\mathbb{P}^n$, Lemmas 9.13 and 9.16 together imply that it suffices to check whether each homogeneous component vanishes at a single representative of that point.

PROOF OF LEMMA 9.16   Let $f \in K[x_0, \ldots, x_n]$ be a polynomial of degree d, which can be written as a sum of its homogeneous components of degree $\leq d$:

(9.17) $$f = \sum_{k=0}^{d} f_k.$$

If each f_k vanishes at every representative of a point a, then it follows from (9.17) that f vanishes at every representative of a.

Conversely, assume that f vanishes at every representative of a. Choose one particular representative $a = [a_0 : \cdots : a_n]$. Then, for any $\lambda \in K \setminus \{0\}$, we have

$$0 = f(\lambda a_0, \ldots, \lambda a_n) = \sum_{k=0}^{d} f_k(\lambda a_0, \ldots, \lambda a_n)$$

$$= \sum_{k=0}^{d} \lambda^k f_k(a_0, \ldots, a_n),$$

where the second equality uses (9.14). In other words, the single-variable polynomial

$$\sum_{k=0}^{d} x^k f_k(a_0, \ldots, a_n) \in K[x]$$

vanishes at infinitely many values of K, so it must be the zero polynomial, implying that $f_k(a_0, \ldots, a_n) = 0$ for all k. Since f_k is homogeneous, Lemma 9.13 then implies that each f_k vanishes at every representative of a. □

With a better understanding of what it means for polynomials to vanish at points of projective space, we now come to the natural definition of a *projective variety*.

9.18 DEFINITION *Projective V-operator*

Let $\mathcal{S} \subseteq K[x_0, \ldots, x_n]$ be a set of polynomials. The *projective vanishing set* of $\mathcal{S}$ is

$$\mathcal{V}_{\mathbb{P}}(\mathcal{S}) = \{a \in \mathbb{P}^n \mid f(a) = 0 \text{ for all } f \in \mathcal{S}\}.$$

We say that a subset $X \subseteq \mathbb{P}^n$ is a *projective variety* if $X = \mathcal{V}_{\mathbb{P}}(\mathcal{S})$ for some set $\mathcal{S} \subseteq K[x_0, \ldots, x_n]$.

> We often write $\mathcal{V}(\mathcal{S}) \subseteq \mathbb{P}^n$ when it is clear from context that we are working in projective space, as opposed to affine space.

Lemma 9.16 implies that every projective variety can be described by a set of homogeneous polynomials, simply by replacing the inhomogeneous polynomials in $\mathcal{S}$ with their homogeneous components. Because of this, it is common in practice to describe a projective variety as the vanishing set of a collection of homogeneous polynomials. Let us consider a few examples of projective varieties.

9.19 EXAMPLE $\emptyset$ and $\mathbb{P}^n$ are projective varieties

As in the affine case, we have $\mathcal{V}(1) = \emptyset$ and $\mathcal{V}(0) = \mathbb{P}^n$, so we see that $\emptyset$ and $\mathbb{P}^n$ are projective varieties.

9.20 EXAMPLE Projective varieties in $\mathbb{P}^1$

In $\mathbb{P}^1$, consider the projective variety

$$X = \mathcal{V}(2x_0 - x_1) = \{[a_0 : a_1] \in \mathbb{P}^1 \mid 2a_0 - a_1 = 0\}.$$

A point $[a_0 : a_1]$ in X cannot have $a_0 = 0$, since then the equation $2a_0 - a_1 = 0$ would force that $a_1 = 0$, as well. Thus, we have

$$X = \{[a : 2a] \in \mathbb{P}^1 \mid a \in K \setminus \{0\}\} = \{[1 : 2]\},$$

since multiplying both coordinates by a^{-1} shows that $[a : 2a] = [1 : 2]$ for any a. More generally (in perfect analogy to the situation for $\mathbb{A}^1$), any projective variety in $\mathbb{P}^1$ is either all of $\mathbb{P}^1$ or a finite (possibly empty) set of points (Exercise 9.2.2).

9.21 EXAMPLE A line in $\mathbb{P}^2$

In $\mathbb{P}^2$, consider the projective variety

$$X = \mathcal{V}(x_0 + x_1 - x_2) = \{[a_0 : a_1 : a_2] \in \mathbb{P}^2 \mid a_0 + a_1 - a_2 = 0\}.$$

A point $[a_0 : a_1 : a_2]$ in X cannot have $a_0 = a_1 = 0$, since then the defining equation would force that $a_2 = 0$, as well. It follows that

$$X = \{[a_0 : a_1 : a_0 + a_1] \in \mathbb{P}^2 \mid a_0, a_1 \in K \text{ not both } 0\},$$

and from here it is not difficult to see that the points of $\mathcal{V}(x_0 + x_1 - x_2)$ are in bijection with $\mathbb{P}^1$.

9.2. THE PROJECTIVE $\mathcal{V}$-OPERATOR

In analogy with the affine case—where the vanishing of a linear polynomial describes a line in $\mathbb{A}^2$—we refer to X as a *line* in $\mathbb{P}^2$. This intuitive terminology should be taken with a grain of salt: as we saw in the previous section, a projective "line" does not really look like our familiar notion of a line from plane geometry, even over the real numbers, where the projective line forms a loop.

As a first step toward utilizing the algebraic structure of polynomial rings to study projective varieties, we note that every projective variety can be defined by an ideal, a result that is parallel to Proposition 1.15 in the affine setting.

9.22 PROPOSITION *Projective varieties are defined by ideals*

If $\mathcal{S} \subseteq K[x_0, \ldots, x_n]$ is a set of polynomials, then

$$\mathcal{V}_\mathbb{P}(\mathcal{S}) = \mathcal{V}_\mathbb{P}(\langle \mathcal{S} \rangle).$$

PROOF Exercise 9.2.3. □

As in the affine case, knowing that any projective variety can be defined by an ideal allows us to leverage the algebraic structure of polynomial rings to deduce that every projective variety can be defined by finitely many polynomials. Moreover, in the projective case we get a little more: using Lemma 9.16, it follows that every projective variety is the vanishing set of a finite set of *homogeneous* polynomials.

9.23 COROLLARY *Projective varieties are finitely generated*

Any projective variety $X \subseteq \mathbb{P}^n$ is of the form $X = \mathcal{V}_\mathbb{P}(f_1, \ldots, f_k)$ where $f_1, \ldots, f_k \in K[x_0, \ldots, x_n]$ are homogeneous polynomials.

PROOF Exercise 9.2.4. □

At this point, we can begin to see the utility of working algebraically with defining ideals of projective varieties, rather than merely sets of polynomials. In particular, it is through passing to an ideal and using Hilbert's Basis Theorem for polynomial rings that one proves Corollary 9.23. Just like in the affine setting, while there may be many defining ideals for a single projective variety, there is always one distinguished ideal among all of its defining ideals—the vanishing ideal. In the next section, we turn to a discussion of vanishing ideals in the projective setting.

Exercises for Section 9.2

9.2.1 Let $f \in K[x_0, x_1, \ldots, x_n]$ be a nonzero polynomial. Prove that f is homogeneous of degree d if and only if, for any $\lambda, a_0, a_1, \ldots, a_n \in K$,

$$f(\lambda a_0, \lambda a_1, \ldots, \lambda a_n) = \lambda^d f(a_0, a_1, \ldots, a_n).$$

9.2.2 Prove that the only projective varieties in $\mathbb{P}^1$ are $\mathbb{P}^1$ and finite sets of points.

9.2.3 Prove Proposition 9.22.

9.2.4 Prove Corollary 9.23.

9.2.5 Prove that finite unions and arbitrary intersections of projective varieties are projective varieties.

9.2.6 Let
$$X = \mathcal{V}(x_0 x_1 - x_2^2) \subseteq \mathbb{P}^2.$$

(a) Let $U = \{[a_0 : a_1 : a_2] \in \mathbb{P}^2 \mid a_0 \neq 0\}$, which, by the results of the previous section, is in natural bijection with $\mathbb{A}^2$. Prove that $X \cap U$ is identified by this bijection with an affine variety in $\mathbb{A}^2$. What is that affine variety? Draw a picture of $X \cap U$ over $\mathbb{R}$.

(b) Compute all points of $X \setminus U$, and describe how these points are approached by points in the affine variety you found in (b).

(c) Describe a bijection between X and $\mathbb{P}^1$.

9.2.7 Complete the previous problem for $X = \mathcal{V}(x_0^2 - x_1 x_2) \subseteq \mathbb{P}^2$.

Section 9.3   The projective $\mathcal{I}$-operator

The projective $\mathcal{V}$-operator allows us to pass from collections of polynomials to subsets of projective space, and we now turn to the projective $\mathcal{I}$-operator, which moves us in the opposite direction. The definition of $\mathcal{I}_\mathbb{P}$ is as one might expect.

9.24 DEFINITION   *Projective $\mathcal{I}$-operator*

Let $X \subseteq \mathbb{P}^n$ be a subset. The *vanishing ideal* of X is

$$\mathcal{I}_\mathbb{P}(X) = \{f \in K[x_0, \ldots, x_n] \mid f(a) = 0 \text{ for all } a \in X\}.$$

We say that a subset of $K[x_0, \ldots, x_n]$ is a *projective vanishing ideal* if it is of the form $\mathcal{I}_\mathbb{P}(X)$ for some $X \subseteq \mathbb{P}^n$.

As for the $\mathcal{V}_\mathbb{P}$-operator, we often write $\mathcal{I}(X)$ when it is clear from context whether we are working in affine or projective space.

Recalling Definition 9.11, when we say that $f(a) = 0$, we are asserting that f vanishes at *every* representative of the point $a \in \mathbb{P}^n$. For example, if $[1:0]$ is a point of X, then the polynomial $f = x_0 - 1$ is *not* an element of $\mathcal{I}(X)$; even though $f(1,0) = 0$, notice that f does not vanish when evaluated at the representative $[2:0] = [1:0]$. The next example elaborates further on this.

9.25 EXAMPLE   Vanishing ideal of a point in $\mathbb{P}^1$

Let $X = \{[1:0]\} \subseteq \mathbb{P}^1$. Then $f = x_1 \in K[x_0, x_1]$ is an element of $\mathcal{I}(X)$, since any representative of the point $[1:0] \in X$ is of the form $[a:0]$ for some a, and $f(a, 0) = 0$ for any choice of a. More generally, we see that $\langle x_1 \rangle \subseteq \mathcal{I}(X)$, and in fact, we claim that there is equality: $\mathcal{I}(X) = \langle x_1 \rangle$.

To prove the remaining inclusion, let $f \in \mathcal{I}(X)$. Then $f(a, 0) = 0$ for all nonzero $a \in K$. It follows that $f(x_0, 0)$ is a single-variable polynomial with infinitely many zeros, so it must be the zero polynomial. Write f as an element of $(K[x_0])[x_1]$:

$$f = \sum_{d \geq 0} f_d(x_0) x_1^d.$$

Using $f_0(x_0) = f(x_0, 0) = 0$, we conclude that

$$f = x_1 \sum_{d \geq 1} f_d(x_0) x_1^{d-1} \in \langle x_1 \rangle.$$

9.26 EXAMPLE   Vanishing ideal of a line in $\mathbb{P}^2$

If X is the line $\mathcal{V}(x_0 + x_1 - x_2) \subseteq \mathbb{P}^2$ of Example 9.21, then

$$\mathcal{I}(X) = \langle x_0 + x_1 - x_2 \rangle.$$

The fact that every element of $\langle x_0 + x_1 - x_2 \rangle$ vanishes at every point of X is essentially immediate, while the reverse inclusion is the content of Exercise 9.3.1.

As in the affine case, projective vanishing ideals are, in fact, ideals, and moreover, they are readily seen to be radical ideals. In the projective setting, though, we get even more. In particular, Lemma 9.16 implies that, for every $f \in \mathcal{I}(X)$, every homogeneous component of f must also be an element of $\mathcal{I}(X)$. This attribute of $\mathcal{I}(X)$ is the defining property of what it means to be a *homogeneous ideal*.

9.27 DEFINITION *Homogeneous ideal*

An ideal $I \subseteq K[x_0, \ldots, x_n]$ is *homogeneous* if, for every $f \in I$, every homogeneous component of f is also in I.

While the above definition of homogeneous ideals is directly motivated by our discussion of vanishing ideals, the following result offers an important alternative characterization of homogeneous ideals that is, perhaps, more straightforward, and that can be quite useful in practice.

9.28 PROPOSITION *Characterizing homogeneous ideals*

An ideal $I \subseteq K[x_0, \ldots, x_n]$ is homogeneous if and only if it admits a set of homogeneous generators.

PROOF   First, suppose that I is a homogeneous ideal, and let $\mathcal{S} \subseteq I$ be the subset consisting of all homogeneous polynomials in I. It suffices to prove that $I = \langle \mathcal{S} \rangle$. The inclusion $\langle \mathcal{S} \rangle \subseteq I$ is because $\mathcal{S} \subseteq I$ and I is an ideal. For the reverse inclusion, suppose that $f \in I$. Then we can express f as a sum of nonzero homogeneous components f_k, and the fact that I is a homogeneous ideal means that $f_k \in I$ for each k. Given that f_k is homogeneous, it follows that $f_k \in \mathcal{S}$. Therefore, f is a sum of elements of $\mathcal{S}$, so $f \in \langle \mathcal{S} \rangle$. We conclude that $I = \langle \mathcal{S} \rangle$, as claimed.

Conversely, suppose $I = \langle \mathcal{S} \rangle$, where $\mathcal{S}$ is a set of homogeneous polynomials. To prove that I is a homogeneous ideal, let $f \in I$. The fact that $I = \langle \mathcal{S} \rangle$ means that

$$f = \sum_{i=1}^{m} g_i h_i$$

for some $g_i \in K[x_0, \ldots, x_n]$ and $h_i \in \mathcal{S}$; in particular, h_i is homogeneous of some degree d_i. For each k, we have

$$f_k = \sum_{i=1}^{m} (g_i h_i)_k,$$

where, in the right-hand side, $(g_i h_i)_k$ denotes the kth homogeneous component of the polynomial $g_i h_i$. By Exercise 9.3.2, we can rewrite the summands as

$$(g_i h_i)_k = \begin{cases} (g_i)_{k-d_i} \cdot h_i & \text{if } d_i \leq k, \\ 0 & \text{if } d_i > k, \end{cases}$$

where $(g_i)_{k-d_i}$ is the $(k - d_i)$th homogeneous component of g_i. This implies that $(g_i h_i)_k \in I$, since it is a multiple of $h_i \in \mathcal{S}$, so f_k is a sum of elements of I, showing that $f_k \in I$. Thus, we have shown that I is a homogeneous ideal.   □

9.3. THE PROJECTIVE $\mathcal{I}$-OPERATOR

Having established an understanding of homogeneous ideals, we now return to the primary topic of this section: vanishing ideals. The next result summarizes the most important algebraic attributes of projective vanishing ideals.

9.29 PROPOSITION $\mathcal{I}_\mathbb{P}(X)$ *is a homogeneous radical ideal*

If $X \subseteq \mathbb{P}^n$ is any subset, then $\mathcal{I}_\mathbb{P}(X) \subseteq K[x_0, \ldots, x_n]$ is a homogeneous radical ideal.

PROOF The fact that $\mathcal{I}_\mathbb{P}(X)$ is a radical ideal follows from the exact same argument as in the affine case (Proposition 1.24), as the reader is encouraged to verify. That $\mathcal{I}(X)$ is homogeneous follows, by definition, from Lemma 9.16.   □

As in the affine case, our primary reason for defining vanishing ideals is to have a distinguished defining ideal for any projective variety. That the vanishing ideal serves this role is verified in the third item of the next result, which is just one of a number of important properties relating the projective $\mathcal{V}$- and $\mathcal{I}$-operators.

9.30 PROPOSITION *Basic properties of $\mathcal{V}_\mathbb{P}$ and $\mathcal{I}_\mathbb{P}$*

Let $\mathcal{S}, \mathcal{T} \subseteq K[x_0, \ldots, x_n]$ and $X, Y \subseteq \mathbb{P}^n$ be subsets.
  1. If $\mathcal{S} \subseteq \mathcal{T}$, then $\mathcal{V}_\mathbb{P}(\mathcal{S}) \supseteq \mathcal{V}_\mathbb{P}(\mathcal{T})$.
  2. If $X \subseteq Y$, then $\mathcal{I}_\mathbb{P}(X) \supseteq \mathcal{I}_\mathbb{P}(Y)$.
  3. $\mathcal{V}_\mathbb{P}(\mathcal{I}_\mathbb{P}(X)) \supseteq X$, with equality if and only if X is a projective variety.
  4. $\mathcal{I}_\mathbb{P}(\mathcal{V}_\mathbb{P}(\mathcal{S})) \supseteq \mathcal{S}$, with equality if and only if $\mathcal{S}$ is a projective vanishing ideal.

PROOF The proofs of these statements are analogous to those of their affine counterparts (Propositions 2.1 and 1.21), as the reader is encouraged to verify.   □

Continuing to parallel the affine situation, we recall that, in the affine case, the relationship between the $\mathcal{V}$- and $\mathcal{I}$-operators was leveraged to prove the existence and uniqueness of irreducible decompositions. We now state the projective analogue, starting with the natural definition of an irreducible projective variety, which carries over verbatim from the affine case.

9.31 DEFINITION *Irreducible projective variety*

A projective variety $X \subseteq \mathbb{P}^n$ is *reducible* if $X = X_1 \cup X_2$ for some projective varieties $X_1, X_2 \subsetneq X$, and X is *irreducible* if it is neither empty nor reducible.

As one might expect, irreducible decompositions always exist and are unique in the projective setting, and the proof of this fact is parallel to the affine situation.

9.32 PROPOSITION/DEFINITION *Irreducible decomposition*

Let $X \subseteq \mathbb{P}^n$ be a nonempty projective variety. Then there exist irreducible projective varieties $X_1, \ldots, X_r \subseteq X$ such that $X_i \not\subseteq X_j$ for any $i \neq j$ and

(9.33) $$X = \bigcup_{i=1}^{r} X_i.$$

Moreover, the irreducible projective varieties $X_1, \ldots, X_r$ are unique up to reordering; we call these the *irreducible components* of X, and refer to (9.33) as the *irreducible decomposition* of X.

PROOF   The proof, which uses the relationship between the projective $\mathcal{V}$- and $\mathcal{I}$-operators, along with the Noetherian property of $K[x_0, \ldots, x_n]$, is analogous to that of the affine statement (Proposition/Definition 2.32). We leave the verification as an exercise to the reader. □

At this point, the structural parallels between projective varieties and affine varieties have begun to emerge, and indeed, many of the results in the projective case have proofs that are identical, or at least analogous, to the affine case. However, our geometric intuition for projective varieties is still lacking; after all, how can we draw pictures of projective varieties when $\mathbb{P}^n$ is so difficult to visualize, even over the real numbers, for $n \geq 2$? The key to answering this question lies in two techniques for moving between the projective setting and the affine setting—affine restrictions and projective closures—to which we devote the next two sections.

Exercises for Section 9.3

9.3.1 Let $X = \mathcal{V}(x_0 + x_1 - x_2) \subseteq \mathbb{P}^2$. We prove that $\mathcal{I}(X) = \langle x_0 + x_1 - x_2 \rangle$. As discussed in Example 9.26, we need only prove $\mathcal{I}(X) \subseteq \langle x_0 + x_1 - x_2 \rangle$.
  (a) Let $f \in \mathcal{I}(X)$. Prove that $f(x_0, x_1, x_0 + x_1) = 0 \in K[x_0, x_1]$.
  (b) Argue that, since $f(x_0, x_1, x_0 + x_1) = 0$, we have $f \in \langle x_0 + x_1 - x_2 \rangle$.

9.3.2 Let $g \in K[x_0, \ldots, x_n]$ be any polynomial and let $h \in [x_0, \ldots, x_n]$ be homogeneous of degree d. Prove that, for each $k \geq 0$, we have

$$(gh)_k = \begin{cases} g_{k-d} \cdot h & \text{if } d \leq k, \\ 0 & \text{if } d > k, \end{cases}$$

where $(gh)_k$ denotes the kth homogeneous component of gh and g_{k-d} is the $(k-d)$th homogeneous component of g.

9.3.3 Which of the following ideals in $K[x, y]$ are homogeneous?
  (a) $\langle x + 1, y^2 \rangle$
  (b) $\langle x + y, x^2 \rangle$
  (c) $\langle x^2 + y, x^2 - y \rangle$
  (d) $\langle x^2 + y^2, xy^2 + y^3 + x^2, y^2 - x^2 \rangle$

9.3.4 Prove that a principal ideal $\langle f \rangle \subseteq K[x_0, \ldots, x_n]$ is homogeneous if and only if the polynomial f is homogeneous.

9.3. THE PROJECTIVE $\mathcal{I}$-OPERATOR

9.3.5 Let $X \subseteq \mathbb{P}^n$ be a projective variety. Prove that X is irreducible if and only if $\mathcal{I}_\mathbb{P}(X)$ is a prime ideal.

9.3.6 Prove that a subset $X \subseteq \mathbb{P}^n$ consists of a single point if and only if $\mathcal{I}_\mathbb{P}(X)$ is a maximal ideal.

9.3.7 Rewrite the proof of Proposition 1.24 in the language and notation of projective varieties, making sure that every step can be carried out in the projective setting, thereby proving that projective vanishing ideals are radical ideals.

9.3.8 Rewrite the proofs of Propositions 2.1 and 1.21 in the language and notation of projective varieties, making sure that every step can be carried out in the projective setting, thereby proving Proposition 9.30.

9.3.9 Rewrite the proof of Proposition/Definition 2.32 in the language and notation of projective varieties, making sure that every step can be carried out in the projective setting, thereby proving Proposition/Definition 9.32.

Section 9.4 Affine restrictions

In order to relate a projective variety to a more easily visualizable affine variety, we recall the second perspective on projective space from Section 9.1, wherein we view $\mathbb{P}^n$ as the result of adding points at infinity to $\mathbb{A}^n$. By ignoring these points, we find for each projective variety an affine restriction, and this restriction gives a helpful—though incomplete—picture of what the projective variety looks like. Before explaining the concept in general, we consider a specific example.

9.34 EXAMPLE Affine restriction of a quadratic curve in $\mathbb{P}^2$

Consider the projective variety

$$X = \mathcal{V}_\mathbb{P}(x_0^2 + x_1^2 - x_2^2) \subseteq \mathbb{P}^2_\mathbb{C}.$$

> *In settings involving both affine and projective varieties, we often use the notation $\mathcal{V}_\mathbb{A}$ and $\mathcal{I}_\mathbb{A}$ for the affine $\mathcal{V}$- and $\mathcal{I}$-operators to distinguish them from their projective counterparts.*

Any point of X that has a nonzero first coordinate can be expressed in homogeneous coordinates as $[1 : a_1 : a_2]$, where $1 + a_1^2 - a_2^2 = 0$. Thus, setting

$$X_0 = \mathcal{V}_\mathbb{A}(1 + x_1^2 - x_2^2) \subseteq \mathbb{A}^2_\mathbb{C},$$

there is a natural bijection

$$X = X_0 \sqcup \{[0 : a_1 : a_2] \mid a_1^2 - a_2^2 = 0\}.$$

There are two points of the second type: a point $[0 : a_1 : a_2] \in \mathbb{P}^2_\mathbb{C}$ satisfying

$$0 = a_1^2 - a_2^2 = (a_1 - a_2)(a_1 + a_2)$$

must be of the form $[0 : a : a]$ or $[0 : a : -a]$, and under the equivalence relation on $\mathbb{P}^2_\mathbb{C}$, this means that it is equal to either $[0 : 1 : 1]$ or $[0 : 1 : -1]$. Thus,

$$X = X_0 \sqcup \{[0 : 1 : 1], [0 : 1 : -1]\}.$$

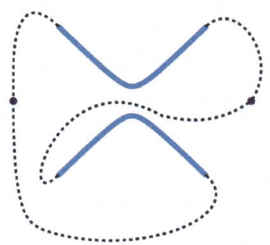

Restricting to the real numbers, we can visualize the affine variety X_0: it is the hyperbola shown at right, which captures almost all of X. The two additional points in X are the points at infinity that one reaches by walking along the two asymptotes of X_0. More specifically, the two asymptotes of X_0 are the lines through the origin of slope 1 and -1, which, as we saw in the discussion following Proposition 9.8, tend toward the points $[0 : 1 : 1]$ and $[0 : 1 : -1]$ in $\mathbb{P}^2$. As you can see in the image by following the path that X makes with X_0 and the two points at infinity, the real points of the full projective variety X form a single loop, just like $\mathbb{P}^1_\mathbb{R}$. In fact, once we have discussed isomorphisms of projective varieties, we will see that there is an isomorphism $X \cong \mathbb{P}^1_\mathbb{C}$.

9.4. AFFINE RESTRICTIONS

Generalizing Example 9.34, if $X = \mathcal{V}_{\mathbb{P}}(\mathcal{S}) \subseteq \mathbb{P}^n$ is a projective variety, then

$$X = X_0 \sqcup \{\text{points at infinity in } X\},$$

where

$$X_0 = \{[a_0 : \cdots : a_n] \in X \mid a_0 \neq 0\},$$

and the points at infinity in X are those with $a_0 = 0$. That is, $X_0 = X \cap \mathbb{A}^n$ under the natural bijection between $\mathbb{A}^n$ and points in $\mathbb{P}^n$ with nonzero first coordinate. Given that points of X_0 can be expressed in the form $[1 : b_1 : \cdots : b_n]$, we see that

$$X_0 = \mathcal{V}_{\mathbb{A}}(\mathcal{S}_0) \subseteq \mathbb{A}^n,$$

where

$$\mathcal{S}_0 = \{f(1, x_1, \ldots, x_n) \mid f \in \mathcal{S}\} \subseteq K[x_1, \ldots, x_n].$$

In particular, X_0 is an affine variety, called the *affine restriction* of X. Let us consider one more class of examples.

9.35 EXAMPLE Restricting lines in $\mathbb{P}^2$

The affine restriction of the projective variety $X = \mathcal{V}_{\mathbb{P}}(x_0 + x_1 - x_2) \subseteq \mathbb{P}^2$ from Example 9.21 is the line $X_0 = \mathcal{V}_{\mathbb{A}}(1 + x_1 - x_2) \subseteq \mathbb{A}^2$, and the only point at infinity of X is $[0 : 1 : 1]$. More generally, the affine restriction of the projective variety

$$L = \mathcal{V}_{\mathbb{P}}(bx_0 + mx_1 - x_2) \subseteq \mathbf{P}^2$$

is

$$L_0 = \mathcal{V}_{\mathbb{A}}(b + mx_1 - x_2) \subseteq \mathbb{A}^2,$$

which is a line with vertical intercept b and slope m. Some reflection should convince the reader that L again contains just one point at infinity: $[0 : 1 : m]$. This is a more precise manifestation of what we saw informally in Section 9.1: the point $[0 : 1 : m]$ is the point at infinity that one reaches by "walking along L_0."

In particular, we see again that the point at infinity reached by walking along L_0 depends only on the slope of L_0. Parallel lines, then, such as

$$\mathcal{V}_{\mathbb{A}}(1 + 3x_1 - x_2), \ \mathcal{V}_{\mathbb{A}}(2 + 3x_1 - x_2) \subseteq \mathbb{A}^2,$$

do not meet in $\mathbb{A}^2$, yet when viewed as the affine restrictions of the projective lines

$$\mathcal{V}_{\mathbb{P}}(x_0 + 3x_1 - x_2), \ \mathcal{V}_{\mathbb{P}}(2x_0 + 3x_1 - x_2) \subseteq \mathbf{P}^2,$$

they meet at the point at infinity $[0 : 1 : 3]$, dictated by their common slope. For this reason, $\mathbb{P}^2$ is sometimes referred to as the setting in which "parallel lines meet." See Exercise 9.4.1 for a more complete exploration of this phenomenon.

The role played by x_0 in the above discussion, as opposed to any other variable, is arbitrary. More generally, restricting a projective variety to the points of $\mathbb{P}^n$ with nonzero ith coordinate yields an affine variety, described in the following definition.

9.36 DEFINITION *Affine patches and affine restrictions*

For each $i \in \{0, 1, \ldots, n\}$, the *ith affine patch* of $\mathbb{P}^n$ is the set

$$\mathbb{A}_i^n = \{[a_0 : a_1 : \cdots : a_n] \in \mathbb{P}^n \mid a_i \neq 0\},$$

and for any set $X \subseteq \mathbb{P}^n$, the intersection $X \cap \mathbb{A}_i^n$ is called the *ith affine restriction* of X.

The reader should pause to convince themselves (Exercise 9.4.2) that for any i, there is a natural bijection $\mathbb{A}_i^n = \mathbb{A}^n$, and under this bijection, the ith affine restriction of $X = \mathcal{V}_{\mathbb{P}}(\mathcal{S})$ is $\mathcal{V}_{\mathbb{A}}(\mathcal{S}_i) \subseteq \mathbb{A}^n$, where

$$\mathcal{S}_i = \{f(x_0, \ldots, x_{i-1}, 1, x_{i+1}, \ldots, x_n) \mid f \in \mathcal{S}\}.$$

9.37 EXAMPLE *Affine restrictions of a cubic curve in $\mathbb{P}^2$*

Consider the three affine restrictions of $X = \mathcal{V}_{\mathbb{P}}(x_0 x_2^2 - 2x_1^3 - 2x_0 x_1^2) \subseteq \mathbb{P}^2$:

$$X_0 = \mathcal{V}_{\mathbb{A}}(x_2^2 - 2x_1^3 - 2x_1^2) \subseteq \mathbb{A}^2,$$
$$X_1 = \mathcal{V}_{\mathbb{A}}(x_0 x_2^2 - 2 - 2x_0) \subseteq \mathbb{A}^2,$$
$$X_2 = \mathcal{V}_{\mathbb{A}}(x_0 - 2x_1^3 - 2x_1^2 x_0) \subseteq \mathbb{A}^2.$$

The full projective variety X is the union of these three subsets, which intersect in points with more than one nonzero coordinate. Thus, one can construct X by "gluing together"—with substantial overlap—three affine varieties. For example, the point $[1 : 1 : 2] \in X$ can be found in each of the three affine restrictions: in X_0, it has affine coordinates $(x_1, x_2) = (1, 2)$; in X_1, it has affine coordinates $(x_0, x_2) = (1, 2)$; and in X_2, it has affine coordinates $(x_0, x_1) = (\frac{1}{2}, \frac{1}{2})$. In the images below, we have depicted the three affine restrictions over the real numbers, marking four color-coded points on each that are identified within X. The arrows suggest the orientation in which these affine restrictions are glued together.

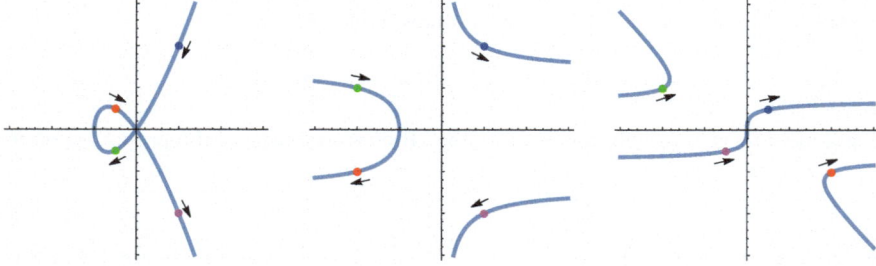

By carefully tracing the curve, moving between affine restrictions when necessary, one sees that, when viewed over the real numbers, X forms a "figure-eight." While the first affine restriction is missing one point (the point at infinity on a vertical line), the second and third are each missing two points, one of which (the point at infinity on a horizontal line) is the point at which the figure-eight crosses itself.

9.4. AFFINE RESTRICTIONS

The passage from a projective variety X to its affine restrictions involves simply setting one of the coordinates equal to 1, but can we reverse this procedure? Namely, starting from an affine variety X, can we find a projective variety $\overline{X}$ such that X is one of the affine restrictions of $\overline{X}$? The answer is "yes," and the associated projective variety is called the *projective closure* of X and is the topic of the next section.

Exercises for Section 9.4

9.4.1 A *line* in $\mathbb{P}^2$ is a projective variety of the form
$$\mathcal{V}_\mathbb{P}(ax_0 + bx_1 + cx_2) \subseteq \mathbb{P}^2$$
where $a, b, c \in K$ are not all zero. For this exercise, let X and Y be a pair of distinct lines in $\mathbb{P}^2$.
  (a) Prove that there is some affine patch $\mathbb{A}_i^2$ such that the affine restrictions X_i and Y_i are both nonempty.
  (b) Without loss of generality, suppose that the affine restrictions X_0 and Y_0 are nonempty in $\mathbb{A}_0^2$. Using the defining equations of X and Y, describe the lines X_0 and Y_0 in $\mathbb{A}^2$.
  (c) Prove that $X \cap Y = X_0 \cap Y_0$ whenever X_0 and Y_0 are not parallel.
  (d) Prove that $X \cap Y$ contains a unique point at infinity when X_0 and Y_0 are parallel.

9.4.2 (a) Show that there is a natural bijection between the affine patch $\mathbb{A}_i^n$ in $\mathbb{P}^n$ and the affine space $\mathbb{A}^n$.
  (b) If $X = \mathcal{V}_\mathbb{P}(\mathcal{S}) \subseteq \mathbb{P}^n$, prove that the bijection in (a) identifies $X \cap \mathbb{A}_i^n$ with $\mathcal{V}_\mathbb{A}(\mathcal{S}_i) \subseteq \mathbb{A}^n$, where
$$\mathcal{S}_i = \{f(x_0, \ldots, x_{i-1}, 1, x_{i+1}, \ldots, x_n) \mid f \in \mathcal{S}\}.$$

9.4.3 Let $X = \mathcal{V}_\mathbb{P}(x_0 x_2 - x_1^2) \subseteq \mathbb{P}^2$.
  (a) Calculate the three affine restrictions X_0, X_1, and X_2, and draw a picture of each over the real numbers.
  (b) Consider the point $(2, 4) \in X_0 \subseteq \mathbb{A}^2$. As an element of X, this is the point $[1 : 2 : 4]$, which also lies in X_1. What are the coordinates of this point in $X_1 \subseteq \mathbb{A}^2$?
  (c) Repeat the reasoning of part (b) for several other points in $X_0 \cap X_1$ to illustrate, visually, how X_0 and X_1 fit together inside X. Then, do the same for X_1 and X_2 and for X_0 and X_2.

9.4.4 Repeat the previous problem where $X = \mathcal{V}(f) \subseteq \mathbb{P}^2$ for your favorite homogeneous polynomial $f \in K[x_0, x_1, x_2]$.

9.4.5 Draw the four affine restrictions of $X = \mathcal{V}(w^2 + x^2 + y^2 - z^2) \subseteq \mathbb{P}_\mathbb{R}^3$, and describe how they fit together.

9.4.6 Draw the four affine restrictions of $X = \mathcal{V}(w^2 + x^2 - y^2 - z^2) \subseteq \mathbb{P}_\mathbb{R}^3$, and describe how they fit together.

Section 9.5 Projective closures

In the previous section, we learned how to view a projective variety as the disjoint union of an affine variety and a collection of points at infinity. In this section, we reverse that process, describing a method for producing a projective variety by adding points at infinity to an affine variety. Let $j_0 : \mathbb{A}^n \to \mathbb{P}^n$ be the function

$$j_0(a_1, \ldots, a_n) = [1 : a_1 : \cdots : a_n],$$

which is a bijection of $\mathbb{A}^n$ onto the affine patch $\mathbb{A}_0^n \subseteq \mathbb{P}^n$. If $X \subseteq \mathbb{A}^n$ is an affine variety, then $j_0(X) \subseteq \mathbb{P}^n$ is a subset of $\mathbb{P}^n$ whose first affine restriction is X by construction, but $j_0(X)$ is not, in general, a projective variety. In order to extend $j_0(X)$ to a projective variety, one must add some additional points. The minimal projective variety obtained in this way is called the *projective closure* of X.

> **9.38 DEFINITION** *Projective closure*
>
> Let $X \subseteq \mathbb{A}^n$ be an affine variety, and let $j_0(X)$ be the image of X in the first affine patch $\mathbb{A}_0^n \subseteq \mathbb{P}^n$. The *projective closure* of X, denoted $\overline{X} \subseteq \mathbb{P}^n$, is the intersection of all projective varieties that contain $j_0(X)$.

That $\overline{X}$ is, itself, a projective variety follows from the fact that intersections of (even infinitely many) projective varieties are projective varieties (Exercise 9.2.5). By definition, $\overline{X}$ is contained within every projective variety containing $j_0(X)$, so it can be thought of as the smallest projective variety containing $j_0(X)$.

9.39 EXAMPLE Projective closure of a line

If $X = \mathcal{V}_\mathbb{A}(1 + x_1 - x_2) = \{(a, 1+a) \mid a \in K\} \subseteq \mathbb{A}^2$, then

$$j_0(X) = \{[1 : a : 1 + a] \mid a \in K\} \subseteq \mathbb{P}^2.$$

In particular, since

$$[1 : a : 1 + a] = [\tfrac{1}{a} : 1 : \tfrac{1}{a} + 1]$$

when $a \neq 0$, we see that $j_0(X)$ contains all points of the form $[b : 1 : b + 1]$ with $b \in K \setminus \{0\}$. But a projective variety that contains all of these points must also contain the corresponding point with $b = 0$ (Exercise 9.5.1). The key point, here, is that any polynomial in $K[x_0, x_1, x_2]$ that vanishes when evaluated at $(b, 1, b+1)$ for all $b \in K \setminus \{0\}$ must also vanish at $(0, 1, 1)$. Thus, given that $[0 : 1 : 1] \notin j_0(X)$, it follows that $j_0(X)$ cannot be a projective variety.

Adding the one missing point $[0 : 1 : 1]$ yields the projective variety

$$\mathcal{V}_\mathbb{P}(x_0 + x_1 - x_2) = j_0(X) \sqcup \{[0 : 1 : 1]\},$$

as one verifies by splitting the points of $\mathcal{V}_\mathbb{P}(x_0 + x_1 - x_2)$ into two disjoint sets depending on whether the first coordinate is zero or nonzero. Since $j_0(X)$ is not, itself, a projective variety, and since the projective variety $\mathcal{V}_\mathbb{P}(x_0 + x_1 - x_2)$ is obtained from $j_0(X)$ by adding just a single point, it follows that $\mathcal{V}_\mathbb{P}(x_0 + x_1 - x_2)$ is the smallest projective variety containing $j_0(X)$, so $\overline{X} = \mathcal{V}_\mathbb{P}(x_0 + x_1 - x_2)$.

9.5. PROJECTIVE CLOSURES

9.40 EXAMPLE *Projective closure of a parabola*

Consider the parabola
$$X = \mathcal{V}_{\mathbb{A}}(x_2 - x_1^2) \subseteq \mathbb{A}^2.$$
Then
$$j_0(X) = \{[1 : a : a^2] \mid a \in K\} \subseteq \mathbb{P}^2.$$
As in the previous example, from the fact that
$$[1 : a : a^2] = [\tfrac{1}{a^2} : \tfrac{1}{a} : 1]$$
for $a \neq 0$, we see that $j_0(X)$ contains $[b^2 : b : 1]$ for any $b \in K \setminus \{0\}$. But again, a projective variety containing all of these points must also contain the point $[0 : 0 : 1]$. Since $j_0(X)$ does not contain this point, it cannot be a projective variety.

By adding the missing point, we obtain the projective variety
$$\mathcal{V}_{\mathbb{P}}(x_0 x_2 - x_1^2) = j_0(X) \sqcup \{[0 : 0 : 1]\},$$
and we conclude that $\overline{X} = \mathcal{V}_{\mathbb{P}}(x_0 x_2 - x_1^2)$.

In each of the previous two examples, notice that the affine variety X was defined by an inhomogeneous polynomial f, and the defining polynomial of $\overline{X}$ could be obtained from f by "homogenizing": multiplying each term of f by a power of x_0 to produce a homogeneous polynomial. To illustrate the idea in another example, let
$$f(x_1, x_2, x_3) = x_1^2 + x_2 + x_1 x_3^5.$$
Then the term of highest degree is the last one, which has degree 5, and we homogenize f by multiplying each term by the necessary power of x_0 to give it degree 5. The result is the homogeneous polynomial
$$\overline{f}(x_0, x_1, x_2, x_3) = x_0^3 x_1^2 + x_0^4 x_2 + x_1 x_3^4.$$
The following definition describes this procedure in general.

9.41 DEFINITION *Homogenization of a polynomial*

Let $f \in K[x_1, \ldots, x_n]$ be a polynomial of degree d. The *homogenization* of f is defined by
$$\overline{f} = x_0^d \cdot f\left(\frac{x_1}{x_0}, \ldots, \frac{x_n}{x_0}\right) \in K[x_0, x_1, \ldots, x_n].$$

The reader should convince themselves that $\overline{f}$ is a homogeneous polynomial of degree d (Exercise 9.5.5), and that this definition agrees with the term-by-term procedure described above. By plugging $x_0 = 1$ into the definition, we have

(9.42) $$\overline{f}(1, x_1, \ldots, x_n) = f(x_1, \ldots, x_n) \in K[x_1, \ldots, x_n].$$

In Examples 9.39 and 9.40, we had $X = \mathcal{V}_{\mathbb{A}}(f)$ and $\overline{X} = \mathcal{V}_{\mathbb{P}}(\overline{f})$, which might lead one to postulate that if $X = \mathcal{V}_{\mathbb{A}}(f_1, \ldots, f_k)$, then $\overline{X} = \mathcal{V}_{\mathbb{P}}(\overline{f}_1, \ldots, \overline{f}_k)$. This would certainly be convenient if it were the case. Unfortunately, the passage from X to $\overline{X}$ is not always quite so simple, as the next example illustrates.

9.43 EXAMPLE The twisted cubic curve

Let $X = \mathcal{V}_{\mathbb{A}}(x_2 - x_1^2, x_3 - x_1^3)$ so that

$$j_0(X) = \{[1 : a : a^2 : a^3] \mid a \in K\} \subseteq \mathbb{P}^3.$$

Similarly to Examples 9.39 and 9.40, any projective variety containing $j_0(X)$ must also contain the point $[0 : 0 : 0 : 1]$. It follows that $j_0(X)$ is not a projective variety, but direct computation (Exercise 9.5.4) shows that $j_0(X) \sqcup \{[0 : 0 : 0 : 1]\}$ is:

$$\overline{X} = \mathcal{V}_{\mathbb{P}}(x_0 x_2 - x_1^2, x_0^2 x_3 - x_1^3, x_1 x_3 - x_2^2) = j_0(X) \cup \{[0 : 0 : 0 : 1]\}.$$

This projective variety is called the *twisted cubic curve*.

The first two defining polynomials of $\overline{X}$ are obtained by homogenizing the defining polynomials of X, but the third is also necessary. Without it, we have

(9.44) $$\mathcal{V}_{\mathbb{P}}(x_0 x_2 - x_1^2,\ x_0^2 x_3 - x_1^3),$$

which contains $j_0(X)$, but also contains the extraneous points $[0 : 0 : b : c] \in \mathbb{P}^3$. Thus, while (9.44) is a projective variety containing $j_0(X)$, it is not $\overline{X}$.

While Example 9.43 shows that the projective closure of $X = \mathcal{V}_{\mathbb{A}}(f_1, \ldots, f_r)$ is not, in general, obtained simply by homogenizing $f_1, \ldots, f_r$, there is a fix: instead of homogenizing only an arbitrarily chosen set of defining polynomials, we should homogenize every polynomial in the vanishing ideal of X. The following notation will be useful.

9.45 DEFINITION *Homogenization of a set*

Let $\mathcal{S} \subseteq K[x_1, \ldots, x_n]$ be a set of polynomials. The *homogenization* of $\mathcal{S}$ is the set

$$\overline{\mathcal{S}} = \{\overline{f} \mid f \in \mathcal{S}\} \subseteq K[x_0, \ldots, x_n].$$

After homogenizing every polynomial in the vanishing ideal of an affine variety, we then obtain enough polynomials to describe its projective closure. This is the statement of the next result.

Careful: even when I is an ideal, its homogenization $\overline{I}$ generally will not be an ideal; can you see why?

9.46 PROPOSITION *Projective closures via homogenization*

If $X \subseteq \mathbb{A}^n$ is an affine variety, then $\overline{X} = \mathcal{V}_{\mathbb{P}}(\overline{\mathcal{I}_{\mathbb{A}}(X)}) \subseteq \mathbb{P}^n$.

PROOF We prove both inclusions.

($\subseteq$) If $[1 : a_1 : \cdots : a_n] \in j_0(X)$ and $g \in \overline{\mathcal{I}_{\mathbb{A}}(X)}$, then $g = \overline{f}$ for some $f \in \mathcal{I}_{\mathbb{A}}(X)$ and hence $g(1, a_1, \ldots, a_n) = f(a_1, \ldots, a_n) = 0$. Thus, $\mathcal{V}_{\mathbb{P}}(\overline{\mathcal{I}_{\mathbb{A}}(X)})$ is a projective variety containing $j_0(X)$, and since $\overline{X}$ is the smallest projective variety containing $j_0(X)$, it follows that $\overline{X} \subseteq \mathcal{V}_{\mathbb{P}}(\overline{\mathcal{I}_{\mathbb{A}}(X)})$.

9.5. PROJECTIVE CLOSURES

($\supseteq$) If we first prove that

(9.47) $$\mathcal{I}_\mathbb{P}(\overline{X}) \subseteq \mathcal{I}_\mathbb{P}\left(\mathcal{V}_\mathbb{P}(\overline{\mathcal{I}_\mathbb{A}(X)})\right).$$

Applying $\mathcal{V}_\mathbb{P}$ to both sides of this containment implies, by Proposition 9.30, that

$$\overline{X} \supseteq \mathcal{V}_\mathbb{P}\left(\overline{\mathcal{I}_\mathbb{A}(X)}\right).$$

To prove (9.47), let $g \in \mathcal{I}_\mathbb{P}(\overline{X})$. Given that $\mathcal{I}_\mathbb{P}(\overline{X})$ is a homogeneous ideal and thus admits a set of homogeneous generators, it suffices to assume for the justification of (9.47) that g is homogeneous.

Even though g is homogeneous, it may not be the homogenization of an element of $K[x_1, \ldots, x_n]$, since it could be the case that every term of g contains x_0. However, if $k \in \mathbb{N}$ is the maximum power of x_0 such that $g = x_0^k h$ for some $h \in K[x_0, x_1, \ldots, x_n]$, then $h = \overline{h_0}$ where $h_0 = h(1, x_1, \ldots, x_n) \in K[x_1, \ldots, x_n]$ (Exercise 9.5.7). Given that g vanishes on $\overline{X} \supseteq j_0(X)$, it follows that h_0 vanishes on X: for any $(a_1, \ldots, a_n) \in X$, we have

$$h_0(a_1, \ldots, a_n) = h(1, a_1, \ldots, a_n) = g(1, a_1, \ldots, a_n) = 0.$$

Thus, $h_0 \in \mathcal{I}_\mathbb{A}(X)$, so $h \in \overline{\mathcal{I}_\mathbb{A}(X)}$, and it follows that

$$g \in \langle \overline{\mathcal{I}_\mathbb{A}(X)} \rangle \subseteq \mathcal{I}_\mathbb{P}\left(\mathcal{V}_\mathbb{P}\left(\overline{\mathcal{I}_\mathbb{A}(X)}\right)\right),$$

where the containment is another application of Proposition 9.30. This completes the proof of (9.47) and hence the proof of the proposition. $\square$

9.48 EXAMPLE *The twisted cubic curve revisited*

In light of Proposition 9.46, we can make further sense of the phenomenon observed in Example 9.43. In that case, in order to compute the projective closure of

$$X = \mathcal{V}_\mathbb{A}(x_2 - x_1^2, x_3 - x_1^3),$$

we observed that it was not enough to just homogenize the two defining polynomials of X. Proposition 9.46 ensures that, after homogenizing *all* of the polynomials in $\mathcal{I}(X)$, we then obtain a set of polynomials defining $\overline{X}$. In this particular case, it turns out that it is sufficient simply to add in the homogenization of one additional polynomial $x_1 x_3 - x_2^2 \in \mathcal{I}(X)$ (which, in this case, happens to be homogeneous).

As we have seen through the example of the twisted cubic curve, it is not generally the case that

$$\overline{\mathcal{V}_\mathbb{A}(f_1, \ldots, f_k)} = \mathcal{V}_\mathbb{P}(\overline{f_1}, \ldots, \overline{f_k}).$$

However, in the special case that $k = 1$, the Nullstellensatz implies that the projective closure is, in fact, obtained by homogenizing the single defining polynomial.

9.49 PROPOSITION *Projective closures of hypersurfaces*

If $f \in K[x_1, \ldots, x_n]$, then $\overline{\mathcal{V}_\mathbb{A}(f)} = \mathcal{V}_\mathbb{P}(\overline{f})$.

PROOF   Let $X = \mathcal{V}_{\mathbb{A}}(f)$ and let $f = q_1^{k_1} \cdots q_\ell^{k_\ell}$ be a distinct irreducible factorization of f. The Nullstellensatz then implies that $\mathcal{I}_{\mathbb{A}}(X) = \langle q_1 \cdots q_\ell \rangle$. Thus, by Proposition 9.46, we have

$$\overline{\mathcal{V}_{\mathbb{A}}(f)} = \mathcal{V}_{\mathbb{P}}(\overline{\mathcal{I}_{\mathbb{A}}(X)}) = \mathcal{V}_{\mathbb{P}}(\overline{\langle q_1 \cdots q_\ell \rangle}).$$

It remains to prove that

$$\mathcal{V}_{\mathbb{P}}(\overline{\langle q_1 \cdots q_\ell \rangle}) = \mathcal{V}_{\mathbb{P}}(\overline{f}).$$

Since $f \in \langle q_1 \cdots q_\ell \rangle$, it follows that $\overline{f} \in \overline{\langle q_1 \cdots q_\ell \rangle}$, so $\mathcal{V}_{\mathbb{P}}(\overline{\langle q_1 \cdots q_\ell \rangle}) \subseteq \mathcal{V}_{\mathbb{P}}(\overline{f})$. To prove the other inclusion, suppose that $a \in \mathcal{V}_{\mathbb{P}}(\overline{f})$, meaning that $\overline{f}(a) = 0$. Using multiplicativity of homogenizations (Exercise 9.5.6), we have

$$\overline{f} = \overline{q}_1^{k_1} \cdots \overline{q}_\ell^{k_\ell},$$

so the assumption that $\overline{f}(a) = 0$ implies that $\overline{q}_i(a) = 0$ for some i. Now given any $g \in \overline{\langle q_1 \cdots q_\ell \rangle}$, there exists some $h \in K[x_1, \ldots, x_n]$ such that

$$g = \overline{h \cdot q_1 \cdots q_\ell} = \overline{h} \cdot \overline{q}_1 \cdots \overline{q}_\ell.$$

Since $\overline{q}_i(a) = 0$ for some i, it follows that $g(a) = 0$, and since g was a general element of $\overline{\langle q_1 \cdots q_\ell \rangle}$, this implies that $a \in \mathcal{V}_{\mathbb{P}}(\overline{\langle q_1 \cdots q_\ell \rangle})$. Thus, we have verified that $\mathcal{V}_{\mathbb{P}}(\overline{f}) \subseteq \mathcal{V}_{\mathbb{P}}(\overline{\langle q_1 \cdots q_\ell \rangle})$, completing the proof. $\square$

The final important result concerning projective closures states that the act of taking projective closures will never add points within affine space, only points at infinity. In other words, taking affine restrictions is, in some sense, inverse to taking projective closures, as we make precise in the next result.

9.50 PROPOSITION $(\overline{X})_0 = X$

If $X \subseteq \mathbb{A}^n$ is an affine variety, then the affine restriction of the projective closure of X is X. More succinctly, $(\overline{X})_0 = X$.

PROOF   Let $X \subseteq \mathbb{A}^n$ be an affine variety and define $\mathcal{S} = \overline{\mathcal{I}_{\mathbb{A}}(X)}$. Combining Proposition 9.46 with Definition 9.36, we see that the affine restriction of the projective closure $\overline{X} = \mathcal{V}_{\mathbb{P}}(\mathcal{S})$ is $\mathcal{V}_{\mathbb{A}}(\mathcal{S}_0)$, where

$$\begin{aligned}\mathcal{S}_0 &= \{g(1, x_1, \ldots, x_n) \mid g \in \overline{\mathcal{I}_{\mathbb{A}}(X)}\} \\ &= \{\overline{f}(1, x_1, \ldots, x_n) \mid f \in \mathcal{I}_{\mathbb{A}}(X)\} \\ &= \mathcal{I}_{\mathbb{A}}(X).\end{aligned}$$

Thus, the affine restriction of $\overline{X}$ is $\mathcal{V}_{\mathbb{A}}(\mathcal{I}_{\mathbb{A}}(X)) = X$, as claimed. $\square$

Affine restrictions and projective closures set up a close correspondence between affine varieties and projective varieties, and as we have seen, the theory in the two settings is essentially parallel. The reader may expect, then, that there is a projective version of the Nullstellensatz. This is indeed the case, and the last section of this chapter is devoted to establishing it.

Exercises for Section 9.5

9.5.1 Let $X \subseteq \mathbb{P}^2$ be a projective variety containing the points $[b : 1 : b + 1]$ for all $b \in K \setminus \{0\}$. Prove that $[0 : 1 : 1] \in X$.

9.5.2 Generalizing the previous exercise, let $f_0, \ldots, f_n \in K[x]$ be single-variable polynomials, not all of which vanish at 0, and let $X \subseteq \mathbb{P}^n$ be a projective variety such that
$$[f_0(b) : \cdots : f_n(b)] \in X$$
for all $b \in K \setminus \{0\}$. Prove that
$$[f_0(0) : \cdots : f_n(0)] \in X.$$

9.5.3 Adapt the arguments of Examples 9.39 and 9.40 to prove that the projective closure of the affine variety
$$X = \mathcal{V}_{\mathbb{A}}(x_1 x_2 - 1) \subseteq \mathbb{A}^2$$
must contain the points $[0 : 0 : 1]$ and $[0 : 1 : 0]$, and conclude that
$$\overline{X} = \mathcal{V}_{\mathbb{P}}(x_1 x_2 - x_0^2).$$
Draw a picture of X over $\mathbb{R}$. What are the asymptotes of X, and how do these relate to the points at infinity of $\overline{X}$?

9.5.4 Let
$$X = \mathcal{V}_{\mathbb{A}}(x_2 - x_1^2, x_3 - x_1^3) \subseteq \mathbb{A}^3$$
be the affine twisted cubic.
 (a) Prove that $\overline{X} = \mathcal{V}_{\mathbb{P}}(x_0 x_2 - x_1^2, x_0^2 x_3 - x_1^3, x_1 x_3 - x_2^2)$.
 (b) Prove that $\overline{X} = \mathcal{V}_{\mathbb{P}}(x_0 x_2 - x_1^2, x_0 x_3 - x_1 x_2, x_1 x_3 - x_2^2)$.
 (The description given in Part (b) is a more common presentation of the twisted cubic curve. Notice that the three equations in (b) are the 2×2 minors of the 2×3 matrix with rows (x_0, x_1, x_2) and (x_1, x_2, x_3).)

9.5.5 Prove that, for any polynomial $f \in [x_1, \ldots, x_n]$ of degree d, the homogenization $\overline{f}$ is a homogeneous polynomial of degree d.

9.5.6 Let $f, g \in K[x_1, \ldots, x_n]$.
 (a) Prove that $\overline{f \cdot g} = \overline{f} \cdot \overline{g}$.
 (b) Is it the case that $\overline{f + g} = \overline{f} + \overline{g}$? Prove or give a counterexample.

9.5.7 Let $h \in K[x_0, \ldots, x_n]$ be a homogeneous polynomial such that $x_0 \nmid h$. Prove that h is the homogenization of
$$h(1, x_1, \ldots, x_n) \in K[x_1, \ldots, x_n].$$

9.5.8 For any ideal $I \subseteq K[x_1, \ldots, x_n]$, prove that $\overline{\mathcal{V}_{\mathbb{A}}(I)} = \mathcal{V}_{\mathbb{P}}(\overline{I})$.
Hint: Using the Nullstellensatz and Proposition 9.46, it suffices to prove that
$$\mathcal{V}_{\mathbb{P}}\left(\overline{\sqrt{I}}\right) = \mathcal{V}_{\mathbb{P}}(\overline{I}).$$

Section 9.6 The projective Nullstellensatz

How might one adapt the statement of the Nullstellensatz to the projective setting? The most straightforward adaptation that one might hope for is that

(9.51) $$\mathcal{I}_\mathbb{P}(\mathcal{V}_\mathbb{P}(I)) = \sqrt{I}$$

for any ideal $I \subseteq K[x_0, \ldots, x_n]$. However, this is not quite correct. First, since vanishing ideals are homogeneous, we probably want to restrict our attention to homogeneous ideals. But even among homogeneous ideals, there still happens to be a case when (9.51) fails:
$$I = \langle x_0, \ldots, x_n \rangle.$$
More explicitly, if $I = \langle x_0, \ldots, x_n \rangle$, then $\mathcal{V}_\mathbb{P}(I)$ consists of all $[a_0 : \cdots : a_n] \in \mathbb{P}^n$ for which each of the polynomials $x_0, \ldots, x_n$ vanishes, which can only be the case if $a_i = 0$ for each i. As no such points exist in $\mathbb{P}^n$, it follows that $\mathcal{V}_\mathbb{P}(I) = \emptyset$. Since every polynomial vacuously vanishes at every point of $\emptyset$, this then implies that $\mathcal{I}_\mathbb{P}(\mathcal{V}_\mathbb{P}(I)) = K[x_0, \ldots, x_n]$. On the other hand, notice that $I = \langle x_0, \ldots, x_n \rangle$ is a radical (in fact, maximal) ideal, so $\sqrt{I} = \langle x_0, \ldots, x_n \rangle$. Tying together these observations, we conclude that

$$\mathcal{I}_\mathbb{P}(\mathcal{V}_\mathbb{P}(I)) = K[x_0, \ldots, x_n] \neq \langle x_0, \ldots, x_n \rangle = \sqrt{I},$$

giving a counterexample to (9.51). To avoid this pesky exception to (9.51), we give the ideal $I = \langle x_0, \ldots, x_n \rangle$ a name that emphasizes our unwillingness to consider it.

9.52 DEFINITION *Irrelevant ideal*

The ideal $\langle x_0, \ldots, x_n \rangle \subseteq K[x_0, \ldots, x_n]$ is called the *irrelevant ideal*. An ideal in $K[x_0, \ldots, x_n]$ that is not the irrelevant ideal is called *relevant*.

In the following statement of the projective Nullstellensatz, the first assertion is the natural analogue of the weak Nullstellensatz in the projective setting, while the second assertion is the natural analogue of the strong Nullstellensatz.

9.53 THEOREM *Projective Nullstellensatz*

Let $I \subseteq K[x_0, \ldots, x_n]$ be a homogeneous ideal.
1. $\mathcal{V}_\mathbb{P}(I) = \emptyset$ if and only if $\sqrt{I} \supseteq \langle x_0, \ldots, x_n \rangle$.
2. If $\sqrt{I}$ is relevant, then $\mathcal{I}_\mathbb{P}(\mathcal{V}_\mathbb{P}(I)) = \sqrt{I}$.

Fortunately, the projective Nullstellensatz can be derived from its affine cousin without re-developing the algebraic machinery. The key idea, setting aside for now the methods we considered in the last two sections, is to leverage another method of passing between affine and projective geometry: that of affine cones.

To motivate the notion of affine cones, notice that a homogeneous ideal can be used in two different ways: we can either use it to define a projective variety $\mathcal{V}_\mathbb{P}(I) \subseteq \mathbb{P}^n$ or an affine variety $\mathcal{V}_\mathbb{A}(I) \subseteq \mathbb{A}^{n+1}$. To begin to understand the relationship between these two perspectives, let us consider an example.

9.6. THE PROJECTIVE NULLSTELLENSATZ

9.54 EXAMPLE $\mathcal{V}_\mathbb{P}(I)$ versus $\mathcal{V}_\mathbb{A}(I)$

Consider $K = \mathbb{R}$ and let $I = \langle -x_0^2 + x_1^2 + x_2^2 \rangle \subseteq \mathbb{R}[x_0, x_1, x_2]$. Then

$$\mathcal{V}_\mathbb{P}(I) = \{[a_0 : a_1 : a_2] \in \mathbb{P}^2 \mid -a_0^2 + a_1^2 + a_2^2 = 0\}.$$

If $a_0 \neq 0$, then by rescaling all three coordinates we can assume that $a_0 = 1$. If, on the other hand, $a_0 = 0$, then the defining equation becomes $a_1^2 + a_2^2 = 0$, implying that $a_1 = a_2 = 0$. Since no point with $a_0 = a_1 = a_2 = 0$ exists in $\mathbb{P}^2$, we then conclude that

$$\mathcal{V}_\mathbb{P}(I) = \{[1 : a_1 : a_2] \in \mathbb{P}^2 \mid -1 + a_1^2 + a_2^2 = 0\},$$

which is in natural bijection with the unit circle in $\mathbb{A}^2$—that is, with the affine variety $\mathcal{V}_\mathbb{A}(-1 + x_1^2 + x_2^2) \subseteq \mathbb{A}^2$.

Alternatively viewing I as defining an affine variety in $\mathbb{A}^3$, we see that $\mathcal{V}_\mathbb{A}(I)$ is the circular cone with vertex at the origin depicted at right. (In the image, x_0 is the vertical coordinate.) The projective variety $\mathcal{V}_\mathbb{P}(I)$—which we have identified with the unit circle—is visible in this image as the intersection of $\mathcal{V}_\mathbb{A}(I)$ with the plane $x_0 = 1$. Furthermore, observe that $\mathcal{V}_\mathbb{A}(I)$ can be viewed as the union of all lines through the origin that correspond, under the bijection of Corollary 9.8, to points of $\mathcal{V}_\mathbb{P}(I)$.

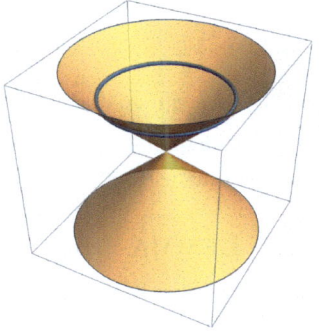

The fact that $\mathcal{V}_\mathbb{A}(I)$ is a cone in the previous example is not a coincidence: a cone $C \subseteq \mathbb{A}^3$ has the property that whenever $(a_0, a_1, a_2) \in C$, the entire line

$$\{(\lambda a_0, \lambda a_1, \lambda a_2) \mid \lambda \in K\}$$

is contained in C. From the projective perspective, this is the statement that membership of $[a_0 : a_1 : a_2]$ in C is well-defined, regardless of how we represent the point $[a_0 : a_1 : a_2] \in \mathbb{P}^2$. With the picture of Example 9.54 in our minds, we now introduce the precise notion of an affine cone over a subset of projective space.

9.55 DEFINITION *Affine cone over a projective variety*

Let $X \subseteq \mathbb{P}^n$ be a subset. The *affine cone* over X is the set

$$C(X) = \{(0, \ldots, 0)\} \cup \{(a_0, \ldots, a_n) \mid [a_0 : \cdots : a_n] \in X\} \subseteq \mathbb{A}^{n+1}.$$

In other words, $C(X)$ is just the origin if $X = \emptyset$, and otherwise, $C(X)$ is the union of all lines through the origin corresponding to points of $X \subseteq \mathbb{P}^n$. For instance, the affine cone over $\mathcal{V}_\mathbb{P}(I) \subseteq \mathbb{P}^2$ in Example 9.54 is equal to $\mathcal{V}_\mathbb{A}(I) \subseteq \mathbb{A}^3$. This is a special case of the following lemma.

> **9.56 LEMMA** $\mathcal{V}_{\mathbb{A}}(I) = C(\mathcal{V}_{\mathbb{P}}(I))$
>
> If $I \subsetneq K[x_0, \ldots, x_n]$ is a homogeneous ideal, then
> $$\mathcal{V}_{\mathbb{A}}(I) = C(\mathcal{V}_{\mathbb{P}}(I)).$$

PROOF First, we note that $(0, \ldots, 0) \in \mathcal{V}_{\mathbb{A}}(I)$ for every homogeneous ideal $I \subsetneq K[x_0, \ldots, x_n]$. To see this, recall from Proposition 9.28 that I has a set of homogeneous generators. If $(0, \ldots, 0) \notin \mathcal{V}_{\mathbb{A}}(I)$, then at least one of these generators must be a homogeneous polynomial f such that $f(0, \ldots, 0) \neq 0$. But the only homogeneous polynomials that do not vanish at $(0, \ldots, 0)$ are the nonzero constant polynomials, and if such a polynomial is among the generators of I, then $I = K[x_0, \ldots, x_n]$, contradicting the assumption that $I \neq K[x_0, \ldots, x_n]$.

Thus, $(0, \ldots, 0) \in \mathcal{V}_{\mathbb{A}}(I)$. Moreover, if at least one coordinate of $(a_0, \ldots, a_n)$ is nonzero, the homogeneity of I implies that

$$(a_0, \ldots, a_n) \in \mathcal{V}_{\mathbb{A}}(I) \Leftrightarrow [a_0 : \cdots : a_n] \in \mathcal{V}_{\mathbb{P}}(I) \Leftrightarrow (a_0, \ldots, a_n) \in C(\mathcal{V}_{\mathbb{P}}(I)).$$

Thus, we have shown that $\mathcal{V}_{\mathbb{A}}(I) = C(\mathcal{V}_{\mathbb{P}}(I))$, as claimed. $\square$

The previous lemma uses affine cones to relate the $\mathcal{V}_{\mathbb{A}}$- and $\mathcal{V}_{\mathbb{P}}$-operators, and the next lemma uses affine cones to relate the $\mathcal{I}_{\mathbb{A}}$- and $\mathcal{I}_{\mathbb{P}}$-operators.

> **9.57 LEMMA** $\mathcal{I}_{\mathbb{P}}(X) = \mathcal{I}_{\mathbb{A}}(C(X))$
>
> If $X \subseteq \mathbb{P}^n$ is a nonempty subset, then
> $$\mathcal{I}_{\mathbb{P}}(X) = \mathcal{I}_{\mathbb{A}}(C(X)).$$

PROOF Since $\mathcal{I}_{\mathbb{P}}(X)$ is a homogeneous ideal, it admits homogeneous generators. None of these generators can be a nonzero constant polynomial, since $X \neq \emptyset$ by assumption, and it follows that they are all homogeneous

> *The exceptions to Lemma 9.56 and 9.57, when $I = K[x_0, \ldots, x_n]$ in the first case or $X = \emptyset$ in the second, are necessary; see Exercise 9.6.2.*

polynomials of positive degree and hence vanish at $(0, \ldots, 0) \in \mathbb{A}^{n+1}$. Thus, we have $f(0, \ldots, 0) = 0$ for all $f \in \mathcal{I}_{\mathbb{P}}(X)$, and with this, the definition of $\mathcal{I}_{\mathbb{P}}(X)$ can be re-expressed as follows:

$$\mathcal{I}_{\mathbb{P}}(X) = \left\{ f \in K[x_0, \ldots, x_n] \;\middle|\; \begin{array}{l} f(a_0, \ldots, a_n) = 0 \text{ if } [a_0 : \cdots : a_n] \in X \text{ and} \\ f(a_0, \ldots, a_n) = 0 \text{ if } (a_0, \ldots, a_n) = (0, \ldots, 0) \end{array} \right\}$$
$$= \mathcal{I}_{\mathbb{A}}(\{(0, \ldots, 0)\} \cup \{(a_0, \ldots, a_n) \mid [a_0 : \cdots : a_n] \in X\})$$
$$= \mathcal{I}_{\mathbb{A}}(C(X)),$$

as claimed. $\square$

Equipped with Lemmas 9.56 and 9.57, the proof of the projective Nullstellensatz is now an application of the affine Nullstellensatz.

9.6. THE PROJECTIVE NULLSTELLENSATZ

PROOF OF THEOREM 9.53   Let $I \subseteq K[x_0, \ldots, x_n]$ be a homogeneous ideal.

We begin by verifying the first assertion of the projective Nullstellensatz: that $\sqrt{I} \supseteq \langle x_0, \ldots, x_n \rangle$ if and only if $\mathcal{V}_{\mathbb{P}}(I) = \emptyset$. First, assume that $\sqrt{I} \supseteq \langle x_0, \ldots, x_n \rangle$. Then

$$\mathcal{V}_{\mathbb{P}}(I) = \mathcal{V}_{\mathbb{P}}(\sqrt{I}) \subseteq \mathcal{V}_{\mathbb{P}}(\langle x_0, \ldots, x_n \rangle) = \emptyset,$$

where the first equality is Exercise 9.6.3. Conversely, assume that $\mathcal{V}_{\mathbb{P}}(I) = \emptyset$. Either it is the case that $I = K[x_0, \ldots, x_n]$—in which case it then follows that $\sqrt{I} = K[x_0, \ldots, x_n] \supseteq \langle x_0, \ldots, x_n \rangle$—or it is the case that $I \subsetneq K[x_0, \ldots, x_n]$, in which case we can apply Lemma 9.56 to obtain

$$\mathcal{V}_{\mathbb{A}}(I) = C(\mathcal{V}_{\mathbb{P}}(I)) = C(\emptyset) = \{(0, \ldots, 0)\},$$

and then the affine Nullstellensatz gives

$$\sqrt{I} = \mathcal{I}_{\mathbb{A}}(\mathcal{V}_{\mathbb{A}}(I)) = \mathcal{I}_{\mathbb{A}}(\{(0, \ldots, 0)\}) = \langle x_0, \ldots, x_n \rangle.$$

In either case, $\sqrt{I} \supseteq \langle x_0, \ldots, x_n \rangle$, and this completes the proof of the first part of the projective Nullstellensatz.

To prove the second assertion of the projective Nullstellensatz, suppose that $\sqrt{I}$ is relevant; we aim to prove that $\mathcal{I}_{\mathbb{P}}(\mathcal{V}_{\mathbb{P}}(I)) = \sqrt{I}$. Consider first the case that $\mathcal{V}_{\mathbb{P}}(I) = \emptyset$. Then $\sqrt{I} \supseteq \langle x_0, \ldots, x_n \rangle$ by the argument above, but the only relevant ideal containing the (maximal) irrelevant ideal is the full ring $K[x_0, \ldots, x_n]$, so we must have $\sqrt{I} = K[x_0, \ldots, x_n]$. Since $\mathcal{I}_{\mathbb{P}}(\mathcal{V}_{\mathbb{P}}(I)) = \mathcal{I}_{\mathbb{P}}(\emptyset) = K[x_0, \ldots, x_n]$, the assertion that $\mathcal{I}_{\mathbb{P}}(\mathcal{V}_{\mathbb{P}}(I)) = \sqrt{I}$ is true in this case. Now consider the case $\mathcal{V}_{\mathbb{P}}(I) \neq \emptyset$. Lemma 9.57 implies that

$$\mathcal{I}_{\mathbb{P}}(\mathcal{V}_{\mathbb{P}}(I)) = \mathcal{I}_{\mathbb{A}}(C(\mathcal{V}_{\mathbb{P}}(I))).$$

Furthermore, since $\mathcal{V}_{\mathbb{P}}(I) \neq \emptyset$, it must be the case that $I \subsetneq K[x_0, \ldots, x_n]$, so from Lemma 9.56 and the affine Nullstellensatz, we obtain

$$\mathcal{I}_{\mathbb{A}}(C(\mathcal{V}_{\mathbb{P}}(I))) = \mathcal{I}_{\mathbb{A}}(\mathcal{V}_{\mathbb{A}}(I)) = \sqrt{I}.$$

Combining the equalities above, it follows that $\mathcal{I}_{\mathbb{P}}(\mathcal{V}_{\mathbb{P}}(I)) = \sqrt{I}$, as desired.   □

As in the affine setting, one application of the projective Nullstellensatz is that the computation of projective vanishing ideals is now significantly simplified.

9.58 **EXAMPLE**   Vanishing ideal of a line in $\mathbb{P}^2$, revisited

Let $X = \mathcal{V}_{\mathbb{P}}(x_0 + x_1 - x_2) \subseteq \mathbb{P}^2$, as in Example 9.26. We outlined a direct computation of $\mathcal{I}_{\mathbb{P}}(X)$ in Exercise 9.3.1, but with the projective Nullstellensatz, we now need only observe that $\langle x_0 + x_1 - x_2 \rangle$ is a radical homogeneous ideal with a nonempty vanishing set. Thus,

$$\mathcal{I}_{\mathbb{P}}(X) = \mathcal{I}_{\mathbb{P}}(\mathcal{V}_{\mathbb{P}}(\langle x_0 + x_1 - x_2 \rangle)) = \langle x_0 + x_1 - x_2 \rangle.$$

Exactly as in the affine setting, the projective Nullstellensatz also clarifies the domains on which the projective $\mathcal{V}$- and $\mathcal{I}$-operators are inverses, allowing us to describe the following dictionary between projective varieties and associated ideals.

> **9.59 PROPOSITION** *Projective varieties and ideals*
>
> The $\mathcal{V}_{\mathbb{P}}$- and $\mathcal{I}_{\mathbb{P}}$-operators are mutually inverse, inclusion-reversing bijections that translate between the following hierarchies of ideals and varieties:
>
> $$\left\{ \begin{array}{c} \text{relevant homogeneous} \\ \text{radical ideals in } K[x_0, \ldots, x_n] \end{array} \right\} \longleftrightarrow \{\text{projective varieties in } \mathbb{P}^n\}$$
>
> $$\cup | \qquad\qquad\qquad\qquad \cup|$$
>
> $$\left\{ \begin{array}{c} \text{relevant homogeneous} \\ \text{prime ideals in } K[x_0, \ldots, x_n] \end{array} \right\} \longleftrightarrow \{\text{irreducible varieties in } \mathbb{P}^n\}$$
>
> $$\cup| \qquad\qquad\qquad\qquad \cup|$$
>
> $$\left\{ \begin{array}{c} \text{relevant homogeneous} \\ \text{maximal ideals in } K[x_0, \ldots, x_n] \end{array} \right\} \longleftrightarrow \{\text{points in } \mathbb{P}^n\}.$$

PROOF   Let $X \subseteq \mathbb{P}^n$ be a projective variety. First, note that $\mathcal{I}_{\mathbb{P}}(X)$ is indeed a homogeneous radical ideal in $K[x_0, \ldots, x_n]$ by Proposition 9.29. Furthermore, $\mathcal{I}_{\mathbb{P}}(X)$ is not the irrelevant ideal. To see why not, suppose to the contrary that

$$\mathcal{I}_{\mathbb{P}}(X) = \langle x_0, \ldots, x_n \rangle.$$

Then Proposition 9.30 implies

$$X = \mathcal{V}_{\mathbb{P}}(\mathcal{I}_{\mathbb{P}}(X)) = \mathcal{V}(\langle x_0, \ldots, x_n \rangle) = \emptyset.$$

But then

$$\mathcal{I}_{\mathbb{P}}(X) = \mathcal{I}_{\mathbb{P}}(\emptyset) = K[x_0, \ldots, x_n],$$

contradicting our assumption above that $\mathcal{I}_{\mathbb{P}}(X) = \langle x_0, \ldots, x_n \rangle$.

Thus, $\mathcal{I}_{\mathbb{P}}$ maps any projective variety in $\mathbb{P}^n$ to a relevant homogeneous radical ideal in $K[x_0, \ldots, x_n]$, and conversely, $\mathcal{V}_{\mathbb{P}}$ maps any such ideal to a projective variety, by definition. The fact that

$$\mathcal{V}_{\mathbb{P}}(\mathcal{I}_{\mathbb{P}}(X)) = X \quad \text{and} \quad \mathcal{I}_{\mathbb{P}}(\mathcal{V}_{\mathbb{P}}(I)) = I$$

on these domains follows from Proposition 9.30 and the projective Nullstellensatz, respectively, justifying the first bijection appearing in the proposition. The other two bijections then follow from the observations that X is irreducible if and only if $\mathcal{I}_{\mathbb{P}}(X)$ is a prime ideal (Exercise 9.3.5), and X is a single point if and only if $\mathcal{I}_{\mathbb{P}}(X)$ is a maximal ideal (Exercise 9.3.6). □

Throughout this chapter, we have begun to develop an intuition for projective varieties, building up to an adaptation of the Nullstellensatz in the projective setting. Now that we have familiarized ourselves with projective varieties, our next task is to introduce and study the structure-preserving maps between them; we turn to this topic in the next chapter.

Exercises for Section 9.6

9.6.1 Draw a picture, over the real numbers, of the affine cone over the projective variety
$$X = \mathcal{V}_{\mathbb{P}}(x^2 - yz) \subseteq \mathbb{P}^2.$$
(Computer graphing software might help.) Where, in your picture, do you see the affine restriction
$$X_0 = \mathcal{V}_{\mathbb{A}}(x^2 - y) \subseteq \mathbb{A}^2?$$

9.6.2 (a) Show that Lemma 9.56 fails if $I = K[x_0, \ldots, x_n]$.
(b) Show that Lemma 9.57 fails if $X = \emptyset$.

9.6.3 Let $I \subseteq K[x_0, \ldots, x_n]$ be an ideal. Prove that
$$\mathcal{V}_{\mathbb{P}}(\sqrt{I}) = \mathcal{V}_{\mathbb{P}}(I).$$

9.6.4 Let $f \in K[x_0, \ldots, x_n]$ be an irreducible homogeneous polynomial and let $X = \mathcal{V}_{\mathbb{P}}(f) \subseteq \mathbb{P}^n$. Prove that $\mathcal{I}_{\mathbb{P}}(X) = \langle f \rangle$.

9.6.5 Let $f \in K[x_0, \ldots, x_n]$ be a nonconstant homogeneous polynomial and let $X = \mathcal{V}_{\mathbb{P}}(f) \subseteq \mathbb{P}^n$. Describe $\mathcal{I}_{\mathbb{P}}(X)$ in terms of the distinct irreducible factors of f.

9.6.6 Prove that $\mathcal{V}_{\mathbb{P}}(f_1, \ldots, f_k) = \emptyset$ if and only if there exists a nonnegative integer d such that every monomial of degree d is contained in $\langle f_1, \ldots, f_k \rangle$.

Chapter 10
Maps of Projective Varieties

LEARNING OBJECTIVES FOR CHAPTER 10

- Define and study regular maps between projective varieties.
- Give various examples of regular maps, including linear maps, polynomial maps, isomorphisms, projective equivalences, and Veronese embeddings.
- Define Segre maps and use them to define products of projective varieties.
- Describe Grassmannians abstractly and realize them, via Plücker maps, as projective varieties.

In the introduction to Chapter 4, the reader was encouraged to ask a key question whenever a new type of mathematical object is introduced: which maps between these objects preserve their relevant structure? For affine varieties, we landed upon polynomial maps as the appropriate notion of structure-preserving maps, and the first goal of this chapter is to define a corresponding notion in the projective setting.

This goal is complicated by a number of crucial differences between affine and projective varieties. First, due to the equivalence relation in the definition of $\mathbb{P}^m$, a polynomial $f \in K[x_0, \ldots, x_m]$ does not give a well-defined function on a projective variety $X \subseteq \mathbb{P}^m$, simply because the value of the polynomial is sensitive to scaling the homogeneous coordinates in $\mathbb{P}^m$. However, if f happens to be homogeneous of degree d, then we have seen that scaling homogeneous coordinates has a fairly simple effect on the value of f:

$$f(\lambda a_0, \ldots, \lambda a_m) = \lambda^d f(a_0, \ldots, a_m) \quad \text{for all} \quad \lambda \in K \setminus \{0\}.$$

It then follows that, given polynomials $f_0, \ldots, f_n \in K[x_0, \ldots, x_m]$ that are all homogeneous *of the same degree*, at least one of which does not vanish at a, we obtain a well-defined value $[f_0(a) : \cdots : f_n(a)] \in \mathbb{P}^n$, which is independent of the choice of homogeneous coordinates for a. Motivated by this observation, such tuples of polynomials form the foundation of our study of maps between projective varieties.

In general, if $X \subseteq \mathbb{P}^m$ and $Y \subseteq \mathbb{P}^n$ are projective varieties, we will say that a function $F : X \to Y$ is a *regular map* if it can be realized—at least locally—by a tuple of polynomials $f_0, \ldots, f_n$ that are homogeneous of the same degree. Regular maps serve as the structure-preserving maps between projective varieties, and our primary aim in the first part of this chapter is to become acquainted with them.

In the latter part of this chapter, we describe two important constructions: products of projective varieties and Grassmannian varieties. At the onset, neither of these two constructions arises as a projective variety in an obvious way, but in each case, we will show that there are natural maps (Segre maps and Plücker maps, respectively), through which we can endow them with the structure of a projective variety.

© The Author(s) 2025
E. Clader, D. Ross, *Beginning in Algebraic Geometry*, Undergraduate Texts in Mathematics, https://doi.org/10.1007/978-3-031-88819-9_11

Section 10.1   Regular maps of projective varieties

Our goal in this section is to familiarize ourselves with the precise notion of a regular map between two projective varieties. As we alluded to in the introduction to this chapter, regular maps between projective varieties are locally modeled by collections of polynomials that are homogeneous of the same degree. Before we introduce the key definitions, a few observations are in order.

Suppose that $f \in K[x_0, \ldots, x_m]$ and $a \in \mathbb{P}^m$. Upon choosing homogeneous coordinates and writing $a = [a_0 : \cdots : a_m]$, we can evaluate f to obtain a value:

$$f(a_0, \ldots, a_m) \in K.$$

However, as we have seen, different choices of homogeneous coordinates for a lead to different values when evaluating f, so the value of f is not well-defined at points of $\mathbb{P}^m$. As a workaround, let us suppose that $f_0, \ldots, f_n \in K[x_0, \ldots, x_m]$ are homogeneous of the same degree d and that at least one of them does not vanish at a. Then we can collect the values together and view them as a point of projective space:

$$[f_0(a_0, \ldots, a_m) : \cdots : f_n(a_0, \ldots, a_m)] \in \mathbb{P}^n.$$

Something quite nice has occurred in doing this: the corresponding point in $\mathbb{P}^n$ is actually independent of the choice of homogeneous coordinates for a. Indeed, if we choose any other homogeneous coordinates $a = [\lambda a_0 : \cdots : \lambda a_m]$, then

$$[f_0(\lambda a_0, \ldots, \lambda a_m) : \cdots : f_n(\lambda a_0, \ldots, \lambda a_m)]$$
$$= [\lambda^d f_0(a_0, \ldots, a_m) : \cdots : \lambda^d f_n(a_0, \ldots, a_m)]$$
$$= [f_0(a_0, \ldots, a_m) : \cdots : f_n(a_0, \ldots, a_m)],$$

where the first equality uses homogeneity of the f_i and the second uses the equivalence relation in $\mathbb{P}^n$. In this situation, we henceforth adopt the shorthand notation

$$[f_0(a) : \cdots : f_n(a)] = [f_0(a_0, \ldots, a_m) : \cdots : f_n(a_0, \ldots, a_m)],$$

which is a slight abuse of notation, given that the individual values $f_i(a)$ are not well-defined. By evaluating at all points of $\mathbb{P}^m$ where at least one of the f_i does not vanish, we thus obtain a function

$$[f_0 : \cdots : f_n] : \mathbb{P}^m \setminus \mathcal{V}(f_0, \ldots, f_n) \to \mathbb{P}^n.$$

Functions arising in this way lead to a natural notion of *polynomial maps* between projective varieties, which serve as a first approximation to the more general notion of *regular maps* that we will introduce later in the section.

10.1 DEFINITION   *Polynomial map between projective varieties*

Let $X \subseteq \mathbb{P}^m$ and $Y \subseteq \mathbb{P}^n$ be projective varieties. A map $F : X \to Y$ is called a *polynomial map* if there exist polynomials $f_0, \ldots, f_n \in K[x_0, \ldots, x_m]$, all homogeneous of the same degree, such that $X \cap \mathcal{V}(f_0, \ldots, f_n) = \emptyset$ and

$$F(a) = [f_0(a) : \cdots : f_n(a)] \quad \text{for all } a \in X.$$

If $f_0, \ldots, f_n$ can be taken to be linear, we say that F is a *linear map*.

10.1. REGULAR MAPS OF PROJECTIVE VARIETIES

The requisite conditions that $f_0, \ldots, f_n$ are homogeneous of the same degree and that $X \cap \mathcal{V}(f_0, \ldots, f_n) = \emptyset$ together ensure that $[f_0(a) : \cdots : f_n(a)]$ is a well-defined point of $\mathbb{P}^n$ for every $a \in X$. Let us consider a few concrete examples.

10.2 EXAMPLE A polynomial map from $\mathbb{P}^1$ to a conic

Let $X = \mathcal{V}(y^2 - xz) \subseteq \mathbb{P}^2$ and consider the function

$$G : \mathbb{P}^1 \to X$$
$$[a : b] \mapsto [a^2 : ab : b^2].$$

This is a polynomial map: in coordinates s, t on the domain, $G = [s^2 : st : t^2]$, and the fact that G is defined at every point of $\mathbb{P}^1$ follows from the observation that

$$\mathcal{V}(s^2, st, t^2) = \emptyset \subseteq \mathbb{P}^1.$$

To verify that the image of G lies in X, note that the defining equation of X, namely $y^2 - xz$, vanishes when evaluated at $[a^2 : ab : b^2]$ for any $[a : b] \in \mathbb{P}^1$.

10.3 EXAMPLE A linear map from a quadric surface to $\mathbb{P}^2$

Let $X = \mathcal{V}(w^2 + x^2 + y^2 - z^2) \subseteq \mathbb{P}^3$. Since $[0:0:0:1] \notin X$, we obtain a linear map

$$H : X \to \mathbb{P}^2$$
$$[a : b : c : d] \mapsto [a : b : c],$$

given by the homogeneous linear polynomials $w, x, y \in K[w, x, y, z]$.

To visualize H, let us consider the restriction of X to the affine patch $\mathbb{A}_0^3$ where $w \neq 0$:

$$X_0 = \mathcal{V}(1 + x^2 + y^2 - z^2) \subseteq \mathbb{A}^3.$$

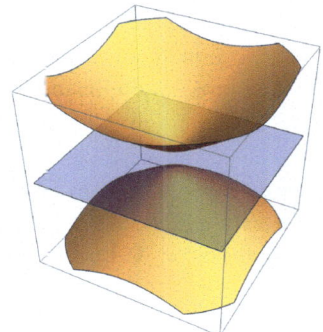

Restricting to this patch, the map H sends $(b, c, d) \in X_0$ to $(b, c) \in \mathbb{A}^2$, which can be visualized over the real numbers as the two-to-one map that vertically projects the two-sheeted hyperboloid depicted at right onto the horizontal coordinate plane.

10.4 EXAMPLE Linear projections

Generalizing the previous example, suppose that $X \subseteq \mathbb{P}^n$ is a projective variety that does not contain the point $[0 : \cdots : 0 : 1]$. The *linear projection of X onto $\mathbb{P}^{n-1}$* is the linear map

$$H : X \to \mathbb{P}^{n-1}$$
$$[a_0 : \cdots : a_{n-1} : a_n] \mapsto [a_0 : \cdots : a_{n-1}].$$

One way to visualize the map H is as follows: given a point $a = [a_0 : \cdots : a_n] \in X$, there is a unique line $L_a \subseteq \mathbb{P}^n$ that passes through both $[0 : \cdots : 0 : 1]$ and a:

$$L_a = \{[ca_0 : \cdots : ca_{n-1} : da_n] \mid [c : d] \in \mathbb{P}^1\}.$$

This line intersects the hyperplane $\mathcal{V}(x_n) \subseteq \mathbb{P}^n$ at the point $[a_0 : \cdots : a_{n-1} : 0]$, and making the natural identification of $\mathcal{V}(x_n)$ with $\mathbb{P}^{n-1}$, this intersection point is $H(a)$. Thus, intuitively, we can view $H(X)$ as the shadow that X casts on $\mathcal{V}(x_n)$ when a light shines from $[0 : \cdots 0 : 1]$. In the affine patch where $x_0 \neq 0$, the lines L_a are "vertical" lines for which only the last coordinate varies, and the hyperplane $\mathcal{V}(x_n)$ is the "horizontal" hyperplane for which the last coordinate is zero; thus, as in Example 10.3, the linear projection in this patch is the vertical projection.

To motivate the more general definition of *regular maps*, we note that the constraint in Definition 10.1 that $X \cap \mathcal{V}(f_0, \ldots, f_n) = \emptyset$ is often too restrictive, and it is useful to allow ourselves the flexibility to work with maps that are described by polynomials only "locally." To better understand this, let us consider an example.

10.5 EXAMPLE A piecewise polynomial function from a conic to $\mathbb{P}^1$

Let $X = \mathcal{V}(y^2 - xz) \subseteq \mathbb{P}^2$ and consider the pair of homogeneous linear polynomials $x, y \in K[x, y, z]$. Note that these polynomials do *not* give rise to a polynomial map from X to $\mathbb{P}^1$, simply because $\mathcal{V}(x, y) \cap X = \{[0:0:1]\} \neq \emptyset$. However, we still obtain a function from a subset of X to $\mathbb{P}^1$:

$$[x:y] : X \setminus \mathcal{V}(x,y) \to \mathbb{P}^1$$
$$[a:b:c] \mapsto [a:b].$$

Importantly, it is not the pair (x, y) that interests us, but the function $[x : y]$ that the pair defines. For example, the function $[5x : 5y]$ is the same as $[x : y]$, simply because points of $\mathbb{P}^1$ are invariant under scaling coordinates. More generally, recalling that two points $[a : b]$ and $[a' : b']$ in $\mathbb{P}^1$ are equal when their cross-multiplications ab' and ba' agree (Exercise 9.1.2), we see that two functions $[f : g]$ and $[f' : g']$ on X are equal—at all points where they are both defined—exactly when their cross-multiplications agree on X:

$$fg' - gf' \in \mathcal{I}(X).$$

For example, the maps $[x : y]$ and $[y : z]$ agree at all points of their common domain within X because the difference of their cross-multiplications is $xz - y^2$, which vanishes on X. Let us consider, then, the second of these two functions:

$$[y:z] : X \setminus \mathcal{V}(y,z) \to \mathbb{P}^1$$
$$[a:b:c] \mapsto [b:c].$$

Observe that the domains of $[x : y]$ and $[y : z]$ collectively cover all of X, since the first omits only $[0:0:1]$ while the second omits only $[1:0:0]$. Since the maps agree at all points where both are defined, we can then combine them to obtain a piecewise-defined function on all of X:

$$F : X \to \mathbb{P}^1$$
$$[a:b:c] \mapsto \begin{cases} [a:b] & \text{if } [a:b:c] \notin \mathcal{V}(x,y) \\ [b:c] & \text{if } [a:b:c] \notin \mathcal{V}(y,z). \end{cases}$$

10.1. REGULAR MAPS OF PROJECTIVE VARIETIES

The function $F : X \to \mathbb{P}^1$ in the previous example is not described by a single pair of polynomials, but it is "locally" polynomial in the following sense: for every point $p \in X$, there exists a pair $f, g \in K[x, y, z]$—possibly different pairs for different points—such that $p \notin \mathcal{V}(f, g)$ and

$$F(a) = [f(a) : g(a)] \quad \text{for all} \quad a \in X \setminus \mathcal{V}(f, g).$$

This characterization of F motivates the definition of regular maps.

10.6 DEFINITION *Regular map between projective varieties*

Let $X \subseteq \mathbb{P}^m$ and $Y \subseteq \mathbb{P}^n$ be projective varieties. A map $F : X \to Y$ is said to be a *regular map* if, for every $p \in X$, there exist polynomials $f_0, \ldots, f_n \in K[x_0, \ldots, x_n]$, all homogeneous of the same degree, such that $p \notin \mathcal{V}(f_0, \ldots, f_n)$ and

$$F(a) = [f_0(a) : \cdots : f_n(a)] \quad \text{for all} \quad a \in X \setminus \mathcal{V}(f_0, \ldots, f_n).$$

In other words, a map $F : X \to Y$ is regular if, for every $p \in X$, we can find a polynomial expression for F that is well-defined at p, even though it may not be well-defined on all of X. It follows from the definitions that every polynomial map is regular, but the converse is not true: not every regular map can be described globally by a single tuple of polynomials. We verify this fact in the next example.

10.7 EXAMPLE A regular map that is not a polynomial map

Let $X = \mathcal{V}(y^2 - xz) \subseteq \mathbb{P}^2$ and consider again the regular map of Example 10.5:

$$F : X \to \mathbb{P}^1$$

$$[a : b : c] \mapsto \begin{cases} [a : b] & [a : b : c] \notin \mathcal{V}(x, y) \\ [b : c] & [a : b : c] \notin \mathcal{V}(y, z). \end{cases}$$

We claim that F is not a polynomial map. To justify this, suppose to the contrary that there exist $f, g \in K[x, y, z]$, homogeneous of the same degree d, such that $\mathcal{V}(f, g) \cap X = \emptyset$ and $F = [f : g]$. Upon precomposing F with the map G from Example 10.2, we obtain a map $F \circ G : \mathbb{P}^1 \to \mathbb{P}^1$. On the one hand, we can readily verify that $F \circ G$ is the identity map on $\mathbb{P}^1$ (see Example 10.9 in the next section), but on the other hand, $F \circ G$ can also be written explicitly in terms of f and g:

$$(F \circ G)([a : b]) = F([a^2 : ab : b^2]) = [f(a^2, ab, b^2) : g(a^2, ab, b^2)].$$

Thus, if we define $f_0(s, t) = f(s^2, st, t^2)$ and $g_0(s, t) = g(s^2, st, t^2)$, then f_0 and g_0 are both homogeneous of degree $2d$ in $K[s, t]$ and they define a polynomial map $[f_0 : g_0] : \mathbb{P}^1 \to \mathbb{P}^1$ that agrees with the identity. This leads to a contradiction, however, because any polynomial map defining the identity on $\mathbb{P}^1$ must be of the form $f_0 = as$ and $g_0 = at$ for some nonzero $a \in K$ (Exercise 10.1.6); in particular, f_0 and g_0 must be linear, so they cannot have degree $2d$ for any d. The contradiction implies that F is not a polynomial map, as claimed.

We close this section by mentioning that a general method for describing a regular map is to define it piecewise, much like we did in Example 10.5. More precisely, a regular map $F : X \to Y$ between projective varieties $X \subseteq \mathbb{P}^m$ and $Y \subseteq \mathbb{P}^n$ can always be described—by definition—by a collection of functions of the form

$$[f_0 : \cdots : f_n] : X \setminus \mathcal{V}(f_0, \ldots, f_n) \to Y.$$

This collection of functions must satisfy the following two properties:

1. If $[f_0 : \cdots : f_n]$ and $[g_0 : \cdots : g_n]$ are both in the collection, then

$$f_i g_j - f_j g_i \in \mathcal{I}(X) \quad \text{for all} \quad i, j.$$

2. Every $p \in X$ must be in the domain of at least one function in this collection.

The first of these two conditions ensures that two functions in this collection agree on their common domain, which allows us to combine them to obtain a well-defined function on the union of their domains, while the second of these two conditions ensures that the union of all the domains covers X. In fact, while it is not obvious from the definition, the reader is encouraged to verify that every regular map can be described not just by a collection of such functions, but by a *finite* collection of such functions (Exercise 10.1.8).

Exercises for Section 10.1

10.1.1 Let $X \subseteq \mathbb{P}^m$ and $Y \subseteq \mathbb{P}^n$ be projective varieties, and let $F : X \to Y$ be a regular map. If $Z \subseteq \mathbb{P}^m$ is any projective variety such that $Z \subseteq X$, explain why the restriction $F|_Z : Z \to Y$ is also a regular map.

10.1.2 Let $X \subseteq \mathbb{P}^\ell$, $Y \subseteq \mathbb{P}^m$, and $Z \subseteq \mathbb{P}^n$ be projective varieties. If $F : X \to Y$ and $G : Y \to Z$ are regular maps, prove that $G \circ F : X \to Z$ is also regular.

10.1.3 Let $X = \mathcal{V}(wz - xy) \subseteq \mathbb{P}^3$.
  (a) Construct a regular map $F : X \to \mathbb{P}^1$ that extends the function

  $$[w : x] : X \setminus \mathcal{V}(w, x) \to \mathbb{P}^1.$$

  (b) Prove that F is surjective.
  (c) For any $b \in \mathbb{P}^1$, compute the preimage

  $$F^{-1}(b) = \{a \in X \mid F(a) = b\}.$$

10.1.4 Prove that every regular map $F : \mathbb{P}^1 \to \mathbb{P}^n$ is a polynomial map.

10.1.5 Prove that the map

$$[x_0 : \cdots : x_{n-1}] : \mathbb{P}^n \setminus \mathcal{V}(x_0, \ldots, x_{n-1}) \to \mathbb{P}^{n-1}$$

cannot be extended to a regular map on all of $\mathbb{P}^n$.

10.1.6 Let $f, g \in K[x, y]$ be homogeneous of the same degree with $\mathcal{V}(f, g) = \emptyset$, and suppose that the polynomial map $[f : g] : \mathbb{P}^1 \to \mathbb{P}^1$ is the identity map. Prove that $f = ax$ and $g = ay$ for some nonzero $a \in K$.

10.1. REGULAR MAPS OF PROJECTIVE VARIETIES

10.1.7 Let $X \subseteq \mathbb{P}^n$ be a projective variety that does not contain $[0 : \cdots : 0 : 1]$ and consider the linear projection

$$H : X \to \mathbb{P}^{n-1}$$
$$[a_0 : \cdots : a_{n-1} : a_n] \mapsto [a_0 : \cdots : a_{n-1}].$$

Prove that H is finite-to-one. In other words, prove that, for any $b \in \mathbb{P}^{n-1}$, the preimage $H^{-1}(b) = \{a \in X \mid H(a) = b\}$ is a finite set.

10.1.8 Let $X \subseteq \mathbb{P}^m$ and $Y \subseteq \mathbb{P}^n$ be projective varieties, and let $F : X \to Y$ be a regular map. Prove that there exists a *finite* collection of functions of the form

$$[f_0 : \cdots : f_n] : X \setminus \mathcal{V}(f_0, \ldots, f_n) \to Y$$

such that

- every function in the collection agrees with F on its domain, and
- every point of X is contained in the domain of at least one function in the collection.

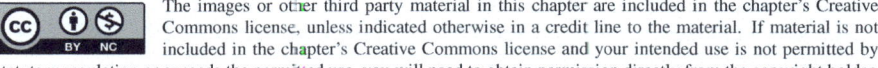

Section 10.2 Isomorphisms of projective varieties

Having developed an appropriate notion of maps between projective varieties, we now turn to a discussion of isomorphisms. We begin with the natural definition.

> **10.8 DEFINITION** *Isomorphism of projective varieties*
>
> An *isomorphism* of projective varieties $X \subseteq \mathbb{P}^m$ and $Y \subseteq \mathbb{P}^n$ is a regular map $F : X \to Y$ for which there exists a regular map $G : Y \to X$ that is inverse to F. If there exists an isomorphism between X and Y, we say that X and Y are *isomorphic* and write $X \cong Y$.

> As in the case of affine varieties, we view the intrinsic nature of a projective variety to be the structure that is preserved under isomorphism.

Some reflection should convince the reader that isomorphisms are an equivalence relation on the set of projective varieties, assuming (as verified in Exercise 10.1.2) that compositions of regular maps are regular. As a first example of an isomorphism, building on the discussion of regular maps in the previous section, we observe that the conic $\mathcal{V}(y^2 - xz) \subseteq \mathbb{P}^2$ is isomorphic to $\mathbb{P}^1$.

10.9 EXAMPLE $\mathcal{V}(y^2 - xz) \cong \mathbb{P}^1$

Let $X = \mathcal{V}(y^2 - xz) \subseteq \mathbb{P}^2$ and consider the regular maps $F : X \to \mathbb{P}^1$ and $G : \mathbb{P}^1 \to X$ that were introduced in Examples 10.7 and 10.2, respectively:

$$F([a:b:c]) = \begin{cases} [a:b] & [a:b:c] \notin \mathcal{V}(x,y) \\ [b:c] & [a:b:c] \notin \mathcal{V}(y,z) \end{cases}$$

and

$$G([a:b]) = [a^2 : ab : b^2].$$

We now show that these two regular maps are inverse to one another. Given a point $[a:b] \in \mathbb{P}^1$, we compute

$$F(G([a:b])) = F([a^2 : ab : b^2]) = \begin{cases} [a^2 : ab] & \text{if } a \neq 0 \\ [ab : b^2] & \text{if } b \neq 0 \end{cases} = [a:b].$$

Conversely, given a point $[a:b:c] \in X$, we compute

$$G(F([a:b:c])) = G\left(\begin{cases} [a:b] & \text{if } (a,b) \neq (0,0) \\ [b:c] & \text{if } (b,c) \neq (0,0) \end{cases}\right)$$

$$= \begin{cases} [a^2 : ab : b^2] & \text{if } (a,b) \neq (0,0) \\ [b^2 : bc : c^2] & \text{if } (b,c) \neq (0,0) \end{cases}$$

$$= [a:b:c],$$

where the final equality uses the fact that $b^2 = ac$ for every $[a:b:c] \in X$.

Since F and G are inverse regular maps, we conclude that $X \cong \mathbb{P}^1$.

10.2. ISOMORPHISMS OF PROJECTIVE VARIETIES

A very special class of isomorphisms arises from linear transformations on projective space. To set up notation, suppose that A is an invertible $(n+1) \times (n+1)$ matrix with entries in K. For each point $p \in \mathbb{P}^n$, we define a new point $Ap \in \mathbb{P}^n$ by choosing homogeneous coordinates for p, viewing them as a column vector, and multiplying the column vector on the left by the matrix A. For example, if

$$A = \begin{pmatrix} 1 & 1 & 0 \\ 1 & 0 & 2 \\ 0 & 1 & -1 \end{pmatrix}$$

and $p = [a_0 : a_1 : a_2]$, then $Ap = [a_0 + a_1 : a_0 + 2a_2 : a_1 - a_2]$.

More generally, if we write $A = (A_{ij})$, the procedure described above gives rise to a linear map $F_A : \mathbb{P}^n \to \mathbb{P}^n$ defined by linear polynomials whose coefficients are the rows of A:

$$F_A = \Big[\sum_{j=0}^{n} A_{0j} x_j : \cdots : \sum_{j=0}^{n} A_{nj} x_j \Big].$$

The invertibility of A implies that F_A is defined at every point of $\mathbb{P}^n$, and, moreover, that the inverse of F_A is the regular map

$$F_{A^{-1}} : \mathbb{P}^n \to \mathbb{P}^n$$

(see Exercise 10.2.3). Thus, F_A is a linear isomorphism from $\mathbb{P}^n$ to itself, and we give isomorphisms arising from invertible matrices in this way a special name.

10.10 DEFINITION *Projective equivalence*

A *projective equivalence* of $\mathbb{P}^n$ is an isomorphism of the form

$$F_A : \mathbb{P}^n \to \mathbb{P}^n$$
$$p \mapsto Ap,$$

where A is an invertible $(n+1) \times (n+1)$ matrix with entries in K. We say that two projective varieties $X, Y \subseteq \mathbb{P}^n$ are *projectively equivalent* if there exists a projective equivalence $F_A : \mathbb{P}^n \to \mathbb{P}^n$ such that $F_A(X) = Y$.

Intuitively, we view a projective equivalence simply as a linear change of coordinates in $\mathbb{P}^n$, similar to a change of basis in a vector space. If $X, Y \subseteq \mathbb{P}^n$ are projectively equivalent, then they are isomorphic; indeed,

> Readers with background in complex analysis may have met projective equivalences of $\mathbb{P}^1$ under the name "Möbius transformations."

if $F_A : \mathbb{P}^n \to \mathbb{P}^n$ is a projective equivalence such that $F_A(X) = Y$, then the restriction $F_A : X \to Y$ is a regular map with regular inverse $F_{A^{-1}} : Y \to X$.

Importantly, projective equivalences of $\mathbb{P}^n$ map projective varieties to projective varieties. More specifically, if $X \subseteq \mathbb{P}^n$ is a projective variety and $F_A : \mathbb{P}^n \to \mathbb{P}^n$ is a projective equivalence, it follows that $F_A(X)$ is also a projective variety in $\mathbb{P}^n$ (Exercise 10.2.4). While the same assertion is true more generally for any regular map $F : \mathbb{P}^m \to \mathbb{P}^n$ (Theorem 11.65), this is a much more difficult fact to prove.

Let us consider a concrete example of a projective equivalence in $\mathbb{P}^2$.

10.11 EXAMPLE Projectively equivalent conics

Consider the two complex projective varieties
$$X = \mathcal{V}(x^2 + y^2 + z^2) \subseteq \mathbb{P}^2_{\mathbb{C}} \quad \text{and} \quad Y = \mathcal{V}(y^2 - xz) \subseteq \mathbb{P}^2_{\mathbb{C}}.$$

We claim that X and Y are projectively equivalent, and we exhibit this through a change of coordinates in the defining equations. More precisely, notice that
$$x^2 + y^2 + z^2 = y^2 - (xi + z)(xi - z).$$

This implies that $[a : b : c] \in X$ if and only if $[ai + c : b : ai - c] \in Y$. In other words, letting A be the invertible matrix
$$A = \begin{pmatrix} i & 0 & 1 \\ 0 & 1 & 0 \\ i & 0 & -1 \end{pmatrix},$$
we see that $F_A(X) = Y$, so X and Y are projectively equivalent.

In fact, the previous example can be generalized in a somewhat surprising way.

10.12 PROPOSITION *All irreducible conics are projectively equivalent*

Let $f, g \in K[x, y, z]$ be irreducible and homogeneous of degree 2. Then $\mathcal{V}(f)$ and $\mathcal{V}(g)$ are projectively equivalent in $\mathbb{P}^2$.

PROOF To prove that all irreducible conics are projectively equivalent, we prove that they are all projectively equivalent to a distinguished conic. More specifically, let $f \in K[x, y, z]$ be irreducible and homogeneous of degree 2; we prove that $\mathcal{V}(f)$ is projectively equivalent to $\mathcal{V}(xy + xz + yz)$.

First, choose two distinct points $[a_1 : b_1 : c_1], [a_2 : b_2 : c_2] \in \mathcal{V}(f)$, and consider the matrix
$$\begin{pmatrix} a_1 & a_2 \\ b_1 & b_2 \\ c_1 & c_2 \end{pmatrix}.$$

Since the columns are not scalar multiples of each other, the rank of this matrix is two, and it follows that there is a unique-up-to-scaling linear relation among the rows. If every choice of third point in $\mathcal{V}(f)$ also satisfied this linear relation, then all of the points of $\mathcal{V}(f)$ would lie on a line in $\mathbb{P}^2$, contradicting the irreducibility of f. Thus, there must exist a third point $[a_3 : b_3 : c_3] \in \mathcal{V}(f)$ that does not satisfy this linear relation, implying that the matrix
$$A = \begin{pmatrix} a_1 & a_2 & a_3 \\ b_1 & b_2 & b_3 \\ c_1 & c_2 & c_3 \end{pmatrix}$$
has linearly independent rows, and is thus invertible.

10.2. ISOMORPHISMS OF PROJECTIVE VARIETIES

The projective equivalence F_A defined by the matrix A maps the three points $[1:0:0], [0:1:0], [0:0:1] \in \mathbb{P}^2$ to our three chosen points of $\mathcal{V}(f)$. Thus, the inverse projective equivalence $F_{A^{-1}}$ identifies $\mathcal{V}(f)$ with an irreducible conic $\mathcal{V}(g) \subseteq \mathbb{P}^2$ such that $[1:0:0], [0:1:0], [0:0:1] \in \mathcal{V}(g)$. But a homogeneous polynomial of degree 2 vanishes at these three points if and only if it has the form

$$g = \alpha xy + \beta xz + \gamma yz,$$

for some $\alpha, \beta, \gamma \in K$, and the irreducibility of g implies that $\alpha, \beta,$ and γ are nonzero. The projective equivalence $[a:b:c] \mapsto [\gamma^{-1}a : \beta^{-1}b : \alpha^{-1}c]$ then identifies $\mathcal{V}(g)$ with $\mathcal{V}(xy + xz + yz)$. Thus, composing the projective equivalences above, we see that $\mathcal{V}(f)$ is projectively equivalent to $\mathcal{V}(xy + xz + yz)$. □

10.13 COROLLARY *All irreducible conics are isomorphic to $\mathbb{P}^1$*

Let $f \in K[x, y, z]$ be irreducible and homogeneous of degree 2. Then

$$\mathcal{V}(f) \cong \mathbb{P}^1.$$

PROOF   Proposition 10.12 implies that all irreducible conics are projectively equivalent, and thus isomorphic. Combining this with Example 10.9, which shows that the irreducible conic $\mathcal{V}(y^2 - xz)$ is isomorphic to $\mathbb{P}^1$, we conclude that all irreducible conics in $\mathbb{P}^2$ are isomorphic to $\mathbb{P}^1$. □

While Corollary 10.13 tells us that the intrinsic nature of irreducible conics in $\mathbb{P}^2$ is rather simplistic—they are all isomorphic to $\mathbb{P}^1$—we caution the reader against being misled into thinking that the intrinsic nature of curves of higher degree in $\mathbb{P}^2$ is just as simple. In fact, upon increasing the degree by one, it can be shown that there are infinitely many distinct isomorphism classes of irreducible cubic curves in $\mathbb{P}^2$, none of which are isomorphic to $\mathbb{P}^1$ (in Example 12.11, we give a specific example of an irreducible cubic that is not isomorphic to $\mathbb{P}^1$). Cubic curves in $\mathbb{P}^2$ form the basis of the study of *elliptic curves*, which is a fascinating branch of mathematics to which a great many researchers have devoted their entire careers, but not a topic that we will touch upon in this book.

The discussion of isomorphisms naturally leads to the question: how might we study the intrinsic nature of projective varieties? In the affine setting, we introduced the coordinate ring of an affine variety as a key tool in this regard: two affine varieties are isomorphic if and only if their coordinate rings are isomorphic, allowing us to study the intrinsic nature of affine varieties by studying the algebraic structure of their coordinate rings. Motivated by the affine setting, we might wonder, then, if there is an algebraic object that captures the intrinsic nature of projective varieties. Following our developments in the affine setting, it is completely natural to introduce the *homogeneous coordinate ring* of a projective variety $X \subseteq \mathbb{P}^n$, defined as the quotient

$$\frac{K[x_0, \ldots, x_n]}{\mathcal{I}(X)}.$$

Unfortunately, unlike in the affine setting, the homogeneous coordinate ring is not preserved by isomorphisms, as the next example illustrates.

10.14 EXAMPLE Isomorphisms and homogeneous coordinate rings

Example 10.9 demonstrated that $X = \mathcal{V}(y^2 - xz) \subseteq \mathbb{P}^2$ is isomorphic to $\mathbb{P}^1$. However, the homogeneous coordinate rings of these two projective varieties are

$$\frac{K[x,y,z]}{\langle y^2 - xz \rangle} \quad \text{and} \quad K[s,t],$$

respectively, and these are not isomorphic rings. In particular, the latter is a unique factorization domain, but the equation $y^2 = xz$ implies that the former is not.

Another way to see that these rings are not isomorphic is that, if they were, then the equivalence of algebra and geometry in the setting of affine varieties would imply that the affine varieties $\mathcal{V}_\mathbb{A}(y^2 - xz)$ and $\mathbb{A}^2$ are isomorphic, but they are not: the former has a singularity at the origin while the latter is nonsingular.

Homogeneous coordinate rings play an important role in projective algebraic geometry, but they are not central to the developments of this text.

While the previous example implies that homogeneous coordinate rings are not complete algebraic invariants of projective varieties, it still leaves open the possibility that there may exist a different ring associated to each projective variety that encodes all of its intrinsic structure. It turns out that this is not the case. As we will see in the next chapter, the intrinsic nature of a projective variety is not determined by a single ring, but by a family of rings, each of which records local geometric information about the projective variety, while the global geometry is determined by homomorphisms between these local rings.

Before getting too far ahead of ourselves, however, we will continue to hone our working knowledge of maps between projective varieties in the rest of this chapter through the introduction of three classical families of maps: the Veronese, Segre, and Plücker maps.

Exercises for Section 10.2

10.2.1 Let $X = \mathcal{V}(x_0x_2 - x_1^2, x_0x_3 - x_1x_2, x_1x_3 - x_2^2) \subseteq \mathbb{P}^3$ be the twisted cubic curve. Consider the map

$$F : \mathbb{P}^1 \to X$$
$$[a:b] \mapsto [a^3 : a^2b : ab^2 : b^3].$$

Prove that F is an isomorphism by constructing a regular inverse.

10.2.2 Let $\ell_1, \ldots, \ell_k \in K[x_0, \ldots, x_n]$ be homogeneous linear polynomials that are linearly independent. Prove that $\mathcal{V}(\ell_1, \ldots, \ell_k) \cong \mathbb{P}^{n-k}$.

10.2.3 Let A be an invertible $(n+1) \times (n+1)$ matrix A with entries in K.
   (a) Prove that $F_A : \mathbb{P}^n \to \mathbb{P}^n$ is a regular map.
   (b) Prove that $F_{A^{-1}}$ is the inverse of F_A.

10.2. ISOMORPHISMS OF PROJECTIVE VARIETIES

10.2.4 Let A be an invertible $(n+1) \times (n+1)$ matrix A with entries in K.
  (a) Let $X = \mathcal{V}(f_1, \ldots, f_k) \in \mathbb{P}^n$. Prove that
  $$F_A(X) = \mathcal{V}(f_1 \circ A^{-1}, \ldots, f_k \circ A^{-1}).$$
  (Here, we view each f_i as a function $f_i : K^{n+1} \to K$ and A^{-1} as a function $A^{-1} : K^{n+1} \to K^{n+1}$, in the natural way.) In particular, if X is a projective variety, then so is $F_A(X)$.
  (b) Let $X, Y \subseteq \mathbb{P}^n$ be projectively equivalent. Prove that X is irreducible if and only if Y is irreducible.

10.2.5 Let $X, Y \subseteq \mathbb{P}^n$ be projective varieties that are projectively equivalent. Prove that
$$\frac{K[x_0, \ldots, x_n]}{\mathcal{I}(X)} \cong \frac{K[x_0, \ldots, x_n]}{\mathcal{I}(Y)}.$$
Thus, while homogeneous coordinate rings are *not* preserved by isomorphisms, they *are* preserved by projective equivalences.

10.2.6 Let $X \subseteq \mathbb{P}^n$ be a projective variety and let $\ell_0, \ldots, \ell_k \in K[x_0, \ldots, x_n]$ be homogeneous linear polynomials such that $\mathcal{V}(\ell_0, \ldots, \ell_k) \cap X = \emptyset$. Consider the linear map
$$L : X \to \mathbb{P}^k$$
$$a \mapsto [\ell_0(a) : \cdots : \ell_k(a)].$$

  (a) Prove that there exist a pair of projective equivalences $F_A : \mathbb{P}^n \to \mathbb{P}^n$ and $F_B : \mathbb{P}^k \to \mathbb{P}^k$ such that, for all $a = [a_0 : \cdots : a_n] \in X$,
  $$F_B(L(F_A(a))) = [a_0 : \cdots : a_j]$$
  for some $j \leq k$. (Hint: Applying row and column operations to the $k \times n$ matrix whose rows are the coefficients of $\ell_1, \ldots, \ell_k$, it can be reduced to a matrix of the form
  $$\begin{pmatrix} I & 0 \\ 0 & 0 \end{pmatrix},$$
  where I represents the $j \times j$ identity matrix for some $j \leq k$ and each 0 represents the zero matrix of the appropriate size.)
  (b) Use (a) to prove that L is finite-to-one. (Hint: See Exercise 10.1.7.)

Section 10.3 Veronese maps

In this section, we introduce the Veronese maps, named in honor of the Italian mathematician Giuseppe Veronese (1854–1917). These are a family of regular maps between projective spaces that can be described by the collection of all monomials of a fixed degree. The utility of Veronese maps in practice is that they allow us to reduce the degree of projective varieties and maps between them. More precisely, as we will see in this section, Veronese maps can be leveraged to prove the following two surprising properties.

1. Up to isomorphism, every projective variety can be realized as the vanishing set of a collection of polynomials of degree at most two (Proposition 10.22).
2. Up to isomorphism, every polynomial map of projective varieties is a linear map (Proposition 10.23).

Let us dive in and begin our discussion of Veronese maps with the definition.

10.15 **DEFINITION** *Veronese maps*

For any $d, n \geq 1$, the *Veronese map* $F_{d,n} : \mathbb{P}^n \to \mathbb{P}^{\binom{d+n}{d}-1}$ is the polynomial map

$$F_{d,n} = [x_0^d : x_0^{d-1}x_1 : x_0^{d-1}x_2 : \cdots : x_n^d],$$

where the monomials appearing in the definition are all of the possible monomials of degree d in the variables $x_0, \ldots, x_n$.

Exercise 10.3.1 outlines a strategy to prove that there are precisely $\binom{d+n}{d}$ monomials of degree d in the variables $x_0, \ldots, x_n$ (which explains the dimension of the projective space that serves as the codomain of $F_{d,n}$). Furthermore, Exercise 10.3.2 asks the reader to verify that these monomials do not simultaneously vanish at any point of $\mathbb{P}^n$, explaining why the domain of $F_{d,n}$ is all of $\mathbb{P}^n$.

As a first example, setting $d = 2$ and $n = 1$, the Veronese map is

$$F_{2,1} : \mathbb{P}^1 \to \mathbb{P}^2$$
$$[a : b] \mapsto [a^2 : ab : b^2],$$

which is the map G that we considered in Examples 10.2 and 10.9. In that particular case, we described the image $F_{2,1}(\mathbb{P}^1)$ as the projective variety $\mathcal{V}(y^2 - xz) \subseteq \mathbb{P}^2$, and our next aim is to generalize this description to $F_{d,n}(\mathbb{P}^n)$ for any d and n. In order to describe the image of any Veronese map as a projective variety, we first establish convenient notation for the coordinates of the codomain of $F_{d,n}$.

Note that degree-d monomials in the variables $x_0, \ldots, x_n$ can be indexed by tuples $D = (d_0, \ldots, d_n)$ of non-negative integers with $d_0 + \cdots + d_n = d$; more specifically, the monomial associated to $D = (d_0, \ldots, d_n)$ is

$$x^D = x_0^{d_0} x_1^{d_1} \cdots x_n^{d_n}.$$

In light of this, we denote the coordinates of $\mathbb{P}^{\binom{d+n}{d}-1}$ by y_D for each such tuple D, so that $F_{d,n}$ is the polynomial map whose y_D-coordinate is x^D.

10.3. VERONESE MAPS

For instance, in the case $(d,n) = (2,1)$ considered previously, we denote the coordinates in $\mathbb{P}^2$ by $y_{(2,0)}, y_{(1,1)}$, and $y_{(0,2)}$, in which case $F_{2,1}$ maps $a = [a_0 : a_1]$ to the point whose $y_{(2,0)}$-coordinate is a_0^2, whose $y_{(1,1)}$-coordinate is $a_0 a_1$, and whose $y_{(0,2)}$-coordinate is a_1^2. From this, one sees that $F_{2,1}$ maps to the variety

$$\mathcal{V}\left(y_{(2,0)} y_{(0,2)} - y_{(1,1)}^2\right) \subseteq \mathbb{P}^2.$$

More generally, images of Veronese maps are described by the following result.

10.16 PROPOSITION *Images of Veronese maps*

For any $d, n \geq 1$, the image of the Veronese map $F_{d,n} : \mathbb{P}^n \to \mathbb{P}^{\binom{d+n}{d}-1}$ is

$$X_{d,n} = \mathcal{V}(\{y_D y_{D'} - y_E y_{E'} \mid D + D' = E + E'\}) \subseteq \mathbb{P}^{\binom{d+n}{d}-1}.$$

PROOF That $F_{d,n}(\mathbb{P}^n) \subseteq X_{d,n}$ is a result of the observation that

$$a^D a^{D'} - a^E a^{E'} = a_0^{d_0+d_0'} \cdots a_n^{d_n+d_n'} - a_0^{e_0+e_0'} \cdots a_n^{e_n+e_n'} = 0$$

for any $a \in \mathbb{P}^n$ and for any tuples D, D', E, E' such that $D + D' = E + E'$.

Conversely, to show that $F_{d,n}(\mathbb{P}^n) \supseteq X_{d,n}$, let $b \in X_{d,n}$, so that the coordinates of b satisfy $b_D b_{D'} = b_E b_{E'}$ for all tuples D, D', E, E' with $D + D' = E + E'$. From the relations $b_D b_{D'} = b_E b_{E'}$, one can show (Exercise 10.3.3) that

(10.17) $$b_{D_1} \cdots b_{D_k} = b_{E_1} \cdots b_{E_k}$$

for all tuples $D_1, \ldots, D_k, E_1, \ldots, E_k$ such that $D_1 + \cdots + D_k = E_1 + \cdots + E_k$. In particular, for $D = (d_0, \ldots, d_n)$, we have

(10.18) $$b_D^d = b_{(d,0,\ldots,0)}^{d_0} b_{(0,d,0,\ldots,0)}^{d_1} \cdots b_{(0,\ldots,0,d)}^{d_n}.$$

This implies that at least one of the coordinates $b_{(d,0,\ldots,0)}, b_{(0,d,0,\ldots,0)}, \ldots, b_{(0,\ldots,0,d)}$ is nonzero, for otherwise (10.18) would yield $b_D = 0$ for all D, which is impossible in projective space.

Suppose, then, without loss of generality, that $b_{(d,0,\ldots,0)}$ is nonzero, so that we can rescale the homogeneous coordinates of b and set $b_{(d,0,\ldots,0)} = 1$. For each $1 \leq i \leq n$, define

$$a_i = b_{(d-1,0,\ldots,0,1,0,\ldots,0)},$$

in which the 1 appears in the ith coordinate of the indexing tuple. We claim that

(10.19) $$F_{d,n}(a) = b.$$

To prove (10.19), note that the definition of $F_{d,n}$ and of a_i implies that, for any $D = (d_0, \ldots, d_n)$, the y_D-coordinate of $F_{d,n}(a)$ is

$$a_0^{d_0} a_1^{d_1} \cdots a_n^{d_n} = b_{(d,0,\ldots,0)}^{d_0} b_{(d-1,1,0,\ldots,0)}^{d_1} \cdots b_{(d-1,0,\ldots,0,1)}^{d_n} = b_{(d,0,\ldots,0)}^{d-1} b_D = b_D,$$

where the second equality is an application of (10.17) and the third follows from the fact that $b_{(d,0,\ldots,0)} = 1$. This proves the assertion (10.19), and thus completes the proof that $X_{d,n} = F_{d,n}(\mathbb{P}^n)$. $\square$

Proposition 10.16 generalizes one aspect of Example 10.9—it gives explicit defining equations for the variety into which $F_{d,n}$ maps—but in the case of Example 10.9, we actually proved more: we showed that the Veronese map $F_{2,1}$ was an isomorphism onto its image. This fact is also true of all Veronese maps.

10.20 PROPOSITION *Veronese maps are embeddings*

For any $d, n \geq 1$, the Veronese map $F_{d,n} : \mathbb{P}^n \to X_{d,n}$ is an isomorphism.

PROOF To prove that $F_{d,n}$ is an isomorphism onto $X_{d,n}$, we construct a regular inverse $G_{d,n} : X_{d,n} \to \mathbb{P}^n$. Given any $b \in X_{d,n}$, define $G_{d,n}(b)$ by the following piecewise description:

In many mathematical contexts, an "embedding" refers to a map that is an isomorphism onto its image.

$$G_{d,n}(b) = \begin{cases} [b_{(d,0,\ldots,0)} : b_{(d-1,1,\ldots,0)} : \cdots : b_{(d-1,0,\ldots,1)}] \\ [b_{(1,d-1,\ldots,0)} : b_{(0,d,\ldots,0)} : \cdots : b_{(0,d-1,\ldots,1)}] \\ \vdots \\ [b_{(1,0,\ldots,d-1)} : b_{(0,1,\ldots,d-1)} : \cdots : b_{(0,0,\ldots,d)}] \end{cases},$$

where each expression is applicable at all points where it gives a well-defined point of $\mathbb{P}^n$. From the defining equations of $X_{d,n}$, we see that the cross-multiplications (see Exercise 9.1.2) of any pair of the expressions for $G_{d,n}$ agree on $X_{d,n}$, so $G_{d,n}$ is well-defined at all points where it is defined. Furthermore, as we saw in the proof of Proposition 10.16, at least one of the coordinates $b_{(d,0,\ldots,0)}, b_{(0,d,\ldots,0)}, \ldots, b_{(0,0,\ldots,d)}$ is nonzero for every $b \in X_{d,n}$, implying that $G_{d,n}$ is defined for all $b \in X_{d,n}$. Thus, $G_{d,n} : X_{d,n} \to \mathbb{P}^n$ is a regular map. It remains to check that $F_{d,n}$ and $G_{d,n}$ are inverse functions, which is left as an exercise to the reader (Exercise 10.3.4). □

The varieties $X_{d,n} \subseteq \mathbb{P}^{\binom{n+d}{d}}$ are called *Veronese varieties*. Up to isomorphism, Proposition 10.20 shows that the Veronese variety $X_{d,n}$ is intrinsically nothing more than a different perspective on projective space $\mathbb{P}^n$. However, by studying the various Veronese models of $\mathbb{P}^n$, it is possible to reduce the maximum degree of the defining polynomials of a projective variety $X \subseteq \mathbb{P}^n$. Before describing how this process works in general, let us consider how it works in a concrete example.

10.21 EXAMPLE Veronese image of a cubic curve

Let $X = \mathcal{V}(x_0^3 + x_1^3 + x_2^3) \subseteq \mathbb{P}^2$, and consider the Veronese map

$$F_{3,2} : X \to \mathbb{P}^{\binom{5}{3}-1} = \mathbb{P}^9.$$

Recall that the coordinates y_D of $\mathbb{P}^9$ are indexed by triples $D = (d_0, d_1, d_2)$ such that $d_0 + d_1 + d_2 = 3$, and observe that, for any $a = [a_0 : a_1 : a_2] \in \mathbb{P}^2$, we have

$$a \in X \iff a_0^3 + a_1^3 + a_2^3 = 0 \iff F_{3,2}(a) \in \mathcal{V}(y_{(3,0,0)} + y_{(0,3,0)} + y_{(0,0,3)}).$$

It then follows that

$$F_{3,2}(X) = X_{3,2} \cap \mathcal{V}\left(y_{(3,0,0)} + y_{(0,3,0)} + y_{(0,0,3)}\right).$$

10.3. VERONESE MAPS

Moreover, by Proposition 10.20, $F_{3,2}$ gives an isomorphism $\mathbb{P}^2 \cong X_{3,2}$, and it follows that the restriction of $F_{3,2}$ gives an isomorphism

$$X \cong X_{3,2} \cap \mathcal{V}\left(y_{(3,0,0)} + y_{(0,3,0)} + y_{(0,0,3)}\right).$$

Since Proposition 10.16 shows that $X_{3,2}$ is defined by quadratic polynomials, this implies that X is isomorphic to a projective variety that can be defined by polynomials of degree at most two, even though our original expression for X described it as the vanishing set of a cubic polynomial.

Building upon the previous example, we now describe the general result.

10.22 PROPOSITION *Projective varieties are defined by quadratics*

Up to isomorphism, every projective variety can be written as the vanishing set of a finite collection of homogeneous polynomials of degree at most two.

PROOF Let $X \subseteq \mathbb{P}^n$ be a projective variety. We first observe (Exercise 10.3.5) that we can write $X = \mathcal{V}(f_1, \ldots, f_k)$ where each $f_i \in K[x_0, \ldots, x_n]$ is homogeneous of the *same* degree d. In other words, we can write each f_i as

$$f_i = \sum_D a_{i,D} x_0^{d_0} \cdots x_n^{d_n}$$

where the sum is over all tuples $D = (d_0, \ldots, d_n)$ of nonnegative integers that sum to d and $a_{i,D} \in K$. Consider the Veronese map

$$F_{d,n} : \mathbb{P}^n \to \mathbb{P}^{\binom{n+d}{d}-1},$$

and define linear polynomials $\ell_1, \ldots, \ell_k$ in the variables of the codomain by

$$\ell_i = \sum_D a_{i,D} y_D.$$

As in Example 10.21, for any $a \in \mathbb{P}^n$, we have

$$a \in X \iff F_{d,n}(a) \in X_{d,n} \cap \mathcal{V}(\ell_1, \ldots, \ell_k),$$

implying that $F_{d,n}(X) = X_{d,n} \cap \mathcal{V}(\ell_1, \ldots, \ell_k)$. Since $F_{d,n} : \mathbb{P}^n \to \mathbb{P}^{\binom{n+d}{d}-1}$ is a regular map, it restricts to a regular map

$$F_{d,n} : X \to X_{d,n} \cap \mathcal{V}(\ell_1, \ldots, \ell_k).$$

Moreover, since $F_{d,n}$ is an isomorphism onto $X_{d,n}$ (Proposition 10.20), it then follows that $F_{d,n}^{-1} : X_{d,n} \to \mathbb{P}^n$ restricts to an inverse regular map

$$F_{d,n}^{-1} : X_{d,n} \cap \mathcal{V}(\ell_1, \ldots, \ell_k) \to X.$$

Thus, $X \cong X_{d,n} \cap \mathcal{V}(\ell_1, \ldots, \ell_k)$, and the result now follows from the fact that $X_{d,n}$ can be defined by quadratics (Proposition 10.16) while each ℓ_i is linear. □

It is worth noting that, while Proposition 10.22 reduces the maximum degree of the defining polynomials of a projective variety, it generally increases the number of defining polynomials quite drastically. For instance, in Example 10.21, one can use a projective analogue of the Fundamental Theorem of Dimension Theory (which will be made precise in the final chapter of the book) to show that we require at least 8 linear and quadratic polynomials to describe the Veronese image of the cubic curve, even though the cubic curve itself only required a single defining polynomial.

As a final application of Veronese maps, the next result says that all polynomial maps are, up to isomorphism on the domain, linear maps.

10.23 **PROPOSITION** *Up to isomorphism, polynomial maps are linear*

If $F : X \to Y$ is a polynomial map of projective varieties, then there exist an isomorphism $G : X \to Z$ and a linear map $L : Z \to Y$ such that $F = L \circ G$.

PROOF   Exercise 10.3.6. □

A surprising consequence of Proposition 10.23 is the following finite-to-one result regarding polynomial maps. This is another indication of just how restrictive polynomial maps are, providing additional justification for the more flexible notion of regular maps that we have been studying, which are only locally polynomial.

10.24 **COROLLARY** *Polynomial maps are finite-to-one*

If $F : X \to Y$ is a polynomial map of projective varieties, then for any $b \in Y$, there are finitely many $a \in X$ such that $F(a) = b$.

PROOF   With notation as in the statement of Proposition 10.23, any polynomial map F can be written as a composition $L \circ G$ where G is an isomorphism and L is linear. The result then follows from the fact that isomorphisms are one-to-one, while linear maps of projective varieties are finite-to-one (Exercise 10.2.6). □

Exercise 10.1.3 gives an example of a regular map that is not finite-to-one, and it then follows from Corollary 10.24 that this regular map is not a polynomial map.

Exercises for Section 10.3

10.3.1 Let $d, n \geq 1$ be integers and let $SB_{d,n}$ be the set of sequences of d "stars" and n "bars." For example, if $(d, n) = (4, 3)$, a few elements of $SB_{d,n}$ are

$$(\star \mid \star \star \mid \mid \star), \quad (\mid \mid \mid \star \star \star \star), \quad \text{and} \quad (\star \mid \star \mid \star \mid \star).$$

(a) Explain why $SB_{d,n}$ has $\binom{d+n}{d}$ elements.
(b) Describe a bijection between $SB_{d,n}$ and the set of monomials of degree d in the variables $x_0, \ldots, x_n$.
(c) Conclude from (a) and (b) that there are exactly $\binom{d+n}{d}$ monomials of degree d in the variables $x_0, \ldots, x_n$.

10.3.2 Prove that the dth Veronese map $F_{d,n}$ is regular at every point of $\mathbb{P}^n$.

10.3. VERONESE MAPS

10.3.3 Let $X_{d,n} \subseteq \mathbb{P}^{\binom{d+n}{d}-1}$ be the Veronese variety and consider a point $b \in X_{d,n}$. This exercise proves that the product of coordinates

$$b_{D_1} \cdots b_{D_k} \in K$$

depends only on $D_1 + \cdots + D_k \in \mathbb{N}^{n+1}$. Fix a tuple $(e_0, \ldots, e_n) \in \mathbb{N}^{n+1}$ such that $e_0 + \cdots + e_n = dk$, and let $\mathcal{S}$ denote the sequences of length dk that contain e_0 entries equal to 0, e_1 entries equal to 1, and so on. For each $\sigma \in \mathcal{S}$, define $b_\sigma \in K$ by setting

$$b_\sigma = b_{D_1} \cdots b_{D_k}$$

where the ith entry of D_j is the number of is in the jth subsequence of σ of length d.

(a) To parse notation, consider the tuple $(e_0, e_1, e_2) = (3, 2, 4)$ where we take $n = 2$ and $d = k = 3$. Write down three examples of sequences $\sigma \in \mathcal{E}$ and the corresponding values $b_\sigma \in K$.

(b) Prove that the function $\varphi : \mathcal{S} \to K$ sending σ to b_σ is a surjection onto

$$\{b_{D_1} \cdots b_{D_k} \mid D_1 + \cdots + D_k = (e_1, \ldots, e_n)\}.$$

(c) Suppose that σ and σ' are sequences in $\mathcal{S}$ that differ by a transposition of two terms. Use the defining equations of $X_{d,n}$ to prove that

$$\varphi(\sigma) = \varphi(\sigma').$$

(d) Using the fact that transpositions generate all permutations, conclude that $b_{D_1} \cdots b_{D_k}$ depends only on $D_1 + \cdots + D_k$.

10.3.4 With notation as in the proof of Proposition 10.20, verify that $F_{d,n}$ and $G_{d,n}$ are inverse functions.

10.3.5 (a) Let $f \in K[x_0, \ldots, x_n]$ and let $d \in \mathbb{N}$. Prove that

$$\mathcal{V}_{\mathbb{P}}(f) = \mathcal{V}_{\mathbb{P}}(x_0^d f, \ldots, x_n^d f).$$

(b) Prove that every projective variety can be defined by a finite set of homogeneous polynomials that all have the same degree.

10.3.6 Prove Proposition 10.23. (Hint: Take G to be a Veronese map.)

Section 10.4   Products and Segre maps

In this section, our primary aim is to study products of projective varieties. The notion of products is quite a bit more involved in the projective case than it is in the affine case. The complications stem from the fact that, while there is a natural identification $\mathbb{A}^m \times \mathbb{A}^n = \mathbb{A}^{m+n}$ in the affine setting, no such identification exists between $\mathbb{P}^m \times \mathbb{P}^n$ and $\mathbb{P}^{m+n}$, as the reader is encouraged to ponder. In fact, it is not even obvious at the onset how to interpret $\mathbb{P}^m \times \mathbb{P}^n$ itself as a projective variety. Our first task is to interpret $\mathbb{P}^m \times \mathbb{P}^n$ as a projective variety by mapping it injectively into a larger projective space. We accomplish this using Segre maps, named in honor of the Italian mathematician Corrado Segre (1863–1924).

10.25 DEFINITION   *Segre maps*

For any $m, n \geq 0$, the *Segre map* $S_{m,n} : \mathbb{P}^m \times \mathbb{P}^n \to \mathbb{P}^{(m+1)(n+1)-1}$ is the map

$$S_{m,n} = [x_0 y_0 : x_0 y_1 : \cdots : x_m y_{n-1} : x_m y_n],$$

where the monomials in the definition are all possible products $x_i y_j$.

To help parse the definition, let us take a look at the first interesting example.

10.26 EXAMPLE   The Segre map $S_{1,1}$

Consider the Segre map in the case $m = n = 1$:

$$S_{1,1} : \mathbb{P}^1 \times \mathbb{P}^1 \to \mathbb{P}^3$$
$$([a_0 : a_1], [b_0 : b_1]) \mapsto [a_0 b_0 : a_0 b_1 : a_1 b_0 : a_1 b_1].$$

Note that scaling the homogeneous coordinates within either $\mathbb{P}^1$ simply results in a uniform scaling of all of the coordinates in the image. Moreover, one checks that

$$a_0 b_0 = a_0 b_1 = a_1 b_0 = a_1 b_1 = 0 \implies a_0 = a_1 = 0 \quad \text{or} \quad b_0 = b_1 = 0.$$

These two observations, together, imply that $S_{1,1}$ is well-defined at every point of $\mathbb{P}^1 \times \mathbb{P}^1$. Generalizing this, the reader is encouraged to verify that $S_{m,n}$ is well-defined at every point of $\mathbb{P}^m \times \mathbb{P}^n$ (Exercise 10.4.1).

Even though Segre maps are well-defined at every point of their domain, it does not make sense to ask whether they are regular maps, simply because we have not yet given the domain the structure of a projective variety. In fact, we will actually leverage the Segre map to endow $\mathbb{P}^m \times \mathbb{P}^n$ with the structure of a projective variety within $\mathbb{P}^{(m+1)(n+1)-1}$. To accomplish this, it will be useful to introduce notation that allows us to conveniently organize the coordinates of $\mathbb{P}^{(m+1)(n+1)-1}$.

Observe that $\mathbb{P}^{(m+1)(n+1)-1}$ has $(m+1)(n+1)$ homogeneous coordinates; we denote these coordinates by z_{ij} with $0 \leq i \leq m$ and $0 \leq j \leq n$. Conveniently, these coordinates naturally organize into an $m \times n$ matrix $A(z)$, whose ij-entry is z_{ij}. With this labeling, we take the convention that the z_{ij}-coordinate of the Segre map $S_{m,n}$ is $x_i y_j$. Using this notation, we now describe the image of the Segre maps.

10.4. PRODUCTS AND SEGRE MAPS

10.27 PROPOSITION *The Segre map injects onto a projective variety*

For any $m, n \geq 0$, the Segre map $S_{m,n}$ is an injection of $\mathbb{P}^m \times \mathbb{P}^n$ onto the projective variety $Z_{m,n} \subseteq \mathbb{P}^{(m+1)(n+1)-1}$ defined by

$$Z_{m,n} = \mathcal{V}(\{z_{ij}z_{k\ell} - z_{i\ell}z_{kj} \mid 0 \leq i, k \leq m, \ 0 \leq j, \ell \leq n\}).$$

PROOF We prove that the image of $S_{m,n}$ is equal to $Z_{m,n}$, and we leave the verification that $S_{m,n}$ is injective as an exercise (Exercise 10.4.2).

> $Z_{m,n}$ is defined by all 2×2 minors of the matrix $A(z)$, so it is an example of a "determinantal variety."

To prove that the image of $S_{m,n}$ is contained in $Z_{m,n}$, it suffices to show that $z_{ij}z_{k\ell} - z_{i\ell}z_{kj}$ vanishes when evaluated at any point in the image of $S_{m,n}$, which follows from the observation that

$$(a_ib_j)(a_kb_\ell) - (a_ib_\ell)(a_kb_j) = 0 \quad \text{for any} \quad (a,b) \in \mathbb{P}^m \times \mathbb{P}^n.$$

Conversely, to prove that $Z_{m,n}$ is contained in the image of $S_{m,n}$, let $c \in Z_{m,n}$. This means that all of the 2×2 minors of $A(c)$ vanish, implying that the rank of $A(c)$ is at most one, or in other words, that every row of $A(c)$ is a multiple of some row. Since there must be at least one nonzero row, assume without loss of generality that the top row is nonzero, and denote it by $b = (b_0, \ldots, b_n)$. Let $a_i \in K$ be the value such that the ith row of $A(c)$ is equal to a_ib. One readily checks that

$$c = S_{m,n}([1 : a_1 : \cdots : a_m], [b_0 : b_1 : \cdots : b_n]),$$

showing that c is in the image of $S_{m,n}$. □

10.28 EXAMPLE *The image of $S_{1,1}$*

Considering again the case $m = n = 1$, denote the coordinates on $\mathbb{P}^3$ by z_{00}, z_{01}, z_{10}, and z_{11}. Proposition 10.27 shows that $S_{1,1}$ is an injection onto the projective variety in $\mathbb{P}^3$ defined by the vanishing of the 2×2 determinant

$$\det \begin{pmatrix} z_{00} & z_{01} \\ z_{10} & z_{11} \end{pmatrix} = z_{00}z_{11} - z_{01}z_{10}.$$

The importance of Proposition 10.27 is that it identifies the product $\mathbb{P}^m \times \mathbb{P}^n$ with a projective variety in $\mathbb{P}^{(m+1)(n+1)-1}$, thereby giving us a natural way to view the product itself as a projective variety. The next result shows that we can use Segre maps to naturally view the product of any two projective varieties—not just projective spaces—as a projective variety.

10.29 PROPOSITION *Products of projective varieties*

If $X \subseteq \mathbb{P}^m$ and $Y \subseteq \mathbb{P}^n$ are projective varieties, then the Segre map $S_{m,n}$ restricts to an injection of $X \times Y$ onto a projective variety in $\mathbb{P}^{(m+1)(n+1)-1}$.

PROOF Suppose that $X = \mathcal{V}(\mathcal{S})$ and $Y = \mathcal{V}(\mathcal{T})$, where $\mathcal{S} \subseteq K[x_0, \ldots, x_m]$ and $\mathcal{T} \subseteq K[y_0, \ldots, y_n]$. For every $f \in \mathcal{S}$ and $g \in \mathcal{T}$ and for each $0 \leq i \leq n$ and $0 \leq j \leq m$, define

$$f_i = f(z_{0i}, z_{1i}, \ldots, z_{mi}) \quad \text{and} \quad g_j = g(z_{j0}, z_{j1}, \ldots, z_{jn}).$$

Let $\mathcal{S}'$ be the collection of all such f_i, and let $\mathcal{T}'$ be the collection of all such g_j. The reader is encouraged to verify (Exercise 10.4.3) that

$$S_{m,n}(X \times Y) = S_{m,n}(\mathbb{P}^m \times \mathbb{P}^n) \cap \mathcal{V}(\mathcal{S}' \cup \mathcal{T}').$$

By Proposition 10.27, we know that $S_{m,n}$ is an injection and that $S_{m,n}(\mathbb{P}^m \times \mathbb{P}^n)$ is a projective variety. Since intersections of projective varieties are themselves projective varieties, we conclude that $S_{m,n}(X \times Y)$ is a projective variety. □

We now use the Segre-map identification of $X \times Y$ with the projective variety $S_{m,n}(X \times Y)$ to define products within the realm of projective varieties.

10.30 DEFINITION *Product of projective varieties*

If $X \subseteq \mathbb{P}^m$ and $Y \subseteq \mathbb{P}^n$ are projective varieties, then their *product* $X \times Y$ (as a projective variety) is the projective variety

$$S_{m,n}(X \times Y) \subseteq \mathbb{P}^{(m+1)(n+1)-1}.$$

10.31 EXAMPLE A doubly-ruled surface

Let us pause to visualize the projective variety $\mathbb{P}^1 \times \mathbb{P}^1$, which, by Definition 10.30 and Example 10.28, is equal to the projective variety $\mathcal{V}(z_{00}z_{11} - z_{01}z_{10}) \subseteq \mathbb{P}^3$. Consider the affine restriction where $z_{00} \neq 0$, which is the affine surface defined by

$$\mathcal{V}_{\mathbb{A}}(z_{11} - z_{01}z_{10}) \subseteq \mathbb{A}^3.$$

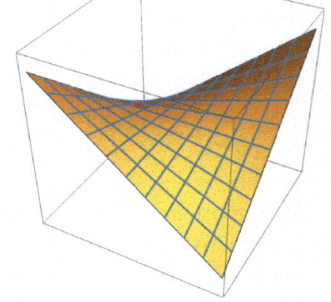

The image to the right is a depiction of this affine restriction over $\mathbb{R}$, from which we can see that the surface is doubly ruled: it can be viewed as a disjoint union of lines in two different ways. The double ruling in this affine restriction reflects the more general fact that the projective surface $\mathbb{P}^1 \times \mathbb{P}^1 \subseteq \mathbb{P}^3$ is also doubly ruled: it can be realized as a disjoint union of (projective) lines in $\mathbb{P}^3$ in two different ways. The two different rulings of $\mathbb{P}^1 \times \mathbb{P}^1$ are given by the subsets in $\mathbb{P}^3$ of the form

$$\{a\} \times \mathbb{P}^1 \quad \text{and} \quad \mathbb{P}^1 \times \{b\},$$

as the reader is encouraged to explore in Exercise 10.4.4.

10.4. PRODUCTS AND SEGRE MAPS

Having realized the product $X \times Y$ as a projective variety, it is now possible to discuss regular maps to and from $X \times Y$. The following result addresses the two most important examples of regular maps from a product: the two projection maps.

10.32 PROPOSITION *Projection maps are regular*

If $X \subseteq \mathbb{P}^m$ and $Y \subseteq \mathbb{P}^n$ are projective varieties, then the projection maps $\pi_1 : X \times Y \to X$ and $\pi_2 : X \times Y \to Y$ are regular.

PROOF We focus on π_1; the argument for π_2 is obtained by symmetry. It suffices to note that, for any $c = (a, b) \in X \times Y \subseteq \mathbb{P}^{(m+1)(n+1)-1}$, we can write $\pi_1(c)$ using the following piecewise expression:

$$\pi_1(c) = \begin{cases} [c_{00} : c_{10} : \cdots : c_{m0}] & c \notin \mathcal{V}(z_{00}, z_{10}, \ldots, z_{m0}) \\ \vdots & \vdots \\ [c_{0n} : c_{1n} : \cdots : c_{mn}] & c \notin \mathcal{V}(z_{0n}, z_{1n}, \ldots, z_{mn}). \end{cases}$$

To verify the above expression for π_1, suppose that $c = (a, b) \in X \times Y$. Then the z_{ij}-coordinate of c is $a_i b_j$. Note that at least one of the coordinates of a and at least one of the coordinates of b are nonzero. If $b_j \neq 0$, it then follows that $c \notin \mathcal{V}(z_{0j}, z_{1j}, \ldots, z_{mj})$, and we compute

$$[c_{0j} : c_{1j} : \cdots : c_{mj}] = [a_0 b_j : a_1 b_j : \cdots : a_m b_j] = [a_0 : a_1 : \cdots : a_m] = \pi_1(c),$$

verifying the above expression for $\pi_1(c)$. $\square$

Just as we can take products of projective varieties, we can also take products of maps to each factor. The next result addresses products of regular maps.

10.33 PROPOSITION *Products of regular maps are regular*

Let $W \subseteq \mathbb{P}^\ell$, $X \subseteq \mathbb{P}^m$, and $Y \subseteq \mathbb{P}^n$ be projective varieties. If $F : W \to X$ and $G : W \to Y$ are regular maps, then the function

$$F \times G : W \to X \times Y$$
$$a \mapsto (F(a), G(a))$$

is a regular map.

PROOF Let $p \in W$. By the definition of regular maps, there exist homogeneous polynomials

$$f_0, \ldots, f_m, g_0, \ldots, g_n \in K[w_0, \ldots, w_\ell]$$

such that $p \notin \mathcal{V}(f_0, \ldots, f_m) \cup \mathcal{V}(g_0, \ldots, g_n)$, and

$$F(a) = [f_0(a) : \cdots : f_m(a)] \quad \text{for all} \quad a \in W \setminus \mathcal{V}(f_0, \ldots, f_m)$$

and

$$G(a) = [g_0(a) : \cdots : g_n(a)] \quad \text{for all} \quad a \in W \setminus \mathcal{V}(g_0, \ldots, g_n).$$

It then follows that $p \notin \mathcal{V}(\{f_i g_j \mid 0 \leq i \leq m,\ 0 \leq j \leq n\})$ and, by the definition of products of projective varieties, we have

$$(F \times G)(a) = [f_0(a)g_0(a) : f_0(a)g_1(a) : \cdots : f_m(a)g_{n-1}(a) : f_m(a)g_n(a)]$$

for all $a \in W \setminus \mathcal{V}(\{f_i g_j \mid 0 \leq i \leq m,\ 0 \leq j \leq n\})$. Thus, $F \times G$ can be locally described by the polynomials $f_i g_j$, verifying that $F \times G$ is regular. $\square$

An important consequence of the previous results is the following, showing that the intrinsic nature of products depends only on the intrinsic nature of each factor.

10.34 COROLLARY *Products preserve isomorphisms*

Let X, X', Y, Y' be projective varieties. If $X \cong X'$ and $Y \cong Y'$, then

$$X \times Y \cong X' \times Y'.$$

PROOF Let $F : X \to X'$ and $G : Y \to Y'$ be isomorphisms. Since the projection maps are regular, precomposing F and G with the projection maps yields regular maps

$$F \circ \pi_1 : X \times Y \to X' \quad \text{and} \quad G \circ \pi_2 : X \times Y \to Y'.$$

The product of these two regular maps is then the regular map

$$X \times Y \to X' \times Y'$$
$$(a, b) \mapsto (F(a), G(b)).$$

Since F and G are isomorphisms, they have regular inverses, and repeating the above procedure with F^{-1} and G^{-1} yields a regular map

$$X' \times Y' \to X \times Y$$
$$(a, b) \mapsto (F^{-1}(a), G^{-1}(b)).$$

The above pair of regular maps between $X \times Y$ and $X' \times Y'$ are inverse to each other, from which we conclude that $X \times Y \cong X' \times Y'$. $\square$

If $X \subseteq \mathbb{P}^m$ and $Y \subseteq \mathbb{P}^n$ are projective varieties, then we have shown that their product $X \times Y$ is a projective variety within $\mathbb{P}^m \times \mathbb{P}^n \subseteq \mathbb{P}^{(m+1)(n+1)-1}$. One might naturally ask: are there other projective varieties within $\mathbb{P}^m \times \mathbb{P}^n$? Indeed, there are, and our final task for this section is to classify them as vanishing sets, with respect to a new notion of vanishing. Given a polynomial $f \in K[x_0, \ldots, x_m, y_0, \ldots, y_n]$, we say that f *vanishes* at a point $(a, b) \in \mathbb{P}^m \times \mathbb{P}^n$ if f vanishes when evaluated at every representative of (a, b). We then define the vanishing operator $\mathcal{V} = \mathcal{V}_{\mathbb{P}^m \times \mathbb{P}^n}$ on products of projective space in the natural way. Using this new notion of vanishing, the next result classifies all of the projective varieties in $\mathbb{P}^m \times \mathbb{P}^n$ in a way that does not require one to explicitly reference the Segre map.

10.35 PROPOSITION *Projective varieties in $\mathbb{P}^m \times \mathbb{P}^n$*

Let Z be a subset of $\mathbb{P}^m \times \mathbb{P}^n$. Then, viewing $\mathbb{P}^m \times \mathbb{P}^n$ as a projective variety via the Segre map, Z is itself a projective variety if and only if Z is a vanishing set $\mathcal{V}(\mathcal{S}) \subseteq \mathbb{P}^m \times \mathbb{P}^n$ for some $\mathcal{S} \subseteq K[x_0, \ldots, x_m, y_0, \ldots, y_n]$.

10.4. PRODUCTS AND SEGRE MAPS

Let us consider an example before the proof.

10.36 EXAMPLE The diagonal in $\mathbb{P}^1 \times \mathbb{P}^1$

Since every projective variety in $\mathbb{P}^1$ is either all of $\mathbb{P}^1$ or a finite set of points, it follows that every product of projective varieties in $\mathbb{P}^1 \times \mathbb{P}^1$ is of one of the following four forms: (i) $\mathbb{P}^1 \times \mathbb{P}^1$, (ii) $\{a_1, \ldots, a_k\} \times \mathbb{P}^1$ for some (possibly empty) finite set $\{a_1, \ldots, a_k\} \subseteq \mathbb{P}^1$, (iii) $\mathbb{P}^1 \times \{b_1, \ldots, b_\ell\}$ for some (possibly empty) finite set $\{b_1, \ldots, b_\ell\} \subseteq \mathbb{P}^1$, or (iv) $\{a_1, \ldots, a_k\} \times \{b_1, \ldots, b_\ell\}$. Proposition 10.35 gives us access to a much richer class of projective varieties in $\mathbb{P}^1 \times \mathbb{P}^1$. For example, the diagonal

$$Z = \{(a,a) \mid a \in \mathbb{P}^1\}$$

is not of one of the four types mentioned above, but it *is* a projective variety in $\mathbb{P}^1 \times \mathbb{P}^1$. To verify this, recall that $a_0 b_1 = a_1 b_0$ if and only if $[a_0 : a_1] = [b_0 : b_1]$, so

$$Z = \mathcal{V}(x_0 y_1 - x_1 y_0) \subseteq \mathbb{P}^1 \times \mathbb{P}^1.$$

Thus, Proposition 10.35 guarantees that Z is a projective variety in $\mathbb{P}^1 \times \mathbb{P}^1$.

We note that, arguing similarly as in the case of a single projective space, any vanishing set in $\mathbb{P}^m \times \mathbb{P}^n$ can be written as

$$\mathcal{V}(f_1, \ldots, f_k),$$

where $f_1, \ldots, f_k$ are *bihomogeneous*, meaning that they are homogeneous in each set of variables separately. Even more, one can take each f_i to be bihomogeneous of the *same degree* in each set of variables. The verifications of these assertions are left as an exercise (Exercise 10.4.6); we use them in the following proof.

PROOF OF PROPOSITION 10.35   Suppose that $Z \subseteq \mathbb{P}^m \times \mathbb{P}^n \subseteq \mathbb{P}^{(m+1)(n+1)-1}$ is a projective variety. By definition, this means that

$$Z = \mathcal{V}(g_1, \ldots, g_\ell)$$

for homogeneous polynomials $g_1, \ldots, g_\ell \in K[z_{ij} \mid 0 \leq i \leq m, \; 0 \leq j \leq n]$ of degrees $d_1, \ldots, d_\ell$, respectively. Under the Segre map, this corresponds to

$$Z = \{(a, b) \in \mathbb{P}^m \times \mathbb{P}^n \mid f_k(a, b) = 0 \text{ for all } k = 1, \ldots, \ell\},$$

where $f_k(x, y)$ is obtained from $g_k(z)$ by substituting $z_{ij} = x_i y_j$ for all i, j. This proves one direction of the proposition.

For the other direction, suppose that Z is a vanishing set in $\mathbb{P}^m \times \mathbb{P}^n$. Then $Z = \mathcal{V}(f_1, \ldots, f_\ell)$ for some polynomials $f_1, \ldots, f_\ell \in K[x_0, \ldots, x_m, y_0, \ldots, y_n]$, each of which can be assumed to be bihomogeneous of the same degree in each set of variables. Thus, in any monomial of f_k, there are the same number of x's as y's, so we can group them into pairs to construct a polynomial $g_k \in K[z_{ij}]$ such that g_k becomes f_k upon making the substitution $z_{ij} = x_i y_j$. Thus, the Segre embedding identifies Z with $(\mathbb{P}^m \times \mathbb{P}^n) \cap \mathcal{V}(g_1, \ldots, g_\ell)$, so Z is a projective variety. □

Exercises for Section 10.4

10.4.1 Let $a = [a_0 : \cdots : a_m] \in \mathbb{P}^m$ and $b = [b_0 : \cdots : b_n] \in \mathbb{P}^n$.
  (a) Prove that $a_i b_j \neq 0$ for some i and j.
  (b) Prove that $[a_0 b_0 : a_0 b_1 : \cdots : a_m b_{n-1} : a_m b_n] \in \mathbb{P}^{(m+1)(n+1)-1}$ does not depend on the choice of homogeneous coordinates for a or b.

10.4.2 Prove that the Segre map $S_{m,n} : \mathbb{P}^m \times \mathbb{P}^n \to \mathbb{P}^{(m+1)(n+1)-1}$ is an injection.

10.4.3 Let $X = \mathcal{V}(\mathcal{S}) \subseteq \mathbb{P}^m$ and $Y = \mathcal{V}(\mathcal{T}) \subseteq \mathbb{P}^n$ where $\mathcal{S} \subseteq K[x_0, \ldots, x_m]$ and $\mathcal{T} \subseteq K[y_0, \ldots, y_n]$. For every $f \in \mathcal{S}$ and $g \in \mathcal{T}$ and for each $0 \leq i \leq n$ and $0 \leq j \leq m$, define

$$f_i = f(z_{0i}, z_{1i}, \ldots, z_{mi}) \quad \text{and} \quad g_j = g(z_{j0}, z_{j1}, \ldots, z_{jn}).$$

Let $\mathcal{S}'$ be the collection of all such f_i, and let $\mathcal{T}'$ be the collection of all such g_j. Prove that

$$S_{m,n}(X \times Y) = S_{m,n}(\mathbb{P}^m \times \mathbb{P}^n) \cap \mathcal{V}(\mathcal{S}' \cup \mathcal{T}').$$

10.4.4 Prove that the subsets of the projective variety $\mathbb{P}^1 \times \mathbb{P}^1 \subseteq \mathbb{P}^3$ of the form $\{a\} \times \mathbb{P}^1$ and $\mathbb{P}^1 \times \{b\}$ are each the image of a linear map $\mathbb{P}^1 \to \mathbb{P}^3$.

10.4.5 Let $X \subseteq \mathbb{P}^m$ and $Y \subseteq \mathbb{P}^n$ be projective varieties. Prove that $X \times Y$ and $Y \times X$ are projectively equivalent in $\mathbb{P}^{(m+1)(n+1)-1}$. In particular, conclude that $X \times Y \cong Y \times X$.

10.4.6 (a) Prove that every polynomial $f \in K[x_0, \ldots, x_m, y_0, \ldots, y_n]$ decomposes uniquely into a sum of bihomogeneous polynomials, and prove that f vanishes at a point $(a, b) \in \mathbb{P}^m \times \mathbb{P}^n$ if and only if every bihomogeneous component of f vanishes at (a, b).
  (b) Let $\mathcal{S} \subseteq K[x_0, \ldots, x_m, y_0, \ldots, y_n]$. Prove that

$$\mathcal{V}(\mathcal{S}) = \mathcal{V}(\langle \mathcal{S} \rangle) \subseteq \mathbb{P}^m \times \mathbb{P}^n,$$

and use this, along with part (a), to prove that $\mathcal{V}(\mathcal{S}) = \mathcal{V}(f_1, \ldots, f_k)$ where each f_i is bihomogeneous.
  (c) Let $\mathcal{S} \subseteq K[x_0, \ldots, x_m, y_0, \ldots, y_n]$. Prove that $\mathcal{V}(\mathcal{S}) = \mathcal{V}(f_1, \ldots, f_k)$ where each f_i is bihomogeneous with the same degree in the x- and y-variables. Hint: For any $d, e \geq 0$, notice that

$$\mathcal{V}(f) = \mathcal{V}(\{x_i^d y_j^e f \mid 0 \leq i \leq m, \ 0 \leq j \leq n\}).$$

Section 10.5   Grassmannians and Plücker maps

Grassmannian varieties, named in honor of Hermann Grassmann (1809–1877), are a family of projective varieties that parametrize linear subspaces of a fixed vector space, or equivalently, linear subvarieties of a fixed projective space. Grassmannians play a central role within many special topics in algebraic geometry. Our aims in this section are to introduce Grassmannians abstractly, describe how to realize them as projective varieties via Plücker maps, and give a hint of how they can be used to answer interesting geometric questions. We begin with the set-theoretic definition.

10.37 DEFINITION   *Grassmannians as sets of linear subspaces*

Let k and n be nonnegative integers with $k \leq n$. The *Grassmannian* $G(k,n)$ is the set of all k-dimensional linear subspaces of the vector space K^n.

For example, $G(0,n)$ is just a single point for any n, because there is a unique zero-dimensional linear subspace of K^n, while $G(1,n)$ has one point for each line through the origin in K^n, so it is in natural bijection with $\mathbb{P}^{n-1}$ (Corollary 9.8).

A common alternative perspective on Grassmannians is in terms of linear subvarieties of projective space, which we now define.

10.38 DEFINITION   *Linear subvariety of projective space*

A subset $L \subseteq \mathbb{P}^n$ is called a *k-dimensional linear subvariety of* $\mathbb{P}^n$ if there exists a $(k+1)$-dimensional linear subspace $V \subseteq K^{n+1}$ such that
$$L = \mathbb{P}(V) = \{[a_0 : \cdots : a_n] \mid (a_0, \ldots, a_n) \in V \setminus \{0\}\}.$$

Linear subvarieties of dimensions one and two are called *lines* and *planes*, while linear subvarieties of codimension one are called *hyperplanes*. Observing that the affine cone over a linear subvariety recovers the linear subspace in $\mathbb{A}^{n+1} = K^{n+1}$ from which it came, it follows that k-dimensional linear subvarieties of $\mathbb{P}^n$ are in natural bijection with $(k+1)$-dimensional linear subspaces of K^{n+1} (Exercise 10.5.1). Thus, we have the following alternative interpretation of Grassmannians.

10.39 DEFINITION   *Grassmannians as sets of linear subvarieties*

Let k and n be positive integers with $k \leq n$. The *Grassmannian* $G(k,n)$ is the set of all $(k-1)$-dimensional linear subvarieties of $\mathbb{P}^{n-1}$.

Grassmannians provide a natural framework for studying geometric questions regarding linear subvarieties in projective space. For example, one of the simplest such questions is: *Given four lines in $\mathbb{P}^3$, does there exist a line that intersects all of them?* As we will see at the end of this section, the answer to this question is "yes," and moreover, under a genericity hypothesis on the four given lines, there are exactly two distinct lines intersecting them all. The key to studying this and many other related questions is to refine our perspective on $G(k,n)$, going beyond a view of it as just an abstract set by endowing it with the structure of a projective variety.

To realize Grassmannians as projective varieties, we leverage matrix representations of linear subspaces. More specifically, given a k-dimensional linear subspace $V \subseteq K^n$—in other words, an element $V \in \mathbf{G}(k, n)$—choose a basis $v_1, \ldots, v_k$ of V and define the *matrix representation* of V with respect to this basis to be the $k \times n$ matrix $M(v_1, \ldots, v_k)$ that has $v_1, \ldots, v_k$ as its rows. Different bases of V produce different matrix representations, but these matrices all share common attributes, as we discover in the next example.

10.40 EXAMPLE Different matrices representing the same subspace

Let $V \subseteq K^4$ be the span of $v_1 = (2, 1, -3, 0)$ and $v_2 = (1, 0, 1, 1)$. Since v_1 and v_2 are linearly independent, they form a basis of V, and we have

$$M(v_1, v_2) = \begin{pmatrix} 2 & 1 & -3 & 0 \\ 1 & 0 & 1 & 1 \end{pmatrix}.$$

However, choosing a different basis of V, such as $w_1 = 2v_1 - v_2 = (3, 2, -7, -1)$ and $w_2 = v_1 + v_2 = (3, 1, -2, 1)$, results in a different matrix representation:

$$M(w_1, w_2) = \begin{pmatrix} 3 & 2 & -7 & -1 \\ 3 & 1 & -2 & 1 \end{pmatrix}.$$

These two matrix representations are not total strangers, of course: they are related by the change-of-basis matrix between v_1, v_2 and w_1, w_2:

$$\begin{pmatrix} 3 & 2 & -7 & -1 \\ 3 & 1 & -2 & 1 \end{pmatrix} = \begin{pmatrix} 2 & -1 \\ 1 & 1 \end{pmatrix} \begin{pmatrix} 2 & 1 & -3 & 0 \\ 1 & 0 & 1 & 1 \end{pmatrix}.$$

In particular, picking out two columns from each of the matrix representations—for instance, consider the first and fourth—multiplicativity of determinants implies that

$$\det \begin{pmatrix} 3 & -1 \\ 3 & 1 \end{pmatrix} = \det \begin{pmatrix} 2 & -1 \\ 1 & 1 \end{pmatrix} \cdot \det \begin{pmatrix} 2 & 0 \\ 1 & 1 \end{pmatrix}.$$

Thus, each 2×2 minor of $M(w_1, w_2)$ is obtained from the corresponding 2×2 minor of $M(v_1, v_2)$ upon scaling by 3. In other words, the six 2×2 minors of these two matrices give rise to the same point of $\mathbb{P}^5$, and the same point of $\mathbb{P}^5$ will be determined by the 2×2 minors of any other matrix representation of V. As we will formalize below, this point of $\mathbb{P}^5$ is the image of V under the *Plücker map*.

Generalizing the previous example, we now aim to describe the Plücker map

$$P_{k,n} : \mathbf{G}(k,n) \to \mathbb{P}^{\binom{n}{k}-1},$$

named in honor of Julius Plücker (1801–1868). To set up notation, let us index the homogeneous coordinates on $\mathbb{P}^{\binom{n}{k}-1}$ by subsets $I \subseteq \{1, \ldots, n\}$ of size k, denoting the corresponding homogeneous coordinate by x_I. Furthermore, given a $k \times n$ matrix M and a subset $I \subseteq \{1, \ldots, n\}$, let M_I be the $k \times |I|$ submatrix of M consisting of the columns indexed by I. The Plücker maps are defined for all nonnegative integers k, n with $k \leq n$ as follows.

10.5. GRASSMANNIANS AND PLÜCKER MAPS

10.41 DEFINITION *Plücker maps*

The *Plücker map*
$$P_{k,n} : \mathbf{G}(k,n) \to \mathbb{P}^{\binom{n}{k}-1}$$
is the function whose value on $V \in \mathbf{G}(k,n)$ is given by choosing a basis of V, letting $M(V)$ denote the $k \times n$ matrix with these basis elements as rows, and setting the value of $P_{k,n}(V) \in \mathbb{P}^{\binom{n}{k}-1}$ in the homogeneous coordinate indexed by I to $\det(M(V)_I)$.

One should pause to convince themselves that $P_{k,n}$ is well-defined; in particular, it should be verified that at least one $k \times k$ minor of $M(V)$ is nonzero for any matrix representation of V, and that the point $P_{k,n}(V)$ does not depend on the choice of basis for V (Exercise 10.5.5). The next example illustrates these observations.

10.42 EXAMPLE *Computing the Plücker map*

Let $V \subseteq K^4$ be the subspace of Example 10.40, and order the coordinates of $\mathbb{P}^{\binom{4}{2}-1} = \mathbb{P}^5$ lexicographically, omitting brackets and commas for brevity:

> We generally order the coordinates of $\mathbb{P}^{\binom{n}{k}-1}$ lexicographically, as one would order them in a dictionary with letters replaced by numbers.

$x_{12}, x_{13}, x_{14}, x_{23}, x_{24}, x_{34}.$

Computing $P_{2,4}(V)$ via the basis $v_1 = (2,1,-3,0)$ and $v_2 = (1,0,1,1)$ gives
$$[-1:5:2:1:1:-3] \in \mathbb{P}^{\binom{4}{2}-1},$$
while the basis $w_1 = (3,2,-7,-1)$ and $w_2 = (3,1,-2,1)$ gives the (same) point
$$[-3:15:6:3:3:-9] \in \mathbb{P}^{\binom{4}{2}-1}.$$

As in the case of Segre maps, it does not make sense to ask whether Plücker maps are regular, because we have not yet endowed the domain with the structure of a projective variety. In fact, similarly to how we defined products as projective varieties in Definition 10.30, we use Plücker maps to endow $\mathbf{G}(k,n)$ with the structure of a projective variety by showing that $P_{k,n}$ is an injection of $\mathbf{G}(k,n)$ onto a projective variety. We begin by arguing injectivity, and then we describe the image.

10.43 PROPOSITION *Plücker maps are injective*

For any integers k and n with $0 \leq k \leq n$, the Plücker map $P_{k,n}$ is injective.

PROOF Suppose that $a \in \mathbb{P}^{\binom{n}{k}-1}$ is a point in the image of $P_{k,n}$. We argue that there is a unique linear subspace $V \in \mathbf{G}(k,n)$ such that $P_{k,n}(V) = a$. Choose a nonzero coordinate $a_I \neq 0$ and rescale so that $a_I = 1$. If $V \in \mathbf{G}(k,n)$ maps to a, then $\det(M(V)_I) \neq 0$, so $M(V)_I$ is invertible. Thus, upon multiplying $M(V)$ by $M(V)_I^{-1}$, we see that any subspace mapping to a admits a matrix representation that has a copy of the identity matrix in the columns indexed by I. We now argue that there is a unique such matrix representation mapping to a.

Let M be a matrix representation for some $V \in \mathbf{G}(k,n)$ such that $P_{k,n}(V) = a$ and such that M_I is the identity matrix. Let $j \in \{1,\ldots,n\} \setminus I$; we argue that the column M_j indexed by j is uniquely determined from the fact that $P_{k,n}(V) = a$. For each $i \in I$, consider the set $I_i = (I \setminus \{i\}) \cup \{j\}$. Since M maps to a, we have

(10.44) $$a_{I_i} = \det(M_{I_i}) \quad \text{for all} \quad i \in I.$$

Notice that M_{I_i} is obtained from the identity matrix by removing the ith column and then adding the column M_j. Some reflection should convince the reader that the determinant of such a matrix is nothing more than the ith entry of M_j (up to a sign). In particular, the relations (10.44) uniquely determine all entries of M_j. Since j was arbitrary, M is uniquely determined from a, completing the proof. $\square$

Our next step is to describe the projective variety in $\mathbb{P}^{\binom{n}{k}-1}$ onto which the Plücker map sends $\mathbf{G}(k,n)$. Given any set $I \subseteq \{1,\ldots,n\}$ and $j \in \{1,\ldots,n\}$, let $I_{<j}$ denote the number of elements of I less than j. Then for each pair of subsets $I, J \subseteq \{1,\ldots,n\}$ of sizes $k-1$ and $k+1$, respectively, we define the corresponding *Plücker polynomial*

(10.45) $$f_{I,J} = \sum_{j \in J} (-1)^{I_{<j} + J_{<j}} x_{I \cup \{j\}} x_{J \setminus \{j\}},$$

where we adopt the convention that $x_I = 0$ if $|I| \neq k$ (in particular, the jth term of the sum is zero whenever $j \in I$). Notice that $f_{I,J}$ is a homogeneous polynomial of degree two in the coordinates of $\mathbb{P}^{\binom{n}{k}-1}$. Let us consider an example.

10.46 EXAMPLE Plücker polynomial in $\binom{4}{2}$ variables

As in Example 10.42, we denote the six homogeneous coordinates of $\mathbb{P}^{\binom{4}{2}-1}$ by $x_{12}, x_{13}, x_{14}, x_{23}, x_{24}, x_{34}$. If we let $I = \{1\}$ and $J = \{2,3,4\}$, then we obtain the Plücker polynomial

$$f_{I,J} = -x_{12}x_{34} + x_{13}x_{24} - x_{14}x_{23}.$$

Choosing any other pair of subsets with $|I| = 1$ and $|J| = 3$, one checks that the corresponding Plücker polynomial is either zero or the same as above, up to a sign. Thus, up to sign, there is a unique nonzero Plücker polynomial in $\binom{4}{2}$ variables.

Evaluating the Plücker polynomial $f_{I,J}$ from Example 10.46 at the point $P_{k,n}(V)$ that we computed in Example 10.42, we obtain

$$-(-1)(-3) + (5)(1) - (2)(1) = 0.$$

This is one instance of the following result describing the images of Plücker maps.

10.47 PROPOSITION *Plücker maps and Plücker polynomials*

For any integers k and n with $0 \leq k \leq n$, we have

$$P_{k,n}(\mathbf{G}(k,n)) = V(f_{I,J} \mid I, J \subseteq \{1,\ldots,n\} \text{ with } |I| = k-1 \text{ and } |J| = k+1).$$

10.5. GRASSMANNIANS AND PLÜCKER MAPS

PROOF   We prove both inclusions.

($\subseteq$) Suppose that $V \in \mathbf{G}(k,n)$ is a k-dimensional linear subspace of K^n and let M be any matrix representation of V. Then for any I and J, we have

$$f_{I,J}(P_{k,n}(V)) = \sum_{j \in J} (-1)^{I_{<j} + J_{<j}} \det(M_{I \cup \{j\}}) \det(M_{J \setminus \{j\}}).$$

Let M_I^i be the matrix obtained by deleting the ith row of the $k \times (k-1)$ matrix M_I, and let M_J^i be the matrix obtained by duplicating the ith row of the $k \times (k+1)$ matrix M_J as the top row. Upon expanding $\det(M_{I \cup \{j\}})$ along the jth column and $\det(M_J^i)$ along the first row, one verifies (Exercise 10.5.6) that

$$\sum_{j \in J} (-1)^{I_{<j} + J_{<j}} \det(M_{I \cup \{j\}}) \det(M_{J \setminus \{j\}}) = \sum_{i=1}^{k} (-1)^i \det(M_I^i) \det(M_J^i).$$

Since M_J^i has two identical rows, $\det(M_J^i) = 0$, so $f_{I,J}$ vanishes at $P_{k,n}(V)$.

($\supseteq$) Suppose that $f_{I,J}(a) = 0$ for all $I, J \subseteq \{1, \ldots, n\}$ with $|I| = k - 1$ and $|J| = k + 1$. Fix I' such that $a_{I'} \neq 0$, and rescale the homogeneous coordinates of a so that $a_{I'} = 1$. Define a $k \times n$ matrix $M = (m_{i,j})$ by

$$m_{i,j} = (-1)^{I'_{<i} + I'_{<j}} a_{(I' \setminus \{i\}) \cup \{j\}},$$

where we set $m_{i,j} = 0$ if $(I' \setminus \{i\}) \cup \{j\}$ does not have size k. Notice that $M_{I'}$ is simply the identity matrix, so the rows of M are linearly independent and span a k-dimensional linear subspace $V \subseteq K^n$. We argue that $P_{k,n}(V) = a$.

For convenience, set $P_{k,n}(V) = b$, and scale the homogeneous coordinates of b so that $b_{I'} = a_{I'} = 1$. For any $i \in I'$ and $j \in \{1, \ldots, n\} \setminus I'$, notice that $M_{(I' \setminus \{i\}) \cup \{j\}}$ differs from the identity matrix by exactly one column. We compute

$$b_{(I' \setminus \{i\}) \cup \{j\}} = \det(M_{(I' \setminus \{i\}) \cup \{j\}}) = (-1)^{I'_{<i} + I'_{<j}} m_{i,j} = a_{(I' \setminus \{i\}) \cup \{j\}},$$

where the first equality is from the definition of b, the second follows from expanding the determinant along the jth column, and the third is from the definition of $m_{i,j}$. It then follows that $b_I = a_I$ for any subset I such that $|I \setminus I'| \leq 1$.

We now proceed to prove that $b_I = a_I$ for all I by induction on $|I \setminus I'|$. The base cases are proved in the previous paragraph. To set up the induction step, fix $\ell \geq 2$ and assume that $b_I = a_I$ whenever $|I \setminus I'| < \ell$; now fix some I such that $|I \setminus I'| = \ell$. Choose $i \in I \setminus I'$ and consider the Plücker polynomial

$$f_{I \setminus \{i\}, I' \cup \{i\}} = \sum_{j \in I' \cup \{i\}} (-1)^{(I \setminus \{i\})_{<j} + (I' \cup \{i\})_{<j}} x_{(I \setminus \{i\}) \cup \{j\}} x_{(I' \cup \{i\}) \setminus \{j\}}.$$

This polynomial vanishes at a (by definition of a) and it also vanishes at $b = P_{k,n}(V)$ (by the proof of ($\subseteq$) above). Plugging a and b into this relation, the terms indexed by $j = i$ are equal to a_I and b_I (up to a sign), while every other term appearing is a product of coordinates indexed by subsets J with $|J \setminus I'| < \ell$. Since $a_J = b_J$ for all such J (by the induction hypothesis), it then follows that $a_I = b_I$. This completes the induction step and thus the proof. $\square$

Proposition 10.47 allows us to realize Grassmannians as projective varieties.

10.48 DEFINITION *Grassmannians as projective varieties*

Let k and n be nonnegative integers with $k \leq n$. The *Grassmannian* $G(k,n)$ (as a projective variety) is the projective variety

$$P_{k,n}(G(k,n)) \subseteq \mathbb{P}^{\binom{n}{k}-1}.$$

As alluded to at the onset of this section, Grassmannians can be used to study many interesting geometric questions about linear subvarieties in projective space. In honor of Hermann Schubert (1848–1911), who made many contributions to the early study of these questions, this branch of algebraic geometry is commonly referred to as *Schubert calculus*. In the next example (with details left to the exercises), we return to the question posed at the beginning of this section.

10.49 EXAMPLE A preview of Schubert calculus

Suppose that we have four fixed lines $L_1, \ldots, L_4$ in $\mathbb{P}^3$, and we want to know if there is a fifth line in $\mathbb{P}^3$ that intersects each of $L_1, \ldots, L_4$. Since lines $L \subseteq \mathbb{P}^3$ correspond to points of $G(2,4)$, we may define four subsets $X_1, \ldots, X_4 \subseteq G(2,4)$ by

$$X_i = \{L \in G(2,4) \mid L \cap L_i \neq \emptyset\}.$$

Notice that a line intersecting every L_i must be in the intersection $X_1 \cap \cdots \cap X_4$; thus, we aim to understand whether $X_1 \cap \cdots \cap X_4$ is nonempty. We now leverage our interpretation of $G(2,4)$ as a projective variety:

$$G(2,4) = \mathcal{V}(-x_{12}x_{34} + x_{13}x_{24} - x_{14}x_{23}) \subseteq \mathbb{P}^5.$$

For each $i = 1, \ldots, 4$, one checks (Exercise 10.5.8) that the condition $L_i \cap L \neq \emptyset$ imposes a linear relation on the Plücker coordinates of L; in other words, for every $i = 1, \ldots, 4$, there is a linear hyperplane $H_i \subseteq \mathbb{P}^5$ such that

$$X_i = G(2,4) \cap H_i.$$

Thus,

$$X_1 \cap \cdots \cap X_4 = G(2,4) \cap (H_1 \cap \cdots \cap H_4).$$

Intersecting four hyperplanes in $\mathbb{P}^5$ gives a linear subvariety of dimension at least one (Exercise 10.5.3), and since any such linear subvariety must have nontrivial intersection with $\mathcal{V}(-x_{12}x_{34} + x_{13}x_{24} - x_{14}x_{23})$ (Exercise 10.5.7), it follows that $X_1 \cap \cdots \cap X_4 \neq \emptyset$. Thus, there must exist at least one line in $\mathbb{P}^3$ intersecting any four fixed lines.

In fact, one can prove that for "almost all" choices of $L_1, \ldots, L_4$, there are exactly two distinct lines in $\mathbb{P}^3$ intersecting all four of these. We do not have the tools to argue this rigorously, but the basic idea is that so long as $H_1 \cap \cdots \cap H_4$ is a "generic" line in $\mathbb{P}^5$, the fact that $G(2,4)$ is defined by a single quadratic polynomial implies that its intersection with this line is exactly two distinct points.

10.5. GRASSMANNIANS AND PLÜCKER MAPS

Throughout this chapter, we have begun to familiarize ourselves with projective varieties and the maps between them. It would be natural at this point for us to turn toward a discussion of attributes of projective varieties, such as dimension and smoothness, parallel to our treatment of affine varieties. However, given how much work we have already devoted to developing these notions in the affine setting, it would certainly be ideal if we could somehow leverage that work, rather than redeveloping those same tools again in this different context. In order to build the necessary tools for importing affine tools into the projective setting, we will turn in the next chapter to the notion of *quasiprojective varieties*, which will simultaneously encompass both the affine and projective settings. This general setting will allow us to view affine varieties as local models for projective varieties, making it possible for us to directly import the notions of dimension and smoothness from the affine setting into the (quasi-)projective setting.

Exercises for Section 10.5

10.5.1 Prove that there is an inclusion-preserving bijection between the set of linear subspaces of K^{n+1} of dimension $k+1$ and the set of linear subvarieties of $\mathbb{P}^n$ of dimension k.

10.5.2 Let $L \subseteq \mathbb{P}^n$ be a subset. Prove that the following are equivalent.
  (i) L is a k-dimensional linear subvariety of $\mathbb{P}^n$.
  (ii) L is the image of an injective linear map $\mathbb{P}^k \to \mathbb{P}^n$.
  (iii) $L = \mathcal{V}(\ell_1, \ldots, \ell_{n-k})$ where $\ell_1, \ldots, \ell_{n-k} \in K[x_0, \ldots, x_n]$ are homogeneous of degree one and linearly independent.

10.5.3 Suppose that L_1 and L_2 are linear subvarieties of $\mathbb{P}^n$ of codimensions d_1 and d_2, respectively. Prove that $L_1 \cap L_2$ is a linear subvariety of codimension at most $d_1 + d_2$.

10.5.4 For any $n \geq 0$, describe a natural bijection between $G(n, n+1)$ and $\mathbb{P}^n$ (without using the Plücker map).

10.5.5 Prove that the Plücker maps are well-defined. In particular, for any linear subspace $V \subseteq K^n$ of dimension k, prove the following:
  (a) If $M(V)$ is any matrix representation of V, then $\det(M(V)_I) \neq 0$ for at least one subset $I \subseteq \{1, \ldots, n\}$ of size k.
  (b) If $M(V)$ and $M(V)'$ are two matrix representations of V, then there exists $\lambda \in K \setminus \{0\}$ such that $\det(M(V)_I) = \lambda \det(M(V)'_I)$ for every subset $I \subseteq \{1, \ldots, n\}$ of size k.

10.5.6 Verify the second displayed equation in the proof of Proposition 10.47.

10.5.7 Let $f \in K[x_0, \ldots, x_n]$ be a homogeneous polynomial and let $L \subseteq \mathbb{P}^n$ be a linear subvariety of positive dimension. Prove that $L \cap \mathcal{V}(f) \neq \emptyset$.
  Hint: Parametrize a line in L and then evaluate f at the points of this line.

10.5.8 Choose a matrix representation $M(V)$ for each linear subspace $V \in \mathbf{G}(2,4)$, and for any $V_1, V_2 \in \mathbf{G}(2,4)$, let $M(V_1, V_2)$ be the 4×4 matrix with $M(V_1)$ above $M(V_2)$.

(a) Prove that $\dim(V_1 \cap V_2) > 0$ if and only if $\det(M(V_1, V_2)) = 0$.

(b) Prove that the 4×4 determinant $\det(M(V_1, V_2))$ can be expressed as

$$\det(M(V_1, V_2)) = \sum_{\substack{I \subseteq \{1,2,3,4\} \\ |I|=2}} s(I) \det(M(V_1)_I) \det(M(V_2)_{I^c}),$$

where I^c denotes the complement $\{1,2,3,4\} \setminus I$ and

$$s(I) = \begin{cases} 1 & \text{if } I = \{1,2\}, \{1,4\}, \{2,3\}, \text{ or } \{3,4\}, \\ -1 & \text{if } I = \{2,4\} \text{ or } \{3,4\}. \end{cases}$$

(c) Fix $V_0 \in \mathbf{G}(2,4)$ and set $X = \{V \in \mathbf{G}(2,4) \mid \dim(V \cap V_0) > 0\}$. Prove that

$$X = \mathbf{G}(2,4) \cap \mathcal{V}(\ell),$$

where ℓ is homogeneous and linear in the coordinates of $\mathbb{P}^{\binom{4}{2}-1}$.

Upon reinterpreting planes in K^4 as lines in $\mathbb{P}^3$, this exercise shows that the set of lines in $\mathbb{P}^3$ that intersect a fixed line is the intersection of $\mathbf{G}(2,4)$ with a hyperplane.

10.5.9 View $\mathbf{G}(k,n)$ as a projective variety via the Plücker map and let I be a subset of $\{1,\ldots,n\}$ of size k.

(a) Prove that the affine restriction

$$\mathbf{G}(k,n) \setminus \mathcal{V}(x_I) \subseteq \mathbb{A}^{\binom{n}{k}-1}$$

is isomorphic as an affine variety to $\mathbb{A}^{k(n-k)}$.

(b) Prove that the projective closure of $\mathbf{G}(k,n) \setminus \mathcal{V}(x_I)$ is $\mathbf{G}(k,n)$.
Hint: Given any $V \in \mathcal{V}(x_I)$, construct a function $f : \mathbb{A}^1 \to \mathbf{G}(k,n)$ such that $f(0) = V$ and $f(a) \notin \mathcal{V}(x_I)$ for $a \neq 0$. Reduced row echelon form may be helpful. Then argue that any polynomial vanishing at $f(a)$ for $a \neq 0$ must also vanish at $f(0)$.

(c) Use Parts (a) and (b) to argue that $\mathbf{G}(k,n)$ is irreducible.

Upon formalizing the notion of dimension for projective varieties, we will also be able to use this exercise to show that $\mathbf{G}(k,n)$ has dimension $k(n-k)$.

Chapter 11
Quasiprojective Varieties

> LEARNING OBJECTIVES FOR CHAPTER 11
> - Define and work with topological concepts such as open and closed sets, closures, denseness, and irreducibility.
> - Define quasiprojective varieties and describe examples of quasiprojective varieties, including examples that are affine, projective, and neither.
> - Describe regular maps and regular functions in the setting of quasiprojective varieties, and understand their continuous and local nature.
> - Understand the locally affine nature of quasiprojective varieties.
> - Define products of quasiprojective varieties and determine whether subsets of products are closed.
> - Understand why regular images of projective varieties are closed.

We have now studied two types of varieties—affine and projective—that behave similarly in some ways and yet appear to occupy two separate worlds. The goal of this chapter is to unify these worlds, introducing a more general notion of "variety" that includes both affine and projective varieties as special cases. The key insight is that, via the inclusion $j_0 : \mathbb{A}^n \to \mathbb{P}^n$, any affine variety X can be viewed as a subset of its projective closure $\overline{X}$. The subsets of projective varieties that arise in this way are special from a topological perspective: they are *open* in the Zariski topology on $\overline{X}$, a topology whose definition we will make precise in the first section of the chapter. On the other hand, the entire projective variety $\overline{X}$ is also open in the Zariski topology on $\overline{X}$, so a natural class of subsets of $\mathbb{P}^n$ that contains both affine and projective varieties is the collection of open subsets of projective varieties; these are what we will refer to as *quasiprojective varieties*.

Studying quasiprojective varieties will require care in several ways: we will have to introduce more topological background than was previously needed, and we will have to ensure that each notion we develop for quasiprojective varieties (such as the definition of "regular map") specializes in the affine and projective settings to the corresponding notion in those cases, when such a notion is defined. One payoff for this work, however, is that we will gain the ability to define many concepts for projective varieties that have been missing from our development thus far. What, for instance, is the dimension of a projective variety, and how can we determine whether a projective variety is smooth? Equipped with the ability to work not just with the variety itself but with any open subset—many of which are affine varieties and are therefore already well-studied in this book—we will be in a position to import these concepts into our projective toolkit.

Section 11.1 Quasiprojective varieties

Up to this point in our journey through algebraic geometry, our treatment has not required a careful study of topology. However, as we work toward unifying the affine and projective settings, topological notions will become more and more important. In this section, we develop basic topological ideas, with a primary focus on the Zariski topology, which is the topology most relevant to the study of algebraic geometry. These topological concepts will allow us to define quasiprojective varieties, which simultaneously generalize both affine and projective varieties.

To motivate the definition of a topology, let us work through a familiar example.

11.1 EXAMPLE Euclidean topology on $\mathbb{R}$

Given two real numbers $a, b \in \mathbb{R}$, consider the intervals

$$(a,b) = \{c \in \mathbb{R} \mid a < c < b\} \quad \text{and} \quad [a,b] = \{c \in \mathbb{R} \mid a \leq c \leq b\}.$$

The first interval (a,b) omits the endpoints and is often referred to as an "open" interval, while the second interval $[a,b]$ includes the endpoints and is often referred to as a "closed" interval. Open and closed intervals in $\mathbb{R}$ motivate the more general notion of open and closed sets in $\mathbb{R}$, as we now describe.

Consider the following property of open intervals: given any number $c \in (a,b)$, all numbers that are sufficiently close to c are also contained in (a,b). For example, given the number $3 \in (1,4)$, every number strictly within a distance of 1 from 3 is also contained in $(1,4)$. Generalizing, we say that a set $U \subseteq \mathbb{R}$ is *open* if it has this property: for every number $c \in U$, there exists a positive number ϵ such that $(c - \epsilon, c + \epsilon) \subseteq U$. In other words, open sets contain the "nearby" points of all of their members, and for this reason, we often call an open set containing c a *neighborhood* of c. We say that a set $Z \subseteq \mathbb{R}$ is *closed* if its complement $\mathbb{R} \setminus Z$ is open. We encourage the reader to convince themselves that the interval $[a,b]$ is, indeed, a closed set, while the half-open interval $[a,b)$ is neither open nor closed.

Open sets in $\mathbb{R}$ satisfy several important properties that are readily verified from the definition above: (i) $\emptyset$ and $\mathbb{R}$ are both open sets in $\mathbb{R}$, (ii) finite intersections of open sets in $\mathbb{R}$ are open, and (iii) arbitrary (even infinite) unions of open sets in $\mathbb{R}$ are open. Equivalently, these properties can be translated to properties of closed sets using De Morgan's law: (i) $\emptyset$ and $\mathbb{R}$ are both closed sets in $\mathbb{R}$, (ii) finite unions of closed sets in $\mathbb{R}$ are closed, and (iii) arbitrary (even infinite) intersections of closed sets in $\mathbb{R}$ are closed. It is these three properties, phrased either in terms of open or closed sets, that we axiomatize to arrive at the notion of a *topological space*.

In general, a topology on a set X is a choice of which subsets of X we will call "open," subject to the condition that the three properties of open sets described in Example 11.1 hold. The key point, philosophically speaking, is that the topology allows us to talk abstractly about "neighborhoods" of points in X, even in the absence of a method for actually measuring distances. This allows us to view the set X not just as a collection of points, but as a structured collection of neighborhoods glued together along their common neighbors. By introducing topology into projective algebraic geometry, we will be able to make precise how a general projective variety can be covered by neighborhoods that are isomorphic to affine varieties, allow-

11.1. QUASIPROJECTIVE VARIETIES

ing us to reduce the study of essential properties of projective varieties—dimension and smoothness, for example—to the study of these properties in the affine setting, where we can leverage the equivalence of algebra and geometry.

As we mentioned in Example 11.1, we can specify a topology either by declaring which sets are open or which are closed. Since topologies in algebraic geometry are most commonly specified using closed sets, we adopt the following definition.

11.2 DEFINITION *Topological space*

Let X be a set. A *topology* on X is a collection of subsets of X, denoted $\mathcal{C}$, whose elements we call *closed sets*, satisfying the following three axioms.

1. The empty set and the whole set are closed: $\emptyset, X \in \mathcal{C}$.

2. Finite unions of closed sets are closed:
$$Z_1, \ldots, Z_n \in \mathcal{C} \implies \bigcup_{i=1}^{n} Z_i \in \mathcal{C}.$$

3. Arbitrary intersections of closed sets are closed:
$$\{Z_\alpha\}_{\alpha \in A} \subseteq \mathcal{C} \implies \bigcap_{\alpha \in A} Z_\alpha \in \mathcal{C}.$$

A set X along with a specified topology $\mathcal{C}$ is called a *topological space*. If X is a topological space, we say that a set $U \subseteq X$ is *open* if $X \setminus U$ is closed.

In addition to Example 11.1, another example of a topology that we met in Section 2.1 is the *Zariski topology on* $\mathbb{A}^n$ whose closed sets are affine varieties. We note that a given set can generally be endowed with multiple topologies; for example, the reader is encouraged to investigate how the Euclidean topology on $\mathbb{R}$ differs from the Zariski topology on $\mathbb{R} = \mathbb{A}^1_\mathbb{R}$ (see Exercise 11.1.1).

The most important topology in our study of algebraic geometry—from which all the other topologies we study will be derived—is the Zariski topology on $\mathbb{P}^n$.

11.3 DEFINITION *Zariski topology on* $\mathbb{P}^n$

The *Zariski topology* on $\mathbb{P}^n$ is the topology whose closed sets (sometimes called *Zariski-closed sets*) are projective varieties.

Some reflection should convince the reader that projective varieties in $\mathbb{P}^n$ satisfy the axioms of a topology (see Exercise 9.2.5). In $\mathbb{P}^1$, they can be described explicitly.

11.4 EXAMPLE *Zariski topology on* $\mathbb{P}^1$

As we have seen in Exercise 9.2.2, the only projective varieties in $\mathbb{P}^1$ are finite sets and all of $\mathbb{P}^1$; this is a complete description of the closed sets in the Zariski topology on $\mathbb{P}^1$. Taking complements, we then see that the Zariski-open sets in $\mathbb{P}^1$ consist of the empty set and sets with finite complement. Notice that nonempty open sets in $\mathbb{P}^1$ are quite large: they can omit only finitely many points.

An important point to ponder upon one's first exposure to topology is the following: while open sets are complements of closed sets, being open in a topological space is not the opposite of being closed. In particular, it is possible for a set to be both open and closed, and it is also possible for a set to be neither open nor closed, as can already be seen in Examples 11.1 and 11.4.

Importantly, once we have described a topology on a set X, there is a natural way to induce topologies on all of the subsets of X, as we now describe.

11.5 DEFINITION *Subspace topology*

Given a topological space Y with closed sets $\mathcal{C}_Y$ and a subset $X \subseteq Y$, the *subspace topology on X (induced from $\mathcal{C}_Y$)* is the topology whose closed sets $\mathcal{C}_X$ are restrictions of closed sets in Y:

$$\mathcal{C}_X = \{Z \cap X \mid Z \in \mathcal{C}_Y\}.$$

The reader is encouraged to verify that the subspace topology is, in fact, a topology (Exercise 11.1.3), and that the open sets in the subspace topology on X are restrictions of open sets in Y (Exercise 11.1.4). Furthermore, it is useful to know that the subspace topology construction is transitive on inclusions (Exercise 11.1.5).

Having already familiarized ourselves with the Zariski topology on projective space, we thus obtain a topology on every subset of projective space.

11.6 DEFINITION *Zariski topology on subsets of projective space*

Let $X \subseteq \mathbb{P}^n$ be a subset. The *Zariski topology on X* is the subspace topology induced from the Zariski topology on $\mathbb{P}^n$. More explicitly, we say $Y \subseteq X$ is *Zariski-closed in X* if

$$Y = Z \cap X$$

for some projective variety $Z \subseteq \mathbb{P}^n$.

Note, here, that the qualifier "in X" is crucial; for instance, if $X \subseteq \mathbb{P}^n$ is not itself a projective variety, then $Y = X$ is Zariski-closed in X (because $Y = \mathbb{P}^n \cap X$ and $\mathbb{P}^n$ is a projective variety) but it is not Zariski-closed in $\mathbb{P}^n$.

To gain a better understanding of the Zariski topology, we look to some familiar examples of subsets of projective space. The next two examples present an explicit description of the closed sets of the Zariski topology on projective and affine varieties. As we shall see, the Zariski-closed subsets of an affine or projective variety are the subsets that are, themselves, affine or projective varieties, respectively.

11.7 EXAMPLE Zariski topology on projective varieties

Let $X \subseteq \mathbb{P}^n$ be a projective variety. By definition, a subset $Y \subseteq X$ is Zariski-closed in X if $Y = Z \cap X$ for some projective variety $Z \subseteq \mathbb{P}^n$. Since intersections of projective varieties are, themselves, projective varieties, this implies that *the Zariski-closed sets of a projective variety $X \subseteq \mathbb{P}^n$ are the subsets $Y \subseteq X$ such that $Y \subseteq \mathbb{P}^n$ is, itself, a projective variety.*

11.1. QUASIPROJECTIVE VARIETIES

11.8 EXAMPLE Zariski topology on affine varieties

Given our definitions, the Zariski topology on affine varieties is subtle, simply because we first need to specify a way to view an affine variety as a subset of projective space. Fortunately, we have a standard way to do this using affine patches.

Let us begin by discussing the subspace topology on the affine patch $\mathbb{A}_0^n \subseteq \mathbb{P}^n$. By definition, a set $X \subseteq \mathbb{A}_0^n$ is Zariski-closed in $\mathbb{A}_0^n$ if $X = Z \cap \mathbb{A}_0^n$ for some projective variety $Z \subseteq \mathbb{P}^n$; in other words, the closed sets in $\mathbb{A}_0^n$ are affine restrictions of projective varieties. We saw in Section 9.4 that affine restrictions of projective varieties are affine varieties, and we saw in Section 9.5 that every affine variety is the affine restriction of its projective closure. Combining these two observations, we conclude that *the Zariski-closed sets of $\mathbb{A}_0^n \subseteq \mathbb{P}^n$ are affine varieties $X \subseteq \mathbb{A}_0^n$*. Thus, after $\mathbb{A}^n$ is identified with the affine patch $\mathbb{A}_0^n \subseteq \mathbb{P}^n$ via the inclusion

$$j_0 : \mathbb{A}^n \to \mathbb{P}^n$$
$$(a_1, \ldots, a_n) \mapsto [1 : a_1 : \cdots : a_n],$$

the above argument shows that the Zariski topology on $\mathbb{A}^n$, as defined in Definition 11.6, is the same as the Zariski topology introduced in Section 2.1.

Now suppose that $X \subseteq \mathbb{A}^n$ is an affine variety (viewed as a subset of $\mathbb{P}^n$ via the inclusion j_0). Since we already know that the Zariski-closed sets in $\mathbb{A}^n$ are affine varieties, it follows (Exercise 11.1.5) that a set $Y \subseteq X$ is Zariski-closed in X precisely when $Y = Z \cap X$ for some affine variety $Z \subseteq \mathbb{A}^n$. Since intersections of affine varieties are, themselves, affine varieties, this implies that *the Zariski-closed sets of an affine variety $X \subseteq \mathbb{A}^n$ are the subsets $Y \subseteq X$ such that $Y \subseteq \mathbb{A}^n$ is, itself, an affine variety.*

We now come to the following definition, the central definition of this chapter, which will unify our study of affine and projective varieties.

11.9 DEFINITION Quasiprojective variety

A subset $X \subseteq \mathbb{P}^n$ is called a *quasiprojective variety* if $X = Z \cap U$ where $Z \subseteq \mathbb{P}^n$ is Zariski-closed and $U \subseteq \mathbb{P}^n$ is Zariski-open.

Notice that both Zariski-closed and Zariski-open sets of $\mathbb{P}^n$ are quasiprojective varieties. However, it is not hard to find examples of quasiprojective varieties that are neither open nor closed; see Exercise 11.1.9 for one such example. We also mention that there are many subsets of projective space that are not quasiprojective varieties. For example, a subset of $\mathbb{P}^1$ is a quasiprojective variety if and only if it is either finite or has a finite complement (Exercise 11.1.6), so any set $X \subseteq \mathbb{P}^1$ such that X and $\mathbb{P}^1 \setminus X$ are both infinite is not a quasiprojective variety.

The key observation we make precise at this point is that this new type of variety simultaneously generalizes both affine and projective varieties.

11.10 EXAMPLE Projective varieties are quasiprojective varieties

Let $X \subseteq \mathbb{P}^n$ be a projective variety. We can write $X = X \cap \mathbb{P}^n$. Since $X \subseteq \mathbb{P}^n$ is closed and $\mathbb{P}^n \subseteq \mathbb{P}^n$ is open, it follows that X is a quasiprojective variety.

11.11 EXAMPLE Affine varieties are quasiprojective varieties

Let $X \subseteq \mathbb{A}^n$ be an affine variety, which we view as a subset of projective space via the inclusion $j_0 : \mathbb{A}^n \to \mathbb{P}^n$. We can write X as the intersection $\overline{X} \cap \mathbb{A}_0^n$, where $\overline{X}$ is the projective closure of X. Since $\overline{X}$ is a projective variety, it is closed. Moreover, since $\mathbb{A}_0^n$ is the complement in $\mathbb{P}^n$ of the closed set $V_\mathbb{P}(x_0)$, it is open. Thus, it follows that X—or, more precisely, $j_0(X)$—is a quasiprojective variety.

While quasiprojective varieties include both affine and projective varieties as special cases, we note that there are many quasiprojective varieties that are neither affine nor projective varieties. For example, the complement of a single point in $\mathbb{A}^2$ is a quasiprojective variety, but—after developing a more robust toolkit for studying quasiprojective varieties—we will argue in Examples 11.43 and 11.75 that it is neither (isomorphic to) an affine variety nor a projective variety.

Subsets of a quasiprojective variety that are, themselves, quasiprojective varieties are particularly important, leading to the following notion.

11.12 DEFINITION Subvariety

Let $X \subseteq \mathbb{P}^n$ be a quasiprojective variety. We say that $Y \subseteq X$ is a *subvariety* of X if $Y \subseteq \mathbb{P}^n$ is, itself, a quasiprojective variety.

Subvarieties of a quasiprojective variety can also be characterized in a topological way that exactly parallels the definition of quasiprojective varieties.

11.13 PROPOSITION Subvarieties of a quasiprojective variety

Let $X \subseteq \mathbb{P}^n$ be a quasiprojective variety. A subset $Y \subseteq X$ is a subvariety of X if and only if $Y = Z \cap U$ where Z is closed in X and U is open in X.

From now on, we generally omit the word "Zariski" from "Zariski-closed" and "Zariski-open."

PROOF Assume first that $Y \subseteq X$ is a subvariety. Then $Y \subseteq \mathbb{P}^n$ is a quasiprojective variety, so it can be written as $Y = Z \cap U$ where Z is closed in $\mathbb{P}^n$ and U is open in $\mathbb{P}^n$. Since $Y \subseteq X$,
$$Y = (X \cap Z) \cap (X \cap U),$$
and by definition of the Zariski topology on X, we see that $X \cap Z$ is closed in X while $X \cap U$ is open in X, proving one direction of the if-and-only-if claim.

Conversely, assume that $Y \subseteq X$ and $Y = Z \cap U$ where Z is closed in X and U is open in X. By definition of the Zariski topology on X, we have
$$Z = X \cap Z' \quad \text{and} \quad U = X \cap U'$$
where Z' is closed in $\mathbb{P}^n$ and U' is open in $\mathbb{P}^n$. Since $X \subseteq \mathbb{P}^n$ is a quasiprojective variety, it can also be written as an intersection of an open and a closed subset of $\mathbb{P}^n$, implying that both Z and U—and, thus, $Z \cap U$—can be written as a finite intersection of open and closed sets in $\mathbb{P}^n$. Since finite intersections of open sets are open and finite intersections of closed sets are closed, this implies that $Y = Z \cap U$ is a quasiprojective variety in $\mathbb{P}^n$, and thus, a subvariety of X. □

11.1. QUASIPROJECTIVE VARIETIES

Having now defined the notion of quasiprojective varieties, in the next section we investigate irreducibility of quasiprojective varieties, which we define topologically, naturally generalizing the notion of irreducibility from the affine and projective settings. The culmination of our discussion of irreducibility will be the proof that, exactly as in the setting of affine and projective varieties, quasiprojective varieties admit unique irreducible decompositions.

Exercises for Section 11.1

11.1.1 Compare the Euclidean and Zariski topologies on $\mathbb{R}$. In other words, investigate whether open sets in the Euclidean topology are open in the Zariski topology, and vice versa.

11.1.2 Every set X admits several somewhat simple topologies. Prove that the following collections of subsets define topologies on any set X.
(a) Indiscrete topology on X: $\mathcal{C} = \{\emptyset, X\}$.
(b) Discrete topology on X: $\mathcal{C}$ contains all subsets of X.
(c) Cofinite topology on X: $\mathcal{C}$ contains X and all finite subsets of X.

11.1.3 Prove that the collection of sets $\mathcal{C}_X$ in Definition 11.5 satisfies the topology axioms.

11.1.4 Let Y be a topological space and $X \subseteq Y$. Prove that the open sets in the subspace topology on X are $U \cap X$ for U an open subset of Y.

11.1.5 Let Z be a topological space and let $X, Y \subseteq Z$ be subsets such that $X \subseteq Y$. Prove that the subspace topology that X inherits from Z is the same as the subspace topology that X inherits from the subspace topology on Y.

11.1.6 Prove that a subset $X \subseteq \mathbb{P}^1$ is a quasiprojective variety if and only if X is finite or $\mathbb{P}^1 \setminus X$ is finite.

11.1.7 Let $X \subseteq \mathbb{P}^n$ be a subset. Prove that the following are equivalent.
(i) $X \subseteq \mathbb{P}^n$ is a quasiprojective variety.
(ii) $X = Y \setminus Z$ where $Y, Z \subseteq \mathbb{P}^n$ are both closed.
(iii) X is an open subset of a projective variety $Y \subseteq \mathbb{P}^n$.

11.1.8 Let $X, Y \subseteq \mathbb{P}^n$ be quasiprojective varieties. For each of the following two assertions, either prove it or give a counterexample.
(a) $X \cap Y$ is a quasiprojective variety.
(b) $X \cup Y$ is a quasiprojective variety.

11.1.9 Prove that the set
$$X = \{[a:b:c] \in \mathbb{P}^2 \mid a \neq 0 \text{ and } b = 0\} \subseteq \mathbb{P}^2$$
is a quasiprojective variety that is neither open nor closed in $\mathbb{P}^2$.

Section 11.2 Closures and irreducibility

The primary aim of this section is to discuss irreducibility and irreducible decompositions of quasiprojective varieties. Given our new understanding of the Zariski topology, we present these ideas in a topological framework. We begin with the general notion of what it means for a topological space to be irreducible.

11.14 DEFINITION *Reducible and irreducible topological spaces*

A topological space X is *reducible* if $X = X_1 \cup X_2$ for some closed subsets $X_1, X_2 \subsetneq X$, and X is *irreducible* if it is neither empty nor reducible.

Given our description of the Zariski-closed sets on affine and projective varieties in Examples 11.7 and 11.8, it follows that Definition 11.14—when applied to the Zariski topology on affine and projective varieties—coincides with our earlier definitions of irreducibility of affine and projective varieties. Thus, Definition 11.14 allows us to extend those earlier notions to the more general context of quasiprojective varieties: a quasiprojective variety is *irreducible* if it is irreducible with respect to its Zariski topology.

We note that irreducibility is quite sensitive to one's choice of topology. For example, $\mathbb{R} = \mathbb{A}^1_{\mathbb{R}}$ is irreducible in the Zariski topology because it is impossible to find affine varieties $X_1, X_2 \subsetneq \mathbb{A}^1_{\mathbb{R}}$ such that $X_1 \cup X_2 = \mathbb{A}^1_{\mathbb{R}}$, but $\mathbb{R}$ is reducible in the Euclidean topology because $(-\infty, 0]$ and $[0, \infty)$ are proper closed subsets whose union is $\mathbb{R}$.

The culmination of this section will be a proof that every quasiprojective variety can be written uniquely as a union of closed irreducible subvarieties. Our proof of this result hinges on the existence and uniqueness of irreducible decompositions in the projective setting (Proposition/Definition 9.32). As a tool for translating between irreducible decompositions in the quasiprojective and projective settings, we now introduce the general topological notions of closures and denseness.

11.15 DEFINITION *Closures and dense subsets*

Let Y be a topological space and $X \subseteq Y$ a subset. The *closure of X in Y*, denoted $\mathrm{cl}_Y(X)$, is the intersection of all closed sets in Y that contain X. We say that X is *dense in Y* if $\mathrm{cl}_Y(X) = Y$.

When the ambient topological space Y is clear from context, we often use the shorthand $\overline{X} = \mathrm{cl}_Y(X)$.

Note that the closure of X in Y is a closed subset of Y—simply because arbitrary intersections of closed sets are closed—and it is contained within every closed subset of Y that contains X. In other words, $\mathrm{cl}_Y(X)$ *is the smallest closed subset of Y containing X*. Unraveling definitions, one also finds that X is dense in Y if and only if every open subset of Y contains at least one element of X (Exercise 11.2.2). In other words, *a set is dense in Y if it is spread out throughout Y, appearing in every neighborhood*. For example, since every open interval of real numbers, no matter how small, contains rational numbers, it follows that $\mathbb{Q}$ is dense in $\mathbb{R}$ with respect to the Euclidean topology.

11.2. CLOSURES AND IRREDUCIBILITY

Let us turn to examples of closures and denseness in the Zariski topology.

11.16 EXAMPLE Closures of sets in $\mathbb{A}^1$

If X is a finite set in $\mathbb{A}^1$, then it is an affine variety and thus closed in $\mathbb{A}^1$, implying that $\overline{X} = X$. If X is infinite, on the other hand, then the only closed set containing X is all of $\mathbb{A}^1$. Thus, it follows that every infinite subset of $\mathbb{A}^1$ is dense in $\mathbb{A}^1$.

11.17 EXAMPLE Projective closures

If $X \subseteq \mathbb{P}^n$ is any subset of projective space, then the closure $\overline{X} \subseteq \mathbb{P}^n$ is called the *projective closure of* X. The specific situation where $X \subseteq \mathbb{A}_0^n$ is an affine variety was discussed at length in Section 9.5. For a general subset $X \subseteq \mathbb{P}^n$, one key fact about projective closures that will be useful in what follows is that $\mathcal{I}_\mathbb{P}(X) = \mathcal{I}_\mathbb{P}(\overline{X})$, as we now justify.

The inclusion $\mathcal{I}_\mathbb{P}(\overline{X}) \subseteq \mathcal{I}_\mathbb{P}(X)$ is simply because any polynomial vanishing on $\overline{X}$ must also vanish on the smaller set X. To prove the other inclusion, it is enough to note that if a polynomial f does not vanish on $\overline{X}$, it cannot vanish on X: otherwise $\overline{X} \cap \mathcal{V}_\mathbb{P}(f)$ would be a closed set with

$$X \subseteq \overline{X} \cap \mathcal{V}_\mathbb{P}(f) \subsetneq \overline{X},$$

contradicting that $\overline{X}$ is the closure of X.

From $\mathcal{I}_\mathbb{P}(X) = \mathcal{I}_\mathbb{P}(\overline{X})$, it follows that the projective closure of X can be expressed as

$$\overline{X} = \mathcal{V}_\mathbb{P}(\mathcal{I}_\mathbb{P}(X)).$$

11.18 EXAMPLE Affine patches are dense in projective space

Consider the affine patch $\mathbb{A}_0^n \subseteq \mathbb{P}^n$. We claim that the projective closure of $\mathbb{A}_0^n$ is $\mathbb{P}^n$. To justify this, we first observe that

$$\mathbb{P}^n = \mathbb{A}_0^n \cup \mathcal{V}_\mathbb{P}(x_0) = \overline{\mathbb{A}_0^n} \cup \mathcal{V}_\mathbb{P}(x_0).$$

Since $\mathbb{P}^n$ is irreducible and both $\overline{\mathbb{A}_0^n}$ and $\mathcal{V}_\mathbb{P}(x_0)$ are closed, it follows that one of them must be equal to $\mathbb{P}^n$. Knowing that $\mathcal{V}_\mathbb{P}(x_0) \neq \mathbb{P}^n$—for example, $\mathcal{V}_\mathbb{P}(x_0)$ does not contain $[1:0:\cdots:0]$—we conclude that $\overline{\mathbb{A}_0^n} = \mathbb{P}^n$, so $\mathbb{A}_0^n$ is dense in $\mathbb{P}^n$.

Generalizing the argument in Example 11.18 to the topological setting, it turns out that every nonempty open set of an irreducible topological space is dense.

11.19 PROPOSITION *Nonempty open sets are dense*

Let X be an irreducible topological space. If $U \subseteq X$ is a nonempty open set, then U is dense in X.

PROOF Let U be a nonempty open subset of an irreducible topological space X. Define $Z = X \setminus U$, which is a closed subset of X, and notice that $X = \overline{U} \cup Z$. By the irreducibility of X, we have $X = \overline{U}$ or $X = Z$, but $X \neq Z$ because Z is the complement of the nonempty set $U \subseteq X$. It follows that $X = \overline{U}$, and we conclude that U is dense in X. □

A useful property of irreducibility in the topological setting is that it can be detected on dense subsets, as the next result verifies.

11.20 PROPOSITION *Denseness and irreducibility*

Let Y be a topological space with a dense subset $X \subseteq Y$. Then X is irreducible in the subspace topology if and only if Y is irreducible.

PROOF Suppose that Y is a topological space and that $X \subseteq Y$ is dense. We prove both directions of the if-and-only-if statement.

Suppose that X is irreducible. To prove that Y is irreducible, let $Y = Y_1 \cup Y_2$ where $Y_1, Y_2 \subseteq Y$ are closed; we must prove that $Y_1 = Y$ or $Y_2 = Y$. Define $X_1 = X \cap Y_1$ and $X_2 = X \cap Y_2$. Notice that X_1 and X_2 are both closed in the subspace topology on X, and $X = X_1 \cup X_2$. Therefore, by the irreducibility of X, we have $X = X_1$ or $X = X_2$. This implies that $X \subseteq Y_1$ or $X \subseteq Y_2$, further implying—as each Y_i is closed in Y—that $\overline{X} \subseteq Y_1$ or $\overline{X} \subseteq Y_2$. But X is dense in Y, so $\overline{X} = Y$, and we conclude that $Y = Y_1$ or $Y = Y_2$, as desired.

Conversely, suppose that Y is irreducible. To prove that X is irreducible, let $X = X_1 \cup X_2$ where $X_1, X_2 \subseteq X$ are closed; we must prove that $X_1 = X$ or $X_2 = X$. Since closures commute with finite unions (Exercise 11.2.3), we have $Y = \overline{X} = \overline{X_1} \cup \overline{X_2}$. As $\overline{X_1}$ and $\overline{X_2}$ are both closed in Y, the irreducibility of Y then implies that $Y = \overline{X_1}$ or $Y = \overline{X_2}$; without loss of generality assume that $Y = \overline{X_1}$. Since X_1 is closed in X, Exercise 11.2.4 implies that $X_1 = \overline{X_1} \cap X$, from which we see that $X_1 = Y \cap X$, implying that $X_1 = X$, as desired. $\square$

Finally, we come to the main result of this section.

11.21 PROPOSITION/DEFINITION *Irreducible decomposition*

Let $X \subseteq \mathbb{P}^n$ be a nonempty quasiprojective variety. Then there exist closed irreducible subvarieties $X_1, \ldots, X_r \subseteq X$ such that $X_i \not\subseteq X_j$ for any $i \neq j$ and

(11.22) $$X = \bigcup_{i=1}^{r} X_i.$$

Moreover, the closed irreducible subvarieties $X_1, \ldots, X_r$ are unique up to reordering; we call these the *irreducible components* of X, and refer to (11.22) as the *irreducible decomposition* of X.

PROOF Let $X \subseteq \mathbb{P}^n$ be a nonempty quasiprojective variety, which we can write as $X = Z \cap U$ where $Z \subseteq \mathbb{P}^n$ is closed and $U \subseteq \mathbb{P}^n$ is open. We prove existence and uniqueness of irreducible decompositions separately.

(Existence) Consider the closure $\overline{X} \subseteq \mathbb{P}^n$. By the existence of irreducible decompositions of projective varieties (Proposition/Definition 9.32), there exist irreducible projective varieties $Z_1, \ldots, Z_r \in \mathbb{P}^n$ such that $Z_i \not\subseteq Z_j$ for any $i \neq j$ and

$$\overline{X} = \bigcup_{i=1}^{r} Z_i.$$

11.2. CLOSURES AND IRREDUCIBILITY

If $Z_i \cap X = \emptyset$ for some i, then the union of the other irreducible components would be a proper closed subset of $\overline{X}$ that contains X, a contradiction. Thus, $Z_i \cap X \neq \emptyset$ for all i, and we define nonempty closed subvarieties of X by $X_i = Z_i \cap X$. By construction, we have

$$X = \bigcup_{i=1}^{r} X_i,$$

and it remains to show that each X_i is irreducible and that $X_i \not\subseteq X_j$ for $i \neq j$.

Since $X = Z \cap U$ and Z is closed, it follows that $\overline{X} \subseteq Z$, and thus $Z_i \subseteq Z$ for every i. Given that $X_i = Z_i \cap X = Z_i \cap (Z \cap U)$, we then see that $X_i = Z_i \cap U$. As U is open, it follows that X_i is a nonempty open subvariety of the irreducible variety Z_i, and therefore dense by Proposition 11.19. By Proposition 11.20, the irreducibility of X_i follows from the irreducibility of Z_i. Moreover, if $X_i \subseteq X_j$ for some $i \neq j$, then taking closures would imply that $Z_i \subseteq Z_j$, which cannot happen as these are distinct irreducible components of $\overline{X}$. Thus, we have shown that irreducible decompositions of quasiprojective varieties exist.

(Uniqueness) Suppose that $X_1, \ldots, X_r$ and $X'_1, \ldots, X'_s$ are two irreducible decompositions of X. Since closures commute with finite unions (Exercise 11.2.3), we obtain two decompositions of the projective variety $\overline{X} \subseteq \mathbb{P}^n$:

(11.23) $$\overline{X} = \bigcup_{i=1}^{r} \overline{X_i} = \bigcup_{i=1}^{s} \overline{X'_i},$$

where the closures are all being taken within $\mathbb{P}^n$. We now argue that each of the decompositions in (11.23) is an irreducible decomposition by showing that each component is irreducible and that none of them is contained in another.

For every i, notice that $\overline{X_i}$ contains the irreducible variety X_i as a dense subset, so Proposition 11.20 implies that each $\overline{X_i}$ is irreducible, and similarly, each $\overline{X'_i}$ is irreducible. Moreover, since X_i is closed in X, it follows that $X_i = \overline{X_i} \cap X$ for every i (Exercise 11.2.4), so it cannot be the case that $\overline{X_i} \subseteq \overline{X_j}$ for some $i \neq j$, as this would imply that $X_i \subseteq X_j$, contradicting the assumption that $X_1, \ldots, X_r$ is an irreducible decomposition of X. Arguing similarly for $\overline{X'_i}$ and $\overline{X'_j}$, we conclude that both decompositions in (11.23) are indeed irreducible decompositions of the projective variety $\overline{X} \subseteq \mathbb{P}^n$.

By uniqueness of irreducible decompositions of projective varieties (Proposition/Definition 9.32), we have that $r = s$ and, after possibly reordering, $\overline{X_i} = \overline{X'_i}$ for all i. Using again the fact that $X_i = \overline{X_i} \cap X$ and $X'_i = \overline{X'_i} \cap X$ (Exercise 11.2.4), we see that $X_i = X'_i$ for all i. Thus, we conclude that irreducible decompositions of quasiprojective varieties are unique. $\square$

Exercises for Section 11.2

11.2.1 Let X be a topological space and $X \subseteq Y$ a subset. Prove that X is closed if and only if $X = \overline{X}$.

11.2.2 Let Y be a topological space and $X \subseteq Y$ a subset. Prove that X is dense in Y if and only if every open set of Y contains at least one element of X.

11.2.3 (a) Let X be a topological space and let $Y_1, \ldots, Y_k \subseteq X$ be subsets. Prove that
$$\overline{Y_1 \cup \cdots \cup Y_k} = \overline{Y_1} \cup \cdots \cup \overline{Y_k}.$$

(b) Can the result in (a) be extended to infinite unions? Does at least one of the inclusions continue to hold? Prove or give a counterexample to justify your answers to these questions.

11.2.4 Let Y be a topological space and let $X \subseteq Y$ be a subset with the subspace topology. For any subset $Z \subseteq X$, prove that $\mathrm{cl}_X(Z) = \mathrm{cl}_Y(Z) \cap X$.

11.2.5 Let $X \subseteq \mathbb{P}^n$ be a quasiprojective variety. Prove that X is an open subset of its projective closure $\overline{X}$.

11.2.6 Let $X \subseteq \mathbb{A}^n$ be a subset with closure $\overline{X} = \mathrm{cl}_{\mathbb{A}^n}(X)$.
(a) Prove that $\mathcal{I}_\mathbb{A}(X) = \mathcal{I}_\mathbb{A}(\overline{X})$.
(b) Deduce that, if $f, g \in K[x_1, \ldots, x_n]$ and $f|_X = g|_X$ as functions on X, then $f|_{\overline{X}} = g|_{\overline{X}}$.

11.2.7 Let X be a quasiprojective variety with irreducible components $X_1, \ldots, X_r$. If $Y \subseteq X$ is a dense open subvariety, prove that $X_1 \cap Y, \ldots, X_r \cap Y$ are the irreducible components of Y.

11.2.8 Prove that X is an irreducible topological space if and only if the intersection of any pair of nonempty open subsets of X is a nonempty open subset of X.

Section 11.3 Regular maps and regular functions

Given that the notion of a quasiprojective variety is a simultaneous generalization of affine and projective varieties, we would like to develop a notion of maps between quasiprojective varieties that specializes in these two settings to our existing notions of polynomial maps (in the affine case) and regular maps (in the projective case). In fact, the correct definition turns out to carry over verbatim from the projective setting, so we continue to refer to it by the same name.

11.24 DEFINITION *Regular map of quasiprojective varieties*

Let $X \subseteq \mathbb{P}^m$ and $Y \subseteq \mathbb{P}^n$ be quasiprojective varieties. A map $F : X \to Y$ is said to be a *regular map* if, for every $p \in X$, there exist polynomials $f_0, \ldots, f_n \in K[x_0, \ldots, x_m]$, all homogeneous of the same degree, such that $p \notin \mathcal{V}(f_0, \ldots, f_n)$ and

$$F(a) = [f_0(a) : \cdots : f_n(a)] \quad \text{for all} \quad a \in X \setminus \mathcal{V}(f_0, \ldots, f_n).$$

In particular, not only can every projective variety be viewed as a quasiprojective variety, but also, the structure-preserving maps between a pair of projective varieties are precisely the same whether one regards them as projective or quasiprojective varieties.

> *In category-theoretic language, this means that the category of projective varieties embeds into the category of quasiprojective varieties.*

It is a significantly more difficult fact—which we relegate to the next section—to draw the same parallel in the affine case: that is, to show that polynomial maps between a pair of affine varieties are the exact same as regular maps between their canonically associated quasiprojective varieties. For now, we simply consider a concrete example of this phenomenon.

11.25 EXAMPLE A regular map of affine patches

Consider the quasiprojective varieties

$$X = \{[a_0 : a_1 : a_2] \in \mathbb{P}^2 \mid a_0 \neq 0\} \quad \text{and} \quad Y = \{[b_0 : b_1] \in \mathbb{P}^1 \mid b_0 \neq 0\},$$

which are the affine patches $\mathbb{A}_0^2$ and $\mathbb{A}_0^1$, respectively. Then

$$F([a_0 : a_1 : a_2]) = [a_0^2 : a_0 a_2 + a_1^2]$$

gives a regular map $F : X \to Y$ because its coordinate functions are polynomials. Under the identification of Y with $\mathbb{A}^1$, this is the map

$$[a_0 : a_1 : a_2] \mapsto \frac{a_0 a_2 + a_1^2}{a_0^2},$$

which appears to be rational, but not polynomial. However, under the further identification of X with $\mathbb{A}^2$, it becomes the map $(a_1, a_2) \mapsto a_2 + a_1^2$, which is indeed a polynomial map between the affine varieties $\mathbb{A}^2$ and $\mathbb{A}^1$.

The difficulty in proving that every regular map between affine varieties is the same as a polynomial map is the possibility that, when one views a pair of affine varieties as quasiprojective, a regular map between them may be only locally described by polynomials, whereas a polynomial map of affine varieties must be given by polynomials globally. The fact that local polynomiality implies global polynomiality for affine varieties is the content of Theorem 11.36 in the next section.

An important special case of regular maps between quasiprojective varieties is the setting in which the target is the quasiprojective variety $\mathbb{A}^1 = K$; we can view such a map as a regular *function* on a quasiprojective variety. Indeed, even if one's primary interest is projective varieties, this gives an important motivation for introducing quasiprojective varieties into the narrative: until now, we have not had a notion of regular functions in the projective setting, but now we can define this notion as a special case of regular maps.

More specifically, consider the specialization of Definition 11.24 to the case where $Y = j_0(\mathbb{A}^1)$. After unwinding the identification of $j_0(\mathbb{A}^1)$ with $\mathbb{A}^1 = K$ via

$$[a_0 : a_1] \in j_0(\mathbb{A}^1) \longleftrightarrow \frac{a_1}{a_0} \in K,$$

we arrive at the following definition.

11.26 DEFINITION *Regular function on a quasiprojective variety*

Let $X \subseteq \mathbb{P}^m$ be a quasiprojective variety. A function $F : X \to K$ is a *regular function* if, for every $p \in X$, there exist polynomials $f, g \in K[x_0, \ldots, x_m]$, homogeneous of the same degree, such that $p \notin \mathcal{V}(g)$ and

$$F(a) = \frac{f(a)}{g(a)} \quad \text{for all} \quad a \in X \setminus \mathcal{V}(g).$$

In other words, a regular function $F : X \to K$ is a function that can be locally described by rational functions (not necessarily polynomials). The next example illustrates a regular function with a description that holds globally.

11.27 EXAMPLE A regular function on $\mathbb{P}^1 \setminus \mathcal{V}(xy)$

Consider the quasiprojective variety $X = \mathbb{P}^1 \setminus \mathcal{V}(xy)$. Notice that X is the complement in $\mathbb{P}^1$ of the two points $[1:0]$ and $[0:1]$. Then

$$F([a:b]) = \frac{a^2 + b^2}{ab}$$

is an example of a regular function on X, simply by virtue of being a quotient of polynomials of the same degree where the denominator never vanishes on X.

Sums, products, and scalar multiples of regular functions are regular (Exercise 11.3.2), and the set of regular functions on a quasiprojective variety naturally forms a K-algebra. We denote this K-algebra with the same notation that we used for the coordinate ring in the affine setting.

11.3. REGULAR MAPS AND REGULAR FUNCTIONS

11.28 DEFINITION *Ring of regular functions*

Let X be a quasiprojective variety. The *ring of regular functions* on X, denoted $K[X]$, is the K-algebra of all regular functions $F : X \to K$.

We warn the reader that $K[X]$ now has a double meaning for affine varieties: it could be its coordinate ring or its ring of regular functions. This tension will be resolved in the next section, where we prove that these two rings are isomorphic. As in the case of regular maps, the main subtlety is that Definition 11.26 is local, allowing for the possibility that a regular function F may be expressed as a different quotient of polynomials f/g around different points $p \in X$. In fact, though, there is a nice class of quasiprojective varieties for which this local-versus-global subtlety does not arise, a class for which we can compute explicit descriptions of rings of regular functions in terms of quotients defined globally.

11.29 LEMMA *Regular functions on open subsets of $\mathbb{P}^m$*

Let $X \subseteq \mathbb{P}^m$ be open. Given any regular function $F : X \to K$, there exist polynomials $f, g \in K[x_0, \ldots, x_m]$ that are homogeneous of the same degree such that $X \cap \mathcal{V}(g) = \emptyset$ and

$$F(a) = \frac{f(a)}{g(a)} \quad \text{for all } a \in X.$$

PROOF Let $p \in X$. By definition of regular functions, there exist polynomials $f, g \in K[x_0, \ldots, x_m]$, homogeneous of the same degree, such that $g(p) \neq 0$ and

$$F(a) = \frac{f(a)}{g(a)} \quad \text{for all} \quad a \in X \setminus \mathcal{V}(g).$$

After canceling factors if necessary (see Exercise 11.3.4), we may assume that the polynomials f and g have no common factors. We aim to prove that $\mathcal{V}(g) \cap X = \emptyset$, so that the quotient description above for F is valid on all of X. To do so, it suffices to prove that g does not have any irreducible factors that vanish at a point of X.

Toward a contradiction, suppose that h is an irreducible factor of g that vanishes at a point $q \in X$. By definition of regular functions, there exist homogeneous polynomials $f_q, g_q \in K[x_0, \ldots, x_m]$ of the same degree such that $g_q(q) \neq 0$ and

$$F(a) = \frac{f_c(a)}{g_c(a)} \quad \text{for all} \quad a \in X \setminus \mathcal{V}(g_q).$$

Notice that $X \setminus \mathcal{V}(g)$ and $X \setminus \mathcal{V}(g_q)$ are both nonempty open sets in $\mathbb{P}^m$, so Exercise 11.2.8 implies that their intersection $X \setminus (\mathcal{V}(g) \cup \mathcal{V}(g_q))$ is also a nonempty open set in $\mathbb{P}^m$. Moreover, for any $a \in X \setminus (\mathcal{V}(g) \cup \mathcal{V}(g_q))$, we have

$$\frac{f(a)}{g(a)} = \frac{f_q(a)}{g_q(a)}.$$

It follows that $fg_q - f_qg \in \mathcal{I}(X \setminus (\mathcal{V}(g) \cup \mathcal{V}(g_q)))$. Since $X \setminus (\mathcal{V}(g) \cup \mathcal{V}(g_q))$ is nonempty and open in $\mathbb{P}^m$, Proposition 11.19 implies that it is dense, and Example 11.17 then implies that

$$\mathcal{I}(X \setminus (\mathcal{V}(g) \cup \mathcal{V}(g_q))) = \mathcal{I}(\mathbb{P}^m) = \{0\}.$$

Therefore, we find that $fg_q - f_qg = 0$. Since $g(q) = 0$ by assumption, whereas $g_q(q) \neq 0$, we must have $f(q) = 0$.

So far, we have shown that $h(q) = 0$ implies $f(q) = 0$, and it follows that $\mathcal{V}(h) \cap X \subseteq \mathcal{V}(f)$. Since $\mathcal{V}(h) \cap X$ is a nonempty open set of the irreducible variety $\mathcal{V}(h)$, it is dense, and it then follows that $\mathcal{V}(h) \subseteq \mathcal{V}(f)$. Since h was assumed to be irreducible, the Nullstellensatz implies that h divides f, contradicting the assumption that f and g have no common factors. Having reached a contradiction, we conclude that the expression $F = f/g$ is the requisite globally-defined ratio. □

Let us consider a few explicit applications of the previous lemma.

11.30 EXAMPLE   The ring of regular functions of $\mathbb{P}^m$

Consider $X = \mathbb{P}^m$. Then Lemma 11.29 implies that a regular function F on X can be expressed as $F = f/g$ where g is nowhere vanishing on $\mathbb{P}^m$. This is only possible if g is constant, and since f and g are homogeneous of the same degree, it follows that f is constant, as well. In other words, every regular function on $\mathbb{P}^m$ is a constant function, so

$$K[\mathbb{P}^m] = K.$$

In fact, this is a special case of a general phenomenon: there are no nonconstant regular functions on any projective variety. The proof of the general case is significantly harder, however, and will take us until the end of the chapter to complete.

11.31 EXAMPLE   The ring of regular functions on $\mathbb{P}^1 \setminus \mathcal{V}(xy)$

As in Example 11.27, let

$$X = \mathbb{P}^1 \setminus \mathcal{V}(xy).$$

Let F be a regular function on X, so by Lemma 11.29, we can express $F = f/g$ where f and g are homogeneous polynomials of the same degree such that g vanishes only if $x = 0$ or $y = 0$. This implies that g is of the form $x^j y^k$ for some $j, k \geq 0$. From here, setting $y = 1$ in f/g yields a polynomial in x and $1/x$. For example, for F as in Example 11.27, we have

$$F(x, 1) = \frac{x^2 + 1}{x} = x + \frac{1}{x} \in K[x, x^{-1}].$$

This dehomogenization process gives a function $K[X] \to K[x, x^{-1}]$. In Exercise 11.3.5, the reader is encouraged to prove that this function is a bijective homomorphism, thereby proving that

> Recall that $K[x, x^{-1}]$ denotes the subalgebra of $K(x)$ generated by x and x^{-1}.

$$K[X] \cong K[x, x^{-1}].$$

11.3. REGULAR MAPS AND REGULAR FUNCTIONS

We conclude this section with a brief discussion of isomorphisms in the quasiprojective setting. The definition is most likely just as the reader would expect.

11.32 DEFINITION *Isomorphism of quasiprojective varieties*

An *isomorphism* of quasiprojective varieties $X \subseteq \mathbb{P}^m$ and $Y \subseteq \mathbb{P}^n$ is a regular map $F : X \to Y$ for which there exists a regular map $G : Y \to X$ that is inverse to F. If there exists an isomorphism between X and Y, we say that X and Y are *isomorphic* and write $X \cong Y$.

11.33 EXAMPLE An isomorphism of quasiprojective varieties

As in Examples 11.27 and 11.31, consider $X = \mathbb{P}^1 \setminus \mathcal{V}(xy)$, and let Y be the quasiprojective variety

$$Y = \mathcal{V}(w^2 - xy) \setminus \mathcal{V}(w) \subseteq \mathbb{P}^2.$$

Then the regular map $F : X \to Y$ given by

$$F([a : b]) = [ab : a^2 : b^2]$$

is an isomorphism because $G([a : b : c]) = [a : c]$ is a regular inverse. We encourage the reader to draw a picture of this isomorphism over $\mathbb{R}$.

As one might hope, rings of regular functions are preserved by isomorphisms. In order to prove this, we require the quasiprojective analogue of a notion we met in the theory of affine varieties: the pullback homomorphism.

11.34 DEFINITION *Pullback homomorphism*

Let X and Y be quasiprojective varieties, and let $F : X \to Y$ be a regular map. The *pullback homomorphism* induced by F is

$$F^* : K[Y] \to K[X]$$
$$F^*(G) = G \circ F.$$

The reader should pause to convince themselves that F^* indeed takes regular functions on Y to regular functions on X (Exercise 11.3.7). Moreover, the basic properties of pullbacks in the affine setting described in Section 4.2 carry over verbatim to the quasiprojective setting. Thus, the same proof as in the affine case gives the following isomorphism-invariance of rings of regular functions.

11.35 PROPOSITION $X \cong Y$ *implies* $K[X] \cong K[Y]$

If $F : X \to Y$ is an isomorphism of quasiprojective varieties, then the pullback homomorphism $F^* : K[Y] \to K[X]$ is an isomorphism of K-algebras.

PROOF Exercise 11.3.8. □

> Unlike the affine setting, the converse of Proposition 11.35 is false: as we saw in Example 11.30, $K[\mathbb{P}^m] = K$ for any m even though $\mathbb{P}^m \not\cong \mathbb{P}^n$ for $m \neq n$.

Having now familiarized ourselves with quasiprojective varieties and the structure-preserving regular maps between them, we are ready to revisit the affine setting, placing the theory of affine varieties and polynomial maps solidly within the context of quasiprojective varieties and regular maps. This is the subject of the next section.

Exercises for Section 11.3

11.3.1 (a) Prove that compositions of regular maps are regular maps.
(b) Prove that isomorphism is an equivalence relation on the set of quasiprojective varieties.

11.3.2 Let $X \subseteq \mathbb{P}^n$ be a quasiprojective variety and let $F, G : X \to K$ be regular functions and $\lambda \in K$. Prove that the sum $F + G : X \to K$, the product $FG : X \to K$, and the scalar multiple $\lambda F : X \to K$ are all regular functions, and verify that these operations satisfy the K-algebra axioms.

11.3.3 Prove that the ring of regular functions on a quasiprojective variety is reduced.

11.3.4 Let $X \subseteq \mathbb{P}^n$ be an irreducible quasiprojective variety and consider homogeneous polynomials $f, g, h \in K[x_0, \ldots, x_n]$ such that f and g have the same degree. Let $F : X \to K$ be a regular function. If $\mathcal{V}(gh) \neq X$ and

$$F(a) = \frac{f(a)h(a)}{g(a)h(a)} \quad \text{for every} \quad a \in X \setminus \mathcal{V}(gh),$$

prove that

$$F(a) = \frac{f(a)}{g(a)} \quad \text{for every} \quad a \in X \setminus \mathcal{V}(g).$$

11.3.5 Let $X = \mathbb{P}^1 \setminus \mathcal{V}(xy)$. Prove that $K[X] \cong K[x, x^{-1}]$.

11.3.6 Using Lemma 11.29, prove that the ring of regular functions on $\mathbb{A}^n$, viewed as the affine patch $\mathbb{A}_0^n \subseteq \mathbb{P}^n$, is $K[x_1, \ldots, x_n]$.

11.3.7 Let $X \subseteq \mathbb{P}^m$ and $Y \subseteq \mathbb{P}^n$ be quasiprojective varieties, and let $F : X \to Y$ be a regular map.
(a) For any $G \in K[Y]$, prove that $G \circ F \in K[X]$.
(b) Prove that $F^* : K[Y] \to K[X]$ is a K-algebra homomorphism.

11.3.8 Prove Proposition 11.35 by verifying that the relevant properties of pullbacks in Section 4.2 extend to regular maps and regular functions of quasiprojective varieties.

11.3.9 Let $Y \subseteq \mathbb{P}^n$ be a quasiprojective variety. Prove that every regular map $F : \mathbb{P}^m \to Y$ is a polynomial map.

Section 11.4 Affine varieties revisited

The primary goal of this section is to prove that, if $X_0 \subseteq \mathbb{A}^n$ is an affine variety, then the polynomial functions on X_0 are the same as the regular functions on the canonically associated quasiprojective variety $j_0(X_0)$. The key difficulty in proving this assertion is that a regular function need only be given locally by a ratio of polynomials, whereas a polynomial function must have a global expression as the restriction of a polynomial. While polynomial functions seem to be much more restrictive than regular functions, the next result shows that these two seemingly different descriptions lead to the exact same set of functions on an affine variety.

> **11.36 THEOREM** *Regular functions on affine varieties are polynomial*
>
> Let $X_0 \subseteq \mathbb{A}^n$ be an affine variety and let $X = j_0(X_0) \subseteq \mathbb{P}^n$ be the associated quasiprojective variety. Then there is a canonical isomorphism
>
> $$K[X] = K[X_0]$$
>
> between the ring of regular functions on X and the coordinate ring of X_0 given by sending $F \in K[X]$ to $F_0 = F \circ j_0 \in K[X_0]$.

PROOF   Let $F : X \to K$ be a regular function and set $F_0 = F \circ j_0 : X_0 \to K$, so that
$$F_0(a_1, \ldots, a_n) = F([1 : a_1 : \cdots : a_n]).$$
The most formidable part of the proof is to show that F_0 is actually a polynomial function on X_0 and thus an element of the coordinate ring. Applying the definition of regular functions and then setting $x_0 = 1$ implies that, for any $p \in X_0$, we may choose polynomials $f_p, g_p \in K[x_1, \ldots, x_n]$ such that $g_p(p) \neq 0$ and

$$F_0(a) = \frac{f_p(a)}{g_p(a)} \quad \text{for all} \quad a \in X_0 \setminus \mathcal{V}_\mathbb{A}(g_p).$$

Furthermore, we may assume that f_p and g_p both vanish on every irreducible component Z of X_0 not containing p: if they do not so vanish, then we may multiply both f_p and g_p by a polynomial vanishing on all of Z but not at p. Choose one such pair (f_p, g_p) for every $p \in X_0$.

We claim that, for any pair $p, q \in X_0$, the polynomials $f_p g_q$ and $f_q g_p$ define the same function $X_0 \to K$. To prove this, we prove that they agree on every irreducible component $Z \subseteq X_0$. If either $p \notin Z$ or $q \notin Z$, then our assumptions imply that $f_p g_q = f_q g_p = 0$ on Z. On the other hand, if $p, q \in Z$, then the equality

$$\frac{f_p(a)}{g_p(a)} = F_0(a) = \frac{f_q(a)}{g_q(a)}$$

holds for all a at which neither g_p nor g_q vanishes. This implies that the polynomial functions $f_p g_q$ and $f_q g_p$ agree on $Z \setminus \mathcal{V}_\mathbb{A}(g_p g_q)$, which is a nonempty open set of the irreducible variety Z. By Proposition 11.19, it follows that $Z \setminus \mathcal{V}_\mathbb{A}(g_p g_q)$ is dense in Z, implying (see Exercise 11.2.6) that $f_p g_q$ and $f_q g_p$ agree on all of Z, justifying the claim.

Now define the ideal

$$J = \langle g_p \mid p \in X_0 \rangle \subseteq K[x_1, \ldots, x_n]$$

generated by the denominators of all the chosen local expressions for F_0. Since every generator of J is nonvanishing at some point of X_0, we have

$$\mathcal{V}_\mathbb{A}(J + \mathcal{I}_\mathbb{A}(X_0)) = \mathcal{V}_\mathbb{A}(J) \cap X_0 = \emptyset,$$

from which the Nullstellensatz implies that $J + \mathcal{I}_\mathbb{A}(X_0) = K[x_1, \ldots, x_n]$. In particular, this means that $1 \in J + \mathcal{I}_\mathbb{A}(X_0)$, so we can write

(11.37) $$1 = h_1 g_{p_1} + \cdots + h_k g_{p_k} + f$$

for some $p_1, \ldots, p_k \in X_0$, $h_1, \ldots, h_k \in K[x_1, \ldots, x_n]$, and $f \in \mathcal{I}_\mathbb{A}(X_0)$.

We now claim that, for every $a \in X_0$, we have

$$F_0(a) = h_1(a) f_{p_1}(a) + \cdots + h_k(a) f_{p_k}(a).$$

To prove this, let $a \in X_0$. Then it follows from (11.37) that $g_{p_i}(a) \neq 0$ for some i; without loss of generality, assume $g_{p_1}(a) \neq 0$. We then have

$$\begin{aligned}
F_0(a) &= \frac{f_{p_1}(a)}{g_{p_1}(a)} \\
&= \left(\sum_{i=1}^{k} h_i(a) g_{p_i}(a) \right) \frac{f_{p_1}(a)}{g_{p_1}(a)} \\
&= \sum_{i=1}^{k} h_i(a) \frac{f_{p_i}(a) g_{p_1}(a)}{g_{p_1}(a)} \\
&= \sum_{i=1}^{k} h_i(a) f_{p_i}(a),
\end{aligned}$$

where the second equality uses equation (11.37) and the third equality uses the fact that $f_{p_1} g_{p_i} = f_{p_i} g_{p_1}$ as functions on X_0. We have thus proven that F_0 can be realized globally on X_0 by the polynomial function $F_0 = [h_1 f_{p_1} + \cdots + h_k f_{p_k}] \in K[X_0]$.

In particular, we now know that $F \mapsto F_0$ defines a function $K[X] \to K[X_0]$, which is manifestly a K-algebra homomorphism. Furthermore, it is a bijection, because if $F_0 \in K[X_0]$ is a polynomial function on X_0, then we can choose a representative $F_0 = [f_0]$ with $f_0 \in K[x_1, \ldots, x_n]$, homogenize it to obtain a homogeneous polynomial $f \in K[x_0, x_1, \ldots, x_n]$ of some degree d, and then set $F = f/x_0^d$ to obtain a regular function on X whose dehomogenization is F_0. Thus, the map $K[X] \to K[X_0]$ is a K-algebra isomorphism, as desired. $\square$

Equipped with Theorem 11.36, we can now compute the ring of regular functions on any quasiprojective variety that arises from the natural embedding of an affine variety. The following example illustrates the computation for a particular quasiprojective variety that we met in the previous section.

11.4. AFFINE VARIETIES REVISITED

11.38 EXAMPLE *The ring of regular functions on $\mathcal{V}(w^2 - xy) \setminus \mathcal{V}(w)$*

Consider the quasiprojective variety
$$Y = \mathcal{V}(w^2 - xy) \setminus \mathcal{V}(w) \subseteq \mathbf{P}^2,$$
which we can identify with $j_0(Y_0)$ where
$$Y_0 = \mathcal{V}_{\mathbb{A}}(1 - xy) \subseteq \mathbb{A}^2.$$
By Theorem 11.36, we have
$$K[Y] = K[Y_0] = \frac{K[x,y]}{\mathcal{I}_{\mathbb{A}}(Y_0)} = \frac{K[x,y]}{\langle 1 - xy \rangle} = K[x, x^{-1}].$$

Now that we know that polynomial functions on affine varieties agree with regular functions on their associated quasiprojective varieties, it can be argued without too much additional work that the same is true of polynomial maps and regular maps.

11.39 COROLLARY *Regular maps on affine varieties are polynomial*

Let $X_0 \subseteq \mathbb{A}^m$ and $Y_0 \subseteq \mathbb{A}^n$ be affine varieties, and let $X = j_0(X_0) \subseteq \mathbf{P}^m$ and $Y = j_0(Y_0) \subseteq \mathbf{P}^n$ be the corresponding quasiprojective varieties. There is a bijection
$$\left\{\begin{array}{c} \text{regular maps} \\ X \to Y \end{array}\right\} \longrightarrow \left\{\begin{array}{c} \text{polynomial maps} \\ X_0 \to Y_0 \end{array}\right\}$$
given by sending $F : X \to Y$ to $F_0 = j_0^{-1} \circ F \circ j_0$.

PROOF If $F : X \to Y$ is a regular map, then it follows from the definitions of regular maps and regular functions that the composition $j_0^{-1} \circ F : X \to Y_0 \subseteq \mathbb{A}^n$ is given by
$$(j_0^{-1} \circ F)(a) = (\widetilde{F}_1(a), \ldots, \widetilde{F}_n(a))$$
in which each $\widetilde{F}_i : X \to K$ is a regular function on X. Theorem 11.36 then implies that
$$\widetilde{F}_1 \circ j_0, \ldots, \widetilde{F}_n \circ j_0 \in K[X_0].$$
In particular, each of these is a polynomial function on X_0. Thus,
$$(j_0^{-1} \circ F \circ j_0)(a) = ((\widetilde{F}_1 \circ j_0)(a), \ldots, (\widetilde{F}_n \circ j_0)(a))$$
is a polynomial map $X_0 \to Y_0$. This shows that the association $F \mapsto F_0$ indeed sends regular maps to polynomial maps, and some reflection should convince the reader that this is, in fact, a bijection (see Exercise 11.4.3). $\square$

At this point, parallel to the discussion of projective varieties in Section 11.3, we have shown that the category of affine varieties embeds into the category of quasiprojective varieties. There is one substantial drawback to the situation, however: although every affine variety can be naturally identified with a quasiprojective variety, the class of quasiprojective varieties that arise in this way is *not* invariant under isomorphism. The following example illustrates this phenomenon.

11.40 EXAMPLE $\mathbb{A}^1 \setminus \{0\}$ is isomorphic to $\mathcal{V}(1 - xy) \subseteq \mathbb{A}^2$

Let $X_0 = \mathbb{A}^1 \setminus \{0\}$. Viewed as a subset of $\mathbb{A}^1$, the set X_0 is not an affine variety; indeed, the only proper affine varieties in $\mathbb{A}^1$ are finite sets of points. However, we can naturally identify X_0 with the quasiprojective variety

$$X = j_0(X_0) = \mathbb{P}^1 \setminus \mathcal{V}(xy),$$

and we have seen in Example 11.33 that X is isomorphic to the quasiprojective variety

$$Y = \mathcal{V}(w^2 - xy) \setminus \mathcal{V}(w) \subseteq \mathbb{P}^2,$$

which *is* a quasiprojective variety that is naturally identified with an affine variety:

$$Y = j_0(Y_0) \quad \text{where} \quad Y_0 = \mathcal{V}(1 - xy) \subseteq \mathbb{A}^2.$$

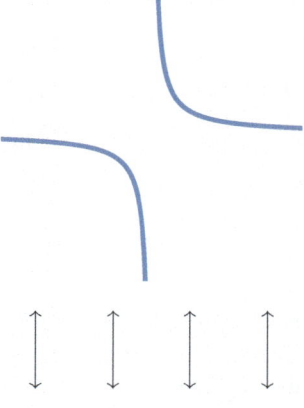

Thus, even though X does not arise naturally from an affine variety, it is isomorphic to a quasiprojective variety that does. The image above depicts the quasiprojective isomorphism between $Y_0 = \mathcal{V}(1 - xy) \subseteq \mathbb{A}^2$ and $X_0 = \mathbb{A}^1 \setminus \{0\} \subseteq \mathbb{A}^1$.

Since we generally prefer properties of varieties that are preserved by isomorphisms, this example demonstrates an unfortunate situation for our terminology. We rectify the matter by a bit of brute force: we simply revise the definition of "affine variety" to insist that it be invariant under isomorphisms of quasiprojective varieties.

11.41 DEFINITION *Affine variety*

A quasiprojective variety $X \subseteq \mathbb{P}^m$ is called an *affine variety* if X is isomorphic to a quasiprojective variety $Y \subseteq \mathbb{P}^n$ such that $Y = j_0(Y_0)$ for some Zariski-closed subset $Y_0 \subseteq \mathbb{A}^n$.

While the quasiprojective variety $\mathbb{P}^1 \setminus \mathcal{V}(xy)$ considered in Example 11.40 is not *equal* to $j_0(X_0)$ for a Zariski-closed subset $X_0 \subseteq \mathbb{A}^1$, we have verified that $\mathbb{P}^1 \setminus \mathcal{V}(xy)$ is *isomorphic* to a quasiprojective variety $Y \subseteq \mathbb{P}^2$ such that $Y = j_0(Y_0)$ for some Zariski-closed subset $Y_0 \subseteq \mathbb{A}^2$. Thus, according to Definition 11.41, the quasiprojective variety $\mathbb{P}^1 \setminus \mathcal{V}(xy)$ will henceforth be referred to as an affine variety.

This terminology creates a new tension, however: now that the term "affine variety" refers to a particular type of quasiprojective variety, how shall we refer to the objects that were formerly known as affine varieties—that is, subsets of $\mathbb{A}^n$ that arise as the vanishing set of a collection of polynomials? As in the previous paragraph, we will henceforth refer to these using the language of the Zariski topology.

11.42 TERMINOLOGY CHANGE

A subset of $\mathbb{A}^n$ of the form $\mathcal{V}_\mathbb{A}(\mathcal{S})$ for some subset $\mathcal{S} \subseteq K[x_1, \ldots, x_n]$ will henceforth be referred to as a *(Zariski-)closed subset* of $\mathbb{A}^n$.

11.4. AFFINE VARIETIES REVISITED

Thus, while $\mathbb{A}^1 \setminus \{0\}$ is not a Zariski-closed subset of $\mathbb{A}^1$, and is therefore not an affine variety according to the terminology that we have used previously in this book, it is now fair to refer to $\mathbb{A}^1 \setminus \{0\}$ as an affine variety, since it is affine as a quasiprojective variety, according to Defini-

> *Note here that we are beginning to elide the difference between a subset $X \subseteq \mathbb{A}^n$ and its corresponding subset $j_0(X) \subseteq \mathbb{P}^n$, a mild abuse of notation to which the reader will become accustomed.*

tion 11.41. This particular example is a special case of a general result that will be proved in the next section: the subset of a closed set $X \subseteq \mathbb{A}^n$ at which a single polynomial does *not* vanish is always an affine variety.

One might wonder whether we require a similar change in terminology regarding projective varieties. As it turns out, it is not necessary in the projective setting because projective varieties are already preserved under isomorphism. More precisely, it will follow from our developments in this chapter that, if $X \subseteq \mathbb{P}^m$ and $Y \subseteq \mathbb{P}^n$ are isomorphic quasiprojective varieties, then one of them is Zariski-closed in its ambient projective space if and only if the other is Zariski-closed (Corollary 11.73).

In the meantime, our final goal of this section is to describe an example of a quasiprojective variety in affine space that is provably *not* an affine variety.

11.43 EXAMPLE $\mathbb{A}^2 \setminus \{(0,0)\}$ is not an affine variety

Consider $X_0 = \mathbb{A}^2 \setminus \{(0,0)\}$, which we identify (Exercise 11.4.4) with the quasiprojective variety
$$X = \mathbb{P}^2 \setminus \mathcal{V}(x_0 x_1, x_0 x_2).$$
Suppose, toward a contradiction, that X is affine, so there exists an isomorphism $F : Y \to X$ where $Y = j_0(Y_0)$ for some closed subset $Y_0 \subseteq \mathbb{A}^n$.

Consider the regular functions $G_1 = x_1/x_0$ and $G_2 = x_2/x_0$ in $K[X]$. Notice that there do not exist any points of X at which both G_1 and G_2 vanish. Thus, F^*G_1 and F^*G_2 do not have any common points of vanishing in Y. Since $K[Y] = K[Y_0]$ (Theorem 11.36), the Nullstellensatz applied to the closed set $Y_0 \subseteq \mathbb{A}^n$ then implies that $K[Y] = \langle F^*G_1, F^*G_2 \rangle$ (see Exercise 11.4.5). But since $F^* : K[X] \to K[Y]$ is an isomorphism (Proposition 11.35), it follows that $K[X] = \langle G_1, G_2 \rangle$.

On the other hand, Lemma 11.29 implies (Exercise 11.4.4) that
$$K[X] \cong K[x_1, x_2].$$
The ideal $\langle G_1, G_2 \rangle$ corresponds under this isomorphism to $\langle x_1, x_2 \rangle \subseteq K[x_1, x_2]$, so it is a proper ideal, contradicting that $\langle G_1, G_2 \rangle = K[X]$. This contradiction implies that it cannot be the case that X is an affine variety.

Exercises for Section 11.4

11.4.1 Let $Y = \mathcal{V}(w^2 - xy) \subseteq \mathbb{P}^2$ be as in Example 11.38, where we observed that
$$K[Y] = K[x, x^{-1}].$$
Describe the regular function $Y \to K$ associated to $x^4 - 3x^{-1} \in K[x, x^{-1}]$ as a quotient of homogeneous polynomials of the same degree in $K[w, x, y]$.

11.4.2 Let $X_0 = \mathcal{V}_\mathbb{A}(x_2 - x_1^2) \subseteq \mathbb{A}^2$, and let $Y_0 = \mathcal{V}_\mathbb{A}(y_1 y_2 - y_3) \subseteq \mathbb{A}^3$. Consider the polynomial map $X_0 \to Y_0$ given by

$$(a_1, a_2) \mapsto (a_1 - a_1^2,\ a_1 + a_2,\ a_1^2 - a_2^2).$$

In terms of homogeneous polynomials of the same degree, describe the corresponding regular map of quasiprojective varieties $X \to Y$ where we set $X = j_0(X_0) \subseteq \mathbb{P}^2$ and $Y = j_0(Y_0) \subseteq \mathbb{P}^3$.

11.4.3 Generalize the construction in the previous exercise and argue that the association in the statement of Corollary 11.39 is invertible, and thus a bijection.

11.4.4 Consider the quasiprojective variety

$$X = \mathbb{P}^2 \setminus \mathcal{V}(x_0 x_1, x_0 x_2).$$

(a) Prove that $X = j_0(\mathbb{A}^2 \setminus \{(0,0)\})$.
(b) Prove that $K[X] \cong K[x_1, x_2]$.

11.4.5 Let $X \subseteq \mathbb{A}^n$ be a closed set, and let $F_1, \ldots, F_k \in K[X]$ be regular functions that do not have any common points of vanishing in X. Use the Nullstellensatz to prove that $K[X] = \langle F_1, \ldots, F_k \rangle$.

11.4.6 Prove that $\mathbb{P}^n$, as a quasiprojective variety, is not affine for any $n > 0$.

Section 11.5 Affine opens and local properties

Now that we have shown that the theory of quasiprojective varieties specializes to the theory of affine varieties, we are prepared to describe one of the most useful features of quasiprojective varieties: every quasiprojective variety is "locally affine," in a sense that we will make precise in this section. The power of this fact is that it allows us to reduce the study of many properties of quasiprojective varieties—properties such as dimension and smoothness—to the corresponding properties of affine varieties, which we have already developed in detail in Part I of the book.

The first step in showing that every quasiprojective variety is locally affine is to study *nonvanishing sets* within closed subsets of affine space.

11.44 PROPOSITION *Nonvanishing sets are affine varieties*

If $X \subseteq \mathbb{A}^n$ is a Zariski-closed set and $f \in K[x_1, \ldots, x_n]$, then the set $U(f) = X \setminus \mathcal{V}(f)$ is an affine variety.

Example 11.40 demonstrates one instance of this result, showing that the nonvanishing set $\mathbb{A}^1 \setminus \mathcal{V}(x)$ is isomorphic to the closed set $\mathcal{V}_\mathbb{A}(xy - 1) \subseteq \mathbb{A}^2$. More generally, a similar argument shows that $\mathbb{A}^n \setminus \mathcal{V}(f)$ is isomorphic to the closed set $\mathcal{V}(x_{n+1}f - 1) \subseteq \mathbb{A}^{n+1}$ for any $f \in K[x_1, \ldots, x_n]$. These special cases provide a model for the general proof, which we now present.

PROOF OF PROPOSITION 11.44   Let $X \subseteq \mathbb{A}^n$ be closed, allowing us to write $X = \mathcal{V}_\mathbb{A}(g_1, \ldots, g_m)$ for some polynomials $g_1, \ldots, g_m \in K[x_1, \ldots, x_n]$. Given $f \in K[x_1, \ldots, x_n]$, define the closed set $Y \subseteq \mathbb{A}^{n+1}$ by

$$Y = \mathcal{V}_\mathbb{A}(g_1, \ldots, g_m, x_{n+1}f - 1).$$

We claim that $U(f)$ and Y are isomorphic as quasiprojective varieties. Since Y is a closed subset of affine space, this will imply that $U(f)$ is an affine variety.

To see that $U(f) \cong Y$, consider the projection map

$$H : Y \to U(f)$$
$$(a_1, \ldots, a_n, a_{n+1}) \mapsto (a_1, \ldots, a_n).$$

Notice (Exercise 11.5.1) that H is a bijection with inverse H^{-1} given by

$$(a_1, \ldots, a_n) \mapsto \left(a_1, \ldots, a_n, \frac{1}{f(a)}\right).$$

It remains to argue that both H and H^{-1} are regular. To do so, identify Y and $U(f)$ with their images in $\mathbb{P}^{n+1}$ and $\mathbb{P}^n$. Let $\bar{f}$ be the homogenization of f and suppose that $\deg(\bar{f}) = d$. Then H corresponds to the regular map

$$[a_0 : \cdots : a_n : a_{n+1}] \mapsto [a_0 : \cdots : a_n],$$

whereas some reflection should convince the reader that H^{-1} corresponds to the regular map

$$[a_0 : \cdots : a_n] \mapsto [a_0 \bar{f}(a) : \cdots : a_n \bar{f}(a) : a_0^{d+1}].$$

Thus, $U(f) \cong Y$, as claimed. □

> The key idea in the previous proof is that, upon passing to projective spaces, we are able to clear $f(a)$ from the denominator.

Proposition 11.44 shows that non-vanishing sets within closed subsets of affine space are, themselves, affine varieties. This is not true of all open subsets of affine space; for instance, the quasiprojective variety $\mathbb{A}^2 \setminus \{(0,0)\}$ encountered in Example 11.43 is open in $\mathbb{A}^2$, but it is not an affine variety.

We now utilize Proposition 11.44 to prove the main result of this section: that every quasiprojective variety is locally isomorphic to an affine variety, in the sense that every point of a quasiprojective variety is contained in an affine open subvariety. In fact, we prove an even stronger result; we prove that every quasiprojective variety can be covered by a *finite* number of affine open subvarieties.

11.45 THEOREM *Quasiprojective varieties are locally affine*

Let X be a quasiprojective variety. Then there exist finitely many affine open subvarieties $U_1, \ldots, U_k \subseteq X$ such that $X = U_1 \cup \cdots \cup U_k$. In particular, for every $p \in X$, there exists an affine open subvariety $U \subseteq X$ with $p \in U$.

PROOF Let $X \subseteq \mathbb{P}^n$ be a quasiprojective variety. By definition, we have

$$X = Y \cap (\mathbb{P}^n \setminus Z) = Y \setminus Z,$$

where $Y, Z \subseteq \mathbb{P}^n$ are projective varieties. Write $Z = \mathcal{V}(f_1, \ldots, f_k)$ for some homogeneous polynomials $f_1, \ldots, f_k \in K[x_0, \ldots, x_n]$. For each $i = 1, \ldots, k$ and each $j = 0, \ldots, n$, define

$$U_{i,j} = (Y \setminus \mathcal{V}(f_i)) \cap \mathbb{A}_j^n.$$

Notice that $Z = \mathcal{V}(f_1, \ldots, f_k) \subseteq \mathcal{V}(f_i)$, implying that $Y \setminus \mathcal{V}(f_i) = X \setminus \mathcal{V}(f_i)$. Thus, $Y \setminus \mathcal{V}(f_i)$ is an open subvariety of X for every i, and since $\mathbb{A}_j^n$ is open in $\mathbb{P}^n$, it follows that $U_{i,j}$ is an open subvariety of X for all i and j. Moreover, observe that

$$U_{i,j} = (Y \cap \mathbb{A}_j^n) \setminus \mathcal{V}(f_{i,j}),$$

where $f_{i,j}$ is the polynomial in n variables obtained from f_i by setting $x_j = 1$. Since $Y \cap \mathbb{A}_j^n$ is a closed subset of $\mathbb{A}_j^n$, it follows from Proposition 11.44 that each $U_{i,j}$ is an affine variety. The theorem now follows from the fact that

$$X = \bigcup_{\substack{1 \leq i \leq k \\ 0 \leq j \leq n}} U_{i,j},$$

which the reader is encouraged to verify in Exercise 11.5.3, thereby showing that every point of X is in (at least) one of the affine open subvarieties $U_{i,j}$. □

Recalling from Section 11.1 that a *neighborhood* of a point $p \in X$ is any open subset of X containing p, Theorem 11.45 can be rephrased as the assertion that *every point of a quasiprojective variety has an affine neighborhood*. We provide a concrete illustration of the theorem in the following example.

11.5. AFFINE OPENS AND LOCAL PROPERTIES

11.46 EXAMPLE $\mathbb{A}^2 \setminus \{(0,0)\}$ is locally affine

Let $X = \mathbb{A}^2 \setminus \{(0,0)\}$, which we view as a quasiprojective variety in $\mathbb{P}^2$ in the usual way. The subset

$$U = X \setminus \mathcal{V}(x) = \mathbb{A}^2 \setminus \mathcal{V}(x)$$

is an affine variety by Proposition 11.44; in particular, $U \cong \mathcal{V}(xz - 1) \subseteq \mathbb{A}^3$ via the isomorphism

$$(x, y) \mapsto \left(x, y, \frac{1}{x}\right).$$

Similarly,

$$V = X \setminus \mathcal{V}(y) = \mathbb{A}^2 \setminus \mathcal{V}(y)$$

is an affine variety. Both U and V are Zariski-open subvarieties of X, and since every point of X must have either its x-coordinate or its y-coordinate nonzero, every point is contained in either U or V. Thus, we have verified that every point of the quasiprojective variety $\mathbb{A}^2 \setminus \{(0,0)\}$ is contained in an affine neighborhood.

As we mentioned at the beginning of this section, the power of Theorem 11.45 lies in its ability to reduce the study of certain properties of quasiprojective varieties to the affine setting. We now give a name to those properties.

11.47 DEFINITION *Local property*

Let $\mathcal{C}$ be a class of topological spaces. A property P on $\mathcal{C}$ is *local* if it satisfies the following condition: if $X \in \mathcal{C}$ and every point $p \in X$ is contained in an open set $U_p \subseteq X$ with property P, then X has property P.

A non-example may help to clarify the definition: the property of being affine is *not* local on the class of quasiprojective varieties, because, for instance, $\mathbb{P}^n$ is not an affine variety (Exercise 11.4.6) but every point in $\mathbb{P}^n$ is contained in some affine patch. To put the situation metaphorically, if poor eyesight prevented one from seeing beyond a small neighborhood of any given point, it would be impossible to determine whether one was standing in the affine variety $\mathbb{A}^n$ or the non-affine variety $\mathbb{P}^n$, somewhat analogously to the way in which early humans were unable to see that they stood not on a flat Earth but a(n approximately) spherical one.

On the other hand, an example of a property that is local on the class of irreducible quasiprojective varieties is the property of being smooth, as we will see in the next chapter. Intuitively, the idea is that if X had a singular point, then even a near-sighted inhabitant of X would know this point was singular if they were standing on it. Similarly, the property of having a particular dimension is a local property of irreducible varieties: the number of independent directions of movement within an irreducible variety can be detected in a neighborhood of any point. We will present an in-depth discussion of both of these properties in the next chapter.

In the meantime, we close this section with two useful results regarding local properties, both of which will be used in the next section. The first lemma says that, to verify that a subset of a topological space is closed, it suffices to check that it is closed within a neighborhood of every point.

11.48 LEMMA *Being closed is local*

Let Y be a topological space and let $X \subseteq Y$ be a subset. If every $p \in Y$ is contained in an open set $U_p \subseteq Y$ such that $X \cap U_p$ is closed in U_p, then X is closed in Y.

PROOF For each $p \in Y$, fix an open set U_p such that $p \in U_p$ and $X \cap U_p$ is closed in U_p. By definition of open sets, there exist closed sets $Z_p \subseteq Y$ for each p such that $U_p = Y \setminus Z_p$, and by definition of the subspace topology, there exist closed sets $Z'_p \subseteq Y$ such that $X \cap U_p = Z'_p \cap U_p$. We now claim that

$$X = \bigcap_{p \in Y} (Z_p \cup Z'_p),$$

from which it follows that X is closed in Y.

($\subseteq$) Suppose that $q \in X$. Then, for each $p \in Y$, we either have $q \in U_p$ or $q \notin U_p$. If $q \in U_p$, then $q \in X \cap U_p = Z'_p \cap U_p$, so $q \in Z'_p$. If $q \notin U_p$, then $q \in Z_p$. In either case, then, we have $q \in Z_p \cup Z'_p$, and this holds for all $p \in Y$.

($\supseteq$) Conversely, suppose that $q \in Z_p \cup Z'_p$ for all $p \in Y$. Since $q \in U_q$, we have $q \notin Z_q$ and therefore the fact that $q \in Z_q \cup Z'_q$ implies that $q \in Z'_q$. Thus, $q \in Z'_q \cap U_q = X \cap U_q$, so $q \in X$. $\square$

The next lemma asserts that regularity can be checked locally on the domain.

11.49 LEMMA *Being regular is local on the domain*

Let $X \subseteq \mathbb{P}^m$ and $Y \subseteq \mathbb{P}^n$ be quasiprojective varieties, and let $F : X \to Y$ be a map. If every $p \in X$ is contained in an open set $U_p \subseteq X$ such that the restriction $F : U_p \to Y$ is regular, then F is regular.

PROOF Assume that $F : X \to Y$ is a map satisfying the hypotheses of the lemma. To prove that F is regular, let $p \in X$. By assumption, there exists an open set $U_p \subseteq X$ and homogeneous polynomials $g_0, \ldots, g_n \in K[x_0, \ldots, x_m]$ of the same degree such that $p \notin \mathcal{V}(g_0, \ldots, g_n)$ and

$$F(a) = [g_0(a) : \cdots : g_n(a)] \quad \text{for all} \quad a \in U_p \setminus \mathcal{V}(g_0, \ldots, g_n).$$

In order for F to satisfy the definition of a regular map, we would need the expression above to hold on all of $X \setminus \mathcal{V}(g_0, \ldots, g_n)$, but it does not necessarily do so.

To remedy this situation, choose a homogeneous polynomial $f \in K[x_0, \ldots, x_n]$ such that f vanishes on the closed set $X \setminus U_p \subseteq X$ and such that $f(p) \neq 0$. Define $f_i = f g_i$ for every $i = 0, \ldots, n$, and notice that the values of the two functions $[f_0 : \cdots : f_n]$ and $[g_0 : \cdots : g_n]$ agree wherever both are defined, but the domain of definition of the first is smaller. In particular, since f vanishes on $X \setminus U_p$, we obtain

$$X \setminus \mathcal{V}(f_0, \ldots, f_n) = U_p \setminus \mathcal{V}(f_0, \ldots, f_n).$$

Thus, by construction, it follows that $p \notin X \setminus \mathcal{V}(f_0, \ldots, f_n)$ and

$$F(a) = [f_0(a) : \cdots : f_n(a)] \quad \text{for all} \quad a \in X \setminus \mathcal{V}(f_0, \ldots, f_n),$$

so F is regular at p. $\square$

Exercises for Section 11.5

11.5.1 Let $g_1, \ldots, g_m, f \in K[x_0, \ldots, x_n]$ and define $X = \mathcal{V}(g_1, \ldots, g_m) \subseteq \mathbb{A}^n$ and $Y = \mathcal{V}(g_1, \ldots, g_m, x_{n+1}f - 1) \subseteq \mathbb{A}^{n+1}$. Prove that the function

$$\varphi : Y \to X \setminus \mathcal{V}(f)$$
$$(a_1, \ldots, a_n, a_{n+1}) \mapsto (a_1, \ldots, a_n)$$

is a bijection and compute its inverse.

11.5.2 Let $X \subseteq \mathbb{A}^n$ be closed and let $f \in K[x_1, \ldots, x_n]$. Setting $F = [f] \in K[X]$, prove that

$$K[X \setminus \mathcal{V}(f)] = \frac{K[X][z]}{\langle zF - 1 \rangle}.$$

(When a is not nilpotent in R, it is common to use the notation $R[a^{-1}]$ for the ring $R[z]/\langle za - 1\rangle$. Thus, we have $K[X \setminus \mathcal{V}(f)] = K[X][F^{-1}]$.)

11.5.3 Let $Y, Z \subseteq \mathbb{P}^n$ be projective varieties with $Z = \mathcal{V}(f_1, \ldots, f_k)$. For each $i = 1, \ldots, k$ and $j = 0, \ldots, n$, define

$$U_{i,j} = (Y \setminus \mathcal{V}(f_i)) \cap \mathbb{A}^n_j.$$

Prove that

$$Y \setminus Z = \bigcup_{\substack{1 \le i \le k \\ 0 \le j \le n}} U_{i,j}.$$

11.5.4 Construct a finite affine open cover of $\mathbb{A}^n \setminus \{(0, \ldots, 0)\}$.

11.5.5 Let $Y = \mathcal{V}(x^2 + y^2 - zw) \subseteq \mathbb{P}^3$, and let $Z = \mathcal{V}(x, y, z) \subseteq \mathbb{P}^3$. Construct a finite affine open cover of $X = Y \setminus Z$.

11.5.6 Let $f \in K[x_0, \ldots, x_r]$ be a nonconstant homogeneous polynomial. Prove that $\mathbb{P}^n \setminus \mathcal{V}(f)$ is an affine variety. (Hint: Use a Veronese map.)

Section 11.6 Continuity of regular maps

Regular maps have been introduced as the "structure-preserving" maps in the context of quasiprojective varieties. Since every quasiprojective variety comes equipped with its Zariski topology, one might naturally ask: Do regular maps preserve the "structure" of this topology? This section is devoted to addressing this question by introducing *continuous maps*, which are the structure-preserving maps in the setting of topology, and proving that regular maps of quasiprojective varieties are continuous with respect to the Zariski topology.

Let us begin by recalling the familiar notion of continuity. One's first encounter with continuity is usually in calculus, studying functions $f : \mathbb{R} \to \mathbb{R}$. In this setting, a continuous function might be introduced as one whose graph can be drawn without lifting your pencil, but this is rather informal. To make it more rigorous, students then grapple with the ϵ-δ definition of continuity: for every $x_0 \in \mathbb{R}$ and every $\epsilon > 0$, there exists $\delta > 0$ such that $|x - x_0| < \delta$ implies $|f(x) - f(x_0)| < \epsilon$.

While the ϵ-δ definition becomes intuitive over time, it remains a bit clunky. Fortunately, there is an elegant topological characterization of continuity: a function $f : \mathbb{R} \to \mathbb{R}$ is continuous (with respect to the ϵ-δ definition) if and only if $f^{-1}(V)$ is Euclidean-open in $\mathbb{R}$ for all Euclidean-open sets $V \subseteq \mathbb{R}$. The importance of this characterization is that it leads to a purely topological formulation of continuity, which makes sense even in the absence of a method for measuring distances, leading to the following general notion of continuity.

11.50 DEFINITION *Continuous map*

Let X and Y be topological spaces. A map $F : X \to Y$ is *continuous* if, for every open set $V \subseteq Y$, the preimage $F^{-1}(V)$ is open in X.

While continuity is most commonly defined in terms of open sets—as we have done here—we can also reformulate continuity in terms of closed sets. In particular (Exercise 11.6.1), a map $F : X \to Y$ is continuous if and only if $F^{-1}(Z)$ is closed in X for every closed set $Z \subseteq Y$. We now state the main result of this section in terms of this second formulation.

11.51 THEOREM *Regular maps are continuous*

Let $F : X \to Y$ be a regular map of quasiprojective varieties. If Z is closed in Y, then $F^{-1}(Z)$ is closed in X.

PROOF Let $X \subseteq \mathbb{P}^m$ and $Y \subseteq \mathbb{P}^n$ be quasiprojective varieties and let $F : X \to Y$ be a regular map. Given a closed subvariety $Z \subseteq Y$, we aim to prove that $F^{-1}(Z)$ is closed in X, and it suffices (by Lemma 11.48) to show that every point $p \in X$ is contained in a neighborhood U_p such that $F^{-1}(Z) \cap U_p$ is closed in U_p.

Consider a point $p \in X$. By definition of regular maps, choose homogeneous $f_0, \ldots, f_n \in K[x_0, \ldots, x_m]$ of the same degree such that $p \notin \mathcal{V}(f_0, \ldots, f_n)$ and

$$F(a) = [f_0(a) : \cdots : f_n(a)] \quad \text{for all} \quad a \in X \setminus \mathcal{V}(f_0, \ldots, f_n).$$

11.6. CONTINUITY OF REGULAR MAPS

Set $U_p = X \setminus \mathcal{V}(f_0, \ldots, f_n)$. Then U_p is open in X and contains p. Consider the regular function $F_p : U_p \to \mathbb{P}^n$ that is globally defined by $F_p = [f_0 : \cdots : f_n]$, and notice that

$$F^{-1}(Z) \cap U_p = F_p^{-1}(Z) = F_p^{-1}(\overline{Z}),$$

where $\overline{Z}$ is the projective closure of Z in $\mathbb{P}^n$ and the last equality holds because Z is closed in Y, so $Z = \overline{Z} \cap Y$. As $\overline{Z}$ is closed in $\mathbb{P}^n$, there exist homogeneous polynomials $g_1, \ldots, g_k \in K[y_1, \ldots, y_n]$ such that $\overline{Z} = \mathcal{V}(g_1, \ldots, g_k)$. Then $g_i(f_0, \ldots, f_n)$ is a homogeneous polynomial in $K[x_1, \ldots, x_m]$ for every $i = 1, \ldots, k$, and some reflection (Exercise 11.6.3) should convince the reader that

$$F_p^{-1}(\overline{Z}) = U_p \cap \mathcal{V}\big(g_1(f_0, \ldots, f_n), \ldots, g_k(f_0, \ldots, f_n)\big).$$

It follows that $F^{-1}(Z) \cap U_p = F_p^{-1}(\overline{Z})$ is closed in U_p, as desired. □

Continuous maps are the structure-preserving maps in topology, so continuous maps with continuous inverses give a natural notion of equivalence among topological spaces. This notion of equivalence is called *homeomorphism*.

11.52 DEFINITION *Homeomorphism*

Let X and Y be topological spaces. A *homeomorphism* from X to Y is a continuous map $F : X \to Y$ for which there is a continuous map $G : Y \to X$ that is inverse to F. If there exists a homeomorphism between X and Y, we say that X and Y are *homeomorphic*.

Since isomorphisms of quasiprojective varieties are regular maps with regular inverses, Theorem 11.51 implies that an isomorphism $F : X \to Y$ of quasiprojective varieties is a homeomorphism with respect to the Zariski topologies on X and Y. This observation is captured by the first two (equivalent) points in the following.

11.53 COROLLARY *Isomorphisms identify subvarieties*

If $F : X \to Y$ is an isomorphism of quasiprojective varieties and $Z \subseteq X$, then the following assertions hold.

1. Z is open in X if and only if $F(Z)$ is open in Y.
2. Z is closed in X if and only if $F(Z)$ is closed in Y.
3. Z is a subvariety of X if and only if $F(Z)$ is a subvariety of Y.

Moreover, if Z is a subvariety of X, then the restriction $F|_Z : Z \to F(Z)$ is an isomorphism of quasiprojective varieties.

PROOF   Exercise 11.6.4. □

In the remainder of this section, we use Theorem 11.51 to develop alternative characterizations of regular maps and isomorphisms. These characterizations are given purely in terms of continuity and regular functions, without reference to the ambient projective spaces in which the quasiprojective varieties live. As we discuss at the end of the section, these characterizations take us one step closer to understanding the true intrinsic nature of quasiprojective varieties.

11.54 PROPOSITION *Characterization of regular maps*

Let $X \subseteq \mathbb{P}^m$ and $Y \subseteq \mathbb{P}^n$ be quasiprojective varieties. A map $F : X \to Y$ is regular if and only if the following two conditions are satisfied:

1. F is continuous with respect to the Zariski topology, and
2. For every open set $V \subseteq Y$ and for every regular function $G \in K[V]$, the function $G \circ F$ is regular on $F^{-1}(V)$.

PROOF First, suppose that the map $F : X \to Y$ is regular. Condition 1 in the statement of the proposition is then a result of Theorem 11.51, while Condition 2 is a result of the observation

See Exercise 11.6.8 for yet another characterization of regular maps in terms of affine open covers.

that the restriction of F to the open set $F^{-1}(V)$ is regular, along with the fact that pullbacks by regular maps are well-defined (Exercise 11.3.7). Therefore, it remains to prove the other direction of the if-and-only-if statement.

Suppose that $F : X \to Y$ satisfies Conditions 1 and 2 in the statement of the proposition; we aim to show that F is regular. By Lemma 11.49, it suffices to prove that every point $p \in X$ is contained in some open set $U_p \subseteq X$ such that the restriction $F : U_p \to Y$ is regular. Let $p \in X$. Without loss of generality, assume that $F(p) \notin \mathcal{V}(y_0)$, and let us start by defining $U = F^{-1}(Y \cap \mathbb{A}_0^n)$, which is an open set of X (by Condition 1) that contains p (since $F(p) \notin \mathcal{V}(y_0)$).

Since $F(a) \notin \mathcal{V}(y_0)$ for every $a \in U$, we can scale the homogeneous coordinates of each $F(a)$ so that the y_0-coordinate is 1. Thus, there exist unique functions $F_1, \ldots, F_n : U \to K$ such that

$$(11.55) \qquad F(a) = [1 : F_1(a) : \cdots : F_n(a)] \quad \text{for every} \quad a \in U.$$

In other words, $F_i = (y_i/y_0) \circ F$. Notice that $y_1/y_0, \ldots, y_n/y_0$ are regular functions on $\mathbb{A}_0^n$, and Condition 2 then tells us that each composition $(y_i/y_0) \circ F = F_i$ is regular on U. Since $p \in U$, the definition of regular functions then implies that, for all $i = 1, \ldots, n$, there exist $g_i, h_i \in K[x_0, \ldots, x_m]$, homogeneous of the same degree, such that $h_i(p) \neq 0$ and

$$(11.56) \qquad F_i(a) = g_i(a)/h_i(a) \quad \text{for all} \quad a \in U \setminus \mathcal{V}(h_i).$$

Define $f_0 = h_1 \cdots h_n$ and $f_i = f_0 g_i / h_i$ for $i = 1, \ldots, n$, and set $U_p = U \setminus \mathcal{V}(f_0)$. Then U_p is an open set of X containing p, and some reflection should convince the reader that the polynomials $f_0, \ldots, f_n \in K[x_0, \ldots, x_m]$ are all homogeneous of the same degree and—using (11.55) and (11.56)—that

$$F(a) = [f_0(a) : \cdots : f_n(a)] \quad \text{for all} \quad a \in U_p.$$

Therefore, we have shown that $F : U_p \to Y$ is regular.

Thus, for each $p \in X$, we have constructed an open set $U_p \subseteq X$ for which the restriction $F : U_p \to Y$ is regular, and we conclude that $F : X \to Y$ is regular. □

Building on the previous result, we now characterize isomorphisms between quasiprojective varieties in terms of open sets and regular functions on open sets.

11.6. CONTINUITY OF REGULAR MAPS

> **11.57 COROLLARY** *Characterization of isomorphisms*
>
> Let X and Y be quasiprojective varieties. A map $F : X \to Y$ is an isomorphism if and only if the following two conditions are satisfied:
>
> 1. F is a homeomorphism with respect to the Zariski topology, and
> 2. For every open set $V \subseteq Y$, the map $G \mapsto G \circ F$ is a well-defined K-algebra isomorphism from $K[V]$ to $K[F^{-1}(V)]$.

PROOF First, suppose that F is an isomorphism. Then F is a regular map and has a regular inverse, which then implies that F is a homeomorphism by Theorem 11.51. Furthermore, since F

> *See Exercise 11.6.9 for yet another characterization of isomorphisms in terms of affine open covers.*

restricts to an isomorphism $F : F^{-1}(V) \to V$ for each open set $V \subseteq Y$, Proposition 11.35 then implies that $F^* : K[V] \to K[F^{-1}(V)]$ is an isomorphism. This proves one direction of the if-and-only-if statement.

Conversely, suppose that $F : X \to Y$ satisfies the two conditions in the statement of the proposition. Since F is continuous and $G \circ F \in K[F^{-1}(V)]$ for every $G \in K[V]$, it follows from Proposition 11.54 that F is regular. And since F is a homeomorphism, it follows that F has a continuous inverse $F^{-1} : Y \to X$. It remains to prove that F^{-1} is also regular.

Let $U \subseteq X$ be an open set. Condition 2 applied to the open set $F(U) \subseteq Y$ implies that $G \mapsto G \circ F$ gives an isomorphism $K[F(U)] \to K[U]$. By surjectivity of this homomorphism, for every $H \in K[U]$, there exists $G \in K[F(U)]$ such that $H = G \circ F$; in particular,

$$H \circ F^{-1} = G \circ F \circ F^{-1} = G \in K[F(U)].$$

> *Our argument does not use the injectivity of $G \mapsto G \circ F$; this injectivity is implied by the surjectivity of F.*

In summary, we have verified that, for every open set $U \subseteq X$ and for every regular function $H \in K[U]$, the composition $H \circ F^{-1}$ is regular on the set $F(U) = (F^{-1})^{-1}(U)$. Since F^{-1} is also continuous, Proposition 11.54 then implies that F^{-1} is regular, as desired. □

Our developments thus far begin to give us a glimpse of the intrinsic nature of quasiprojective varieties. Unlike the case of affine varieties, where the intrinsic nature of a variety was completely determined by one K-algebra (its coordinate ring), the intrinsic nature of a general quasiprojective variety has both a topological aspect and an algebraic aspect that must be considered. More specifically, the essential data of a quasiprojective variety X includes both the information of the open sets in X (a topology on X) and the information of the ring of regular functions on each open set $U \subseteq X$ (a "sheaf" of K-algebras on X). By Corollary 11.57, a function $F : X \to Y$ between quasiprojective varieties is an isomorphism if and only if F is a homeomorphism—meaning that F preserves the topological structure—and $F^* : K[V] \to K[F^{-1}(V)]$ is an isomorphism of K-algebras for every open set $V \subseteq Y$—meaning that F preserves the algebraic structure.

In the affine setting, the situation is much simpler: in order to know whether a map is an isomorphism in the affine setting, we simply need to check whether its pullback is an isomorphism of K-algebras, and we then get the topological homeomorphism for free. In other words, the algebra completely governs the topology (and all other intrinsic data) in the affine setting. However, as we have seen (in Example 11.30, for instance), this is not the case in the more general quasiprojective case, where the K-algebra of global regular functions does not typically capture the full intrinsic nature of the variety. When studying general quasiprojective varieties, we need the topological structure, as it essentially keeps track of how the local affine neighborhoods (where the algebra completely governs the topology) are glued together to form a more global object.

Exercises for Section 11.6

11.6.1 Let X and Y be topological spaces and let $F : X \to Y$ be a map. Prove that $F^{-1}(V)$ is open in X for every open set V in Y if and only if $F^{-1}(Z)$ is closed in X for every closed set Z in Y.

11.6.2 Let $F : X \to Y$ be a continuous map of topological spaces.
  (a) Let Z be a subspace of X, prove that restricting the domain gives a continuous map $F : Z \to Y$.
  (b) Let Z be a topological space such that Y is a subspace of Z. Prove that extending the codomain gives a continuous map $F : X \to Z$.
  (c) Let Z be subspace of Y containing $F(X)$. Prove that restricting the codomain gives a continuous map $F : X \to Z$.

11.6.3 Let $X \subseteq \mathbb{P}^m$ be a quasiprojective variety and let $f_0, \ldots, f_n \in K[x_1, \ldots, x_m]$ be homogeneous polynomials of the same degree such that

$$\mathcal{V}(f_0, \ldots, f_n) \cap X = \emptyset.$$

If $F = [f_0 : \cdots : f_n] : X \to \mathbb{P}^n$ and $Z = \mathcal{V}(g_1, \ldots, g_k) \subseteq \mathbb{P}^n$ for some $g_1, \ldots, g_k \in K[y_1, \ldots, y_n]$, prove that

$$F^{-1}(Z) = X \cap \mathcal{V}(g_1(f_0, \ldots, f_n), \ldots, g_k(f_0, \ldots, f_n)).$$

11.6.4 Prove Corollary 11.53.

11.6.5 Prove that irreducibility is an intrinsic property of quasiprojective varieties. In other words, if X and Y are isomorphic quasiprojective varieties, prove that X is irreducible if and only if Y is irreducible.

11.6.6 Let $F : X \to Y$ be a continuous map of topological spaces. If Z is an irreducible subspace of X, prove that $F(Z)$ is an irreducible subspace of Y.

11.6.7 Let X be a quasiprojective variety and let $\mathcal{S} \subseteq K[X]$ be a set of regular functions. Define

$$\mathcal{V}_X(\mathcal{S}) = \{a \in X \mid F(a) = 0 \text{ for all } F \in \mathcal{S}\}.$$

Prove that $\mathcal{V}_X(\mathcal{S})$ is closed in X.

11.6. CONTINUITY OF REGULAR MAPS

11.6.8 Let $F: X \to Y$ be a map of quasiprojective varieties. Prove that F is regular if and only if there exist affine open covers

$$X = U_1 \cup \cdots \cup U_k \quad \text{and} \quad Y = V_1 \cup \cdots \cup V_\ell$$

such that, for every $i \in \{1, \ldots, k\}$, there exists $j \in \{1, \ldots, \ell\}$ such that $U_i \subseteq F^{-1}(V_j)$ and the restriction

$$F|_{U_i} : U_i \to V_j$$

is a regular map of affine varieties.

11.6.9 Let $F: X \to Y$ be a bijective map of quasiprojective varieties. Prove that F is an isomorphism if and only if there exists an affine open cover

$$X = U_1 \cup \cdots \cup U_k$$

such that, for every i, the set $F(U_i)$ is open in Y and F restricts to give an isomorphism $U_i \cong F(U_i)$.

Section 11.7 Products and graphs

We now turn toward the topic of products of quasiprojective varieties, with an eye toward studying graphs of regular maps, which will play a central role in the next section. In order to make sense of products in the quasiprojective setting, we first recall from Section 10.4 that the Segre map

$$S_{m,n} : \mathbb{P}^m \times \mathbb{P}^n \to \mathbb{P}^{(m+1)(n+1)-1}$$

is an injection that takes the product $X \times Y$ of a pair of projective varieties to a projective variety in $\mathbb{P}^{(m+1)(n+1)-1}$. This allows us to identify $X \times Y$ with its image under $S_{m,n}$ in order to endow it with the structure of a projective variety. A similar result holds for quasiprojective varieties, for essentially the same reason.

> **11.58 PROPOSITION** *Products of quasiprojective varieties*
>
> If $X \subseteq \mathbb{P}^m$ and $Y \subseteq \mathbb{P}^n$ are quasiprojective varieties, then the Segre map $S_{m,n}$ restricts to an injection of $X \times Y$ onto a quasiprojective variety in $\mathbb{P}^{(m+1)(n+1)-1}$.

PROOF Proposition 10.27 shows that $S_{m,n}$ is an injection, so it remains to show that $S_{m,n}(X \times Y)$ is a quasiprojective variety. To do so, we start by using the assumption that X is quasiprojective to write $X = Z \setminus Z'$ and $Y = W \setminus W'$, where Z and Z' are closed in $\mathbb{P}^m$ and W and W' are closed in $\mathbb{P}^n$. Set-theoretically, notice that

$$X \times Y = (Z \times W) \setminus ((Z \times W') \cup (Z' \times W)),$$

and since $S_{m,n}$ is an injection, it then follows that

$$S_{m,n}(X \times Y) = S_{m,n}(Z \times W) \setminus (S_{m,n}(Z \times W') \cup S_{m,n}(Z' \times W)).$$

By Proposition 10.29, the sets $S_{m,n}(Z \times W)$, $S_{m,n}(Z \times W')$, and $S_{m,n}(Z' \times W)$ are all closed, showing that $S_{m,n}(X \times Y)$ is the complement of a closed set in a closed set, or in other words, a quasiprojective variety. $\square$

In light of this proposition, we can identify $X \times Y$ with $S_{m,n}(X \times Y)$ in order to define what we mean by a product within the realm of quasiprojective varieties.

> **11.59 DEFINITION** *Product of quasiprojective varieties*
>
> If $X \subseteq \mathbb{P}^m$ and $Y \subseteq \mathbb{P}^n$ are quasiprojective varieties, then their *product* $X \times Y$ (as a quasiprojective variety) is the quasiprojective variety
>
> $$S_{m,n}(X \times Y) \subseteq \mathbb{P}^{(m+1)(n+1)-1}.$$

We should pause to convince ourselves that this general definition of products agrees with our prior notions of products in the affine and projective settings. In the projective setting, this agreement is immediate from the definition. In the affine setting, this amounts to proving that $S_{m,n}$ restricts to an isomorphism (of quasiprojective varieties) from $\mathbb{A}^{m+n} = \mathbb{A}_0^m \times \mathbb{A}_0^n$ onto its image (Exercise 11.7.2).

11.7. PRODUCTS AND GRAPHS

We can now make sense of such objects as the product of a projective variety with an affine variety—within the realm of quasiprojective varieties—and we can speak of closed and open subsets of the resulting product.

11.60 EXAMPLE *The product $\mathbb{P}^1 \times \mathbb{A}^2$*

Consider the product $\mathbb{P}^1 \times \mathbb{A}^2$, which sits inside of $\mathbb{P}^1 \times \mathbb{P}^2$. By definition, the Segre map $S_{1,2} : \mathbb{P}^1 \times \mathbb{P}^2 \to \mathbb{P}^5$ is given by

$$S_{1,2}([a_0 : a_1], [b_0 : b_1 : b_2]) = [a_0b_0 : a_0b_1 : a_0b_2 : a_1b_0 : a_1b_1 : a_1b_2],$$

with coordinates on $\mathbb{P}^5$ accordingly denoted $[z_{00} : z_{01} : z_{02} : z_{10} : z_{11} : z_{12}]$. We have $\mathbb{A}^2 = \mathbb{A}_0^2 = \mathbb{P}^2 \setminus \mathcal{V}(y_0)$ and the reader can verify that

$$S_{1,2}(\mathbb{P}^1 \times \mathcal{V}(y_0)) = S_{1,2}(\mathbb{P}^1 \times \mathbb{P}^2) \cap \mathcal{V}(z_{00}, z_{10}).$$

Combining this with the expression for $S_{1,2}(\mathbb{P}^1 \times \mathbb{P}^2)$ in Proposition 10.27, we find that $S_{1,2}(\mathbb{P}^1 \times \mathbb{A}^2)$ is the quasiprojective variety in $\mathbb{P}^5$ given by

(11.61) $\quad \mathcal{V}(z_{00}z_{11} - z_{10}z_{01}, z_{00}z_{12} - z_{10}z_{02}, z_{01}z_{12} - z_{11}z_{02}) \setminus \mathcal{V}(z_{00}, z_{10}).$

In accordance with Definition 11.59, let us denote this quasiprojective variety by $\mathbb{P}^1 \times \mathbb{A}^2$. An example of a closed subset of $\mathbb{P}^1 \times \mathbb{A}^2$, then, is the set

$$Z = \{([a_0 : a_1], (b_1, b_2)) \mid a_0 - a_1 = 0, \; b_1 - 1 = 0\},$$

since $S_{1,2}$ identifies Z with the intersection of (11.61) with the projective variety $\mathcal{V}(z_{00} - z_{01}, z_{01} - z_{10})$; see Exercise 11.7.3.

If $X \subseteq \mathbb{P}^m$ and $Y \subseteq \mathbb{P}^n$ are quasiprojective varieties, then the Zariski topology on $X \times Y$ is the subspace topology inherited from $\mathbb{P}^m \times \mathbb{P}^n$. As we saw in Proposition 10.35, the closed sets of $\mathbb{P}^m \times \mathbb{P}^n$ are exactly the vanishing sets of collections of polynomials in $K[x_0, \ldots, x_m, y_0, \ldots, y_n]$. Thus, it follows that the closed subsets of $X \times Y$ correspond to intersections of the form

$$(X \times Y) \cap \mathcal{V}(\mathcal{S}),$$

where $\mathcal{S} \subseteq K[x_0, \ldots, x_m, y_0, \ldots, y_n]$.

In preparation for our developments in the next section, we now turn toward a discussion of *graphs* of regular maps, naturally generalizing the notion of graphs of real-valued functions that one studies in calculus.

11.62 DEFINITION *Graph of a regular map*

Let $F : X \to Y$ be a regular map of quasiprojective varieties. The *graph* of F is the set
$$\Gamma_F = \{(a, b) \in X \times Y \mid b = F(a)\} \subseteq X \times Y.$$

For example, if $F : \mathbb{A}_{\mathbb{R}}^1 \to \mathbb{A}_{\mathbb{R}}^1$ is the polynomial map defined by $F(a) = a^2$, the graph Γ_F is the usual parabola in $\mathbb{A}_{\mathbb{R}}^1 \times \mathbb{A}_{\mathbb{R}}^1 = \mathbb{A}_{\mathbb{R}}^2$.

Let us consider a slightly more general example.

11.63 EXAMPLE *The graph of a regular map*

Let $X = \mathbb{P}^1 \setminus \mathcal{V}(x_0 x_1)$, and let $Y = \mathbb{P}^1$. Then the map $F : X \to Y$ given by

$$F([a_0 : a_1]) = [a_0 a_1 : a_0^2 + a_1^2]$$

is regular, and its graph can be described as

$$\Gamma_F = \{([a_0 : a_1], [b_0 : b_1]) \in X \times Y \mid [b_0 : b_1] = [a_0 a_1 : a_0^2 + a_1^2]\}.$$

The condition that $[b_0 : b_1] = [a_0 a_1 : a_0^2 + a_1^2]$ is equivalent to the equality of the cross-multiplications $b_0(a_0^2 + a_1^2) = b_1(a_0 a_1)$, so we can write

$$\Gamma_F = (X \times Y) \cap \mathcal{V}(y_0(x_0^2 + x_1^2) - y_1(x_0 x_1)) \subseteq \mathbb{P}^1 \times \mathbb{P}^1.$$

Since Γ_F is the intersection of $X \times Y$ with a closed set in $\mathbb{P}^1 \times \mathbb{P}^1$, it follows that Γ_F is a closed subset of the quasiprojective variety $X \times Y$.

The final observation made at the end of Example 11.63 illustrates a property that is true of all graphs of regular maps, as the following proposition verifies.

11.64 PROPOSITION *Graphs of regular maps are closed*

Let $X \subseteq \mathbb{P}^m$ and $Y \subseteq \mathbb{P}^n$ be quasiprojective varieties, and let $F : X \to Y$ be a regular map. Then Γ_F is closed in $X \times Y$.

PROOF Let $(p,q) \in X \times Y$. We will construct a neighborhood U_p of (p,q) (depending only on p) such that $\Gamma_F \cap U_p$ is closed in U_p. By Lemma 11.48, it then follows that Γ_F is closed in $X \times Y$.

Since F is a regular map, choose polynomials $f_0, \ldots, f_n \in K[x_0, \ldots, x_m]$, homogeneous of the same degree, such that $p \notin \mathcal{V}(f_0, \ldots, f_n)$ and

$$F(a) = [f_0(a) : \cdots : f_n(a)] \quad \text{for all} \quad a \in X \setminus \mathcal{V}(f_0, \ldots, f_n).$$

Viewing $f_0, \ldots, f_n$ as polynomials in $K[x_0, \ldots, x_m, y_0, \ldots, y_n]$ that depend only on the x-variables, define the open set

$$U_p = (X \times Y) \setminus \mathcal{V}(f_0, \ldots, f_n),$$

which contains the point (p,q). Define

$$Z_p = \mathcal{V}(\{y_i f_j - y_j f_i \mid 0 \leq i, j \leq n\}) \subseteq \mathbb{P}^m \times \mathbb{P}^n,$$

and notice that $\Gamma_F \cap U_p = Z_p \cap U_p$ since the condition $b_i f_j(a) = b_j f_i(a)$ for all i, j is equivalent to the requirement that

$$[b_0 : \cdots : b_n] = [f_0(a) : \cdots : f_n(a)] = F(a)$$

for every $(a,b) \in U_p$. Since Z_p is closed in $\mathbb{P}^m \times \mathbb{P}^n$ by Proposition 10.35, we then conclude that $\Gamma_F \cap U_p = Z_p \cap U_p$ is closed in U_p, as desired. $\square$

11.7. PRODUCTS AND GRAPHS

The importance of Proposition 11.64 in what follows is that, if $F : X \to Y$ is a regular map, we can identify the image $F(X) \subseteq Y$ with the image of $\Gamma_F \subseteq X \times Y$ under the projection $X \times Y \to Y$. Thus, an understanding of Γ_F, together with a careful study of the projection map, will allow us to make statements about the images of regular maps. This is the content of the final section of this chapter.

Exercises for Section 11.7

11.7.1 Prove that, if $U \subseteq \mathbb{P}^m$ and $V \subseteq \mathbb{P}^n$ are open, then the image

$$S_{m,n}(U \times V) \subseteq \mathbb{P}^{(m+1)(n+1)-1}$$

is open in $S_{m,n}(\mathbb{P}^m \times \mathbb{P}^n)$.

11.7.2 (a) Prove that the Segre map $S_{m,n}$ restricts to an isomorphism between the affine space $\mathbb{A}^{m+n} = \mathbb{A}_0^m \times \mathbb{A}_0^n$ and its image in $\mathbb{P}^{(m+1)(n+1)-1}$.
(b) Let $X \subseteq \mathbb{A}^m$ and $Y \subseteq \mathbb{A}^n$ be closed sets. Prove that the closed set $X \times Y \subseteq \mathbb{A}^{m+n}$ is isomorphic to the quasiprojective variety $X \times Y$ defined in Definition 11.59. (Hint: Use Part (a) and Corollary 11.53.)

11.7.3 Consider the Segre map $S_{1,2} : \mathbb{P}^1 \times \mathbb{P}^2 \to \mathbb{P}^5$, with coordinates denoted as in Example 11.60. Fill in the details of Example 11.60, as follows.
(a) Verify that $S_{1,2}(\mathbb{P}^1 \times \mathcal{V}(y_0)) = S_{1,2}(\mathbb{P}^1 \times \mathbb{P}^2) \cap \mathcal{V}(z_{00}, z_{10})$.
(b) Verify that the set Z defined in Example 11.60 is mapped by $S_{1,2}$ to the set $S_{1,2}(\mathbb{P}^1 \times \mathbb{A}^2) \cap \mathcal{V}(z_{00} - z_{01}, z_{01} - z_{10})$.

11.7.4 Let X_1 and X_2 be quasiprojective varieties, and let $Y_1 \subseteq X_1$ and $Y_2 \subseteq X_2$ be closed subvarieties. Prove that $Y_1 \times Y_2$ is closed in $X_1 \times X_2$.

11.7.5 Let X and Y be quasiprojective varieties. Prove that the projection maps $\pi_1 : X \times Y \to X$ and $\pi_2 : X \times Y \to Y$ are regular.

11.7.6 Prove that, if $F : Z \to X$ and $G : Z \to Y$ are regular maps of quasiprojective varieties, then the map $F \times G : Z \to X \times Y$ defined by

$$(F \times G)(a) = (F(a), G(a))$$

is a regular map.

11.7.7 Suppose that X_1, X_2, Y_1, Y_2 are quasiprojective varieties such that $X_1 \cong X_2$ and $Y_1 \cong Y_2$. Prove that

$$X_1 \times X_2 \cong Y_1 \times Y_2.$$

11.7.8 Let X and Y be quasiprojective varieties. Prove that $X \times Y$ is irreducible if and only if X and Y are irreducible.

11.7.9 Let $F : X \to Y$ be a regular map of quasiprojective varieties. Prove that the graph $\Gamma_F \subseteq X \times Y$ is isomorphic to X.

Section 11.8 Images of projective varieties

We have now arrived at the final section of this chapter, in which our goal is to use our prior developments to prove the following fundamental theorem.

> **11.65 THEOREM** *Regular images of projective varieties are closed*
>
> Let $X \subseteq \mathbb{P}^m$ be closed and let $Y \subseteq \mathbb{P}^n$ be a quasiprojective variety. If $F : X \to Y$ is a regular map, then $F(X)$ is closed in Y.

This result is yet another illustration of the fact that projective varieties are complete, in the sense that they are not missing any points: no matter how we map a projective variety into another quasiprojective variety, the image will always be closed. Theorem 11.65 stands in stark contrast to the affine setting; as we have seen in the case of the affine hyperbola—pictured at right—the projection down to the affine line is a regular map whose image is not closed. Intuitively, the missing point in the image detects the asymptote in the hyperbola where the two branches wander off to infinity.

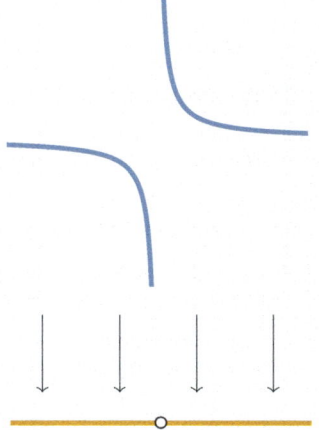

To put the distinction between affine and projective varieties another way, recall that in Section 11.4 we were careful to re-define "affine variety" to mean not just a closed subset of $\mathbb{A}^n$ but a quasiprojective variety isomorphic to such a subset. However, we can safely define "projective variety" to mean simply a closed subset of $\mathbb{P}^n$, and this notion is already invariant under isomorphism, because Theorem 11.65 implies (Corollary 11.73) that any quasiprojective variety $X \subseteq \mathbb{P}^n$ that is isomorphic to a closed subset of some other projective space must itself be closed in $\mathbb{P}^n$.

Yet another way to describe the divergence between affine and projective varieties that we will see in this section is given by Corollary 11.74, which says that—in contrast to the wealth of polynomial functions on an affine variety—there are no nonconstant regular functions on an irreducible projective variety. The key idea here is to view a regular function on X as a regular map $F : X \to \mathbb{A}^1$ and to use the fact that the only closed subsets of $\mathbb{A}^1$ are $\mathbb{A}^1$ itself and finite sets of points. The irreducibility and projectivity of X imply that, among these options, $F(X)$ must be single a point.

With this motivation in place, let us begin our journey toward proving Theorem 11.65. By Exercise 11.8.1, a somewhat different perspective on the image $F(X)$ of a regular map $F : X \to Y$ is that $F(X) = \pi(\Gamma_F)$, where $\Gamma_F \subseteq X \times Y$ is the graph of F and π is the projection onto the second factor:

$$\pi : X \times Y \to Y$$
$$\pi(x, y) = y.$$

Since Γ_F is closed in $X \times Y$ (Proposition 11.64), it suffices to prove the following.

11.8. IMAGES OF PROJECTIVE VARIETIES 353

> **11.66 LEMMA** *Projections from projective varieties are closed*
>
> If $X \subseteq \mathbb{P}^m$ is closed and $Y \subseteq \mathbb{P}^n$ is a quasiprojective variety, then the projection $\pi : X \times Y \to Y$ takes closed sets of $X \times Y$ to closed sets of Y.

PROOF   This is a rather formidable proof, beginning with a series of reductions. We first reduce to the case where $X = \mathbb{P}^m$, then we reduce to the case where Y is affine, and we lastly reduce to the case where Y is $\mathbb{A}^n$. We then prove the result in the case $X \times Y = \mathbb{P}^m \times \mathbb{A}^n$.

(**Reduction to** $X = \mathbb{P}^m$) Suppose the lemma is true when $X = \mathbb{P}^m$, and let $X \subseteq \mathbb{P}^m$ be closed. Suppose that $Z \subseteq X \times Y$ is closed. Since X is closed in $\mathbb{P}^m$, it follows that $X \times Y$ is closed in $\mathbb{P}^m \times Y$ (Exercise 11.7.4), so the closed inclusions

$$Z \subseteq X \times Y \subseteq \mathbb{P}^m \times Y$$

show that Z is closed in $\mathbb{P}^m \times Y$. If $\pi : X \times Y \to Y$ and $\overline{\pi} : \mathbb{P}^m \times Y \to Y$ denote the respective projection maps, then our assumption implies that $\overline{\pi}(Z)$ is closed in Y. But $\pi(Z) = \overline{\pi}(Z)$, so $\pi(Z)$ is also closed in Y.

Thus, the special case $X = \mathbb{P}^m$ implies the general case of the lemma, so we assume from now on that $X = \mathbb{P}^m$.

(**Reduction to** Y **affine**) Suppose the lemma is true when Y is affine, and let Y be any quasiprojective variety. Let $Z \subseteq \mathbb{P}^m \times Y$ be closed. Since quasiprojective varieties are locally affine (Theorem 11.45), every $p \in Y$ is contained in an open set $Y_p \subseteq Y$ for which Y_p is affine. Intersecting the closed set $Z \subseteq \mathbb{P}^m \times Y$ with $\mathbb{P}^m \times Y_p$ shows that

$$Z \cap (\mathbb{P}^m \times Y_p) \subseteq \mathbb{P}^m \times Y_p$$

is closed in $\mathbb{P}^m \times Y_p$ for each p, and therefore, our assumption implies that under the projection map $\pi_p : \mathbb{P}^m \times Y_p \to Y_p$, the image

$$\pi_p(Z \cap (\mathbb{P}^m \times Y_p)) \subseteq Y_p$$

is closed in Y_p. But $\pi(Z) \cap Y_p = \pi_p(Z \cap (\mathbb{P}^m \times Y_p))$, so we have shown that $\pi(Z) \cap Y_p$ is closed in Y_p. As this is true for each $p \in Y$, we see that $\pi(Z)$ is locally closed in Y, and thus is closed in Y by Lemma 11.48.

Thus, the special case in which Y is affine implies the general case of the lemma, so we furthermore assume from now on that Y is affine.

(**Reduction to** $Y = \mathbb{A}^n$) Finally, suppose that the lemma holds when $Y = \mathbb{A}^n$, and let $Y \subseteq \mathbb{A}^n$ be closed. Let $Z \subseteq \mathbb{P}^m \times Y$ be closed, and note that Z is also closed as a subset of $\mathbb{P}^m \times \mathbb{A}^n$. Let $\pi : \mathbb{P}^m \times Y \to Y$ and $\overline{\pi} : \mathbb{P}^m \times \mathbb{A}^n \to \mathbb{A}^n$ denote the respective projections. Since $\pi(Z) = \overline{\pi}(Z)$ and because $\overline{\pi}(Z)$ is closed in $\mathbb{A}^n$ by assumption, it follows that $\pi(Z)$ is closed in Y.

Thus, we have reduced the entire lemma to the special case in which $X = \mathbb{P}^m$ and $Y = \mathbb{A}^n$. We now move on to the proof of the lemma in this special case, but we caution the reader that the argument is quite involved; it may be worth working through Example 11.72 below either before or alongside reading the proof to help digest the strategy and notation.

(Proof for $X = \mathbb{P}^m$ and $Y = \mathbb{A}^n$) Let $Z \subseteq \mathbb{P}^m \times \mathbb{A}^n$ be closed, and write

$$Z = \mathcal{V}(f_1, \ldots, f_\ell)$$

for some $f_1, \ldots, f_\ell \in K[x_0, \ldots, x_m, y_1, \ldots, y_n]$. We may assume that each f_k is homogeneous in the x-variables, and we let d_k denote the degree of f_k in the x-variables. We note that

(11.67) $$\pi(Z) = \{b \in \mathbb{A}^n \mid \pi^{-1}(b) \cap Z \neq \emptyset\}.$$

The heart of what remains is to unpack the condition that $\pi^{-1}(b) \cap Z \neq \emptyset$.

Given $b = (b_1, \ldots, b_n) \in \mathbb{A}^n$, define

(11.68) $$f_k^b(x_0, \ldots, x_m) = f_k(x_0, \ldots, x_m, b_1, \ldots, b_n) \in K[x_0, \ldots, x_m].$$

for each $k = 1, \ldots, \ell$. We naturally identify the set

$$\pi^{-1}(b) \cap Z = \{(a,b) \in \mathbb{P}^m \times \{b\} \mid f_k^b(a) = 0 \text{ for all } k = 1, \ldots, \ell\}$$

with the vanishing set $\mathcal{V}(f_1^b, \ldots, f_\ell^b) \subseteq \mathbb{P}^m$. Thus, we obtain the following characterization for when $\pi^{-1}(b) \cap Z$ is empty:

$$\pi^{-1}(b) \cap Z = \emptyset \iff \mathcal{V}(f_1^b, \ldots, f_\ell^b) = \emptyset \subseteq \mathbb{P}^m.$$

In light of the projective Nullstellensatz, the condition that $\mathcal{V}(f_1^b, \ldots, f_\ell^b) = \emptyset$ is equivalent to the condition that there exists some $r \geq 0$ such that every monomial x^α of degree r is contained in $\langle f_1^b, \ldots, f_\ell^b \rangle$ (Exercise 9.6.6). For convenience, let M_r index the set of degree-r monomials in $x_0, \ldots, x_m$, so each $\alpha \in M_r$ corresponds to a degree-r monomial x^α. Thus, $\pi^{-1}(b) \cap Z = \emptyset$ if and only if there exists $r \geq 0$ such that, for every $\alpha \in M_r$, we have

(11.69) $$x^\alpha = \sum_{k=1}^{\ell} g_k f_k^b$$

for some $g_1, \ldots, g_\ell \in K[x_0, \ldots, x_m]$. We can write $g_k = \sum_\beta a_{k,\beta} x^\beta$ for monomials x^β, and it follows that (11.69) is equivalent to the existence of $a_{k,\beta} \in K$ such that

(11.70) $$x^\alpha = \sum_{k=1}^{\ell} \sum_\beta a_{k,\beta} x^\beta f_k^b.$$

Since $\alpha \in M_r$ and f_k^b is homogeneous of degree d_k, we can assume that each β appearing in the kth summand of (11.70) lies in M_{r-d_k}, as all other terms must sum to zero. For each $k \in \{1, \ldots, \ell\}$ and each $\beta \in M_{r-d_k}$, we now write $x^\beta f_k^b$ as a linear combination of monomials:

(11.71) $$x^\beta f_k^b = \sum_\alpha c^b_{(k,\beta),\alpha} x^\alpha,$$

and we organize the coefficients of these linear combinations into a matrix

$$C_r^b = \left(c^b_{(k,\beta),\alpha} \right).$$

11.8. IMAGES OF PROJECTIVE VARIETIES

Notice that the columns of C_r^b are indexed by M_r while the rows are indexed by pairs (k, β) with $\beta \in M_{r-d_k}$.

Observe that the row of C_r^b indexed by (k, β) records the coefficients of $x^\beta f_k^b$ as a linear combination of $\{x^\alpha \mid \alpha \in M_r\}$, so the existence of (11.70) for every $\alpha \in M_r$ is equivalent to the condition that the standard basis vectors are in the row span of C_r^b, which is equivalent to the condition that the rank of C_r^b is $|M_r|$ (the number of columns of C_r^b). Summarizing, we have argued that

$$\pi^{-1}(b) \cap Z = \emptyset \iff \text{there exists } r \geq 0 \text{ such that } \mathrm{rk}(C_r^b) = |M_r|.$$

Negating the previous condition, we obtain

$$\pi^{-1}(b) \cap Z \neq \emptyset \iff \text{for every } r \geq 0, \text{ we have } \mathrm{rk}(C_r^b) < |M_r|.$$

It now follows that $\pi^{-1}(b) \cap Z \neq \emptyset$ if and only if all $|M_r| \times |M_r|$ minors of C_r^b vanish for every $r \geq 0$. Since these minors are polynomial expressions in the entries of C_r^b, and since each entry of C_r^b is a polynomial expression in $b_1, \ldots, b_n$—as can be seen from their definition in (11.68) and (11.71)—it follows that the points $b \in \mathbb{A}^n$ for which $\pi^{-1}(b) \cap Z \neq \emptyset$ form a closed subset of $\mathbb{A}^n$. We then conclude from (11.67) that $\pi(Z)$ is closed, finishing the proof. □

In the next example, we give a concrete illustration of some of the notation used in the proof of Lemma 11.66.

11.72 EXAMPLE Projecting a closed set

To illustrate the strategy in the proof of Lemma 11.66, consider the closed subset

$$Z = \mathcal{V}(x_0 + y_1 x_1, x_1^2 + y_2 x_0 x_1) \subseteq \mathbb{P}^1 \times \mathbb{A}^2$$

and its image under the projection $\pi : \mathbb{P}^1 \times \mathbb{A}^2 \to \mathbb{A}^2$. First, we note, for instance, that $(1, 0) \notin \pi(Z)$, since setting $y_1 = 1$ and $y_2 = 0$ in the defining equations of Z gives

$$\pi^{-1}(1,0) \cap Z = \left\{ [a_0 : a_1] \in \mathbb{P}^1 \;\middle|\; \begin{smallmatrix} a_0 - a_1 = 0, \\ a_1^2 = 0 \end{smallmatrix} \right\} = \emptyset.$$

On the other hand, for instance, $(1, 1) \in \pi(Z)$, since setting $y_1 = y_2 = 1$ in the defining equations of Z gives

$$\pi^{-1}(1,1) \cap Z = \left\{ [a_0 : a_1] \in \mathbb{P}^1 \;\middle|\; \begin{smallmatrix} a_0 + a_1 = 0, \\ a_1^2 + a_0 a_1 = 0 \end{smallmatrix} \right\} = \{[1 : -1]\}.$$

In general, the proof of Lemma 11.66 shows that $b = (b_1, b_2) \in \pi(Z)$ if and only if, for every $r \geq 0$, the $|M_r| \times |M_r|$ minors of the matrix C_r^b are all zero. To help us parse this statement, and especially to help parse the notation, let us compute the matrix explicitly for $r = 3$.

To calculate the matrix C_3^b for a generic $b = (b_1, b_2)$, we start by considering the polynomials

$$f_1^b = x_0 + b_1 x_1 \quad \text{and} \quad f_2^b = x_1^2 + b_2 x_0 x_1.$$

Multiplying each of these polynomials by all possible monomials in x_0 and x_1 that give a product of degree 3 in the x-variables, and recording the coefficients of the monomials x_0^3, $x_0^2 x_1$, $x_0 x_1^2$, and x_1^3 in the result, we obtain:

$$f_1^b \cdot x_0^2 = \boxed{1} \cdot x_0^3 + \boxed{b_1} \cdot x_0^2 x_1 + \boxed{0} \cdot x_0 x_1^2 + \boxed{0} \cdot x_1^3$$
$$f_1^b \cdot x_0 x_1 = \boxed{0} \cdot x_0^3 + \boxed{1} \cdot x_0^2 x_1 + \boxed{b_1} \cdot x_0 x_1^2 + \boxed{0} \cdot x_1^3$$
$$f_1^b \cdot x_1^2 = \boxed{0} \cdot x_0^3 + \boxed{0} \cdot x_0^2 x_1 + \boxed{1} \cdot x_0 x_1^2 + \boxed{b_1} \cdot x_1^3$$
$$f_2^b \cdot x_0 = \boxed{0} \cdot x_0^3 + \boxed{b_2} \cdot x_0^2 x_1 + \boxed{1} \cdot x_0 x_1^2 + \boxed{0} \cdot x_1^3$$
$$f_2^b \cdot x_1 = \boxed{0} \cdot x_0^3 + \boxed{0} \cdot x_0^2 x_1 + \boxed{b_2} \cdot x_0 x_1^2 + \boxed{1} \cdot x_1^3.$$

Then C_3^b is the 5×4 matrix consisting of the boxed entries, and the vanishing of the 4×4 minors of this matrix is necessary in order for b to be an element of $\pi(Z)$. One of these minors is the determinant of the matrix given by deleting the fourth row of C_3^b, which the reader can check is

$$1 - b_1 b_2.$$

In particular, then, an element $b = (b_1, b_2) \in \pi(Z)$ must satisfy $b_1 b_2 = 1$, which is consistent with our calculations that $(1, 0) \notin \pi(Z)$ but $(1, 1) \in \pi(Z)$.

More conditions on b_1, b_2 result from the vanishing of other minors of C_3^b, and of the minors of C_r^b for other $r \geq 1$, but in fact, they are all implied by the single condition that $b_1 b_2 = 1$. This can be checked directly from the definition of $\pi(Z)$: it consists of all points $(b_1, b_2) \in \mathbb{A}^2$ for which

$$a_0 + b_1 a_1 = 0 = a_1^2 + b_2 a_0 a_1 \quad \text{for some} \quad [a_0 : a_1] \in \mathbb{P}^1,$$

a system of equations that can readily be solved to yield $b_1 b_2 = 1$. Thus, either by this direct calculation or by the method of minors illustrated above, we find that $\pi(Z)$ is, indeed, a closed set: $\pi(Z) = \mathcal{V}(y_1 y_2 - 1) \subseteq \mathbb{A}^2$.

The proof of Theorem 11.65 now follows quickly from Lemma 11.66.

PROOF OF THEOREM 11.65 Since $F(X) = \pi(\Gamma_F)$ and $\Gamma_F \subseteq X \times Y$ is closed in $X \times Y$ by Proposition 11.64, Lemma 11.66 implies that $F(X)$ is closed in Y. □

We now verify the two important consequences of Theorem 11.65 that were mentioned at the beginning of this section.

11.73 COROLLARY *Being closed in projective space is intrinsic*

If $X \subseteq \mathbb{P}^m$ and $Y \subseteq \mathbb{P}^n$ are isomorphic quasiprojective varieties, then X is closed in $\mathbb{P}^m$ if and only if Y is closed in $\mathbb{P}^n$.

PROOF Suppose that X is closed in $\mathbb{P}^m$ and $F : X \to Y$ is an isomorphism. We can view F as a regular map from X to $\mathbb{P}^n$, and by Theorem 11.65, it follows that $F(X)$ is closed in $\mathbb{P}^n$. But $F(X) = Y$ since F is an isomorphism, showing that Y is closed in $\mathbb{P}^n$. The if-and-only-if statement then follows by symmetry. □

11.8. IMAGES OF PROJECTIVE VARIETIES

11.74 COROLLARY *Regular functions on projective varieties*

If X is an irreducible projective variety, then every regular function on X is constant. In other words, $K[X] = K$.

PROOF A regular function $F : X \to K$ is the same as a regular map $F : X \to \mathbb{A}^1$, so Theorem 11.65 implies that $F(X) \subseteq \mathbb{A}^1$ is closed. The only nonempty closed subsets of $\mathbb{A}^1$ are all of $\mathbb{A}^1$ and finite sets of points. It cannot be the case that $F(X) = \mathbb{A}^1$, since if so, then composing F with the inclusion $j_0 : \mathbb{A}^1 \to \mathbb{P}^1$ would yield a regular map $X \to \mathbb{P}^1$ with image equal to $\mathbb{A}^1 \subseteq \mathbb{P}^1$, which is not closed. Thus, the image of F must be a finite collection of points $a_1, \ldots, a_\ell \in \mathbb{A}^1$, meaning that
$$X = F^{-1}(a_1) \sqcup \cdots \sqcup F^{-1}(a_\ell).$$
Each of the sets in this disjoint union is closed in X (Theorem 11.51), so the fact that X is irreducible implies that we must have $\ell = 1$. In other words, $F(X)$ consists of a single element, so F is a constant function. $\square$

We close this section by returning to a familiar example.

11.75 EXAMPLE $\mathbb{A}^2 \setminus \{(0,0)\}$ *is not a projective variety*

As we saw in Example 11.43, the quasiprojective variety $\mathbb{A}^2 \setminus \{(0,0)\}$ is not isomorphic to an affine variety. We can also now verify that it is not isomorphic to a projective variety. We give two arguments, each hinging on one of the two corollaries that we just proved regarding projective varieties.

Our first argument for why $\mathbb{A}^2 \setminus \{(0,0)\}$ is not isomorphic to a projective variety is a consequence of the fact that $\mathbb{A}^2 \setminus \{(0,0)\} \subseteq \mathbb{P}^2$ is not closed. Indeed, $\mathbb{A}^2 \setminus \{(0,0)\}$ is a nonempty open, and thus dense, subset of $\mathbb{P}^2$, so the only way it could be closed is if it were all of $\mathbb{P}^2$, which it is not. Since $\mathbb{A}^2 \setminus \{(0,0)\}$ is not closed in $\mathbb{P}^2$, Corollary 11.73 implies that it is not isomorphic to a projective variety.

Our second argument for why $\mathbb{A}^2 \setminus \{(0,0)\}$ is not isomorphic to a projective variety is a consequence of the observation made in Example 11.43 that
$$K[\mathbb{A}^2 \setminus \{(0,0)\}] = K[x,y].$$
Since the ring of regular functions on $\mathbb{A}^2 \setminus \{(0,0)\}$ is not simply the constant functions K, it follows from Corollary 11.74 that $\mathbb{A}^2 \setminus \{(0,0)\}$ cannot be isomorphic to a projective variety.

Throughout this chapter, we have begun to familiarize ourselves with the theory of quasiprojective varieties. Importantly, we have now made precise how both affine varieties and projective varieties sit within this more general framework. In the next—and final—chapter of the book, we turn to the study of key attributes of quasiprojective varieties, like dimension and smoothness. Building on the work that we have carried out in this chapter, we will see that the fundamental properties of these attributes can be imported directly from the affine setting, without too much additional work, by utilizing the locally affine nature of quasiprojective varieties.

Exercises for Section 11.8

11.8.1 Let $F : X \to Y$ be a function of sets, and let $\Gamma_F \subseteq X \times Y$ be the graph of F. Prove that $F(X) = \pi(\Gamma_F)$ where $\pi : X \times Y \to Y$ is the projection map.

11.8.2 Let $Z \subseteq \mathbb{P}^1 \times \mathbb{A}^2$ be as in Example 11.72, and calculate the matrix C_2^b. What conditions on $b \in \pi(Z)$ result from the vanishing of the minors of this matrix? Verify that these conditions are satisfied when $b_1 b_2 = 1$, as claimed in Example 11.72.

11.8.3 Let $Z = \mathcal{V}(x_0 + y_1 x_1, x_0 + y_2 x_1) \subseteq \mathbb{P}^1 \times \mathbb{A}^2$, and let $\pi : \mathbb{P}^1 \times \mathbb{A}^2 \to \mathbb{A}^2$ denote the projection.
  (a) Calculate the image $\pi(Z)$ directly.
  (b) Calculate the matrices C_1^b and C_2^b, and confirm that all of their maximal minors vanish for $b \in \pi(Z)$.

11.8.4 Prove that every regular map from an irreducible projective variety to an affine variety is constant.

11.8.5 Prove that $\mathbb{P}^1 \times \mathbb{A}^1$ is (isomorphic to) neither an affine nor a projective variety.

11.8.6 Prove that $\mathbb{P}^2 \setminus \{[1:0:0]\}$ is (isomorphic to) neither an affine nor a projective variety.

11.8.7 Prove that every regular map $\mathbb{P}^1 \to \mathbb{P}^1$ is either constant or surjective.

Chapter 12

Culminating Topics

LEARNING OBJECTIVES FOR CHAPTER 12
- Describe rational functions and rational maps of quasiprojective varieties.
- Define dimension, tangent spaces, and smoothness of quasiprojective varieties, and import key results from the theory of affine varieties.
- Study the blow-up of $\mathbb{A}^n$ at a point, and explore examples of how it can be used to resolve singularities.
- Calculate the dimensions of fibers of regular maps.
- Use Grassmannians and dimension theory to study lines on surfaces in $\mathbb{P}^3$.
- Count lines on smooth cubic surfaces.

In this last chapter of the book, we discuss several culminating topics, tying together our prior developments of affine, projective, and quasiprojective varieties. In the first three sections, we further develop the parallelism between the affine and quasiprojective settings, extending the notions of dimension and smoothness to the quasiprojective setting. In some sense, these extensions require little additional work: dimension and smoothness are local properties, and since every quasiprojective variety is locally affine, our understanding of dimension and smoothness for affine varieties can be imported directly to the quasiprojective context.

In the fourth section, we explore the notion of a "blow-up," which is a construction that replaces a point in a variety with the projective space parametrizing the tangent directions at that point. As we will see through examples, a key application of blow-ups is to resolve the singular points of a variety. An important take-away from this section is that, even if one is initially interested only in affine varieties, the quest to resolve singularities naturally leads one into the quasiprojective setting.

We then turn toward a key theorem of regular maps—a vast generalization of the Rank-Nullity Theorem—which describes the dimensions of the "fibers" of a regular map in terms of the dimensions of the domain and codomain. As we will see, a geometric application of this result is that every surface in $\mathbb{P}^3$ defined by a homogeneous polynomial of degree at most three contains at least one line.

The culmination of our efforts—presented in the final section—is one of the most famous classical results in algebraic geometry: every smooth cubic surface in $\mathbb{P}^3$ contains exactly 27 lines. Remarkably, neither the number of lines nor the combinatorics of which pairs of lines meet each other depends on the ground field or on the defining equation of the smooth cubic surface. This classical result provides just a glimpse of the rich and beautiful structure that one will become more familiar with as they continue their studies in algebraic geometry.

Section 12.1   Rational functions

In preparation for defining dimension of quasiprojective varieties in the next section, we devote this section to the introduction of rational functions. In the context of affine varieties, rational functions on X were defined as elements of $\text{Frac}(K[X])$; that is, as ratios of polynomial functions on X. When $X \subseteq \mathbb{P}^n$ is quasiprojective, we again consider ratios of polynomials, but the key new feature is that a ratio of two homogeneous polynomials of the same degree $f, g \in K[x_0, \ldots, x_n]$ can now be viewed as defining a regular function on the open set $X \setminus \mathcal{V}(g)$. It is with this observation in mind that we define rational functions on quasiprojective varieties in terms of regular functions on open sets.

12.1 DEFINITION   *Rational function on a quasiprojective variety*

Let X be a quasiprojective variety. A *rational function on* X is an equivalence class of pairs (U, F), where $U \subseteq X$ is a dense open subset of X and F is a regular function on U, under the equivalence relation

$$(U, F) \sim (V, G) \iff F(a) = G(a) \ \forall a \in U \cap V.$$

Intuitively, a rational function on X is a regular function defined "almost everywhere," in much the same way that a ratio like $1/x$ is a regular function on $\mathbb{A}^1 \setminus \{0\}$, which is "almost all" of $\mathbb{A}^1$. One should check that the relation in Definition 12.1 is an equivalence relation (Exercise 12.1.1), and this hinges on the assumption that the domains are dense, not just open, implying that the domains of two representatives of a rational function must intersect in a dense open subset. The motivation for the equivalence relation is simply that different representatives of a rational function may have different domains, but we want to consider them equivalent if they agree "almost everywhere." The following example illustrates this notion of equivalence.

12.2 EXAMPLE   Rational functions on $\mathbb{P}^1$

Let $U = \mathbb{P}^1 \setminus \{[1:0]\}$ and $V = \mathbb{P}^1 \setminus \{[1:0], [0:1]\}$, and consider the regular functions $F \in K[U]$ and $G \in K[V]$ defined by

$$F([a:b]) = \frac{a}{b} \quad \text{and} \quad G([a:b]) = \frac{a^2}{ba}.$$

Note that both U and V are dense open subsets of $\mathbb{P}^1$, so both (U, F) and (V, G) give rise to rational functions on $\mathbb{P}^1$. Moreover, since F and G agree on $U \cap V$, we conclude that $(U, F) \sim (V, G)$. Intuitively, F and G define the same rational function simply because the quotient defining G reduces to the quotient defining F.

A rational function on X is generally denoted $F : X \dashrightarrow K$, where the dashed arrow indicates that F may not be defined on all of X. By the *domain* of a rational function $F : X \dashrightarrow K$, we mean the union of the domains of all of its representatives. Since dense open subsets are closed under unions, the domain of a rational function is a dense open subset of X. Moreover, by the local nature of regular functions, a rational function $F : X \dashrightarrow K$ is regular on its domain. Thus, we may view the domain of a rational function as the largest set on which a rational function is regular.

12.1. RATIONAL FUNCTIONS

Like the set of regular functions, the set of rational functions naturally forms a K-algebra (Exercise 12.1.2), leading to the following definition.

12.3 DEFINITION *Ring of rational functions*

Let X be a quasiprojective variety. The *ring of rational functions* on X, denoted $K(X)$, is the K-algebra of all rational functions $F : X \dashrightarrow K$.

Using that isomorphisms of quasiprojective varieties identify open sets and the rings of regular functions on them (Corollary 11.57), it follows that rings of regular functions on quasiprojective varieties are intrinsic (Exercise 12.1.3).

Let us consider a few concrete examples of rings of regular functions.

12.4 EXAMPLE $K(\mathbb{P}^n) = K(x_1, \ldots, x_n)$

Let us first argue that the rational functions on $\mathbb{P}^n$ can all be represented by ratios of homogeneous polynomials in $K[x_0, \ldots, x_n]$ of the same degree. Such a ratio manifestly defines a rational function on $\mathbb{P}^n$: given $f, g \in K[x_0, \ldots, x_n]$, homogeneous of the same degree, f/g defines a regular function on the open set $U = \mathbb{P}^n \setminus \mathcal{V}(g)$, which is dense because $\mathbb{P}^n$ is irreducible. Conversely, given any rational function on $\mathbb{P}^n$ represented by a regular function $F : U \to K$ on an open subset $U \subseteq \mathbb{P}^n$, Lemma 11.29 shows that there exist $f, g \in K[x_0, \ldots, x_n]$, homogeneous of the same degree, such that $\mathcal{V}(g) \cap U = \emptyset$ and $F = f/g$ on U. It follows that

$$(U, F) \sim (\mathbb{P}^n \setminus \mathcal{V}(g), f/g),$$

showing that any rational function on $\mathbb{P}^n$ can be represented by a ratio of homogeneous polynomials of the same degree, as claimed.

Some reflection should convince the reader that two ratios f_1/g_1 and f_2/g_2 define the same rational function on $\mathbb{P}^n$ if and only if $f_1 g_2 - f_2 g_1 = 0$, which is the same as the relation defining elements of $K(x_0, \ldots, x_n) = \text{Frac}(K[x_0, \ldots, x_n])$. In other words, we can identify $K(\mathbb{P}^n)$ with the subring of $K(x_0, \ldots, x_n)$ comprised of quotients whose numerator and denominator are both homogeneous of the same degree. Upon dehomogenizing at x_0, we then obtain $K(\mathbb{P}^n) = K(x_1, \ldots, x_n)$.

12.5 EXAMPLE Rational functions on $\mathcal{V}(xy) \subseteq \mathbb{P}^2$

Let X be the reducible variety $\mathcal{V}(xy) \subseteq \mathbb{P}^2$, comprised of the two components $\mathcal{V}(x)$ and $\mathcal{V}(y)$. Define rational functions $F : X \dashrightarrow K$ and $G : X \dashrightarrow K$ by

$$F([a:b:c]) = a/c \quad \text{and} \quad G([a:b:c]) = b/c,$$

both of which are defined on the dense open set $U = X \setminus \{[1:0:0], [0:1:0]\}$. Since neither F nor G vanishes on its domain, neither of them is zero in $K(X)$. However, the product $F \cdot G$ vanishes on U, from which we see that the ring of rational functions on X is not an integral domain, and in particular, it is not a field.

In the previous examples, we observed that the ring of rational functions on the irreducible variety $\mathbb{P}^n$ is not just a ring, but a field, while the ring of rational functions on the reducible variety $\mathcal{V}(xy) \subseteq \mathbb{P}^2$ is not a field, as it has zero divisors. The next result naturally generalizes these observations.

12.6 PROPOSITION *Rational functions on irreducible varieties*

Let X be a quasiprojective variety. Then $K(X)$ is a field if and only if X is irreducible.

PROOF   Exercise 12.1.4.   □

As a consequence of Proposition 12.6, we typically refer to $K(X)$ as the *field of rational functions* when X is an irreducible quasiprojective variety.

As we did for regular functions, we should resolve the double use of the notation $K(X)$ in the affine setting by confirming that, when X is an irreducible affine variety, its function field—which we defined as the fraction field of its coordinate ring—is the same as the field of rational functions on the associated quasiprojective variety.

12.7 PROPOSITION *Field of rational functions on affine varieties*

Let $X_0 \subseteq \mathbb{A}^n$ be an irreducible closed set, and let $X = j_0(X_0) \subseteq \mathbb{P}^n$ be the corresponding quasiprojective variety. Then there is a canonical K-algebra isomorphism

$$K(X) = K(X_0)$$

between the field of rational functions on X and the function field of X_0.

PROOF   Recall that $K(X_0) = \mathrm{Frac}(K[X_0])$, so an element of $K(X_0)$ can be expressed as F_0/G_0 with $F_0, G_0 \in K[X_0]$ and $G_0 \neq 0$. Choose polynomial representatives $f_0, g_0 \in K[x_1, \ldots, x_n]$ so that $F_0 = [f_0]$ and $G_0 = [g_0]$, and let $f, g \in K[x_0, \ldots, x_n]$ be homogeneous polynomials of the same degree that dehomogenize at x_0 to f_0 and g_0, respectively. As f/g is regular on $X \setminus \mathcal{V}(g)$, we obtain a rational function

$$f/g : X \dashrightarrow K.$$

We begin by verifying that the map $F_0/G_0 \mapsto f/g$ is well-defined on $K(X_0)$. If F_0'/G_0' is an equivalent quotient, then $F_0 G_0' = F_0' G_0 \in K[X_0]$. Therefore, any polynomial representatives $F_0' = [f_0']$ and $G_0' = [g_0']$ satisfy $f_0 g_0' - f_0' g_0 \in \mathcal{I}(X_0)$. Thus, if f' and g' are homogeneous polynomials of the same degree that dehomogenize at x_0 to f_0' and g_0', then f/g and f'/g' agree on the dense open subset $X \setminus \mathcal{V}(gg') \subseteq X$, so they represent the same rational functions on X. In light of this, we obtain a well-defined map $\varphi : K(X_0) \to K(X)$ given by the above construction:

$$\varphi(F_0/G_0) = f/g.$$

As φ is readily seen to be a homomorphism, it remains to check that φ is bijective.

To prove that φ is injective, assume that $\varphi(F_0/G_0) = f/g = 0 \in K(X)$. Then f/g must vanish on a dense open subset of X, implying that f vanishes X. But this can only happen if F_0 vanishes on X_0, so $F_0/G_0 = 0 \in K(X_0)$, proving injectivity.

To prove that φ is surjective, let $F : X \dashrightarrow K$ be a rational function. Given any point p in the domain of F, regularity of F at p implies that there exist homogeneous polynomials f and g such that $p \notin \mathcal{V}(g)$ and $F = f/g$ on $X \setminus \mathcal{V}(g)$. Upon dehomogenizing at x_0, we obtain polynomials $f_0, g_0 \in K[x_1, \ldots, x_n]$ such that $\varphi([f_0]/[g_0]) = F$, proving surjectivity and thereby completing the proof.   □

12.1. RATIONAL FUNCTIONS

Proposition 12.7 tells us that the field of rational functions on an irreducible affine variety is simply the fraction field of its coordinate ring. In fact, knowing how to compute $K(X)$ in this very special case allows us to compute $K(X)$ on any irreducible quasiprojective variety X. This is a consequence of the next result.

12.8 PROPOSITION $K(X)$ *is determined on dense open subvarieties*

Let X be a quasiprojective variety, and let $U \subseteq X$ be a dense open subvariety of X. Then
$$K(X) = K(U).$$

PROOF Any dense open subset of U is also a dense open subset of X, so a rational function $F : U \dashrightarrow K$ defined on $V \subseteq U$ also gives rise to a rational function $F : X \dashrightarrow K$ defined on the same set $V \subseteq X$. This leads to a canonical inclusion $K(U) \subseteq K(X)$.

Conversely, if $G : X \dashrightarrow K$ is a rational function defined on a dense open set $W \subseteq X$, then $W \cap U$ is a dense open set in X and $(U, G) \sim (U \cap W, G|_{U \cap W})$. Since $U \cap W$ is a dense open set in U, it follows that $G \in K(X)$ is equivalent to a rational function coming from $K(U)$, and we conclude that $K(U) = K(X)$. □

The combination of Propositions 12.7 and 12.8, along with the fact that every quasiprojective variety is locally affine, implies that the field of rational functions on any irreducible quasiprojective variety can be computed by reducing to the function field of an affine variety. The following examples illustrate this strategy.

12.9 EXAMPLE $K(\mathbb{P}^n) = K(x_1, \ldots, x_n)$, revisited

Consider the affine patch $\mathbb{A}_0^n \subseteq \mathbb{P}^n$. Then we have
$$K(\mathbb{P}^n) = K(\mathbb{A}_0^n) = \operatorname{Frac}(K[x_1, \ldots, x_n]) = K(x_1, \ldots, x_n),$$
where the first equality is Proposition 12.8 and the second is Proposition 12.7.

12.10 EXAMPLE *The field of rational functions on* $\mathcal{V}(wx^2 - y^3)$

Consider the projective variety $X = \mathcal{V}(wx^2 - y^3) \subseteq \mathbb{P}^2$, which contains the affine patch $X_0 = \mathcal{V}_\mathbb{A}(x^2 - y^3) \subseteq \mathbb{A}^2$ as a dense open subset. We calculated $K(X_0)$ in Example 6.7, yielding
$$K(X) = K(X_0) = \operatorname{Frac}\left(\frac{K[x, y]}{\langle x^2 - y^3 \rangle}\right) \cong K(t),$$
where the isomorphism is given by identifying $[x]$ with t^3 and $[y]$ with t^2.

Since $K(X) \cong K(\mathbb{P}^1)$, one might be led to suspect that $X \cong \mathbb{P}^1$. This is false, but something weaker is true. While the two varieties are not isomorphic everywhere, the map depicted to the right becomes an isomorphism upon removing the cusp point $[1 : 0 : 0]$ from X and its preimage $[1\ 0]$ from $\mathbb{P}^1$. We will return to this discussion in Example 12.14 below.

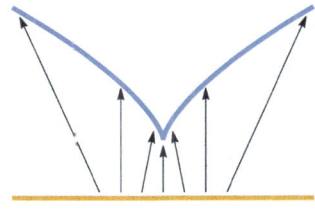

12.11 EXAMPLE The field of rational functions on $\mathcal{V}(w^3 + x^3 + y^3)$

Let $X = \mathcal{V}(w^3 + x^3 + y^3) \subseteq \mathbb{P}^2$. Similarly to the previous example, we can calculate $K(X)$ by calculating the function field of the affine piece X_0, yielding

$$K(X) = K(X_0) = \operatorname{Frac}\left(\frac{K[x,y]}{\langle 1 + x^3 + y^3 \rangle}\right).$$

In Example 6.8, we proved that $K(X_0) \not\cong K(t_1, \ldots, t_d)$ for any d. As isomorphic varieties have isomorphic rings of rational functions (Exercise 12.1.3), it follows that $X \not\cong \mathbb{P}^d$ for any d. In particular, $X \not\cong \mathbb{P}^1$, and this provides a concrete example of a phenomenon discussed in Section 10.2: unlike the case of irreducible conics, irreducible cubics in $\mathbb{P}^2$ need not be isomorphic to $\mathbb{P}^1$.

Given that regular functions are a special case of regular maps, one might expect that rational functions are a special case of a more general notion of rational maps, and indeed this is the case. In fact, the definition of rational maps mimics the definition of rational functions almost verbatim.

12.12 DEFINITION *Rational map*

Let X and Y be quasiprojective varieties. A *rational map* from X to Y is an equivalence class of pairs (U, F), where $U \subseteq X$ is a dense open subset of X and $F : U \to Y$ is a regular map, under the equivalence relation

$$(U, F) \sim (V, G) \iff F(a) = G(a) \ \forall a \in U \cap V.$$

We denote a rational map by $F : X \dashrightarrow Y$. The *domain* of a rational map $F : X \dashrightarrow Y$, denoted $U_F \subseteq X$, is the union of the domains of all of the representatives of F, and the *image* of F is $F(U_F) \subseteq Y$.

Intuitively, a rational map is a map between quasiprojective varieties that is defined "almost everywhere." The analogue of "isomorphism" in the study of rational maps is the notion of *birational equivalence*.

12.13 DEFINITION *Birational equivalence*

Let X and Y be quasiprojective varieties. A *birational equivalence* is a rational map $F : X \dashrightarrow Y$ for which there exists a rational map $G : Y \dashrightarrow X$ such that $G \circ F$ is a rational map equivalent to id_X and $F \circ G$ is a rational map equivalent to id_Y. If there exists a birational equivalence between X and Y, we say that X and Y are *birational*.

Notice that the composition $G \circ F$ is defined on the open set of points in U_F that map into U_G, so the requirement in Definition 12.13 that $G \circ F$ is a rational map means that this set must be dense in X. In essence, birational varieties are isomorphic "almost everywhere," giving rise to a relation that is strictly weaker than that of isomorphism. The next example gives a concrete illustration of a birational equivalence: the cuspidal curve of Example 12.10 is birational to the projective line.

12.1. RATIONAL FUNCTIONS

12.14 EXAMPLE $\mathcal{V}(wx^2 - y^3)$ is birational to $\mathbb{P}^1$

Let $X = \mathcal{V}(wx^2 - y^3) \subseteq \mathbb{P}^2$, as in Example 12.10. Then X is birational to $\mathbb{P}^1$. To see this, consider the regular map $F : \mathbb{P}^1 \to X$ given by

$$F([a:b]) = [a^3 : b^3 : ab^2].$$

While F does not have an inverse, there is a rational map $G : X \dashrightarrow \mathbb{P}^1$ given by

$$G([a:b:c]) = [c:b],$$

which is defined on the dense open subset of X where x and y are not both zero, and which is inverse to G as rational maps. Visually, this is the statement that the maps F and G become isomorphisms when the cusp at $[1:0:0]$ (shown in Example 12.10) is removed from X and its preimage $[1:0]$ is removed from $\mathbb{P}^1$.

In much the same way that isomorphic quasiprojective varieties have isomorphic rings of regular functions, birational quasiprojective varieties have isomorphic rings of rational functions; for instance, for X as in Example 12.14, the result of Example 12.10 shows that

$$K(X) \cong K(t) \cong K(\mathbb{P}^1),$$

reflecting the birational equivalence of Example 12.14. In fact, in contrast to the situation for regular functions, the converse of the above statement is also true. We state this result here, but since birational equivalence will not play a central role in the remainder of this text, we relegate the proof to the exercises.

12.15 PROPOSITION *Birationality and rational functions*

Let X and Y be quasiprojective varieties. Then X and Y are birational if and only if $K(X)$ and $K(Y)$ are isomorphic K-algebras.

PROOF   Exercises 12.1.8 and 12.1.9.   □

There is a large and active subfield of algebraic geometry, known as *birational geometry*, devoted to classifying varieties according to whether they are or are not birational. As birational varieties are isomorphic "almost everywhere," they share certain attributes: their rings of rational functions are the same, by Proposition 12.15, and therefore (as we will see in the next section), they have the same dimension. However, birational varieties can also differ in key respects; for instance, one may be smooth (a notion we will meet for quasiprojective varieties in the section after next) while the other may be singular, as is the case for the pair of varieties in Example 12.14. The quest to find smooth varieties birational to a given singular one, known as *desingularization* and based on the notion of *blowing up* singularities, is a fundamental topic in algebraic geometry, a notion that we will get a closer glimpse of later in this chapter.

Exercises for Section 12.1

12.1.1 (a) Prove that the relation in Definition 12.1 is an equivalence relation.
   (b) Give an example to show that the relation in Definition 12.1 can fail to be transitive if the open sets are not required to be dense.

12.1.2 Let X be a quasiprojective variety. Given dense open subsets $U, V \subseteq X$, regular functions $F : U \to K$ and $G : V \to K$, and a scalar $\lambda \in K$, define sums, products, and scalar multiples in the natural way:

$$F + G : U \cap V \to K, \quad FG : U \cap V \to K, \quad \text{and} \quad \lambda F : U \to K.$$

Explain why these operations are well-defined on equivalence classes of rational functions, and prove that they endow the set of rational functions with the structure of a K-algebra.

12.1.3 Let X and Y be isomorphic quasiprojective varieties. Prove that their rings of rational functions are isomorphic: $K(X) \cong K(Y)$.

12.1.4 Prove Proposition 12.6.

12.1.5 Let X be a quasiprojective variety with irreducible components $X_1, \ldots, X_k$. Prove that
$$K(X) = K(X_1) \oplus \cdots \oplus K(X_k).$$

12.1.6 Prove that $\mathbb{A}^n$ is birational to $\mathbb{P}^n$.

12.1.7 Let X and Y be quasiprojective varieties. Prove that X is birational to Y if and only if there exist dense open subsets $U \subseteq X$ and $V \subseteq Y$ such that U and V are isomorphic quasiprojective varieties.

12.1.8 Let $F : X \dashrightarrow Y$ be a rational map of quasiprojective varieties.
   (a) Suppose that the image of F is dense in Y. Construct a pullback homomorphism $F^* : K(Y) \to K(X)$.
   (b) Suppose that F is a birational equivalence. Prove that the image of F is dense in Y and that F^* is an isomorphism.

12.1.9 Let $\varphi : K(Y) \to K(X)$ be a homomorphism of K-algebras, where X and Y are quasiprojective varieties.
   (a) Construct a rational map $F : X \dashrightarrow Y$ such that $F^* = \varphi$, where F^* is as in Exercise 12.1.8. (Hint: Prove that Y contains a dense open affine subvariety V, and consider the images under φ of the coordinate functions on V.)
   (b) Prove that if φ is an isomorphism, then F is a birational equivalence.

Section 12.2 Dimension of quasiprojective varieties

In this section, we extend the notion of dimension from the setting of affine varieties to that of quasiprojective varieties, and we then use the local nature of dimension to import key results from the affine setting to the more general quasiprojective setting.

To begin, recall that we defined the dimension of an irreducible affine variety as the transcendence degree of its function field (Section 6.5). Moreover, in the previous section, we introduced rational functions as a generalization of the notion of function fields to the quasiprojective setting. From this perspective, the following definition naturally generalizes our definition of dimension from the affine setting.

12.16 DEFINITION *Dimension of a quasiprojective variety*

The *dimension* of an irreducible quasiprojective variety X is the transcendence degree of its field of rational functions over K:

$$\dim(X) = \operatorname{trdeg}_K(K(X)).$$

The *dimension* of a nonempty reducible quasiprojective variety is the maximum of the dimensions of its irreducible components.

An essential property of dimension in this more general quasiprojective setting is that it is local in nature. In particular, the dimension of a quasiprojective variety can be determined on any dense open subset, as the next result verifies.

12.17 PROPOSITION *Dimension and dense open subvarieties*

If X is a quasiprojective variety and $U \subseteq X$ is a dense open subset, then

$$\dim(X) = \dim(U).$$

PROOF   Assume first that X is irreducible. Then for any nonempty (and thus dense) open subset $U \subseteq X$, we have $K(X) = K(U)$ (Proposition 12.8), and it follows that $\operatorname{trdeg}(K(X)) = \operatorname{trdeg}(K(U))$, implying that $\dim(X) = \dim(U)$.

To prove the general case, suppose that $X = X_1 \cup \cdots \cup X_k$ is the irreducible decomposition of X. If $U \subseteq X$ is open and dense, it then follows that

$$U = (X_1 \cap U) \cup \cdots \cup (X_k \cap U)$$

is the irreducible decomposition of U (Exercise 11.2.7). Since each X_i is irreducible and $X_i \cap U$ is a nonempty open subset of X_i, the argument above then implies that $\dim(X_i) = \dim(X_i \cap U)$. It follows that the maximum dimension of an irreducible component of X is the same as the maximum dimension of an irreducible component of U, from which we conclude that $\dim(X) = \dim(U)$. □

Proposition 12.17 provides us with a useful way of reducing the dimension theory of quasiprojective varieties to the affine setting: given any irreducible quasiprojective variety X whose dimension we would like to know, we can simply choose an affine open subset $U \subseteq X$ and compute the dimension of U. The next several examples leverage this strategy.

12.18 EXAMPLE Dimension of projective space

Since $\mathbb{P}^n$ contains $\mathbb{A}_0^n = \mathbb{A}^n$ as a dense open subset, Proposition 12.17 implies that

$$\dim(\mathbb{P}^n) = \dim(\mathbb{A}^n) = n.$$

This conclusion also follows directly from Definition 12.16 and the observation made in Example 12.9 that $K(\mathbb{P}^n) = K(x_1, \ldots, x_n)$.

12.19 EXAMPLE Dimension of projective hypersurfaces

Let $f \in K[x_0, \ldots, x_n]$ be an irreducible homogeneous polynomial and consider the irreducible projective variety

$$X = \mathcal{V}_\mathbb{P}(f) \subseteq \mathbb{P}^n.$$

We would like to compute the dimension of X. Without loss of generality, assume that $f \neq x_0$, so that

$$X_0 = \mathcal{V}_\mathbb{P}(f) \cap \mathbb{A}_0^n \neq \emptyset.$$

Since X_0 is a dense open subset of X, Proposition 12.17 implies that X and X_0 have the same dimension. Furthermore, since X_0 is an affine variety, we can utilize tools that were developed to study dimension in the affine setting. In particular, since

$$X_0 = \mathcal{V}_\mathbb{A}(f_0) \quad \text{where} \quad f_0(x_1, \ldots, x_n) = f(1, x_1, \ldots, x_n),$$

the Fundamental Theorem of Dimension Theory (Theorem 6.45) implies that X_0 has dimension $n - 1$, and we conclude that $\dim(X) = n - 1$, as well.

Moreover, as the vanishing set of any nonconstant homogeneous polynomial (not necessarily irreducible) is the union of the vanishing sets of its distinct irreducible factors, we can remove the hypothesis that f is irreducible and conclude that *every irreducible component of the vanishing set of a single nonconstant homogeneous polynomial in projective space has codimension one.*

12.20 EXAMPLE Dimension of Grassmannians

Consider the Grassmannian $\mathbb{G}(k, n) \subseteq \mathbb{P}^{\binom{n}{k}-1}$, introduced in Section 10.5. In Exercise 10.5.9, the reader was tasked with showing that $\mathbb{G}(k, n)$ is irreducible and that all of its affine restrictions are isomorphic to $\mathbb{A}^{k(n-k)}$. Thus, $\mathbb{G}(k, n)$ contains a dense open subvariety isomorphic to an affine space of dimension $k(n - k)$, and it follows that

$$\dim(\mathbb{G}(k, n)) = k(n - k).$$

To understand this dimension more intuitively, recall that elements of $\mathbb{G}(k, n)$ can be represented by $k \times n$ matrices with linearly independent rows. Moreover, up to a change of basis, we can always choose a matrix representation that contains a $k \times k$ submatrix equal to the identity. Since there are then $k(n - k)$ remaining entries that can vary independently, we see that there are $k(n - k)$ independent parameters within $\mathbb{G}(k, n)$, consistent with the dimension computed above.

12.2. DIMENSION OF QUASIPROJECTIVE VARIETIES

To further illustrate the utility of working affine-locally, we now develop a generalization of the Fundamental Theorem of Dimension Theory in the quasiprojective setting. To state it in its most general form, we introduce the notion of *local hypersurfaces*, a new type of closed subvariety that can be locally described by the vanishing of a single regular function. In the next definition, we adopt the notation $\mathcal{V}(F) = \{a \in U \mid F(a) = 0\}$ for any regular function $F \in K[U]$, which is a slight, but natural, generalization of how we have been using the notation $\mathcal{V}(-)$ thus far.

12.21 DEFINITION *Local hypersurface*

Let X be a quasiprojective variety. We say that a nonempty subset $Y \subseteq X$ is a *local hypersurface in* X if every $p \in X$ is contained in an open set $U \subseteq X$ such that $Y \cap U = \mathcal{V}(F)$ for a regular function $F \in K[U]$ that does not vanish on any irreducible component of U.

The key point of Definition 12.21 is that it gives a uniform framework for considering two types of subvarieties of a quasiprojective variety $X \subseteq \mathbb{P}^n$: those of the form $Y = X \cap \mathcal{V}(F)$ for a regular function $F \in K[X]$ not vanishing on any irreducible component of X, and those of the form $Y = X \cap \mathcal{V}(f)$ for a homogeneous polynomial $f \in K[x_0, \ldots, x_n]$ not vanishing on any irreducible component of X. We note that local hypersurfaces are closed subvarieties of X. To justify this, observe that $\mathcal{V}(F) \subseteq U$ is the preimage of the single point $\{0\}$ under the regular map $F: U \to \mathbb{A}^1$, so it is closed in U by continuity of regular maps (Theorem 11.51). From here, Lemma 11.48 then implies that local hypersurfaces are closed.

We now state the Fundamental Theorem of Dimension Theory in this context.

12.22 THEOREM *Fundamental Theorem of Dimension Theory*

Let X be a quasiprojective variety whose irreducible components all have the same dimension, and let $Y \subseteq X$ be a local hypersurface in X. Then every irreducible component of Y has dimension $\dim(X) - 1$.

PROOF Let Z be an irreducible component of Y, and let $p \in Z$. By definition of local hypersurfaces, there exists an open subset $U \subseteq X$ containing p and a regular function $F \in K[U]$ that does not vanish on any irreducible component of U such that
$$Y \cap U = \mathcal{V}(F).$$
By restricting to a smaller open subset containing p if necessary, we may further assume that U is affine. Since $Z \cap U$ is irreducible, it must be contained in some irreducible component of U; call this component V. Then V is an irreducible affine variety and $Z \cap V$ is an irreducible component of $Y \cap V = \mathcal{V}(F|_V)$. Since $\mathcal{V}(F|_V)$ contains p and $F|_V \neq 0$, it follows that $\mathcal{V}(F|_V)$ is neither empty nor all of V, so the strong form of the Fundamental Theorem of Dimension Theory in the affine setting implies that
$$\dim(Z \cap V) = \dim(V) - 1.$$
Irreducibility of Z gives $\dim(Z) = \dim(Z \cap V)$, while irreducibility of V implies that V is dense in some component of X, all of which have the same dimension, so $\dim(V) = \dim(X)$. Thus, we obtain $\dim(Z) = \dim(X) - 1$, as claimed. □

By iteratively applying the previous result, we obtain the following upper bound on the dimension of vanishing sets within quasiprojective varieties.

12.23 COROLLARY *Dimension of vanishing sets*

Let $X \subseteq \mathbb{P}^n$ be a quasiprojective variety, all of whose irreducible components have the same dimension, and let $Y = \mathcal{V}(f_1,\ldots,f_k,F_1,\ldots,F_\ell) \subseteq X$ where $f_1,\ldots,f_k \in K[x_0,\ldots,x_n]$ are homogeneous and $F_1,\ldots,F_\ell \in K[X]$. If $Y \neq \emptyset$, then
$$\dim(Y) \geq \dim(X) - k - \ell.$$

PROOF Exercise 12.2.3. □

Moreover, exactly as in the affine setting, the Fundamental Theorem of Dimension Theory leads to the following alternative characterization of dimension, which, in many topological contexts, is taken as the definition of dimension.

12.24 COROLLARY *Chains of inclusions and dimension*

If X is a quasiprojective variety, then $\dim(X)$ is equal to the maximum d such that there exist closed irreducible subvarieties $X_0,\ldots,X_d \subseteq X$ with
$$X_0 \subsetneq X_1 \subsetneq \cdots \subsetneq X_d \subseteq X.$$

PROOF Exercise 12.2.4. □

To close this section, we specialize to the study of solutions of homogeneous systems of polynomials. Recall that one of our original motivations for studying $\mathbb{P}^n$ was to recover solutions of polynomial systems that may have been "lost to infinity." For example, while two lines may not intersect in $\mathbb{A}^2$, we have seen that a pair of lines will always intersect within $\mathbb{P}^2$; in essence, the intersection of two parallel lines in $\mathbb{A}^2$ occurs at a "point at infinity" in $\mathbb{P}^2$. The next result is a vast generalization of this observation, guaranteeing that the intersection of k hypersurfaces in $\mathbb{P}^n$ will always have dimension at least $n - k$; in particular, the intersection is guaranteed to be nonempty as long as $k \leq n$.

12.25 PROPOSITION *Dimension of homogeneous systems*

If $f_1,\ldots,f_k \in K[x_0,\ldots,x_n]$ are nonconstant homogeneous polynomials with $k \leq n$, then $\mathcal{V}_\mathbb{P}(f_1,\ldots,f_k) \neq \emptyset$ and
$$\dim(\mathcal{V}_\mathbb{P}(f_1,\ldots,f_k)) \geq n - k.$$

One should compare the statement of Proposition 12.25 with the statement of Corollary 6.47 in the affine setting. The key difference in the two results is that, in the projective setting, we are able to remove the assumption of nonemptiness that was required as a hypothesis in the affine setting.

12.2. DIMENSION OF QUASIPROJECTIVE VARIETIES

PROOF OF PROPOSITION 12.25   If $\mathcal{V}_{\mathbb{P}}(f_1, \ldots, f_k) \neq \emptyset$, then Corollary 12.23 implies that
$$\dim(\mathcal{V}_{\mathbb{P}}(f_1, \ldots, f_k)) \geq \dim(\mathbb{P}^n) - k = n - k.$$
Thus, we need only show that $\mathcal{V}_{\mathbb{P}}(f_1, \ldots, f_k) \neq \emptyset$. To do so, consider the affine vanishing set
$$\mathcal{V}_{\mathbb{A}}(f_1, \ldots, f_k) \subseteq \mathbb{A}^{n+1}.$$
Since $f_1, \ldots, f_k$ are nonconstant homogeneous polynomials, they all vanish at the origin. Thus, $\mathcal{V}_{\mathbb{A}}(f_1, \ldots, f_k) \neq \emptyset$, and by Corollary 6.47, it follows that
$$\dim(\mathcal{V}_{\mathbb{A}}(f_1, \ldots, f_k)) \geq n + 1 - k > 0.$$
Since a positive-dimensional variety must contain more than one point, we see that $f_1, \ldots, f_k$ must vanish at some nonzero point $(a_0, \ldots, a_n) \in \mathbb{A}^{n+1}$, implying that $\mathcal{V}_{\mathbb{P}}(f_1, \ldots, f_k)$ contains the point $[a_0 : \cdots : a_n] \in \mathbb{P}^n$. Thus, we conclude that $\mathcal{V}_{\mathbb{P}}(f_1, \ldots, f_k) \neq \emptyset$, as desired. □

Exercises for Section 12.2

12.2.1 Prove the following results by using locality of dimension to reduce them to the appropriate results from the affine setting.

(a) If $X, Y \subseteq \mathbb{P}^n$ are quasiprojective varieties with $X \subseteq Y$, then
$$\dim(X) \leq \dim(Y).$$

(b) If, furthermore, Y is irreducible and $X \subsetneq Y$ is closed in Y, then
$$\dim(X) < \dim(Y).$$

12.2.2 Let $X \subseteq \mathbb{P}^n$ be a projective variety such that every irreducible component of X has dimension $n - 1$. Prove that $\mathcal{I}(X)$ is principal, generated by any homogeneous polynomial of minimal degree in $\mathcal{I}(X)$.

12.2.3 Prove Corollary 12.23.

12.2.4 Prove Corollary 12.24.

12.2.5 Let $F : X \to Y$ be a regular map of quasiprojective varieties such that $F(X)$ is dense in Y. Prove that $\dim(Y) \leq \dim(X)$. (Hint: Show that the pullback—see Exercise 12.1.8—gives an injection $F^* : K(Y) \to K(X)$.)

12.2.6 Let X and Y be quasiprojective varieties. Prove that
$$\dim(X \times Y) = \dim(X) + \dim(Y).$$

12.2.7 Prove that every polynomial map $\mathbb{P}^n \to \mathbb{P}^m$ is constant when $n > m$.

12.2.8 Prove that every regular map $\mathbb{P}^n \to \mathbb{P}^m$ is constant when $n > m$.

Section 12.3   Local rings and tangent spaces

We now turn to the task of generalizing the concepts of tangent spaces and smoothness from the affine setting to the quasiprojective setting. As a first attempt at defining the tangent space at a point of a quasiprojective variety, one might simply restrict to an affine neighborhood of that point and then use the tangent space of the corresponding affine variety at that point. While this gives the correct notion, one drawback of this approach is that it is not immediately clear that different affine restrictions give rise to the same (canonically isomorphic) tangent space. To avoid this subtlety, we develop an alternative approach to tangent spaces using local rings.

Roughly speaking, the *local ring* of a quasiprojective variety X at a point $a \in X$ behaves like the coordinate ring of an infinitesimal neighborhood of a, a neighborhood that is contained in every other neighborhood. The local ring is an algebraic tool that captures geometric information about X near a, and in particular, can be used to define the tangent space of X at a in an intrinsic way, naturally generalizing the intrinsic description of tangent spaces in Theorem 7.18. As we will see, this leads to a definition of the tangent space at a point of a quasiprojective variety that recovers our affine notion upon restricting to any affine neighborhood.

We begin our discussion with a formal definition of the local ring at a point.

12.26   DEFINITION   *Local ring of a quasiprojective variety at a point*

Let X be a quasiprojective variety with $a \in X$. The *local ring of X at a*, denoted $K[X]_a$, is the set of equivalence classes of pairs (U, F) where $U \subseteq X$ is an open subset of X containing a and $F : U \to K$ is a regular function. Two pairs (U, F) and (V, G) are equivalent if there exists an open set $W \subseteq U \cap V$ containing a such that $F(b) = G(b)$ for all $b \in W$.

In other words, an element of $K[X]_a$ can be represented by a regular function on a neighborhood of a, and two regular functions on two possibly different neighborhoods give rise to the same element if and only if they agree on some possibly smaller neighborhood. Thus, an element of the local ring of X at a can be viewed as a regular function defined on an arbitrarily small neighborhood of a. For this reason, we refer to elements of the local ring $K[X]_a$ as *local functions at a*.

There are several aspects of Definition 12.26 that require further clarification. For example, one should pause to understand why the relation on pairs (U, F) is, in fact, an equivalence relation, and one should also carefully define the natural algebraic operations of addition, multiplication, and scalar multiplication on $K[X]_a$ that make it a K-algebra (see Exercise 12.3.1).

Before further developing local rings, let us consider a few concrete examples.

12.27   EXAMPLE   Local functions given by quotients of polynomials

Let $X \subseteq \mathbb{P}^n$ be a quasiprojective variety with $a \in X$. Given two homogeneous polynomials $f, g \in K[x_0, \ldots, x_n]$ that have the same degree, their quotient defines a regular map $f/g : X \setminus \mathcal{V}(g) \to K$. In particular, if $g(a) \neq 0$, then f/g gives rise to a local function at a defined on the neighborhood $U = X \setminus \mathcal{V}(g)$. In fact, we will see below that every local function can be realized as a quotient of polynomials.

12.3. LOCAL RINGS AND TANGENT SPACES

12.28 EXAMPLE *Different quotients giving rise to the same local function*

Consider the reducible projective variety $X = \mathcal{V}(xy) = \mathcal{V}(x) \cup \mathcal{V}(y) \subseteq \mathbb{P}^2$, and let $a = [1:0:0]$. Note that a is in the irreducible component $\mathcal{V}(y)$ but is not in the irreducible component $\mathcal{V}(x)$. Consider the rational function

$$\frac{x+y}{x-y} : X \setminus \{[0:0:1]\} \to K.$$

This rational function is *not* constant, because it takes value 1 at $[1:0:0]$ and value -1 at $[0:1:0]$. However, the local function at a that this rational function defines is the constant function 1, simply because its numerator and denominator agree upon restricting to the neighborhood $U = X \setminus \mathcal{V}(x)$. This example illustrates the local nature of local functions: a local function at a does not "see" the irreducible component $\mathcal{V}(x) \subseteq X$ because this irreducible component does not contain a.

Building on the observation of the previous example, the next result shows that the local ring of a quasiprojective variety X at a point a truly is local to $a \in X$: in other words, it is insensitive to restricting to smaller neighborhoods of a.

12.29 PROPOSITION *Local rings are local*

Let X be a quasiprojective variety and $a \in X$. If $U \subseteq X$ is an open subvariety containing a, then
$$K[U]_a = K[X]_a.$$

PROOF Let X be a quasiprojective variety with $a \in X$, and let $U \subseteq X$ be an open subvariety containing a. There is a canonical restriction homomorphism on local rings
$$K[X]_a \to K[U]_a$$
that sends a local function $(V, F) \in K[X]_a$ to $(U \cap V, F|_{U \cap V}) \in K[U]_a$. Note that a local function is 0 if and only if any regular function representing it vanishes on a neighborhood of a; this implies that the restriction map is injective. The restriction map is surjective because every element of $K[U]_a$ can, by definition, be viewed as an element of $K[X]_a$, so it is the restriction of itself. Thus, $K[X]_a = K[U]_a$. □

Since quasiprojective varieties are locally affine, Proposition 12.29 reduces the study of local rings to the affine setting. Thus, it will be useful to have a concrete description of local rings of affine varieties, which we now provide.

12.30 LEMMA *Local rings of affine varieties*

Let X be an affine variety with $a \in X$. Every element of $K[X]_a$ can be represented as a quotient F/G where $F, G \in K[X]$ are regular functions and $G(a) \neq 0$. Two quotients F/G and F'/G' give rise to the same local function if and only if there is an $H \in K[X]$ such that $H(a) \neq 0$ and
$$H(FG' - F'G) = 0 \in K[X].$$

PROOF   Since open sets and rings of regular functions on open sets are preserved by isomorphisms, it suffices to assume that X is a closed subset of $\mathbb{A}_0^n \subseteq \mathbb{P}^n$. Let $a \in X$ and suppose that (U, H) represents an element of $K[X]_a$, so that $H \in K[U]$. By definition of regular functions, there exist $f, g \in K[x_0, \ldots, x_n]$, homogeneous of the same degree d, such that $g(a) \neq 0$ and

$$H(b) = \frac{f(b)}{g(b)} \quad \text{for all} \quad b \in U \setminus \mathcal{V}(g).$$

Define $F = f(x_0, x_1, \ldots, x_n)/x_0^d$ and $G = g(x_0, \ldots, x_n)/x_0^d$. As F and G are each quotients of homogeneous polynomials of the same degree with denominators that do not vanish on X, it follows that $F, G \in K[X]$, and by construction, we have

$$H(b) = \frac{F(b)}{G(b)} \quad \text{for all} \quad b \in U \setminus \mathcal{V}(G).$$

This shows that (U, H) is equivalent to $(X \setminus \mathcal{V}(G), F/G)$ in $K[X]_a$, so every local function can be realized as a quotient of two regular functions on X.

Now suppose that F/G and F'/G' are quotients of regular functions on X that give rise to the same local function in $K[X]_a$. Then $FG' - F'G$ must vanish on a neighborhood of a, and since open subsets of irreducible components are dense and the vanishing set of $FG' - F'G$ is closed in X, this implies that $FG' - F'G$ vanishes on every irreducible component of X that contains a. Let $H \in K[X]$ be any regular function such that $H(a) \neq 0$ and such that H vanishes on every irreducible component of X that does not contain a. Then $H(FG' - F'G) = 0 \in K[X]$.

Conversely, suppose that there exist $F, F', G, G', H \in K[X]$ such that none of the latter three vanish at a and $H(FG' - F'G) = 0 \in K[X]$. It then follows that

$$F(b)/G(b) = F'(b)/G'(b) \quad \text{for all} \quad b \in X \setminus \mathcal{V}(HGG').$$

Since $X \setminus \mathcal{V}(HGG')$ is an open neighborhood of a, we conclude that F/G and F'/G' give rise to the same local function in $K[X]_a$. □

> The description in Lemma 12.30 is the motivation for a more general algebraic process of "localization;" see Exercise 12.3.9 for more details.

We now revisit tangent spaces of affine varieties from a local-ring perspective. Recall from Chapter 7 that the tangent space $T_a X$ of an affine variety X at a point $a \in X$ was first defined extrinsically in terms of tangent vectors and gradients at a, but was later described intrinsically in terms of the maximal ideal of regular functions on X that vanish at a. The second intrinsic perspective has a natural analogue in the setting of local rings, which we now describe.

Let X be a quasiprojective variety with $a \in X$, and let $J_a \subseteq K[X]_a$ be the ideal of local functions that vanish at a. We note that J_a is a maximal ideal in $K[X]_a$; in fact, it is the *unique* maximal

> Motivated by this geometric setting, rings with a unique maximal ideal are commonly called "local rings."

ideal in $K[X]_a$ (Exercise 12.3.2). Specializing to the affine setting, where tangent spaces have already been defined in Chapter 7, we obtain the following alternative description of tangent spaces in terms of the ideal $J_a \subseteq K[X]_a$.

12.3. LOCAL RINGS AND TANGENT SPACES

> **12.31 PROPOSITION** *Tangent spaces of affine varieties, revisited*
>
> If X is an affine variety with $a \in X$ and $J_a \subseteq K[X]_a$ is the ideal of local functions that vanish at a, then there is a canonical vector-space isomorphism
>
> $$T_a X = \left(J_a / J_a^2\right)^\vee.$$

PROOF As in Section 7.3, let $I_a \subseteq K[X]$ denote the maximal ideal of regular (that is, polynomial) functions that vanish at a. By Theorem 7.18, there is a canonical vector-space isomorphism

$$T_a X = \left(I_a / I_a^2\right)^\vee.$$

Thus, we must prove that there is a canonical vector-space isomorphism

$$I_a / I_a^2 = J_a / J_a^2.$$

Consider the restriction homomorphism $K[X] \to K[X]_a$. Some reflection should convince the reader that this map sends I_a into J_a and I_a^2 into J_a^2. Thus, we obtain a canonical linear map

$$(12.32) \qquad I_a / I_a^2 \to J_a / J_a^2,$$

and it remains to show that this linear map is bijective.

To prove surjectivity of (12.32), first note that, by Lemma 12.30, every element of $K[X]_a$ can be written as F/G for some $F, G \in K[X]$ with $G(a) \neq 0$. Suppose that F/G gives rise to an element of J_a, or equivently, that $F \in I_a$. After possibly scaling both F and G, we can assume that $G(a) = 1$. It then follows that both F and $1/G - 1$ give rise to elements of J_a, so

$$F/G - F = F(1/G - 1) \in J_a^2.$$

This implies that the local functions defined by F/G and F become equivalent in the quotient J_a / J_a^2, and it follows that $[F/G] \in J_a / J_a^2$ is the image of $[F] \in I_a / I_a^2$, proving surjectivity of (12.32).

To prove injectivity, suppose that $F \in I_a$ restricts to a local function that lies in J_a^2. Elements in J_a^2 are finite sums of products of pairs of elements in J_a. Thus, by Lemma 12.30, there exist $m \geq 1$ and $G_i, G_i', H_i, H_i' \in K[X]$ for $i = 1, \ldots, m$ such that $G_i(a) = 0 = G_i'(a)$ and $H_i(a) \neq 0 \neq H_i'(a)$ for all $i = 1, \ldots, m$, and a neighborhood U of a such that

$$F(b) = \sum_{i=1}^m \frac{G_i(b)}{H_i(b)} \frac{G_i'(b)}{H_i'(b)} \quad \text{for all} \quad b \in U.$$

Choose $H \in K[X]$ such that $H(a) \neq 0$ and such that H vanishes on $X \setminus U$. Then

$$H(b) F(b) = H(b) \sum_{i=1}^m \frac{G_i(b)}{H_i(b)} \frac{G_i'(b)}{H_i'(b)} \quad \text{for all} \quad b \in X.$$

Upon clearing denominators, we see that $(H_1 \cdots H_m)(H_1' \cdots H_m') H F \in I_a^2$, and since F is the only factor in this product that lies in I_a, it follows that $F \in I_a^2$ (Exercise 12.3.3). This proves injectivity of (12.32), completing the proof. $\square$

The previous result motivates the following intrinsic definition of tangent spaces in the general setting of quasiprojective varieties.

12.33 DEFINITION *Tangent spaces of quasiprojective varieties*

Let X be a quasiprojective variety with $a \in X$, and let J_a denote the ideal in $K[X]_a$ of local functions that vanish at a. The *tangent space of X at a*, denoted $T_a X$, is the vector space
$$T_a X = \left(J_a / J_a^2\right)^{\vee}.$$

By the local nature of Definition 12.33, it follows that $T_a X = T_a U$ whenever X is a quasiprojective variety with $a \in X$ and U is a neighborhood of a. In particular, tangent spaces of quasiprojective varieties can always be computed by first restricting to an affine open neighborhood, then computing as in Chapter 7.

It also follows from the intrinsic nature of tangent spaces in the affine setting that tangent spaces are intrinsic in the quasiprojective setting. More specifically, if $F : X \to Y$ is an isomorphism of quasiprojective varieties such that $F(a) = b$, then F restricts to an isomorphism between any affine neighborhood of a and the image of that neighborhood—which is an affine neighborhood of b—in Y. By the intrinsic nature of tangent spaces in the affine context, it follows that $T_a X = T_b Y$.

Equipped with the notion of tangent spaces in the quasiprojective setting, we may proceed naturally to compare the vector-space dimension of the tangent space to the dimension of the variety. A first observation that one can make by reducing to the affine setting is that
$$\dim(T_a X) \geq \dim(X)$$
whenever X is irreducible (Exercise 12.3.4). As in the affine setting, the points where equality is obtained are particularly important.

12.34 DEFINITION *Smoothness of quasiprojective varieties*

Let X be an irreducible quasiprojective variety with $a \in X$. We say that X is *smooth at a* if $\dim(T_a X) = \dim(X)$; otherwise, we say that X is *singular at a*. We say that X is *smooth* if it is smooth at every point $a \in X$; otherwise, we say that X is *singular*.

Since our development of tangent spaces and smoothness in the general quasiprojective setting is local in nature, one's intuition for these notions can be derived entirely from the affine setting. In particular, whether a quasiprojective variety is smooth or singular at a point can be determined by restricting to an affine neighborhood of that point. As such, the results that were developed in the affine setting—such as the Jacobian criterion for smoothness and the fact that singular points are closed—naturally carry over to the quasiprojective setting. We direct the reader to the exercises to explore projective analogues of these results.

We note that the Jacobian criterion in the projective case generally involves one more partial derivative than we might expect. To close this section, we discuss a special case of the Jacobian criterion—concerning hypersurfaces in $\mathbb{P}^n$—where we see how homogeneity helps us account for this extra partial derivative.

12.3. LOCAL RINGS AND TANGENT SPACES

12.35 EXAMPLE Singularities of projective hypersurfaces

Let $f \in K[x_0, \ldots, x_n]$ be an irreducible homogeneous polynomial, and consider the projective hypersurface $X = \mathcal{V}(f) \subseteq \mathbb{P}^n$. We show that X is singular at a point $a \in X$ if and only if all partial derivatives of f vanish at a:

$$\frac{\partial f}{\partial x_0}(a) = \cdots = \frac{\partial f}{\partial x_n}(a) = 0.$$

Without loss of generality, assume that $a \notin \mathcal{V}(x_0)$, so $a = [1 : a_1 : \cdots : a_n]$. Restrict to the affine neighborhood $X_0 = X \cap \mathbb{A}_0^n$, making the natural identification

$$X_0 = \mathcal{V}(f_0) \subseteq \mathbb{A}^n \quad \text{where} \quad f_0(x_1, \ldots, x_n) = f(1, x_1, \ldots, x_n).$$

Set $\underline{a} = (a_1, \ldots, a_n) \in \mathbb{A}^n$. Since f, and thus f_0, is irreducible, $\mathcal{I}(X_0) = \langle f_0 \rangle$. By the Jacobian criterion in the affine setting (see Example 7.34), we see that X_0 is singular at $\underline{a}$—and thus X is singular at a—if and only if

$$\frac{\partial f_0}{\partial x_1}(\underline{a}) = \cdots = \frac{\partial f_0}{\partial x_n}(\underline{a}) = 0.$$

Some reflection on derivatives should convince the reader that

$$\frac{\partial f}{\partial x_i}(1, a_1, \ldots, a_n) = \frac{\partial f_0}{\partial x_i}(a_1, \ldots, a_n) \quad \text{for all} \quad i = 1, \ldots, n,$$

and we conclude that X is singular at $a \in X$ if and only if

$$\frac{\partial f}{\partial x_1}(a) = \cdots = \frac{\partial f}{\partial x_n}(a) = 0.$$

But what about the partial derivative at x_0? As a consequence of the fact that f is a nonconstant homogeneous polynomial, Euler's identity for homogeneous polynomials (Exercise 12.3.6) gives

$$\deg(f) f = x_0 \frac{\partial f}{\partial x_0} + \cdots + x_n \frac{\partial f}{\partial x_n},$$

and since $f(a) = 0$ and $a \notin \mathcal{V}(x_0)$, it then follows that all of the partial derivatives vanish at a if and only if all but $\partial f/\partial x_0$ vanish at a. Thus, we conclude that X is singular at $a \in X$ if and only if all of the partial derivatives of f vanish at $a \in X$.

Exercises for Section 12.3

12.3.1 (a) Prove that the relation among pairs (U, F) in Definition 12.26 is an equivalence relation.
(b) Carefully define sums, products, and scalar multiples of local functions.
(c) Verify that the local ring $K[X]_a$ is a K-algebra.

12.3.2 Let X be a quasiprojective variety with $a \in X$. Prove that $K[X]_a$ has a unique maximal ideal $J_a \subseteq K[X]_a$, comprised of all local functions that vanish at a.

12.3.3 Let I be a maximal ideal in a ring R and let $a, b \in R$ be elements such that $a \in I$, $b \notin I$, and $ab \in I^2$. Prove that $a \in I^2$.

12.3.4 Let X be an irreducible quasiprojective variety with $a \in X$. Prove that
$$\dim(T_a X) \geq \dim(X).$$
(Hint: Reduce to the appropriate result from Chapter 7.)

12.3.5 (a) Let $f \in K[x, y, z]$ be an irreducible homogeneous polynomial of degree two. Prove that $\mathcal{V}(f) \subseteq \mathbb{P}^2$ is smooth.
 (b) Given an example to show that (a) does not generalize to degree three.

12.3.6 Let $f \in K[x_0, \ldots, x_n]$ be a nonconstant homogeneous polynomial. Prove that
$$\deg(f) f = x_0 \frac{\partial f}{\partial x_0} + \cdots + x_n \frac{\partial f}{\partial x_n}.$$

12.3.7 Suppose that $X \subseteq \mathbb{P}^n$ is an irreducible projective variety with vanishing ideal $\mathcal{I}(X) = \langle f_1, \ldots, f_m \rangle$. Prove that X is singular at $a \in X$ if and only if
$$\mathrm{rk}(\mathrm{Jac}_{f_1,\ldots,f_m}(a)) < \mathrm{codim}(X).$$
(Note: $\mathrm{Jac}_{f_1,\ldots,f_m}(a)$ is not well-defined because scaling the coordinates of a will scale the entries of the Jacobian matrix, but the rank of the Jacobian *is* well-defined because this scaling does not affect the rank.)

12.3.8 Prove that the singular points of an irreducible quasiprojective variety form a closed subvariety.

12.3.9 Let R be a ring and let $S \subseteq R$ be a subset closed under multiplication.
 (a) Consider the relation on $R \times S$ defined by $(r, s) \sim (r', s')$ if there exists $t \in S$ such that $t(rs' - r's) = 0$. Prove that $\sim$ is an equivalence relation. The set of equivalence classes is called the *localization of R at S*, denoted R_S.
 (b) Prove that R_S is a ring with operations defined by
$$(r, s) + (r', s') = (rs' + r's, ss') \quad \text{and} \quad (r, s)(r', s') = (rr', ss').$$
Moreover, if R is a K-algebra, prove that R_S is a K-algebra with scalar multiplication
$$a(r, s) = (ar, s).$$
 (c) If R is an integral domain and $S = R \setminus \{0\}$, prove that
$$R_S = \mathrm{Frac}(R).$$
 (d) If R is a ring and $S = \{1, a, a^2, \ldots\}$ for some $a \in R$ that is not nilpotent, prove that
$$R_S = R[x]/\langle ax - 1 \rangle.$$
 (e) Let $R = K[X]$ for some affine variety X and let $I_a \subseteq R$ be the maximal ideal of functions that vanish at $a \in X$. If $S = R \setminus I_a$, prove that
$$R_S = K[X]_a.$$

Section 12.4 Blow-ups

In this section, we present an examples-driven introduction to blow-ups, a construction that has many applications throughout algebraic geometry. Throughout the examples, we emphasize one particular application, which is the use of blow-ups to resolve singularities. We begin with a very specific type of blow-up: the blow-up of $\mathbb{A}^n$ at 0. To help parse notation, we will consistently use $x_1, \ldots, x_n$ for the coordinates of $\mathbb{A}^n$ and $y_1, \ldots, y_n$ for the homogeneous coordinates of $\mathbb{P}^{n-1}$ (take note that we are starting with y_1, not y_0).

12.36 DEFINITION *Blow-up of affine space at the origin*

The *blow-up of $\mathbb{A}^n$ at the origin* is the quasiprojective variety

$$\mathrm{Bl}_0(\mathbb{A}^n) = \mathcal{V}(x_i y_j - x_j y_i \mid 1 \leq i < j \leq n) \subseteq \mathbb{A}^n \times \mathbb{P}^{n-1},$$

equipped with the *blow-up map* $\pi : \mathrm{Bl}_0(\mathbb{A}^n) \to \mathbb{A}^n$, which is the restriction of the projection of $\mathbb{A}^n \times \mathbb{P}^{n-1}$ onto its first factor.

Before delving too much into the details of the definition, let us illustrate it in the first nontrivial example, when $n = 2$.

12.37 EXAMPLE *Blow-up of $\mathbb{A}^2$ at the origin*

Consider the blow-up of $\mathbb{A}^2$ at the origin,

$$\mathrm{Bl}_0(\mathbb{A}^2) = \mathcal{V}(x_1 y_2 - x_2 y_1) \subseteq \mathbb{A}^2 \times \mathbb{P}^1,$$

which is depicted to the right (over $\mathbb{R}$), along with the (downward) blow-up map

$$\pi : \mathrm{Bl}_0(\mathbb{A}^2) \to \mathbb{A}^2.$$

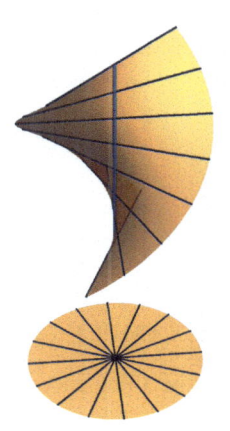

To gain some understanding for this depiction of the blow-up, consider a point

$$((a_1, a_2), [b_1, b_2]) \in \mathrm{Bl}_0(\mathbb{A}^2).$$

By definition, we must have $a_1 b_2 = a_2 b_1$, and some reflection should convince the reader that this is equivalent to the condition that $(a_1, a_2) \in \mathbb{A}^2$ lies on the line determined by $b = [b_1 : b_2] \in \mathbb{P}^1$. In other words, the points of $\mathrm{Bl}_0(\mathbb{A}^2)$ with a fixed $\mathbb{P}^1$-coordinate b correspond canonically to the points on the line in $\mathbb{A}^2$ parametrized by the point $b \in \mathbb{P}^1$.

Thus, in the image above, we have depicted $\mathbb{P}^1$ as the vertical coordinate, in which case each line in $\mathbb{A}^2$ is lifted in $\mathrm{Bl}_0(\mathbb{A}^2) \subseteq \mathbb{A}^2 \times \mathbb{P}^1$ to that same line at the height in the $\mathbb{P}^1$-coordinate determined by that line. This image is incomplete, however; in order to complete it, the top of $\mathrm{Bl}_0(\mathbb{A}^2)$ should loop around and connect to the bottom—forming a Möbius band—because, as we rotate the lines around the origin, we ultimately return back to where we started.

We note that every point in $\mathbb{A}^2$ except the origin has a unique preimage in $\mathrm{Bl}_0(\mathbb{A}^2)$—this is just to say that each of these points determines a unique line through the origin—whereas the preimage of the origin consists of the entire $\mathbb{P}^1$ parametrizing all lines through the origin in $\mathbb{A}^2$. In other words, "blowing up" $\mathbb{A}^2$ at the origin can be viewed as the process of replacing the origin in $\mathbb{A}^2$ with a copy of the projective space $\mathbb{P}^1$.

The intuition for $\mathrm{Bl}_0(\mathbb{A}^2)$ that we developed in the previous example generalizes quite naturally to $\mathrm{Bl}_0(\mathbb{A}^n)$. In particular, a point $(a, b) \in \mathbb{A}^n \times \mathbb{P}^{n-1}$ lies on $\mathrm{Bl}_0(\mathbb{A}^n)$ if and only if the point $a \in \mathbb{A}^n$ lies on the line through the origin determined by $b \in \mathbb{P}^{n-1}$ (Exercise 12.4.1). In particular, $\mathrm{Bl}_0(\mathbb{A}^n)$ can be viewed as the result of replacing the origin in $\mathbb{A}^n$ with the entire projective space $\mathbb{P}^{n-1}$ parametrizing all possible tangent directions along which one might approach the origin. In addition to this intuitive description of the blow-up, we also note that $\mathrm{Bl}_0(\mathbb{A}^n)$ is irreducible, has dimension n, and is smooth (Exercise 12.4.3).

> The term "blow-up" might best be viewed as a metaphor for inflating a balloon or zooming in on an image (not as a metaphor for an explosion).

Having described the blow-up of $\mathbb{A}^n$ at the origin, we can now define the blow-up of an affine variety $X \subseteq \mathbb{A}^n$ at the origin.

12.38 DEFINITION *Blow-up of an affine variety at the origin*

Let $X \subseteq \mathbb{A}^n$ be a Zariski-closed set containing the origin. The *blow-up of X at the origin* is the quasiprojective variety

$$\mathrm{Bl}_0(X) = \overline{\pi^{-1}(X \setminus \{0\})} \subseteq \mathrm{Bl}_0(\mathbb{A}^n).$$

In other words, to compute the blow-up of $X \subseteq \mathbb{A}^n$ at the origin, we first remove the origin and consider the preimage of $X \setminus \{0\}$ within $\mathrm{Bl}_0(\mathbb{A}^n)$, and then we take the Zariski closure. To gain some intuition, let us consider a concrete example.

12.39 EXAMPLE *Blow-up of a nodal curve in $\mathbb{A}^2$*

Consider the affine curve

$$X = \mathcal{V}(x_1^2 + x_1^3 - x_2^2) \subseteq \mathbb{A}^2,$$

depicted in the lower image at right along with its blow-up $\mathrm{Bl}_0(X)$ in the upper image. To justify this depiction of $\mathrm{Bl}_0(X)$, we first note (Exercise 12.4.4) that we can parametrize X as

$$X = \{(a^2 - 1, a^3 - a) \mid a \in \mathbb{A}^1\},$$

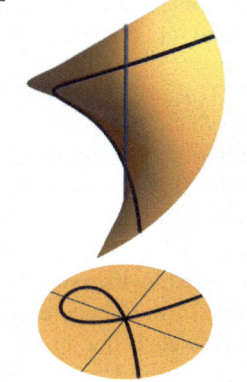

where both $a = 1$ and $a = -1$ correspond to the origin, where X loops back on itself. Since $[a^2 - 1 : a^3 - a] = [1 : a]$ for all $a \neq \pm 1$, it then follows that

$$\pi^{-1}(X \setminus \{0\}) = \{((a^2 - 1, a^3 - a), [1 : a]) \mid a \in \mathbb{A}^1 \setminus \{\pm 1\}\} \subseteq \mathbb{A}^2 \times \mathbb{P}^1.$$

12.4. BLOW-UPS

In order to compute the closure of $\pi^{-1}(X \setminus \{0\})$, we simply insert the two missing values $a = \pm 1$ in the above expression (Exercise 12.4.5), obtaining

$$\mathrm{Bl}_0(X) = \{((a^2 - 1, a^3 - a), [1 : a]) \mid a \in \mathbb{A}^1\}.$$

Notice that the preimage of the origin in $\mathrm{Bl}_0(X)$ contains two points $[1 : \pm 1]$, corresponding to the lines in $\mathbb{A}^2$ with slopes ± 1, which can be viewed as the tangent lines of the two "branches" of X at the origin. Thus, blowing up X at the origin pulls apart the self-intersection point according to the slopes of the two branches.

The curve X in Example 12.39 is singular at the origin, as the Jacobian criterion readily verifies, but its blow-up $\mathrm{Bl}_0(X)$ is isomorphic to $\mathbb{A}^1$ (Exercise 12.4.5), so the blow-up is smooth. In other words, the blow-up has *resolved* the singularity in X. More generally, a *resolution of the singularities* of a quasiprojective variety X is a smooth quasiprojective variety Y and a regular map $\pi : Y \to X$ that restricts to an isomorphism on the preimage of the smooth locus in X. (In particular, this means that X and Y are birational.) Blow-ups are an essential tool in constructing resolutions of singularities; the next example presents another illustration of how a blow-up can be used to resolve a curve singularity.

12.40 EXAMPLE Blow-up of a cuspidal curve in $\mathbb{A}^2$

Consider the affine curve

$$X = \mathcal{V}(x_1^2 - x_2^3) \subseteq \mathbb{A}^2,$$

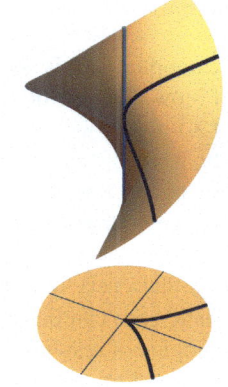

depicted in the lower image at right, along with its blow-up $\mathrm{Bl}_0(X)$ in the upper image. As in the previous example, we consider a parametrization of X:

$$X = \{(a^3, a^2) \mid a \in \mathbb{A}^1\},$$

where $a = 0$ corresponds to the singular point at the origin. Since $[a^3 : a^2] = [a : 1]$ for all $a \neq 0$, it follows that

$$\pi^{-1}(X \setminus \{0\}) = \{((a^3, a^2), [a : 1]) \mid a \in \mathbb{A}^1 \setminus \{0\}\} \subseteq \mathbb{A}^2 \times \mathbb{P}^1.$$

Arguing similarly to the previous example, we then compute the closure by including the value $a = 0$ in the above expression, obtaining

$$\mathrm{Bl}_0(X) = \{((a^3, a^2), [a : 1]) \mid a \in \mathbb{A}^1\}.$$

As in the previous example, we have $\mathrm{Bl}_0(X) \cong \mathbb{A}^1$. Thus, again, we see that the blow-up $\mathrm{Bl}_0(X)$ is a resolution of the cusp singularity in X. Intuitively, the blow-up has allowed us to replace the sharp change in direction of X at the origin by a smooth path in a higher-dimensional space.

Thus far, we have only seen examples of blow-ups of curves in $\mathbb{A}^2$. Stepping up the dimension by one, we now consider the blow-up of a surface in $\mathbb{A}^3$.

12.41 EXAMPLE   Blow-up of an affine cone

Consider the affine surface

$$X = \mathcal{V}(x_1^2 + x_2^2 - x_3^2) \subseteq \mathbb{A}^3.$$

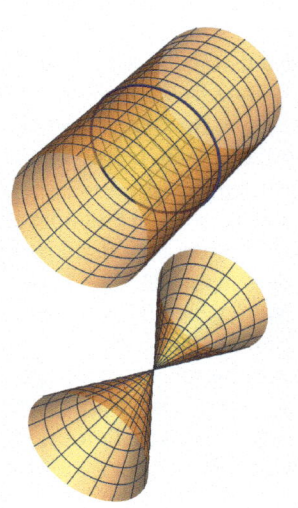

Over $\mathbb{R}$, this is a circular cone with vertex at the origin, depicted in the lower image at right. The blow-up $\mathrm{Bl}_0(X)$ is depicted in the upper image. Intuitively, to understand what is happening here, we note that X is a union of lines in $\mathbb{A}^3$, all of which meet at the origin. Blowing up at the origin pulls apart all of these lines where they meet, allowing us to view $\mathrm{Bl}_0(X)$ as a *disjoint* union of (the blow-ups of) these lines within $\mathbb{A}^3 \times \mathbb{P}^2$. When working over $\mathbb{R}$, this disjoint union of (blow-ups of) lines can naturally be identified with the cylinder depicted above, though we warn the reader that the global geometry of $\mathrm{Bl}_0(X)$ is not as straightforward when working over an algebraically closed field like $\mathbb{C}$.

The examples in this section have just begun to scratch the surface of the rich theory of blow-ups. We note that there is a much more general (intrinsic) construction $\mathrm{Bl}_Y(X)$ that blows up a quasiprojective variety X along a closed subvariety Y. Intuitively, $\mathrm{Bl}_Y(X)$ can be constructed from X by replacing each point of Y with the projective space parametrizing the tangent directions at that point that are normal to Y. The most celebrated result in the theory of blow-ups is the Resolution of Singularities Theorem, proved by the Japanese mathematician Heisuke Hironaka in 1964, which says that every quasiprojective variety over an algebraically closed field of characteristic zero can be resolved to a smooth variety by a sequence of blow-ups. In the examples of resolutions discussed throughout this section, only one blow-up was ever needed to resolve the singularities; for an example of a curve singularity that requires two blow-ups to resolve, we direct the reader to Exercise 12.4.6. In positive characteristic, it is still an open question whether or not every variety can even be resolved.

Exercises for Section 12.4

12.4.1   Prove that a point $(a, b) \in \mathbb{A}^n \times \mathbb{P}^{n-1}$ lies on $\mathrm{Bl}_0(\mathbb{A}^n)$ if and only if a lies on the line in $\mathbb{A}^n$ determined by the point $b \in \mathbb{P}^{n-1}$.

12.4.2   Consider the blow-up map $\pi : \mathrm{Bl}_0(\mathbb{A}^n) \to \mathbb{A}^n$.
   (a) Prove that the preimage of a point $(a_1, \ldots, a_n) \in \mathbb{A}^n$ is a single point if at least one of the coordinates a_i is nonzero.
   (b) Prove that the preimage of the origin is isomorphic to $\mathbb{P}^{n-1}$.

12.4. BLOW-UPS

12.4.3 (a) Prove that $\text{Bl}_0(\mathbb{A}^n)$ has a dense open subset isomorphic to $\mathbb{A}^n \setminus \{0\}$.
 (b) Prove that $\text{Bl}_0(\mathbb{A}^n)$ is irreducible.
 (c) Prove that $\text{Bl}_0(\mathbb{A}^n)$ has dimension n.
 (d) Prove that $\text{Bl}_0(\mathbb{A}^n)$ is smooth.

12.4.4 Let $X = \mathcal{V}(x_1^2 + x_1^3 - x_2^2) \subseteq \mathbb{A}^2$. Prove that
$$X = \{(a^2 - 1, a^3 - a) \mid a \in \mathbb{A}^1\}.$$

12.4.5 Consider the subset
$$Y = \{((a^2 - 1, a^3 - a), [1 : a]) \mid a \in \mathbb{A}^1 \setminus \{\pm 1\}\} \subseteq \mathbb{A}^2 \times \mathbb{P}^1.$$

 (a) Suppose that $f \in K[x_1, x_2, y_1, y_2]$ vanishes at every point of Y. Prove that f also vanishes at the two points $((0,0), [1 : \pm 1])$.
 (b) Find three polynomials $f_1, f_2, f_3 \in K[x_1, x_2, y_1, y_2]$ such that
$$\mathcal{V}(f_1, f_2, f_3) = \{((a^2 - 1, a^3 - a), [1 : a]) \mid a \in \mathbb{A}^1\} \subseteq \mathbb{A}^2 \times \mathbb{P}^1.$$
 (c) Conclude that
$$\overline{Y} = \{((a^2 - 1, a^3 - a), [1 : a]) \mid a \in \mathbb{A}^1\}.$$
 (d) Prove that $\overline{Y} \cong \mathbb{A}^1$.

12.4.6 Let $X = \mathcal{V}(x_1^2 - x_2^5) \subseteq \mathbb{A}^2$. This exercise shows that X must be blown up twice in order to resolve its singularity.
 (a) Prove that X is singular at the origin and smooth elsewhere.
 (b) Prove that $\text{Bl}_0(X) \subseteq \mathbb{A}^2 \times \mathbb{P}^1$ is singular at the point $((0,0), [0:1])$.
 (c) Let $U \subseteq \mathbb{A}^2 \times \mathbb{P}^1$ be the affine open subset where $y_2 \neq 0$ and set $Y = U \cap \text{Bl}_0(X)$, so that Y is an affine open subset of $\text{Bl}_0(X)$ with singular point at the origin in $U \cong \mathbb{A}^3$. Prove that $\text{Bl}_0(Y)$ is smooth.

12.4.7 Consider the rational map $F : \mathbb{A}^n \dashrightarrow \mathbb{P}^{n-1}$ given by
$$F(a_1, \ldots, a_n) = [a_1 : \cdots : a_n],$$
which is regular on $\mathbb{A}^n \setminus \{(0, \ldots, 0)\}$.
 (a) Prove that there does not exist a regular map $\tilde{F} : \mathbb{A}^n \to \mathbb{P}^{n-1}$ that agrees with F on $\mathbb{A}^n \setminus \{(0, \ldots, 0)\}$.
 (b) Construct a regular map $G : \text{Bl}_0(\mathbb{A}^n) \to \mathbb{P}^{n-1}$ with $G(a, b) = F(a)$ for all $(a, b) \in \text{Bl}_0(\mathbb{A}^n)$ with $a \neq (0, \ldots, 0)$.

This exercise illustrates another application of blow-ups: by blowing up the domain of a rational map, we can often replace our domain with a birational domain on which the rational map can be extended to a regular map.

Section 12.5 Theorem on Fiber Dimensions

We now come to one of the most central results in all of algebraic geometry: the Theorem on Fiber Dimensions. To motivate our developments, let us briefly take a step back into the context of linear algebra. Arguably the most fundamental result of linear algebra is the Rank-Nullity Theorem, which says that, for any linear map $F : V \to W$ of finite-dimensional vector spaces, we have

$$\dim(F^{-1}(0)) = \dim(V) - \dim(F(V)).$$

In other words, the dimension of the fiber over zero—which is just the null space of F—is given by the difference of the dimension of the domain and the dimension of the image of F. While a

> A "fiber" of a map generally refers to the preimage of a single point:
> $$F^{-1}(b) = \{a \mid F(a) = b\}.$$

general fiber $F^{-1}(w)$ is not a linear subspace of V, every fiber is closely related to the null space $F^{-1}(0)$. In particular, given any $w \in F(V)$ and $v_0 \in F^{-1}(w)$, we can write

$$F^{-1}(w) = F^{-1}(0) + v_0 = \{v + v_0 \mid v \in F^{-1}(0)\},$$

showing that every fiber is simply a translation of the null space. Thus, defining the dimension of a translate of a subspace as the dimension of the subspace itself, we arrive at the following fundamental result about dimensions of fibers of linear maps:

$$\dim(F^{-1}(w)) = \dim(V) - \dim(F(V)) \quad \text{for all} \quad w \in F(V).$$

The image below depicts one instance of this result, in which a three-dimensional vector space is mapped onto a one-dimensional vector space and several two-dimensional fibers are highlighted in the domain.

The goal of this section is to generalize this fundamental relationship from linear algebra to polynomial algebra, replacing vector spaces by quasiprojective varieties and linear maps by regular maps. There is an important concession that we must make in order for this generalization to hold: in algebraic geometry, while the fibers have the expected dimension almost everywhere, there may be a closed subset of the image over which the fibers are bigger—but never smaller—than expected. We illustrate this phenomenon with an example from last section.

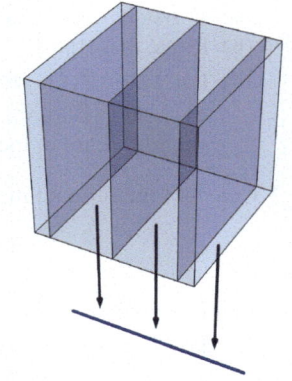

12.42 EXAMPLE Fibers of blow-ups

Let $\pi : \text{Bl}_0(\mathbb{A}^n) \to \mathbb{A}^n$ be the blow-up map with $n > 1$. As we discussed in the previous section, the fiber over any nonzero point of $\mathbb{A}^n$ is a single point, while the fiber over the origin is isomorphic to $\mathbb{P}^{n-1}$. Since $\dim(\mathbb{A}^n) = \dim(\text{Bl}_0(\mathbb{A}^n)) = n$ and π is surjective, we should expect the fibers to have dimension zero, and this expectation is fulfilled for any nonzero point of $\mathbb{A}^n$, but the fiber over the special point 0 has dimension $n - 1$, which is larger than expected.

12.5. THEOREM ON FIBER DIMENSIONS

We now state the first part of the Theorem on Fiber Dimensions, which says that the fibers are never smaller than expected.

12.43 THEOREM *Theorem on Fiber Dimensions, Part 1*

Let $F : X \to Y$ be a regular map of quasiprojective varieties where X is irreducible. Then for any $b \in F(X)$, we have

$$\dim(F^{-1}(b)) \geq \dim(X) - \dim(\overline{F(X)}).$$

We take the closure of $F(X) \subseteq Y$ because $F(X)$ may not, itself, be a subvariety of Y. See Exercise 12.5.1.

As we will see, Theorem 12.43 is a consequence of the Fundamental Theorem of Dimension Theory. The next lemma is the bridge that connects these two results, essentially asserting that every point of an irreducible quasiprojective variety can locally be defined by the expected number of equations.

12.44 LEMMA *Every point in Y is locally defined by $\dim(Y)$ equations*

If Y is an irreducible quasiprojective variety of dimension d and $b \in Y$, then there exists an affine open subset $U \subseteq Y$ with $b \in U$ and regular functions $F_1, \ldots, F_d \in K[U]$ such that $\mathcal{V}(F_1, \ldots, F_d) = \{b\}$.

PROOF If $d = 0$, then $Y = \{b\}$, and $U = Y$ satisfies the desired conclusion. Thus, suppose that $d > 0$. Begin by choosing an affine open subset $V \subseteq Y$ with $b \in V$. Since Y is irreducible, V is also irreducible and $\dim(V) = d > 0$. Up to isomorphism, we may represent V as a closed subset of affine space: $V \subseteq \mathbb{A}^n$. We will recursively construct polynomials $g_1, \ldots, g_d \in K[y_1, \ldots, y_n]$, all vanishing at b, such that every irreducible component of $V \cap \mathcal{V}(g_1, \ldots, g_k)$ has dimension $d - k$.

To begin the recursion, choose a polynomial $g_1 \in K[y_1, \ldots, y_n]$ such that g_1 vanishes at b but not on all of V (for example, if $b' \in V \setminus \{b\}$ differs from b in the ith coordinate, the polynomial $g_1 = y_i - b_i$ suffices). Since V is irreducible of dimension d, the Fundamental Theorem of Dimension Theory then implies that every irreducible component of $V \cap \mathcal{V}(g_1)$ has dimension $d - 1$.

Suppose, now, that we have found $g_1, \ldots, g_k \in K[y_1, \ldots, y_n]$, all of which vanish at b, such that every irreducible component of $V \cap \mathcal{V}(g_1, \ldots, g_k)$ has dimension $d - k > 0$. By Exercise 12.5.2, we may choose $g_{k+1} \in K[y_1, \ldots, y_n]$ such that g_{k+1} vanishes at b but not on an entire irreducible component of $V \cap \mathcal{V}(g_1, \ldots, g_k)$. By the Fundamental Theorem of Dimension Theory again, every irreducible component of $V \cap \mathcal{V}(g_1, \ldots, g_k, g_{k+1})$ has dimension $d - k - 1$, as desired.

By construction, $V \cap \mathcal{V}(g_1, \ldots, g_d)$ is zero-dimensional, and thus a finite set $\{b, b_1, \ldots, b_\ell\}$. Let $U \subseteq V$ be any affine open neighborhood of b contained within the open set $V \setminus \{b_1, \ldots, b_\ell\}$, and set $F_k = g_k|_U$ for $k = 1, \ldots, d$. Then $F_1, \ldots, F_d$ are regular functions on U and $\mathcal{V}(F_1, \ldots, F_d) = \{b\}$, completing the proof of the lemma. $\square$

We now prove the first part of the Theorem on Fiber Dimensions.

PROOF OF THEOREM 12.43   As the statement of the theorem does not reference Y outside of $\overline{F(X)}$, we may assume that $Y = \overline{F(X)}$, so that $F : X \to Y$ is a regular map with dense image. The irreducibility of X then implies the irreducibility of Y (Exercise 12.5.3). Let $b \in F(X)$ and set $d = \dim(Y)$. By Lemma 12.44, choose an affine open subset $U \subseteq Y$ and regular functions $F_1, \ldots, F_d \in K[U]$ such that

$$\mathcal{V}(F_1, \ldots, F_d) = \{b\}.$$

Set $V = F^{-1}(U)$, which is an open subvariety of X by continuity of regular maps. Since X is irreducible, V is also irreducible and $\dim(V) = \dim(X)$. Denote by $G : V \to U$ the restriction of F to V, and note that $F^{-1}(b) = G^{-1}(b)$. Since $G(a) = b$ if and only if $F_i(G(a)) = 0$ for all $i = 1, \ldots, d$, it follows that

$$G^{-1}(b) = \mathcal{V}(G^*(F_1), \ldots, G^*(F_d)).$$

Since $G^{-1}(b)$ is the vanishing set of d regular functions on V, Corollary 12.23 then implies that

$$\dim(G^{-1}(b)) \geq \dim(V) - d.$$

By construction, this inequality translates to the assertion in the statement of the theorem:

$$\dim(F^{-1}(b)) \geq \dim(X) - \dim(\overline{F(X)}). \qquad \square$$

We now proceed to the second—and more challenging—part of the Theorem on Fiber Dimensions, which essentially says that the expected dimension of fibers is always realized over a nonempty open subset of the image.

12.45 THEOREM   *Theorem on Fiber Dimensions, Part 2*

Let $F : X \to Y$ be a regular map of quasiprojective varieties where X is irreducible. Then there exists a nonempty open subset $U \subseteq Y$ such that for every $b \in U$, we have $F^{-1}(b) = \emptyset$ or

$$\dim(F^{-1}(b)) = \dim(X) - \dim(\overline{F(X)}).$$

PROOF   As in the proof of Theorem 12.43, we assume that $Y = \overline{F(X)}$ and note that the irreducibility of X thus implies the irreducibility of Y. We now reduce to the case where X and Y are both affine.

(Reduction to Y affine) Assume that the theorem holds when the codomain is affine, and let $F : X \to Y$ be a regular map where X is irreducible and $\overline{F(X)} = Y$. For any nonempty affine open set $V \subseteq Y$, we can apply our assumption to the restriction of F to the (open) preimage $F^{-1}(V) \subseteq X$. Thus, by assumption, there exists a nonempty open $U \subseteq V$ such that, for all $b \in U$,

$$F^{-1}(b) = \emptyset \quad \text{or} \quad \dim(F^{-1}(b)) = \dim(F^{-1}(V)) - \dim(V).$$

Since X and Y are irreducible, $F^{-1}(V) \subseteq X$ and $V \subseteq Y$ are dense open sets and thus have the same dimensions as X and Y, respectively, from which we conclude that, for all $b \in U$,

$$F^{-1}(b) = \emptyset \quad \text{or} \quad \dim(F^{-1}(b)) = \dim(X) - \dim(Y).$$

12.5. THEOREM ON FIBER DIMENSIONS

As U is a nonempty open subset of the open set $V \subseteq Y$, it is also a nonempty open set of Y, and we have shown that the conclusion of the theorem for $F : X \to Y$ holds for this choice of U. Since the general form of the theorem follows from the special case where the codomain is affine, we henceforth assume that Y is affine.

(Reduction to X affine) Assume that the theorem holds when the domain is affine, and let $F : X \to Y$ be a regular map where X is irreducible and $\overline{F(X)} = Y$. By Theorem 11.45, choose a finite set of affine open subsets $V_1, \ldots, V_k \subseteq X$ such that $X = V_1 \cup \cdots \cup V_k$. As X is irreducible, each V_i is dense in X, and it follows that $F(V_i)$ is dense in Y (Exercise 12.5.4). Let $F_i : V_i \to Y$ denote the restriction of F to V_i. Since V_i is affine and irreducible and since $\overline{F_i(V_i)} = Y$, our assumption implies that there exists a nonempty open subset $U_i \subseteq Y$ such that, for every $b \in U_i$,

$$F_i^{-1}(b) = \emptyset \quad \text{or} \quad \dim(F_i^{-1}(b)) = \dim(V_i) - \dim(Y) = \dim(X) - \dim(Y).$$

Consider the nonempty open set $U = U_1 \cap \cdots \cap U_k \subseteq Y$, and let $b \in U$. If Z is an irreducible component of $F^{-1}(b)$, then $V_i \cap Z$ is nonempty for at least one i, since $X = V_1 \cup \cdots \cup V_k$. Therefore, we have

$$\dim(Z) = \dim(V_i \cap Z) \leq \dim(F_i^{-1}(b)) = \dim(X) - \dim(Y),$$

where the first equality follows from the assumption that Z is irreducible, implying that $V_i \cap Z$ is dense, the inequality is because $V_i \cap Z \subseteq F_i^{-1}(b)$, and the second equality follows from our assumption that $b \in U_i$. Since we have found a uniform upper bound on the dimension of every irreducible component of $F^{-1}(b)$, it follows that $\dim(F^{-1}(b)) \leq \dim(X) - \dim(Y)$. The other inequality follows from Theorem 12.43. Thus, we have constructed a nonempty open set $U \subseteq Y$ such that, for every $b \in U$,

$$F^{-1}(b) = \emptyset \quad \text{or} \quad \dim(F^{-1}(b)) = \dim(X) - \dim(Y).$$

Since the general form of the theorem follows from the setting where the domain is affine, we henceforth assume that X is affine.

(Proof for X and Y affine) Let $F : X \to Y$ be a regular map of irreducible affine varieties such that $\overline{F(X)} = Y$. Since $F(X)$ is dense in Y, the pullback homomorphism $F^* : K[Y] \to K[X]$ is an injection (Exercise 4.2.10). Via this injection, we view $K[Y]$ as a subalgebra of $K[X]$, and upon taking fraction fields, we also view $K(Y)$ as a subfield of $K(X)$. Regarding transcendence degrees, we have

(12.46) $$\operatorname{trdeg}_{K(Y)} K(X) = \operatorname{trdeg}_K K(X) - \operatorname{trdeg}_K K(Y),$$

which follows from the fact that every transcendence basis of $K(Y)$ over K extends to a transcendence basis of $K(X)$ over K, and the elements that we extend by give rise to a transcendence basis of $K(X)$ over $K(Y)$ (Exercise 12.5.5).

Writing X as a closed subset of affine space $X \subseteq \mathbb{A}^m$, define $A \subseteq K(X)$ to be the $K(Y)$-algebra generated by the coordinate functions $[x_1], \ldots, [x_m]$:

$$A = K(Y)[[x_1], \ldots, [x_m]] \subseteq K(X).$$

Since A is a finitely generated over $K(Y)$ by $[x_1], \ldots, [x_m]$, the Noether Normalization Theorem guarantees the existence of a finite Noether basis $\{H_1, \ldots, H_e\} \subseteq A$.

More specifically, $H_1, \ldots, H_e$ are elements of A that are algebraically independent over $K(Y)$ and such that A is integral over $K(Y)[H_1, \ldots, H_e]$. In particular, for each $i = 1, \ldots, m$, there exists a monic polynomial

$$g_i(z) \in \bigl(K(Y)[H_1, \ldots, H_e]\bigr)[z]$$

such that $g_i([x_i]) = 0 \in A$. Since $K(X) = K([x_1], \ldots, [x_m])$ and each $[x_i]$ is algebraic over $K(Y)[H_1, \ldots, H_e]$, it follows that the Noether basis $\{H_1, \ldots, H_e\}$ gives a transcendence basis of $K(X)$ over $K(Y)$, and (12.46) then implies that

$$e = \dim(X) - \dim(Y).$$

By definition of A, for each $i = 1, \ldots, e$, we can write $H_i = F_i / G_i$ for some $F_i \in K[X]$ and $G_i \in K[Y] \setminus \{0\}$. Moreover, each coefficient of $g_i(z)$—as a polynomial in $H_1, \ldots, H_e$ and z—is an element of $K(Y)$, and thus can be represented as a quotient of regular functions on Y. Let $G \in K[Y] \setminus \{0\}$ be the product of $G_1, \ldots, G_e$ and all regular functions on Y appearing as denominators in the coefficients of $g_1(z), \ldots, g_e(z)$, and define the nonempty open set

$$U = Y \setminus \mathcal{V}(G).$$

We now argue that U satisfies the conclusion of the theorem.

Let $b \in U$ and assume that $F^{-1}(b) \neq \emptyset$. The construction of U implies that $G_i(b) \neq 0$ for all $i = 1, \ldots, e$, and the restriction of the function $H_i = F_i / G_i$ to the fiber $F^{-1}(b)$ gives a regular function $H_i^b \in K[F^{-1}(b)]$ defined by

$$H_i^b(a) = \frac{F_i(a)}{G_i(a)} \quad \text{for every} \quad a \in F^{-1}(b).$$

Similarly, since all of the coefficients of $g_i(z)$ are well-defined functions on U, we can restrict all of the coefficients of the polynomial $g_i(z)$ to the fiber $F^{-1}(b)$, obtaining a polynomial

$$g_i^b(z) \in K[H_1^b, \ldots, H_e^b][z].$$

Importantly, as $g_i(z)$ is monic, $g_i^b(z)$ is also monic. Since $F^{-1}(b)$ is closed within the closed set $X \subseteq \mathbb{A}^m$, the coordinate functions $[x_1], \ldots, [x_m]$ generate $K[F^{-1}(b)]$. Restricting the relations $g_i([x_i]) = 0$ to the fiber $F^{-1}(b)$, we obtain relations $g_i^b([x_i]) = 0 \in K[F^{-1}(b)]$, showing that each coordinate function is integral over $K[H_1^b, \ldots, H_e^b]$. Thus, Theorem 5.27 implies that $K[F^{-1}(b)]$ is integral over $K[H_1^b, \ldots, H_e^b]$, and from Part 1 of Corollary 6.31, we conclude that

$$\dim(F^{-1}(b)) \leq e = \dim(X) - \dim(Y).$$

Combining this with Theorem 12.43, it follows that

$$\dim(F^{-1}(b)) = \dim(X) - \dim(Y),$$

completing the proof of the theorem. $\square$

12.5. THEOREM ON FIBER DIMENSIONS

The next example provides a small illustration of just a few of the key ideas used in the previous proof.

12.47 EXAMPLE Illustration of the proof of Theorem 12.45

Consider $X = \mathcal{V}(xy_1 - y_2) \subseteq \mathbb{A}^3$ and $Y = \mathbb{A}^2$, and let $F : X \to \mathbb{A}^2$ be the regular map defined by $F(a, b_1, b_2) = (b_1, b_2)$. Some reflection (Exercise 12.5.6) should convince the reader that $\overline{F(X)} = Y$, that $\dim(X) = \dim(Y) = 2$, that the fiber over any nonzero point of $F(X)$ is a single point, and that the fiber over $(0,0)$ is isomorphic to $\mathbb{A}^1$. Our aim in this example is to work through the details in the proof of Theorem 12.45 for this particular map of affine varieties, culminating in the construction of a nonempty open set $U \subseteq Y$ over which the fibers of F, when nonempty, have the expected dimension of zero.

Notice that $K(Y) = K(y_1, y_2)$ and the algebra A appearing in the proof of Theorem 12.45 is

$$A = K(y_1, y_2)[[x]] \subseteq K(X)$$

The empty set is a Noether basis for A over $K(y_1, y_2)$, as the single generator $[x]$ is a solution of the monic polynomial

$$g(z) = z - \frac{y_2}{y_1} \in K(y_1, y_2)[z],$$

which follows from the observation that $xy_1 - y_2 \in \mathcal{I}(X)$. Following the proof of Theorem 12.45, we let $G \in K[Y] = K[y_1, y_2]$ be the product of all denominators appearing in our Noether basis and in our monic polynomials. In this case, $G = y_1$, so that $U = \mathbb{A}^2 \setminus \mathcal{V}(y_1)$. Thus, for any $(b_1, b_2) \in U$, we have $b_1 \neq 0$, and for any $(a, b_1, b_2) \in F^{-1}(b_1, b_2)$, we obtain a well-defined monic relation

$$0 = g(a) = a - \frac{b_2}{b_1}.$$

In this example, this relation uniquely determines a from (b_1, b_2), showing that the fibers over U have dimension zero, as expected.

The final result of this section is an application of the Theorem on Fiber Dimensions, providing a useful method for verifying that a variety is irreducible.

12.48 THEOREM *Equidimensional irreducible fibers $\Rightarrow$ irreducible*

Let $F : X \to Y$ be a surjective regular map where X is projective and Y is irreducible. If $F^{-1}(b)$ is irreducible of the same dimension for every $b \in Y$, then X is irreducible.

PROOF Let $X_1, \ldots, X_k$ be the irreducible components of X; our aim is to show that $X = X_i$ for some i. Let $F_i : X_i \to Y$ denote the restriction of F to X_i. Since X is projective, Theorem 11.65 implies that $F(X_i)$ is closed in Y for each i. Furthermore, because F is surjective, it follows that

$$Y = F(X) = \bigcup_{i=1}^{k} F(X_i).$$

Notice that the irreducibility of Y implies that $Y = F(X_i)$ for at least one i. For each i such that $F(X_i) = Y$, Theorem 12.45 implies that there exists a nonempty open set $U_i \subseteq Y$ such that
$$\dim(F_i^{-1}(b)) = \dim(X_i) - \dim(Y) \quad \text{for all} \quad b \in U_i.$$
For each i such that $F(X_i) \neq Y$, set $U_i = Y \setminus F(X_i)$. Let $U = U_1 \cap \cdots \cap U_k$, which is a nonempty open subset of Y.

Pick some $b_0 \in U$. Then the irreducibility of $F^{-1}(b_0)$ implies that $F^{-1}(b_0)$ is contained in some X_i. Without loss of generality, suppose that $F^{-1}(b_0) \subseteq X_1$, so that $F^{-1}(b_0) = F_1^{-1}(b_0)$. By definition of U, we know that F_1 is surjective and
$$\dim(F^{-1}(b_0)) = \dim(F_1^{-1}(b_0)) = \dim(X_1) - \dim(Y).$$
Now let $b \in Y$ be any element. Since all fibers of F have the same dimension,

(12.49) $$\dim(F^{-1}(b)) = \dim(F^{-1}(b_0)) = \dim(X_1) - \dim(Y).$$

Furthermore, by Theorem 12.43, we have

(12.50) $$\dim(F_1^{-1}(b)) \geq \dim(X_1) - \dim(Y).$$

Combining (12.49) and (12.50), we conclude that $\dim(F^{-1}(b)) \leq \dim(F_1^{-1}(b))$. However, since $F_1^{-1}(b) \subseteq F^{-1}(b)$, we also get the other inequality, implying that $\dim(F_1^{-1}(b)) = \dim(F^{-1}(b))$. Since $F^{-1}(b)$ is irreducible by hypothesis and since $F_1^{-1}(b) = F^{-1}(b) \cap X_1$ is a closed subset of $F^{-1}(b)$ of the same dimension, we conclude (Exercise 12.2.1) that $F^{-1}(b) = F_1^{-1}(b)$. Since $F^{-1}(b) = F_1^{-1}(b)$ for all $b \in Y$, it follows that $X = X_1$, as desired. □

Exercises for Section 12.5

12.5.1 Let $F : \mathbb{A}^2 \to \mathbb{A}^2$ be defined by $F(a_1, a_2) = (a_1, a_1 a_2)$.
  (a) Compute $F(\mathbb{A}^2) \subseteq \mathbb{A}^2$.
  (b) Prove that $F(\mathbb{A}^2)$ is not a subvariety of $\mathbb{A}^2$.
  (c) Compute the fiber $F^{-1}(b_1, b_2)$ for any $(b_1, b_2) \in \mathbb{A}^2$.

12.5.2 (a) Let $a_0, a_1, \ldots, a_k \subseteq \mathbb{A}^n$ be distinct points in $\mathbb{A}^n$. Prove that there exists $f \in K[x_1, \ldots, x_n]$ such that $f(a_i) = 0$ if and only if $i = 0$.
  (b) Let $a \in \mathbb{A}^n$ and let $X_1, \ldots, X_k \subseteq \mathbb{A}^n$ be nonempty subsets, none of which is equal to $\{a\}$. Prove that there exists $f \in K[x_1, \ldots, x_n]$ such that $f(a) = 0$ and $f \notin \mathcal{I}(X_i)$ for all $i = 1, \ldots, k$.

12.5.3 Let $F : X \to Y$ be a continuous map of topological spaces such that X is irreducible. Prove that $F(X)$ and $\overline{F(X)}$ are irreducible subspaces of Y.

12.5.4 Suppose that $F : X \to Y$ is a continuous map of topological spaces such that $F(X)$ is dense in Y. Prove that $F(U)$ is dense in Y if U is dense in X.

12.5.5 Let $K \subseteq L \subseteq M$ be fields such that L is finitely generated over K and M is finitely generated over L. Prove that
$$\operatorname{trdeg}_K(M) = \operatorname{trdeg}_K(L) + \operatorname{trdeg}_L(M).$$

12.5. THEOREM ON FIBER DIMENSIONS

12.5.6 Set $X = \mathcal{V}(xy_1 - y_2) \subseteq \mathbb{A}^3$ and $Y = \mathbb{A}^2$, and let $F : X \to \mathbb{A}^2$ be the regular map defined by $F(a, b_1, b_2) = (b_1, b_2)$.

(a) Prove that $\overline{F(X)} = Y$.
(b) Prove that $\dim(X) = \dim(Y) = 2$.
(c) Prove that the fiber of F over any nonzero point of $\mathbb{A}^2$ is either empty or a single point.
(d) Prove that $F^{-1}(0,0)$ is isomorphic to $\mathbb{A}^1$.

12.5.7 Let $F : X \to Y$ be a surjective regular map of projective varieties. Prove that the sets
$$Y_k = \{b \in Y \mid \dim(F^{-1}(b)) \geq k\}$$
are closed in Y.

Open Access This chapter is licensed under the terms of the Creative Commons Attribution-NonCommercial 4.0 International License (http://creativecommons.org/licenses/by-nc/4.0/), which permits any noncommercial use, sharing, adaptation, distribution and reproduction in any medium or format, as long as you give appropriate credit to the original author(s) and the source, provide a link to the Creative Commons license and indicate if changes were made.

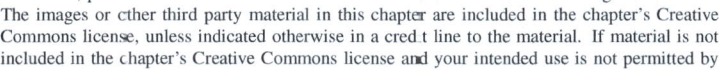

The images or other third party material in this chapter are included in the chapter's Creative Commons license, unless indicated otherwise in a credit line to the material. If material is not included in the chapter's Creative Commons license and your intended use is not permitted by statutory regulation or exceeds the permitted use, you will need to obtain permission directly from the copyright holder.

Section 12.6 Lines on surfaces

In the final two sections of the book, we study lines on surfaces, with the ultimate aim of proving the classical fact that every smooth cubic surface contains exactly 27 lines. In fact, one of the trickiest parts of proving this classical result is proving that every cubic surface contains at least one line; we prove this fact in this section as an application of the Theorem on Fiber Dimensions. Toward this end, let us take a step back and begin by making precise what we mean by the degree of a surface in $\mathbb{P}^3$.

For the purposes of the next two sections, a *surface in* $\mathbb{P}^3$ is any irreducible projective variety $X \subseteq \mathbb{P}^3$ of dimension two. If $X \subseteq \mathbb{P}^3$ is a surface, it follows that $\mathcal{I}(X)$ is a principal ideal, generated by any irreducible polynomial that vanishes on X (Exercise 12.2.2). As any two generators of a principal ideal differ by a scalar, the following notion is well-defined.

12.51 **DEFINITION** *Degree of a surface in* $\mathbb{P}^3$

Let $X \subseteq \mathbb{P}^3$ be a surface. The *degree of* X *in* $\mathbb{P}^3$ is the polynomial degree of any generator of $\mathcal{I}(X)$.

More generally, we note that it is possible to define the *degree* of any projective variety $X \subseteq \mathbb{P}^n$. For a hypersurface X in $\mathbb{P}^n$, the degree is simply the degree of any generator of $\mathcal{I}(X)$, but for projective varieties in $\mathbb{P}^n$ of higher codimension, the notion of degree is a bit more complicated to make precise.

> *Intuitively, the degree of a variety in $\mathbb{P}^n$ is a measure of its "curviness."*

Since two polynomials generating the vanishing ideal of a surface $X \subseteq \mathbb{P}^3$ differ by a scalar, we now shift our perspective from thinking about surfaces to working directly with polynomials up to scaling.

12.52 **NOTATION** $K[x_0, x_1, x_2, x_3]_d$ *and* $\mathbb{P}_{3,d}$

Let $K[x_0, x_1, x_2, x_3]_d$ be the linear subspace of $K[x_0, x_1, x_2, x_3]$ consisting of homogeneous polynomials of degree d, and define

$$\mathbb{P}_{3,d} = \frac{K[x_0, x_1, x_2, x_3]_d \setminus \{0\}}{\sim},$$

where $f \sim g$ iff $f = \lambda g$ for some $\lambda \in K \setminus \{0\}$.

We naturally identify $\mathbb{P}_{3,d}$ with the projective space whose coordinates are indexed by the coefficients of polynomials in $K[x_0, x_1, x_2, x_3]_d$:

$$\mathbb{P}_{3,d} = \mathbb{P}^{\binom{d+3}{3}-1}.$$

Each surface of degree d corresponds to a point of the projective space $\mathbb{P}_{3,d}$ in a natural way: if $\mathcal{I}(X) = \langle f \rangle$, then we identify X with $[f] \in \mathbb{P}_{3,d}$, which is well-defined by our remarks above. However, not every point of $\mathbb{P}_{3,d}$ corresponds to a surface of degree d in $\mathbb{P}^3$, because some of the points of $\mathbb{P}_{3,d}$—such as $[x_0^d]$—are reducible polynomials, so they do not define (irreducible) surfaces.

12.6. LINES ON SURFACES

Even though $\mathbb{P}_{3,d}$ contains extraneous points from the perspective of studying surfaces of degree d, adding these extra points carries the benefit of allowing us to work with a projective space, as opposed to a subset of a projective space. Uniform statements about points in $\mathbb{P}_{3,d}$, once proven, will specialize to give us conclusions about all surfaces of degree d.

As a primer for working with lines on surfaces, we remind the reader that there are several ways to view a line $L \subseteq \mathbb{P}^3$, and we will use them interchangeably:

- as a vanishing set $\mathcal{V}(\ell_0, \ell_1) \subseteq \mathbb{P}^3$ where $\ell_0, \ell_1 \in K[x_0, x_1, x_2, x_3]_1$ is a pair of linearly independent homogeneous linear polynomials;
- as $\mathbb{P}(V) = \dfrac{V \setminus \{0\}}{\sim}$ where $V \subseteq K^4$ is a two-dimensional linear subspace;
- as the image of an injective linear map $\mathbb{P}^1 \to \mathbb{P}^3$.

We also recall that lines in $\mathbb{P}^3$ are parametrized by the Grassmannian $G(2,4)$, which we view as a projective hypersurface in $\mathbb{P}^5$ via the Plücker map (Section 10.5).

We now turn to the main question of this section: given $[f] \in \mathbb{P}_{3,d}$, does the surface $\mathcal{V}(f) \subseteq \mathbb{P}^3$ contain a line? Let us begin with a concrete example for $d = 3$.

12.53 EXAMPLE A line on the Fermat cubic surface

Let $K = \mathbb{C}$ and consider the irreducible cubic polynomial $f = x_0^3 + x_1^3 + x_2^3 + x_3^3$, commonly referred to as the "Fermat cubic." Let $L \subseteq \mathbb{P}^3$ be the line defined by

$$L = \{[a : b : -a : -b] \mid [a : b] \in \mathbb{P}^1\} \subseteq \mathbb{P}^3.$$

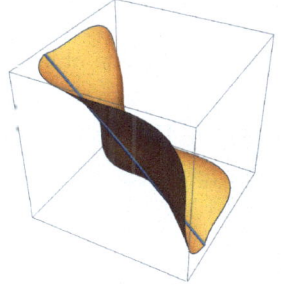

Since f vanishes at every point of L, as one readily verifies, it then follows that $L \subseteq \mathcal{V}(f)$.
The affine restrictions of (the real points of) $\mathcal{V}(f)$ and L are depicted above.

While the previous example exhibits a particular cubic surface that contains a line, we aim to prove that *every* cubic surface contains a line, which is a significantly more challenging assertion to verify. As a warmup to this more general claim, we first discuss the more elementary settings of lines in planes (where $d = 1$) and lines in quadric surfaces (where $d = 2$).

In the case where $d = 1$, a polynomial $f \in K[x_0, x_1, x_2, x_3]_1$ defines a plane $P = \mathcal{V}(f) \subseteq \mathbb{P}^3$. As every plane contains a line—in fact, each plane contains infinitely many lines parametrized by $G(2,3) \cong \mathbb{P}^2$—we conclude that for any $[f] \in \mathbb{P}_{3,1}$, the projective variety $\mathcal{V}(f) \subseteq \mathbb{P}^3$ contains infinitely many lines, and that these lines can be parametrized by $\mathbb{P}^2$.

The case $d = 2$ is slightly more challenging, and we address it in the next result.

12.54 PROPOSITION *Quadric surfaces are covered by lines*

Given $[f] \in \mathbb{P}_{3,2}$ and $a \in \mathcal{V}(f)$, there exists at least one line $L \subseteq \mathbb{P}^3$ such that $a \in L \subseteq \mathcal{V}(f)$. In particular, $\mathcal{V}(f)$ contains infinitely many lines.

PROOF   Let $f \in K[x_0, x_1, x_2, x_3]_2$ be a homogeneous polynomial of degree two and let $a \in \mathcal{V}(f) \subseteq \mathbb{P}^3$. Write $a = [a_0 : a_1 : a_2 : a_3]$ and assume, without loss of generality, that $a_0 = 1$. Restricting to the affine patch $\mathbb{A}_0^3 = \mathbb{A}^3$, any line in $\mathbb{P}^3$ containing a is uniquely determined by its affine restriction, which is an affine line of the form
$$L_b = \{(a_1, a_2, a_3) + t(b_1, b_2, b_3) \mid t \in K\}$$
for some $b = (b_1, b_2, b_3) \neq (0, 0, 0)$. Since the line L_b is independent of scaling b, we view $b = [b_1 : b_2 : b_3]$ as an element of $\mathbb{P}^2$.

If L is a line containing a, then f vanishes on L if and only if $f_0 = f(1, x_1, x_2, x_3)$ vanishes on the affine restriction of L. Evaluating f_0 at every point of such an affine restriction L_b, and collecting the powers of t, we obtain an expression of the form

(12.55) $f_0(a_1 + tb_1, a_2 + tb_2, a_3 + tb_3) = g_1(b_1, b_2, b_3)t + g_2(b_1, b_2, b_3)t^2,$

where g_i is homogeneous of degree i, and the t^0 term is zero as $f_0(a_1, a_2, a_3) = 0$. Thus, f_0 vanishes on L_b if and only if $b \in \mathcal{V}(g_1, g_2)$. Since any two homogeneous polynomials have nonempty common vanishing set in $\mathbb{P}^2$ (Proposition 12.25), there must be at least one solution $b \in \mathcal{V}(g_1, g_2) \in \mathbb{P}^2$, from which we conclude that there exists at least one line $L = \overline{L}_b \subseteq \mathbb{P}^3$ such that $a \in L \subseteq \mathcal{V}(f)$.

The final assertion in the proposition follows: if there were only finitely many lines on $\mathcal{V}(f)$, then every point of $\mathcal{V}(f)$ would lie on one of these lines, so $\mathcal{V}(f)$ would be a finite union of lines, contradicting the two-dimensionality of $\mathcal{V}(f)$. □

12.56 EXAMPLE   Lines on quadric surfaces

Below, we have depicted the (real points of the) affine restrictions of three quadric surfaces: $\mathcal{V}(2x_0^2 - 2x_1^2 - 2x_2^2 + x_3^2)$, $\mathcal{V}(x_0 x_3 - x_1 x_2)$, and $\mathcal{V}(x_1^2 + x_2^2 - x_3^2)$.

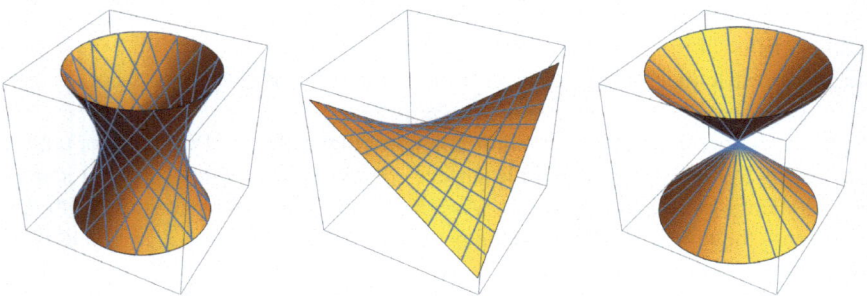

We have included a set of lines on each surface to give a sense of how the infinitely many lines are arranged. We note that the first two surfaces are smooth, whereas the third is singular, and this is reflected in how the lines are arranged on the surfaces. More specifically, Exercise 12.7.6 argues that a quadric surface in $\mathbb{P}^3$ is smooth if and only if every point lies on exactly two lines on the surface. Such a surface is called "doubly ruled," and we have previously seen an example: the Segre embedding $\mathbb{P}^1 \times \mathbb{P}^1 \subseteq \mathbb{P}^3$, which is the second surface depicted above. In fact, this is essentially the only example: Exercise 12.7.7 argues that every smooth quadric surface in $\mathbb{P}^3$ is projectively equivalent to the Segre surface. We defer these two exercises to the next section as some of the methods introduced there will be helpful.

12.6. LINES ON SURFACES

> *The standing hypothesis that K is algebraically closed is essential in Proposition 12.54: the unit sphere, viewed as a quadric surface over $\mathbb{R}$, does not contain any lines (over $\mathbb{R}$).*

Notice that, for $d \geq 3$, the right-hand side of (12.55) acquires additional coefficients that would need to vanish in order for X to contain a line passing through $a \in X$, but three or more homogeneous polynomials generally do not have common solutions in $\mathbb{P}^2$. This reflects the fact that surfaces of degree $d \geq 3$ do not generally contain lines passing through all of their points. Consequently, the method used in the proof of Proposition 12.54 breaks down for $d \geq 3$, and we require more sophisticated techniques.

In order to study lines on surfaces with $d \geq 3$, we now introduce a new tool: the *incidence variety*, which parametrizes all lines on all surfaces of degree d.

12.57 DEFINITION *Incidence variety of lines on surfaces*

For any $d \geq 1$, the *incidence variety of lines on degree-d surfaces* is the subset of $\mathbb{P}_{3,d} \times \mathbf{G}(2,4)$ defined by
$$\Omega_d = \{([f], L) \mid L \subseteq \mathcal{V}(f)\} \subseteq \mathbb{P}_{3,d} \times \mathbf{G}(2,4).$$

In words, a point $([f], L)$ in Ω_d corresponds to a nonzero homogeneous polynomial $f \in K[x_0, x_1, x_2, x_3]_d$ (up to scaling) along with a line $L \subseteq \mathbb{P}^3$ such that L happens to lie on the vanishing set $\mathcal{V}(f) \subseteq \mathbb{P}^3$. In order to use the tools of algebraic geometry to study incidence varieties, we first verify that Ω_d is, in fact, a variety.

12.58 PROPOSITION *Lines on surfaces form a projective variety*

For $d \geq 1$, the incidence variety Ω_d of lines on degree-d surfaces is a closed subvariety of $\mathbb{P}_{3,d} \times \mathbf{G}(2,4)$. In particular, Ω_d is a projective variety.

PROOF Consider $[f] \in \mathbb{P}_{3,d}$ and $L \in \mathbf{G}(2,4)$. To show that Ω_d is closed in $\mathbb{P}_{3,d} \times \mathbf{G}(2,4)$, we aim to characterize the condition that $([f], L) \in \Omega_d$ in terms of polynomial constraints on the coordinates of $\mathbb{P}_{3,d}$ (which correspond to the coefficients of the polynomial f) and the Plücker coordinates of $\mathbf{G}(2,4)$ (which correspond to the coordinates on the ambient $\mathbb{P}^5$ that $\mathbf{G}(2,4)$ sits within).

Write $L = \mathbb{P}(V)$ for some two-dimensional subspace $V \subseteq K^4$ and let $\{v, w\}$ be a basis of V. In Exercise 12.6.3, the reader is encouraged to prove that there is a surjective map $K^4 \to V$ given by
$$u \mapsto (u \cdot v)w - (u \cdot w)v,$$
where "$\cdot$" denotes the usual dot product on K^4. If we write $u = (a_0, a_1, a_2, a_3)$, $v = (b_0, b_1, b_2, b_3)$, and $w = (c_0, c_1, c_2, c_3)$, every element of V can be written as
$$(u \cdot v)w - (u \cdot w)v = (a_0 p_{00} + \cdots + a_3 p_{30}, \ldots, a_0 p_{03} + \cdots + a_3 p_{33}),$$
where $p_{ij} = b_i c_j - b_j c_i$. By definition of the Plücker map $P_{2,4}$, notice that each nonzero p_{ij} is (up to a sign) a Plücker coordinate of the line $L \in \mathbf{G}(2,4) \subseteq \mathbb{P}^5$.

The condition that $([f], L) \in \Omega_d$ is equivalent to f vanishing on V, and by our expressions for points of V in terms of Plücker coordinates p_{ij}, this is equivalent to

$$f(a_0 p_{00} + \cdots + a_3 p_{30}, \ldots, a_0 p_{03} + \cdots + a_3 p_{33}) = 0$$

for every $a_0, \ldots, a_3 \in K$. Expand as a polynomial in $a_0, \ldots, a_3$: this vanishing then occurs if and only if every coefficient of every monomial in $a_0, \ldots, a_3$ vanishes. As these coefficients are polynomials in the Plücker coordinates of L and the coefficients of f, we conclude that Ω_d is closed in $\mathbb{P}_{3,d} \times \mathbf{G}(2,4)$. □

Let us describe Ω_d explicitly as a vanishing set in the simplest case: $d = 1$.

12.59 EXAMPLE Defining equations of Ω_1

Any point $[f] \in \mathbb{P}_{3,1}$ corresponds to a linear polynomial $f = \alpha_0 x_0 + \cdots + \alpha_3 x_3$, up to scaling. Furthermore, via the Plücker map $P_{2,4}$, any point $L \in \mathbf{G}(2,4)$ can be written in coordinates as

$$L = [p_{01} : p_{02} : p_{03} : p_{12} : p_{13} : p_{23}] \subseteq \mathbb{P}^5.$$

Set $p_{ji} = -p_{ij}$ if $i < j$, and set $p_{ii} = 0$. Following the proof of Proposition 12.58, one checks that f vanishes on $L \in \mathbf{G}(2,4)$ if and only if

$$\alpha_0 p_{i0} + \alpha_1 p_{i1} + \alpha_2 p_{i2} + \alpha_3 p_{i3} = 0 \quad \text{for} \quad i = 0, 1, 2, 3.$$

Thus, $\Omega_1 \subseteq \mathbb{P}_{3,1} \times \mathbf{G}(2,4)$ is the vanishing set of four polynomials that are linear separately in the four variables $\alpha_0, \ldots, \alpha_3$ coming from $\mathbb{P}_{3,1} = \mathbb{P}^3$ and the six variables $p_{01}, \ldots, p_{23}$ coming from $\mathbf{G}(2,4) \subseteq \mathbb{P}^5$.

As Ω_d is a subvariety of a product $\mathbb{P}_{3,d} \times \mathbf{G}(2,4)$, it comes equipped with a projection map to each factor of the product. In what follows, we play these two projection maps against each other. As it turns out, the first projection is what we actually want to study; after all, the first projection is surjective if and only if $\mathcal{V}(f)$ contains a line for every $[f] \in \mathbb{P}_{3,d}$. On the other hand, as we will see in the next result, it is only through an understanding of the second projection map that we can glean structural information about Ω_d, such as its dimension, which will be necessary in order to conclude useful results regarding the first projection map.

12.60 PROPOSITION Dimension of the space of lines on surfaces

For any $d \geq 1$, the incidence variety Ω_d is irreducible, and

$$\dim(\Omega_d) = \binom{d+3}{3} + 2 - d = \dim(\mathbb{P}_{3,d}) + 3 - d.$$

The $3 - d$ appearing in this result is a hint as to the reason why cubic surfaces (for which $d = 3$) are special.

PROOF Let $\pi_2 : \Omega_d \to \mathbf{G}(2,4)$ be the restriction of the projection onto the second factor of $\mathbb{P}_{3,d} \times \mathbf{G}(2,4)$. In order to prove the theorem, we study the fibers of the map π_2.

12.6. LINES ON SURFACES

Let $L \in \mathbf{G}(2,4)$ be a line in $\mathbb{P}^3$, which we can write as $L = \mathcal{V}(\ell_0, \ell_1)$ for some linearly independent $\ell_0, \ell_1 \in K[x_0, x_1, x_2, x_3]_1$. Notice that $([f], L) \in \pi_2^{-1}(L)$ if and only if $f \in \mathcal{I}(L) = \langle \ell_0, \ell_1 \rangle$. In other words, $\pi_2^{-1}(L)$ is in natural bijection with the points in $\mathbb{P}_{3,d}$ represented by elements of

$$\langle \ell_0, \ell_1 \rangle \cap K[x_0, x_1, x_2, x_3]_d \subseteq K[x_0, x_1, x_2, x_3]_d.$$

Since $\langle \ell_0, \ell_1 \rangle \cap K[x_0, x_1, x_2, x_3]_d$ is a linear subspace of $K[x_0, x_1, x_2, x_3]_d$, it follows that $\pi_2^{-1}(L)$ is a linear subvariety of $\mathbb{P}_{3,d}$ of dimension one less than the vector-space dimension of $\langle \ell_0, \ell_1 \rangle \cap K[x_0, x_1, x_2, x_3]_d$. Let us compute that dimension.

Extend ℓ_0, ℓ_1 to a basis $\{\ell_0, \ell_1, \ell_2, \ell_3\}$ of the four-dimensional vector space $K[x_0, x_1, x_2, x_3]_1$, and notice that there is a degree-preserving isomorphism of K-algebras $K[y_0, y_1, y_2, y_3] \to K[x_0, x_1, x_2, x_3]$ that sends y_i to ℓ_i. This isomorphism restricts to a vector-space isomorphism

$$\langle y_0, y_1 \rangle \cap K[y_0, y_1, y_2, y_3]_d \cong \langle \ell_0, \ell_1 \rangle \cap K[x_0, x_1, x_2, x_3]_d,$$

so we can now turn our focus to computing the dimension of the former. A basis of $\langle y_0, y_1 \rangle \cap K[y_0, y_1, y_2, y_3]_d$ is given by the monomials that are divisible by either y_0 or y_1. Thus, upon subtracting the $d+1$ monomials of degree d that are divisible by neither y_0 nor y_1, we compute

$$\dim\left(\langle y_0, y_1 \rangle \cap K[y_0, y_1, y_2, y_3]_d\right) = \binom{d+3}{3} - d - 1.$$

Summarizing the previous two paragraphs, we have now argued that, for any $L \in \mathbf{G}(2,4)$, the fiber $\pi_2^{-1}(L)$ is a linear subvariety of $\mathbb{P}_{3,d}$ of dimension

$$(12.61) \qquad \dim(\pi_2^{-1}(L)) = \binom{d+3}{3} - d - 2.$$

Therefore, since $\mathbf{G}(2,4)$ is irreducible and $\pi_2 : \Omega_d \to \mathbf{G}(2,4)$ is surjective with irreducible fibers all of the same dimension, Theorem 12.48 implies that Ω_d is irreducible. Knowing that Ω_d is irreducible then allows us to apply Theorem 12.45, which guarantees the existence of a nonempty open set $U \subseteq \mathbf{G}(2,4)$ such that

$$\dim(\Omega_d) = \dim(\pi_2^{-1}(L)) + \dim(\mathbf{G}(2,4)) \quad \text{for every} \quad L \in U.$$

Since $\dim(\mathbf{G}(2,4)) = 4$ and all fibers of π_2 have the same dimension given by (12.61), we conclude that

$$\dim(\Omega_d) = \binom{d+3}{3} - d + 2. \qquad \square$$

We are now prepared to prove the main result of this section.

12.62 THEOREM *Every cubic surface contains a line*

For any homogeneous cubic polynomial $f \in K[x_0, x_1, x_2, x_3]_3$, there exists at least one line $L \subseteq \mathbb{P}^3$ such that $L \subseteq \mathcal{V}(f)$.

PROOF Let $\pi_1 : \Omega_3 \to \mathbb{P}_{3,3}$ be the restriction of the projection onto the first factor of $\mathbb{P}_{3,3} \times G(2,4)$. The theorem is equivalent to the assertion that π_1 is surjective. Suppose, toward a contradiction, that π_1 is not surjective. Since Ω_3 is a projective variety, $\pi_1(\Omega_3)$ is closed in $\mathbb{P}_{3,3}$, and our assumption then implies that $\pi_1(\Omega_3)$ is a proper closed subvariety of the irreducible variety $\mathbb{P}_{3,3}$. Thus (Exercise 12.2.1), we have
$$\dim(\pi_1(\Omega_3)) < \dim(\mathbb{P}_{3,3}) = 19.$$
However, since $\dim(\Omega_3) = 19$ (Proposition 12.60) while $\dim(\pi_1(\Omega_3)) < 19$, Theorem 12.43 then implies that every nonempty fiber of π_1 has dimension at least one. On the other hand, Exercise 12.6.6 describes an explicit example of a cubic polynomial $f \in K[x_0, x_1, x_2, x_3]_3$ such that $\mathcal{V}(f)$ contains a finite, nonzero number of lines, showing that the fiber $\pi_1^{-1}([f])$ has dimension zero, a contradiction. This contradiction implies that π_1 is surjective, as claimed. □

So far in this section, we have proved that there exists a line $L \subseteq \mathcal{V}(f)$ for any $[f] \in \mathbb{P}_{3,d}$ with $d \leq 3$. We now close the section by stating what happens when $d > 3$, leaving the proof as an exercise to the reader.

12.63 PROPOSITION *Most high-degree surfaces do not contain lines*

For $d > 3$, the set of points $[f] \in \mathbb{P}_{3,d}$ for which $\mathcal{V}(f) \subseteq \mathbb{P}^3$ contains a line is a closed subvariety of $\mathbb{P}_{3,d}$ of codimension at least $d - 3$.

PROOF Exercise 12.6.7. □

In other words, for $d > 3$, there is a nonempty open subset of points $[f] \in \mathbb{P}_{3,d}$ such that the corresponding surfaces $\mathcal{V}(f) \subseteq \mathbb{P}^3$ do not contain a single line.

Intuitively, the curvier a surface is, the less likely it is to contain a line.

To put it yet another way, if $f \in K[x_0, x_1, x_2, x_3]$ is a homogeneous polynomial of degree $d > 3$, then in order for $\mathcal{V}(f) \subseteq \mathbb{P}^3$ to contain a line, its coefficients must satisfy certain nontrivial polynomial relations. Moreover, as d increases, the set of $[f] \in \mathbb{P}_{3,d}$ for which $\mathcal{V}(f)$ contains a line becomes relatively smaller and smaller compared to the dimension of $\mathbb{P}_{3,d}$, meaning that the coefficients of f must satisfy more and more relations in order to guarantee that $\mathcal{V}(f)$ contains a line.

Exercises for Section 12.6

12.6.1 Consider the quadric surface $X = \mathcal{V}(x_0^2 - x_1^2 - x_2^2 - x_3^2) \subseteq \mathbb{P}_\mathbb{C}^3$, whose real points are the unit sphere within $\mathbb{A}_0^3$. Construct two distinct lines in X that contain $a = [1:0:0:1]$.

12.6.2 Consider the quadric surface $X = \mathcal{V}(x_1^2 + x_2^2 - x_3^2) \subseteq \mathbb{P}_\mathbb{C}^3$, whose real points restrict to the circular cone with vertex at the origin in the affine patch $\mathbb{A}_0^3$. Prove that every line in X contains $[1:0:0:0]$, and use this to explain why every point of X other than $[1:0:0:0]$ is contained in a unique line in X.

12.6.3 Let $v, w \in K^4$. Prove that the map $K^4 \to K^4$ defined by sending u to $(u \cdot v)w - (u \cdot w)v$ is a surjection onto the span of v and w.

12.6. LINES ON SURFACES

12.6.4 Write down defining equations of Ω_2 as a subvariety of $\mathbb{P}_{3,2} \times \mathbf{G}(2,4)$.

12.6.5 Give an example of a point $[f] \in \mathbb{P}_{3,2}$ for which the fiber $\pi_1^{-1}([f]) \subseteq \Omega_2$ has dimension one, and an example for which the fiber has dimension two.

12.6.6 Consider the cubic surface $X = \mathcal{V}(x_0^3 - x_1 x_2 x_3) \subseteq \mathbb{P}^3$.
  (a) Prove that $X \cap \mathcal{V}(x_0)$ is the union of three lines in $\mathbb{P}^3$.
  (b) Prove that the affine restriction
  $$X_0 = \mathcal{V}(1 - x_1 x_2 x_3) \subseteq \mathbb{A}^3$$
  does not contain any (affine) lines. (Hint: Any affine line can be parametrized as $\{a + tb \mid t \in K\}$ for some $a, b \in \mathbb{A}^3$ with $b \neq 0$.)
  (c) Conclude that X contains exactly three lines.

12.6.7 Prove Proposition 12.63. (Hint: The set in question is $\pi_1(\Omega_d) \subseteq \mathbb{P}_{3,d}$.)

Section 12.7 Lines on smooth cubic surfaces

We have arrived at the final section of the book, where we discuss the following classical result, which was originally proved in 1849 by Arthur Cayley (1821–1895) and George Salmon (1819–1904) and builds on many of our prior developments.

> **12.64 THEOREM** *Lines on smooth cubic surfaces*
>
> Let $X \subseteq \mathbb{P}^3$ be a smooth cubic surface. Then X contains exactly 27 lines.

One reason Theorem 12.64 is such a beautiful result is the uniform nature of the conclusion: the number of lines on a smooth cubic surface depends neither on the underlying (algebraically closed) field, nor on the defining cubic equation. Moreover, unlike the case of smooth quadric surfaces, which are all projectively equivalent to one another (Exercise 12.7.7), it can be shown that there are infinitely many distinct smooth cubic surfaces, even after accounting for projective equivalence; that every surface in this infinite family has the same number of lines is quite remarkable! As a warm-up example, the reader is encouraged in Exercise 12.7.1 to find all 27 lines on the Fermat cubic surface, which is the most symmetric example.

It is difficult to visualize the lines on a general cubic surface over $\mathbb{C}$, because a complex line generally intersects the real points of the surface in a single point. However, there are specific examples of cubic surfaces over $\mathbb{C}$ for which all of the complex lines contain real lines. The most famous such example is the Clebsch cubic surface, named in honor of Alfred Clebsch (1833–1872).

12.65 EXAMPLE Lines on the Clebsch cubic surface

The Clebsch cubic surface is defined by

$$X = \mathcal{V}(w^3 + x^3 + y^3 + z^3 - (w + x + y + z)^3) \subseteq \mathbb{P}^3.$$

Below, we have depicted—from three different angles—an affine restriction of the real points of X along with the 27 lines on X, all of which are visible as real lines on this affine restriction.

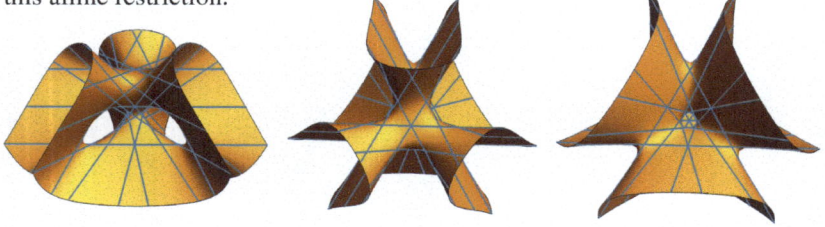

In working toward a proof of Theorem 12.64, we prove a number of preparatory lemmas concerning lines on smooth cubic surfaces. Throughout, we make liberal use of projective equivalences to transform lines and planes to convenient positions, noting that these projective equivalences transform smooth cubic surfaces to smooth cubic surfaces. In each instance, the justification that a projective equivalence with the desired properties exists can be reduced to a linear-algebra statement by replacing lines and planes in $\mathbb{P}^2$ and $\mathbb{P}^3$ with linear subspaces in K^3 and K^4; we leave these linear-algebra justifications to the reader (Exercises 12.7.2 and 12.7.3).

12.7. LINES ON SMOOTH CUBIC SURFACES

The proof of the next lemma concerning lines on smooth cubic surfaces is a prime example of how a projective equivalence can be used to simplify an argument.

12.66 LEMMA *Lines meeting on a smooth cubic surface are coplanar*

Let $X \subseteq \mathbb{P}^3$ be a smooth cubic surface. If L_1, L_2, and L_3 are lines on X such that $L_1 \cap L_2 \cap L_3 \neq \emptyset$, then L_1, L_2, and L_3 lie on a common plane.

PROOF Assume that $X \subseteq \mathbb{P}^3$ is an irreducible cubic surface that contains three intersecting lines that are not contained in a plane; we prove that X is singular at the point where these three lines meet.

Given three intersecting lines in $\mathbb{P}^3$ not contained in a plane, Exercise 12.7.3 shows that there is a projective equivalence of $\mathbb{P}^3$ taking these three lines to the three lines $\mathcal{V}(x,y)$, $\mathcal{V}(x,z)$, and $\mathcal{V}(y,z)$, which meet at the point $[1:0:0:0]$. Thus, working up to projective equivalence, we can assume that X contains these three specific lines. Let $\mathcal{I}(X) = \langle f \rangle$. Since f vanishes on $\mathcal{V}(x,y)$, $\mathcal{V}(x,z)$, and $\mathcal{V}(y,z)$, some reflection should convince the reader that every term of f must be divisible by at least two of the three variables x, y, and z, and it follows that

$$f = xy\ell_1 + xz\ell_2 + yz\ell_3 \quad \text{for some} \quad \ell_1, \ell_2, \ell_3 \in K[w, x, y, z]_1.$$

Furthermore, by irreducibility, ℓ_1, ℓ_2, and ℓ_3 are all nonzero. Computing partial derivatives, we have

$$\frac{\partial f}{\partial w} = xy\frac{\partial \ell_1}{\partial w} + xz\frac{\partial \ell_2}{\partial w} + yz\frac{\partial \ell_3}{\partial w},$$

$$\frac{\partial f}{\partial x} = y\ell_1 + xy\frac{\partial \ell_1}{\partial x} + z\ell_2 + xz\frac{\partial \ell_2}{\partial x} + yz\frac{\partial \ell_3}{\partial x},$$

$$\frac{\partial f}{\partial y} = x\ell_1 + xy\frac{\partial \ell_1}{\partial y} + xz\frac{\partial \ell_2}{\partial y} + z\ell_3 + yz\frac{\partial \ell_3}{\partial y},$$

$$\frac{\partial f}{\partial z} = xy\frac{\partial \ell_1}{\partial z} + x\ell_2 + xz\frac{\partial \ell_2}{\partial z} + y\ell_3 + yz\frac{\partial \ell_3}{\partial z}.$$

Since these partial derivatives all vanish at $[1:0:0:0] \in X$, the Jacobian criterion for smoothness (see Example 12.35) implies that X is singular at this point. □

The next lemma concerns the types of curves that can arise by intersecting a smooth cubic surface with a plane. In preparation for the lemma, recall that any plane $P \subseteq \mathbb{P}^3$ is the image of an injective linear map $\mathbb{P}^2 \to \mathbb{P}^3$, allowing us to view any closed subvariety of P as a projective variety in $\mathbb{P}^2$, simply by taking its preimage along such an injection. Furthermore, any two injective linear maps $\mathbb{P}^2 \to \mathbb{P}^3$ mapping to the same plane $P \subseteq \mathbb{P}^3$ differ by a projective equivalence of $\mathbb{P}^2$, and since projective equivalences of $\mathbb{P}^2$ preserve the degree of irreducible curves $C \subseteq \mathbb{P}^2$, it makes sense to refer to the "degree" of a closed irreducible curve within a plane $P \subseteq \mathbb{P}^3$. With this terminology in place, the next result characterizes all possible configurations of curves that arise from intersecting a smooth cubic surface with a plane.

> **12.67 LEMMA** *Linear slices of smooth cubic surfaces*
>
> Let $X \subseteq \mathbb{P}^3$ be a smooth cubic surface and let $P \subseteq \mathbb{P}^3$ be a plane. Then $X \cap P$ is exactly one of the following:
>
> (i) an irreducible curve of degree three in P,
>
> (ii) a union of a line and an irreducible curve of degree two in P, or
>
> (iii) a union of three distinct lines in P.

PROOF Let $X \subseteq \mathbb{P}^3$ be a smooth cubic surface with $\mathcal{I}(X) = \langle f \rangle$, and let $P \subseteq \mathbb{P}^3$ be a plane. After translating both X and P by a projective equivalence, we may assume that $P = \mathcal{V}(z)$, which we identify with $\mathbb{P}^2$ with coordinates w, x, and y. The restriction of f to P is given by $g = f(w, x, y, 0)$, which is a homogeneous polynomial of degree three in $K[w, x, y]$. Notice that g must be nonzero, as otherwise f would be divisible by z, contradicting the irreducibility of X.

If g is irreducible, then $X \cap P$ is an irreducible curve of degree three in P, and if g factors as a product of a linear and an irreducible quadratic polynomial, then $X \cap P$ is a union of a line and an irreducible curve of degree two in P. To finish the proof, assume that g factors as a product of three linear polynomials; we must prove that $X \cap P$ is a union of three distinct lines (as opposed to one or two lines).

Toward a contradiction, assume that $X \cap P$ contains fewer than three lines; we show that X is singular. Since g factors as a product of three linear polynomials but $\mathcal{V}(g) = X \cap P$ contains fewer than three lines, g must have a repeated linear factor. Up to projective equivalence, we may assume that the line where this repeated linear factor vanishes is $L = \mathcal{V}(y, z)$. In other words, $g = f(w, x, y, 0) = y^2 \ell$ for some nonzero $\ell \in K[w, x, y, z]_1$, and thus,

$$f = y^2 \ell + zq \quad \text{for some nonzero} \quad q \in K[w, x, y, z]_2.$$

By computing partial derivatives, one then verifies (Exercise 12.7.4) that X is singular at any point $[a : b : 0 : 0] \in L$ where $q(a, b, 0, 0) = 0$, and there must be at least one such point because K is algebraically closed. Thus, X is singular if $X \cap P$ is a union of fewer than three lines, completing the proof. □

> *In Case* (iii) *of Lemma 12.67, there are two possible arrangements of the three lines: either they meet in pairs at three distinct points of P, or all three meet at a common point of P.*

Given a smooth cubic surface X and a line $L \subseteq X$, our next aim is to determine all of the other lines in X that intersect L, and we accomplish this by studying the intersection of X with all planes containing L. More precisely, if P is a plane containing a line L, then Lemma 12.67 implies that $X \cap P$ is either the union of L and an irreducible conic, or $X \cap P$ contains two more lines on X (in addition to L). Since every line on X that meets L lies in exactly one of the latter planes, understanding the lines on X that intersect L amounts to characterizing the planes containing L for which the residual conic in $X \cap P$ is reducible. The next lemma gives a numerical tool for determining when a conic is reducible.

12.7. LINES ON SMOOTH CUBIC SURFACES

12.68 LEMMA *Determining when a conic is reducible*

If $g = ax^2 + by^2 + cz^2 + dxy + exz + fyz \in K[x,y,z]$, then g is reducible if and only if
$$4abc + def - af^2 - be^2 - cd^2 = 0.$$

PROOF   Assume that the characteristic of K is not two (the characteristic-two case is left to Exercise 12.7.5). Given g as in the lemma, define the matrix

$$M_g = \begin{pmatrix} 2a & d & e \\ d & 2b & f \\ e & f & 2c \end{pmatrix}.$$

Notice that $2g$ can be written as the matrix product
$$2g = (x,y,z) M_g (x,y,z)^T.$$

Upon observing that
$$\det(M_g) = 2(4abc + def - af^2 - be^2 - cd^2),$$

we see that the assertion in the lemma is equivalent to the assertion that g is reducible if and only if $\det(M_g) = C$.

Let $A = (a_{i,j})$ be an invertible 3×3 matrix, and consider the polynomial $\tilde{g}(x,y,z) = g((x,y,z)A^T)$; in other words, $\tilde{g}$ is obtained from g by the linear change of variables determined by A:
$$\tilde{g}(x,y,z) = g(a_{1,1}x + a_{1,2}y + a_{1,3}z, \ldots, a_{3,1}x + a_{3,2}y + a_{3,3}z).$$

Notice that
$$2\tilde{g} = (x,y,z) A^T M_g A (x,y,z)^T,$$

from which it follows that $M_{\tilde{g}} = A^T M_g A$. Since $\det(A) = \det(A^T) \neq 0$, multiplicativity of determinants then implies that $\det(M_g) = 0$ if and only if $\det(M_{\tilde{g}}) = 0$. Thus, the vanishing of $\det(M_g)$—as well as the irreducibility of g—is invariant under any invertible linear change of variables.

As we saw in the proof of Proposition 10.12, up to an invertible linear change of variables, we may assume that $g = axy + bxz + cyz$. The lemma now follows from the observation that such g is reducible if and only if one of a, b, or c is zero, which is true if and only if $\det(M_g) = 2abc = 0$.   □

The culmination of the previous two lemmas is the following result, which is the key tool that will help us count lines on smooth cubic surfaces.

12.69 LEMMA *Every line on a smooth cubic surface meets 10 others*

Let $X \subseteq \mathbb{P}^3$ be a smooth cubic surface and $L \subseteq X$ a line. Then there are exactly five planes P containing L for which $P \cap L$ is a union of three lines. As a consequence, L intersects exactly ten other lines in X.

PROOF Up to projective equivalence, we may assume that $L = \mathcal{V}(y,z)$. Suppose that $\mathcal{I}(X) = \langle f \rangle$. Since f vanishes on L, every term of f is divisible by either y or z, and separating the terms according to their degrees in y and z, we can write

$$f = \ell_1 w^2 + \ell_2 wx + \ell_3 x^2 + q_1 w + q_2 x + c, \tag{12.70}$$

with $\ell_1, \ell_2, \ell_3 \in K[y,z]_1$ linear, $q_1, q_2 \in K[y,z]_2$ quadratic, and $c \in K[y,z]_3$ cubic.
For any $a = (a_0, a_1) \in K^2 \setminus \{0\}$, define a plane $P_a \subseteq \mathbb{P}^3$ by

$$P_a = \{[\alpha : \beta : a_0 \gamma : a_1 \gamma] \mid [\alpha : \beta : \gamma] \in \mathbb{P}^2\} \subseteq \mathbb{P}^3.$$

Notice that P_a contains L (where $\gamma = 0$), and moreover, every plane in $\mathbb{P}^3$ containing L is equal to P_a for some a. Additionally, observe that $P_a = P_b$ if and only if $a = \lambda b$ for some $\lambda \in K \setminus \{0\}$. In other words, the planes in $\mathbb{P}^3$ containing L are in one-to-one correspondence with the points of $\mathbb{P}^1$.

Let t, u, and v denote the coordinates on the $\mathbb{P}^2$ in the definition of P_a. In these coordinates, the restriction of f to P_a is $f_a = f(t, u, a_0 v, a_1 v) \in K[t, u, v]$, which by (12.70) is equal to

$$f_a = (\ell_1(a)t^2 + \ell_2(a)tu + \ell_3(a)u^2 + q_1(a)tv + q_2(a)uv + c(a)v^2)v.$$

Lemma 12.67 implies that $X \cap P_a$ is a union of three lines if and only if the quadratic factor in $f_a \in K[t, u, v]$ is reducible. By Lemma 12.68, this occurs if and only if

$$4\ell_1 \ell_3 c + \ell_2 q_1 q_2 - \ell_1 q_2^2 - \ell_3 q_1^2 - c\ell_2^2 \in K[y,z] \tag{12.71}$$

vanishes when evaluated at a. Thus, to prove the lemma, we must prove that (12.71) has exactly five zeros in $\mathbb{P}^1$. Upon observing that (12.71) is a homogeneous polynomial of degree five in $K[y,z]$, we know that it factors into five linear polynomials, and we must show that these linear factors are distinct.

Suppose that $a \in \mathbb{P}^1$ is a zero of (12.71), or in other words, that $P_a \cap X$ is a union of three distinct lines. Up to projective equivalence on $\mathbb{P}^3$, we may assume that $P_a = \mathcal{V}(z)$, or equivalently, that z is a linear factor of (12.71). To prove that this linear factor is distinct, we must prove that (12.71) is not divisible by z^2. Let L_1 and L_2 be the other two lines in $X \cap P_a$, aside from L. We consider two cases, depending on whether the three lines in $X \cap P_a$ have nonempty intersection.

(Case 1: $L \cap L_1 \cap L_2 = \emptyset$) Since the three lines L, L_1, and L_2 lie in the common plane $P_a = \mathcal{V}(z)$ and do not have a common point of intersection, then Exercise 12.7.2 implies that a projective equivalence can transform the lines to $L = \mathcal{V}(y,z)$, $L_1 = \mathcal{V}(w,z)$ and $L_2 = \mathcal{V}(x,z)$. Since f vanishes on L, L_1, and L_2, it then follows that every term in f is divisible by either z or wxy. From (12.70), this implies that z divides ℓ_1, ℓ_3, q_1, q_2, and c, and by the irreducibility of f, we then see that z does not divide ℓ_2. Therefore, every term in (12.71) is divisible by z^2 except possibly the final term, and to prove that z^2 does not divide (12.71), it suffices to prove that z^2 does not divide the cubic term c. To prove this final assertion, note that if z^2 divided c, then $X = \mathcal{V}(f)$ would be singular at $[0:0:1:0]$, as the reader can readily verify by computing partial derivatives of f via (12.70), and using that z divides q_1 and q_2 and that z^2 divides c. Thus, the smoothness of X implies that z is not a repeated linear factor of (12.71) in this case.

(Case 2: $L \cap L_1 \cap L_2 \neq \emptyset$) In this case, Exercise 12.7.2 implies that up to projective equivalence, $L = \mathcal{V}(y, z)$, $L_1 = \mathcal{V}(x, z)$, and $L_2 = \mathcal{V}(x + y, z)$. Thus, the part of f that is not divisible by z must be divisible by $xy(x + y)$. From (12.70), this implies that z divides ℓ_1, ℓ_2, q_1, and c, and, furthermore, that the coefficient of y in ℓ_3 is the same as the coefficient of y^2 in q_2. In particular, from the last point, it follows that z divides ℓ_3 if and only if z divides q_2, and by the irreducibility of f, it follows that z cannot divide either, as otherwise z would divide every term in (12.70). Since z does not divide q_2, then z^2 does not divide $\ell_1 q_2^2$, and since every other term in (12.71) is divisible by z^2, it follows that (12.71) is *not* divisible by z^2. Thus, z is not a repeated linear factor of (12.71) in this case.

Having verified that (12.71) has distinct linear factors, we conclude that it has exactly five zeros in $\mathbb{P}^1$, giving the five special planes in the statement of the lemma. The final assertion in the lemma follows: each line on X that intersects L lies in a unique plane containing L, and since the intersection of X with this plane contains two lines, Lemma 12.67 implies that the intersection must contain a third line. Therefore, every line on X that intersects L must lie in one of the five special planes containing L, and (aside from L) there are exactly ten such lines. □

We are finally prepared to count lines on smooth cubic surfaces.

PROOF OF THEOREM 12.64 Let $X \subseteq \mathbb{P}^3$ be a smooth cubic surface and choose any line L_1 on X; such a line exists by Theorem 12.62. Choose a plane $P \subseteq \mathbb{P}^3$ such that $L_1 \subseteq P$ and such that $X \cap P$ is a union of three lines; such a plane exists by Lemma 12.69. Call the other two lines in this plane L_2 and L_3. Now let L be any other line on X. Notice that L cannot lie on P, as the only lines in $X \cap P$ are L_1, L_2, and L_3. Since the intersection of a plane in $\mathbb{P}^3$ and a line not lying on it is a single point, it follows that $L \cap P$ is a single point of $X \cap P = L_1 \cup L_2 \cup L_3$. Thus, L meets $L_1 \cup L_2 \cup L_3$ at a unique point. Lemma 12.66 implies that this point of intersection cannot occur at a point where two or more of L_1, L_2, and L_3 intersect, and we conclude that every line on X other than L_1, L_2, and L_3 must intersect exactly one of these three lines. Lemma 12.69 tells us that there are exactly ten lines on X intersecting each of L_1, L_2, and L_3, and since each L_i meets the other two lines in $X \cap P$, it follows that there are eight remaining lines on X that meet each L_i. Since each line on X is exactly one of these, we conclude that X contains $3 + 8 + 8 + 8 = 27$ lines □

Having counted the lines on any smooth cubic surface, we conclude with a discussion of the combinatorics of how the lines intersect. Remarkably, as we will see below, the combinatorics of which pairs of lines intersect is identical for every smooth cubic surface. As a tool for helping us visualize which pairs of lines intersect, we introduce the *line-incidence graph* of a smooth cubic surface $X \subseteq \mathbb{P}^3$, which we define as the (combinatorial) graph with one vertex for each line on X and an edge between two vertices if and only if the two lines intersect. The next result asserts that this graph does not depend on the choice of smooth cubic surface X.

12.72 THEOREM *Configuration of lines on smooth cubic surfaces*

Every smooth cubic surface has the same line-incidence graph.

PROOF   Let G be the graph pictured below. We claim that G is the line-incidence graph of any smooth cubic surface. To prove this, suppose that $X \subseteq \mathbb{P}^3$ is a smooth cubic surface; we describe how to label the vertices of G by the lines on X such that two vertices are adjacent in G if and only if the corresponding lines intersect in X. We encourage the reader to actually label the vertices of G as they work through the proof. Depending on the choices one makes, there are many ways to label G, reflecting the many symmetries of this graph. As a first check, the reader can verify that G has 27 vertices, representing the 27 lines on X, and that each vertex is adjacent to 10 others, reflecting the conclusion of Lemma 12.69.

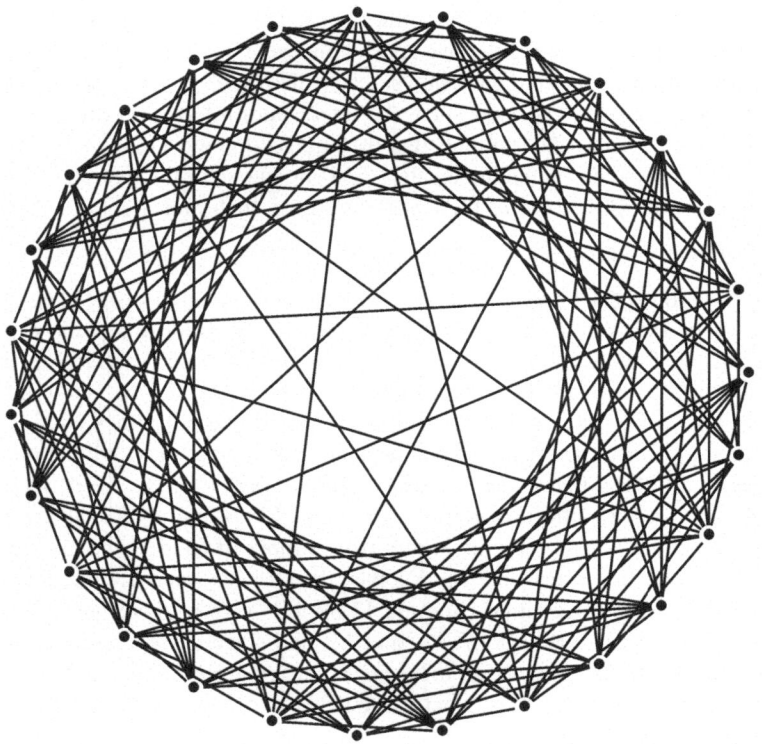

Let $L_1 \subseteq X$ be any line on X. (*Pick any vertex of G and label it L_1.*) Since L_1 meets only ten of the remaining 26 lines on X, there exists a line $L_1' \subseteq X$ that does not meet L_1. (*Pick any vertex of G that is not adjacent to L_1 and label it L_1'.*) Note that L_1 lies in five special planes, each containing two additional lines on X, and L_1' meets exactly one of each pair of such lines; let $L_{1,2}, \ldots, L_{1,6}$ denote the five lines that meet both L_1 and L_1'. (*Find the five vertices adjacent to both L_1 and L_1' and label them $L_{1,2}, \ldots, L_{1,6}$.*) Since L_1 meets $L_{1,i}$ for each i, the plane determined by these two intersecting lines contains a third line of X; call it L_i'. (*Find the unique vertex adjacent to both L_1 and $L_{1,i}$ and label it L_i'.*) Similarly, let L_i be the third line in the plane determined by the intersecting lines L_1' and $L_{1,i}$. (*Find the unique vertex adjacent to both L_1' and $L_{1,i}$ and label it L_i.*) So far, we have found 17 lines on X:

$$L_1, \ldots, L_6, L_1', \ldots, L_6', L_{1,2}, \ldots, L_{1,6}.$$

12.7. LINES ON SMOOTH CUBIC SURFACES

By construction, any triple of lines of the form $\{L_1, L_{1,i}, L'_i\}$ or $\{L'_1, L_{1,i}, L_i\}$ with $i \geq 2$ are coplanar, so their corresponding vertices form triangles in the line-incidence graph. Furthermore, for $1 < i < j \leq 6$, notice that L_i was chosen to be disjoint from L_1, and L_i is also disjoint from $L_{1,j}$, as otherwise it would need to lie in the same plane as the intersecting lines $L_{1,j}$ and L'_1, but we already know that the third line in this plane is $L_j \neq L_i$. As L_i does not meet either of the intersecting lines L_1 and $L_{1,j}$, it must meet the third line in the plane determined by them, which is L'_j. Similarly, for $1 < i < j \leq 6$, the line L'_i meets the line L_j. Thus, the vertices corresponding to pairs $\{L_i, L'_j\}$ are adjacent in the line-incidence graph of X for all distinct $i, j \in \{1, \ldots, 6\}$. The 50 edges in the line-incidence graph described in this paragraph correspond to all edges in G connecting the 17 vertices that have thus far been labeled, as the reader can readily verify.

For each $1 < i < j \leq 6$, notice that none of the 17 lines described so far intersect both of the intersecting lines L_i and L'_j. Thus, the third line on the plane determined by L_i and L'_j is not yet in our list; let $L_{i,j}$ denote this line. (*Find the unique vertex adjacent to L_i and L'_j and label it $L_{i,j}$.*) Since there are 10 such new lines, we have now accounted for all 27 lines on X. Since $L_{i,j}$ is disjoint from L_1 for $1 < i < j \leq 6$—because we have already accounted for the 10 lines meeting L_1—and since $L_{i,j}$ is also disjoint from $L_{1,i}$—because it meets a different line in the triple $\{L'_1, L_{1,i}, L_i\}$—it follows that $L_{i,j}$ must meet the triple $\{L_1, L_{1,i}, L'_i\}$ in L'_i. Similarly, the line $L_{i,j}$ must meet L_j. Thus, in the line-incidence graph, the vertices corresponding to pairs $\{L_{i,j}, L_i\}$, $\{L_{i,j}, L'_i\}$, $\{L_{i,j}, L_j\}$, and $\{L_{i,j}, L'_j\}$ are all adjacent, giving 40 additional edges in the line-incidence graph.

Thus far, we have accounted for all 10 lines intersecting any L_i or L'_i, so it only remains to describe which lines of the form $L_{i,j}$ intersect one another. If i, j, k, ℓ are distinct, then the fact that $L_{i,j}$ intersects neither L_k nor L'_ℓ implies that $L_{i,j}$ intersects the triple $\{L_k, L_{k,\ell}, L'_\ell\}$ in $L_{k,\ell}$. Thus, $L_{i,j}$ meets $L_{k,\ell}$ whenever $\{i, j\} \cap \{k, \ell\} = \emptyset$, giving 45 additional edges in the line-incidence graph. As we have now described 10 vertices adjacent to each $L_{i,j}$, we can be certain that we have successfully accounted for all possible edges in the line-incidence graph of X.

In summary, the 27 lines on X can be organized into six disjoint lines $L_1, \ldots, L_6$, six disjoint lines $L'_1, \ldots, L'_6$, and 15 lines $L_{i,j}$ with $1 \leq i < j \leq 6$ such that

- L_i meets L'_j if and only if $i \neq j$,
- L_i meets $L_{j,k}$ if and only if $i \in \{j, k\}$,
- L'_i meets $L_{j,k}$ if and only if $i \in \{j, k\}$, and
- $L_{i,j}$ meets $L_{k,\ell}$ if and only if $\{i, j\} \cap \{k, \ell\} = \emptyset$.

The reader is encouraged to check that all of these intersections are consistent with the 135 edges in their labeling of the graph G. □

While Theorem 12.72 shows that the line-incidence graph of every smooth cubic surface is the same, we mention in closing that not *all* of the combinatorics of how lines intersect remains uniform among all smooth cubic surfaces. In particular, an *Eckardt point* of a cubic surface is one where three lines on the surface meet. Some special cubic surfaces have Eckardt points (for example, the image in Example 12.65 shows that the Clebsch surface does, and Exercise 12.7.1 shows that the Fermat cubic surface does), but a general cubic surface has no Eckardt points.

Exercises for Section 12.7

12.7.1 Let $X = V(x_0^3 + x_1^3 + x_2^3 + x_3^3) \subseteq \mathbb{P}^3$ be the Fermat cubic surface.
   (a) Prove that X is smooth.
   (b) Give a parametrization of each of the 27 lines on X.
   (c) Let ζ be a third root of unity and suppose that $0 \leq i < j \leq 3$. Prove that the plane $V(x_i + \zeta x_j)$ contains three lines on X, and that these three lines have nonempty intersection.
   (d) Find 18 points on X where three lines on X meet.

12.7.2 Prove that there exists a projective equivalence of $\mathbb{P}^2$ satisfying each of the following properties.
   (a) Taking any line L to the line $V(x)$.
   (b) Taking any pair of lines L_1, L_2 to the lines $V(x), V(y)$.
   (c) Taking any three lines L_1, L_2, L_3 with $L_1 \cap L_2 \cap L_3 = \emptyset$ to the lines $V(x), V(y), V(z)$.
   (d) Taking any three lines L_1, L_2, L_3 with $L_1 \cap L_2 \cap L_3 \neq \emptyset$ to the lines $V(x), V(y), V(x+y)$.

12.7.3 Prove that there exists a projective equivalence of $\mathbb{P}^3$ satisfying each of the following properties.
   (a) Taking any line L to the line $V(y,z)$.
   (b) Taking any plane P to the plane $V(z)$.
   (c) Taking any three noncoplanar lines L_1, L_2, L_3 with $L_1 \cap L_2 \cap L_3 \neq \emptyset$ to the lines $V(x,y), V(x,z), V(y,z)$.

12.7.4 Suppose that f is an irreducible polynomial of the form $f = y^2 \ell + zq$ for some nonzero $\ell \in K[w,x,y,z]_1$ and some nonzero $q \in K[w,x,y,z]_2$. Prove that $V(f) \subseteq \mathbb{P}^3$ is singular at any point $[a:b:0:0]$ where $q(a,b,0,0) = 0$.

12.7.5 Suppose that the characteristic of K is two, and consider a homogeneous quadratic polynomial $g = ax^2 + by^2 + cz^2 + dxy + exz + fyz \in K[x,y,z]$. Prove that g is reducible if and only if $def - af^2 - be^2 - cd^2 = 0$.

12.7.6 Let $X \subseteq \mathbb{P}^3$ be a quadric surface.
   (a) Suppose that X is smooth. Prove that every point of X is contained in exactly two lines on X.
   (b) Suppose that some point of X is contained in a unique line on X. Prove that X is singular.
   (c) Suppose that some point of X is contained in more than two lines on X. Prove that X is singular.

12.7.7 Let $X \subseteq \mathbb{P}^3$ be a smooth quadric surface.
   (a) Prove that X contains three disjoint lines.
   (b) Prove that any three disjoint lines in $\mathbb{P}^3$ are projectively equivalent to the three lines $V(w,x), V(y,z)$, and $V(w-y, x-z)$.
   (c) Prove that X is projectively equivalent to $V(wz - xy) \subseteq \mathbb{P}^3$.

Coda: Where to from here?

We have covered an enormous amount of mathematics in this book, and yet it is not an exaggeration to say that we have barely scratched the surface of the subject of algebraic geometry. This book is a doorway. If you, the reader, have followed it carefully, then you have entered the house of algebraic geometry and are prepared to explore many of its various rooms, but which of these rooms you choose to explore first is in some sense a matter of taste and opportunity. In this coda, we describe several complementary directions in which one could proceed after studying this book. The first two directions center around focusing on specific types of varieties that are particularly well-behaved and well-understood (algebraic curves and toric varieties), while the third direction revolves around generalizing the notions of varieties entirely and entering the wide world of scheme theory. We hope that this coda will give the reader a flavor of what might be to come and a bit of direction for further pursuing their mastery of the vast and beautiful subject of algebraic geometry.

Algebraic curves

Many of the tools of projective algebraic geometry are perhaps best learned by first focusing deeply on the case of one-dimensional projective varieties—otherwise known as *algebraic curves*—or even more specifically on *projective plane curves*, which are projective varieties defined by the vanishing of a single polynomial in $\mathbb{P}^2$. While this may seem like a very specialized setting, it is already a rich and fascinating world that provides, among other things, an excellent introduction to the tools of intersection theory.

The prototypical example of an intersection-theoretic statement about projective plane curves is something we have briefly mentioned in this book: Bézout's Theorem. Roughly speaking, Bézout's Theorem is the statement that if $C_1 = \mathcal{V}(f_1)$ and $C_2 = \mathcal{V}(f_2)$ are projective plane curves that do not share an irreducible component, and if $\deg(f_i) = d_i$, then C_1 and C_2 intersect in $d_1 \cdot d_2$ points. This statement requires qualification, however, when tangency is involved. For instance, the curves $C_1 = \mathcal{V}(y)$ and $C_2 = \mathcal{V}(yz - x^2)$ are defined by polynomials of degrees 1 and 2, respectively, so Bézout's Theorem should tell us that they intersect in two points, when in fact their intersection is the single point $[0:0:1]$. The resolution is to observe that C_1 and C_2 are tangent at $[0:0:1]$, and to view this point as an intersection of "multiplicity" two; in general, Bézout's Theorem holds only when points of intersection are counted with the appropriate multiplicity. Defining intersection multiplicity precisely, and proving that it satisfies properties like Bézout's Theorem, is where the subject of intersection theory begins. A classic reference for a student wishing to learn about intersection multiplicities and Bezout's Theorem in the setting of algebraic curves is Fulton's book [5].

In addition to providing an introduction to intersection theory, another reason that curves may be of interest is that, when the ground field is $\mathbb{C}$, they provide the opportunity for a complex-analytic approach to algebraic geometry. The key point, here, is that a smooth one-dimensional projective variety over $\mathbb{C}$ can be viewed as a one-dimensional compact complex manifold, also known as a *Riemann surface*.

A similar statement is true for smooth projective varieties of any dimension, but in dimension one, miraculously, the converse is also true: any compact Riemann surface is isomorphic, as a complex manifold, to a smooth one-dimensional projective variety. Thus, the study of curves can be approached either via algebraic geometry (by viewing curves as projective varieties), or via complex analysis (by viewing curves as Riemann surfaces). There is a beautiful interplay between the algebraic and analytic perspectives, and readers who find the analytically-minded approach to curves illuminating may wish to study Cavalieri and Miles's book [1] or Miranda's [7].

Beyond geometry, there is also a rich and important interplay between algebraic curves and number theory. To get just a hint of this connection, we recall that the infamous "Fermat's Last Theorem" is the number-theoretic statement that, for any $n \geq 3$, there do not exist positive integers x, y, z such that $x^n + y^n = z^n$. This assertion can be rephrased in the realm of algebraic geometry as the statement that, for $n \geq 3$, the projective plane curve $\mathcal{V}(x^n + y^n - z^n) \subseteq \mathbb{P}^2_\mathbb{C}$ has no points whose coordinates are all nonzero rational numbers. This is just a first indication of the close tie between number theory and the study of rational points on algebraic curves. For an introduction to the theory of rational points on curves from a number-theoretic perspective, the reader may wish to study Silverman and Tate's book [9].

Toric varieties

Another potential avenue for further study is toric geometry, which is a concrete and hands-on gateway into algebraic geometry in dimensions greater than one. Briefly, a *toric variety* is an algebraic variety X over $\mathbb{C}$ that contains a dense open set isomorphic to the "algebraic torus" $\mathbb{T} \cong (\mathbb{C}^*)^n$, such that the group action of $\mathbb{T}$ on itself by componentwise multiplication extends to an action of $\mathbb{T}$ on all of X. For example, projective space $\mathbb{P}^n$ is a toric variety, where the dense open set consists of the points whose homogeneous coordinates are all nonzero.

The key idea behind the study of toric geometry is that the mere existence of the dense $\mathbb{T}$-action on X provides X with a great deal of combinatorial structure, and this allows one to reinterpret the geometry of X entirely in terms of discrete geometry. For example, an important piece of algebro-geometric information about a projective variety X is its *degree*, which roughly measures how "curved" X is by counting the number of points of intersection between X and a linear space of complementary dimension. When X is a toric variety, its degree is equal to the volume of an associated polytope P_X, and in this way, toric geometry translates an artifact of algebraic geometry (degree) into one of discrete geometry (volume).

The beauty of the translation to the discrete setting that toric geometry provides is that it allows a student to get a taste of many of the common tools of algebraic geometry—from divisor theory to resolutions of singularities to cohomology and intersection theory—in a concrete context that is conducive to computational examples. One should be forewarned, however, that toric varieties make up a very small slice of the vast landscape of general algebraic varieties; for example, the only smooth projective toric variety of dimension one is $\mathbb{P}^1$. A student interested in learning more about toric geometry may wish to consult the comprehensive textbook [3] by Cox, Little, and Schenck.

Scheme theory

The final topic for further study that we mention here is an enormous one, and is really a doorway of its own into all of the rest of modern algebraic geometry: the notion of a scheme.

To motivate the idea of a scheme, recall that there is a bijection between affine varieties and finitely-generated reduced K-algebras given by $X \mapsto K[X]$. One way in which to understand the inverse of this bijection is to recall that the points of X are in bijection with the maximal ideals of $K[X]$, so we can recover X—at least as a set—as the *maximal spectrum*

$$\mathrm{maxSpec}(K[X]) = \{\text{maximal ideals } M \subseteq K[X]\}.$$

In fact, this can be used to recover X not merely as a set but as a topological space: the Zariski-closed subsets of an affine variety X are precisely the sets $\mathcal{V}(I) \subseteq X$ where $I \subseteq K[X]$ is an ideal, and if we unwind this in terms of the bijection with $\mathrm{maxSpec}(K[X])$, it yields a Zariski topology on $\mathrm{maxSpec}(K[X])$ whose closed sets are precisely the sets $Z_I = \{\text{maximal ideals } M \supseteq I\}$ for any ideal $I \subseteq K[X]$.

Thus, we have described an injection

(C.1)
$$\{\text{finitely-generated reduced } K\text{-algebras}\} \to \{\text{topological spaces}\}$$
$$R \mapsto \mathrm{maxSpec}(R)$$

whose image consists precisely of affine varieties with their Zariski topology. The left-hand side is a rather special class of rings, however, so a natural question is whether we can expand (C.1) to a bijection from arbitrary commutative rings R to some generalization of affine varieties. We have already seen a hint as to why such a generalization might be useful: the affine curves $\mathcal{V}(y) \subseteq \mathbb{A}^2$ and $\mathcal{V}(y - x^2) \subseteq \mathbb{A}^2$ (which are affine restrictions of the projective plane curves considered earlier in this coda) intersect in

$$X = \mathcal{V}(y) \cap \mathcal{V}(y - x^2) = \mathcal{V}(y, y - x^2).$$

If we were to forget that $\langle y, y - x^2 \rangle$ is not a radical ideal, then we would accidentally come to the conclusion that $K[X]$ is the ring

$$\frac{K[x,y]}{\langle y, y - x^2 \rangle} \cong \frac{K[x]}{\langle x^2 \rangle},$$

which is not reduced. The fact that this ring is not reduced—and, in particular, that it is generated by an element whose square is zero—is precisely what captures the fact that $\mathcal{V}(y)$ and $\mathcal{V}(y - x^2)$ intersect with "multiplicity two." Thus, although $K[x, y]/\langle y, y - x^2 \rangle$ is not the coordinate ring of the variety X, it may be worthwhile to view it as the coordinate ring of some enhancement of X that remembers that X arose as a point of tangency.

Perhaps, then, we should simply replace finitely-generated reduced K-algebras with arbitrary commutative rings in (C.1). Unfortunately, this cannot be the right answer: for arbitrary commutative rings, the association $R \mapsto \mathrm{maxSpec}(R)$ is not injective. As an example to illustrate the problem, the reader is encouraged to verify that the rings $K[x]/\langle x \rangle$ and $K[x]/\langle x^2 \rangle$ each have a single maximal ideal, but they are not isomorphic rings.

The solution is to consider a somewhat larger and better-behaved set than the maximal spectrum, which is the *prime spectrum*

$$\text{Spec}(R) = \{\text{prime ideals } P \subseteq R\}.$$

This set can be equipped with a Zariski topology in exactly the same way as we did for the maximal spectrum. Furthermore, $\text{Spec}(R)$ inherits additional structure from the ring R, known as a "structure sheaf" (which we describe in more detail below). A topological space along with a sheaf of rings on it is called a *ringed space*, and we obtain a map

(C.2)
$$\{\text{commutative rings}\} \to \{\text{ringed spaces}\}$$
$$R \mapsto \text{Spec}(R)$$

that is, at last, an injection. An *affine scheme*, by definition, is a ringed space in the image of this map—that is, a set of the form $\text{Spec}(R)$ for some commutative ring R, along with its Zariski topology and structure sheaf. The starting point of scheme theory is that, upon restricting the codomain of (C.2) to affine schemes, it becomes an equivalence of categories, vastly generalizing the equivalence of algebra and geometry that was central to the developments in this book.

So far, our discussion of schemes has taken place entirely in the affine setting, which leads to a natural question: is there a notion that relates to (quasi)projective varieties in the same way that affine schemes relate to affine varieties? This will require substantially more care because, while the affine variety X is fully determined by its ring $K[X]$ of regular functions, the corresponding statement for quasiprojective varieties is false.

Instead, as we have hinted at in the end of Section 11.6, what determines a quasiprojective variety X is the data of the rings $K[U]$ on *all* open sets $U \subseteq X$, which are related to one another by way of restriction maps $K[U] \to K[U \cap V]$ whenever two open sets U and V intersect. This data, essentially, is what is known as a *sheaf of rings* on X. Somewhat more precisely, a sheaf of rings on any topological space X is an assignment of a ring $\mathcal{R}(U)$ to each open set $U \subseteq X$, and a ring homomorphism $\mathcal{R}(U) \to \mathcal{R}(U')$ whenever $U' \subseteq U$, subject to certain compatibility conditions.

In particular, when X is a quasiprojective variety (viewed as a topological space with the Zariski topology), there is a sheaf of rings $\mathcal{O}_X$ on X known as the *structure sheaf*, defined by

$$\mathcal{O}_X(U) = K[U]$$

for any open set $U \subseteq X$. In fact, the quasiprojective variety X is determined by the ringed space $(X, \mathcal{O}_X)$, so in this sense, quasiprojective varieties can be viewed as a special case of ringed spaces. Not every ringed space is a quasiprojective variety, though: a key feature of the ringed space $(X, \mathcal{O}_X)$ is that it is "locally affine" in the sense that it admits an open cover by sets $U_\alpha \subseteq X$ for which the restriction $(U_\alpha, \mathcal{O}_X|_{U_\alpha})$ is isomorphic to an affine variety with its structure sheaf. A ringed space satisfying this condition (which may or may not be a quasiprojective variety) is referred to as an *abstract algebraic variety*. Having recast the notion of a variety in the language of sheaves, we may press on to extend the notion of schemes beyond the affine context: first, as mentioned above, one describes a natural way to construct a structure sheaf on the affine scheme $\text{Spec}(R)$ for any commutative ring R, and then, one defines a *scheme* to be a ringed space with an open cover by sets isomorphic as ringed spaces to affine schemes.

While the language of schemes may be initially intimidating, any student interested in further study in algebraic geometry must grapple with it. We encourage the reader to consult such references as Eisenbud and Harris's book [4], Hartshorne's classic [6], or Vakil's [11] to begin the journey toward making these concepts part of their mathematical repertoire.

Index

affine cones, 273
affine linear transformations, 109
affine patches, 264
affine restrictions, 262–264
    of projective closures, 270
affine space, 38
    dimension of, 178
    is irreducible, 77
    is smooth, 210
affine twisted cubic, 45
affine vanishing ideals, 47
    are radical, 51
    of products, 217
affine vanishing sets, 39
affine varieties, 43
    are defined by ideals, 45
    are finitely generated, 71
    as quasiprojective varieties, 334
    inclusions of, 62
    intersections of, 63, 64, 66
    irreducibility of, 61, 73
    irreducible components of, 61, 79
    irreducible decompositions of, 79
    unions of, 63, 64, 66
algebraic (in)dependence, 145
algebraic closures, 56
algebraic elements of ring extensions, 139
algebraic extensions of rings, 140
algebraic field extensions
    are transitive, 170
    characterization of, 168
algebraically closed fields, 56, 151
    are infinite, 56
algebras over fields, 85, 90
    as ring extensions of fields, 91
    finite generation, 96
    First Isomorphism Theorem, 93
    generators, 95
    homomorphisms of, 91
    polynomial combinations, 95
    presentations of, 97
    quotients of, 92
    subalgebras of, 93
algebras over rings, 127, 131
ascending chain condition, 69

bases of vector spaces, 158, 168
bilinear maps, 232
birational geometry, 365
birationality
    of irreducible affine varieties, 164
    of quasiprojective varieties, 364
blow-ups
    of affine space, 379
    of affine varieties, 380
Blue Marble, The, 158

Cartesian products, 215, 216
categories, 112
    equivalence of, 124
Cayley, Arthur, 400
characteristic of a field, 59
Clebsch cubic surface, 400
Clebsch, Alfred, 400
closed embedding
    of affine varieties, 117
closures (topological), 320
codimension, 178
continuous maps, 342
coordinate functions, 86, 97
coordinate rings, 85, 87
    and coordinate functions, 97
    as K-algebras, 91, 94
    as polynomial functions, 87
    as quotients, 88, 94
    as regular functions, 331
    characterization of, 100
    detect isomorphisms, 122
    of products, 238
Cramer's Rule, 142, 207
cross-multiplications, 250
curves, 178

De Morgan's law, 314
dense subset
    of an affine variety, 117
dense subsets (topological), 320
derivatives, 192
desingularization, 365
determinantal variety, 299
determinants

and Cramer's Rule, 142
and minimal polynomials, 184
dimension of affine varieties, 157
    and inclusions, 180
    and Noether bases, 149, 179
    axioms, 159
    bounded by defining equations, 188
    bounds on, 179
    Fundamental Theorem, 157, 160, 182, 207
        proof of, 185
        strong form of, 187
    key idea, 159
    of a product, 222
dimension of homogeneous systems, 370
dimension of quasiprojective varieties, 367
    and dense open subsets, 367
    and inclusions, 371
    Fundamental Theorem, 369
dimension of vector spaces, 158
direct sums, 219
disjoint unions, 247
distinct irreducible factorizations, 53
    and irreducible decompositions, 82
distinct irreducible factors, 53
    and irreducible components, 82
division algorithm, 23
domains of rational functions, 360
domains of rational maps, 364
dominant map
    of affine varieties, 117
double dual, 199
doubly-ruled surface, 300
dual
    of a linear map, 200
    of a vector space, 197
    of a vector-space basis, 197

Earth, 158
Eckardt point, 407
Eisenstein's Criterion, 32, 101
elliptic curves, 289
empty function, 86
equivalence of algebra & geometry, 124
Euclid's Lemma
    for integers, 9

    for polynomials, 27
Euclidean topology, 46
    on $\mathbb{R}$, 314
Euler's identity, 377
exchange lemma, 174
extrinsic property, 125

factorization domains, 8
fiber, 384
field generators, 167
field of rational functions, 362
finitely-generated
    algebras, 96
    field extensions, 167
    ideals, 68
    modules, 135
First Isomorphism Theorem
    for K-algebras, 93
    for algebras over rings, 131, 146
    for modules, 131, 133
    for rings, 14
formal linear combinations, 225
fraction fields, 29
free modules, 226
function fields
    as a fraction field, 163
    as rational functions, 362
    of irreducible affine varieties, 157
functor, 116
    fully faithful, 119
Fundamental Theorem of Algebra, 56
Fundamental Theorem of Dimension Theory, 185, 187, 369

generic smoothness, 213
Gröbner bases, 82
gradient vectors, 195
graphs of regular maps, 349
Grassmann, Hermann, 305
Grassmannians, 305
    as projective varieties, 310
ground field, 38
    assumptions on, 59

Hilbert's Basis Theorem, 70, 255
Hilbert, David, 70
Hironaka, Heisuke, 365, 382
homeomorphic, 343

INDEX 417

homeomorphism, 343
homogeneous coordinate rings, 289
homogeneous ideals, 258
homogeneous polynomials, 251
homogenizations
    of polynomials, 267
    of sets, 268
hypersurfaces, 178

ideals, 12
    cosets, 13
    finite generation, 68
    generating sets, 12
    products of, 64
    sums of, 64
images of rational maps, 364
incidence varieties, 395
indexing set, 65
integral closure, 144
integral elements of ring extensions, 139
integral extensions of rings, 140
    and finitely-generated modules, 141
intrinsic property, 125
irreducible affine varieties, 73
    as vanishing of prime ideals, 75
    characterized by vanishing ideal, 74
    products of, 220
irreducible components
    and distinct irreducible factors, 82
    of affine varieties, 79
    of products, 222
    of projective varieties, 260
    of quasiprojective varieties, 322
irreducible decompositions
    and irreducible factorizations, 82
    of affine varieties, 79
    of products, 222
    of projective varieties, 260
    of quasiprojective varieties, 322
irreducible elements of rings, 7
irreducible polynomials, 7
    characterizations of, 32
irreducible projective varieties, 259
irreducible quasiprojective varieties, 320
irrelevant ideal, 272
isomorphism classes, 124
isomorphisms

    of affine varieties, 108
    of projective varieties, 286
    of quasiprojective varieties, 329

Jacobian criterion, 376
    for affine hypersurfaces, 212
    for affine varieties, 211
    for projective hypersurfaces, 377
Jacobian matrix, 211

Krull dimension, 189

line-incidence graphs, 405
linear combinations, 134
linear maps
    of projective varieties, 280
        are finite-to-one, 291
    of vector spaces, 130
linear maps of vector spaces
    determinants of, 184
linear projections, 281
linear subspaces, 131
linear subvarieties of $\mathbb{P}^n$, 305
linearizations, 191
    as linear approximations, 192
    at a point, 192
    of polynomials, 192
    of products, 223
local dimension, 210
local functions, 372
local hypersurface, 369
local property, 339
local rings, 372
    are local, 373
    of affine varieties, 373

matrix minors, 212
maximal ideals, 18
    and points, 77
    are prime, 19
minimal polynomials, 182
    and Noether bases, 183
modules, 127–129
    finite generation, 135
    generators of, 134
    homomorphisms of, 130
    quotients of, 131
    submodule generated by a set, 134

submodules of, 130
monic polynomials, 139
monomials, 2
    degree of, 5
multiplicative maps, 235

nilpotents, 99
Noether Bases, 146
Noether bases
    and dimension, 179
    and minimal polynomials, 183
    and transcendence bases, 172
Noether Normalization Theorem, 127, 145, 146, 180, 387
    proof of, 148
Noether, Emmy, 68
Noetherian rings, 61, 69, 260
    are factorization domains, 72
    are not hereditary, 72
    are preserved by quotients, 72
    ascending chain condition, 69
    Hilbert's Basis Theorem, 70
nonvanishing sets, 337
Nullstellensatz, 37, 48, 55, 57–59, 75, 77, 82, 100, 101, 127, 186, 188, 206, 269, 270, 328, 332
    projective version, 272
    proof of, 153
    statement of, 57
    translation of, 56

partial derivatives, 192
Plücker maps, 307
    are injective, 307
    images of, 308
Plücker polynomials, 308
Plücker, Julius, 306
polynomial combinations, 95
polynomial functions, 85, 86
polynomial maps
    of affine varieties, 105, 106
    of projective varieties, 280
        are finite-to-one, 296
        are linear, 296
polynomial rings, 3
    are factorization domains, 9
    are integral domains, 5
    are Noetherian, 71, 80

are unique factorization domains, 31
    as K-algebras, 91
    recursive nature, 4, 131, 132
polynomials, 3
    additivity of degree, 6
    degree of, 5
    homogeneous components, 252
    homogenizations of, 267
    irreducibility of, 7
    monic, 139
    vanishing of, 38
    versus functions, 41
prime elements of rings, 9
    are irreducible, 10
prime ideals, 18
    and irreducible affine varieties, 77
    are radical, 54
principal ideal domains, 21
    are unique factorization domains, 21
principal ideals, 12
products
    of affine varieties, 216
        and dimension, 222
        and smoothness, 223
        are affine varieties, 216
        preserve irreducibility, 220
    of projective varieties, 299, 300
    of quasiprojective varieties, 348
projective closures, 266, 321
    of hypersurfaces, 269
    via homogenization, 268
projective equivalences, 287
projective Nullstellensatz, 272
projective space, 245
    as a quotient, 246
    as an extension of affine space, 248
    as lines through the origin, 249
    points at infinity, 248
projective vanishing, 251
projective vanishing ideals, 257
    are homogeneous, 259
    are radical, 259
projective vanishing sets, 254
projective varieties, 245, 254
    are defined by ideals, 255
    are defined by quadratics, 295
    are finitely generated, 255

INDEX

irreducibility of, 259
irreducible components of, 260
irreducible decompositions of, 260
pullbacks of polynomial maps, 110, 112
    and compositions, 114
    and isomorphisms, 115
    and the identity function, 114
    are K-algebra homomorphisms, 113
pullbacks of regular maps, 329
pulling back polynomial functions
    is a bijection, 119
    is functorial, 114

quadric surfaces, 394
quasiprojective varieties, 313, 317
    are locally affine, 338
quotients
    by maximal ideals, 19
    by prime ideals, 19
    by radical ideals, 100
    of algebras, 92
    of modules, 131, 133
    of rings, 13

radical ideals, 51
    characterization of, 52
radicals of ideals, 51
    are radical ideals, 52
radicals of principal ideals, 53
Rank-Nullity Theorem, 111, 160, 384
rational functions
    on irreducible affine varieties, 163, 165
    on quasiprojective varieties, 360
rational maps, 364
rational varieties, 164
reduced rings, 85, 99
    and quotients by radical ideals, 100
regular functions
    on affine varieties, 331
    on quasiprojective varieties, 326
regular maps
    are continuous, 342
    of projective varieties, 283
    of quasiprojective varieties, 325
ring of rational functions, 361
ring of regular functions, 327

Salmon, George, 400
Schubert calculus, 310
Schubert, Hermann, 310
Segre maps, 298
    images of, 299
Segre, Corrado, 298
single-variable polynomial rings, 23
    are principal ideal domains, 25
    division algorithm, 23, 182
    factor theorem, 24
    finite zeros theorem, 25
singular points
    of affine varieties, 191, 210
        are closed, 212
    of quasiprojective varieties, 376
smooth points
    of affine varieties, 191, 210
    of quasiprojective varieties, 376
smoothness
    of affine varieties, 210
        and products, 223
    of projective hypersurfaces, 377
    of quasiprojective varieties, 376
square-free, 53
subspace topology, 316
subvarieties, 318
surfaces, 178

tangent spaces
    of affine varieties, 191, 194
        and gradient vectors, 195
        and isomorphisms, 204
        and Jacobian matrices, 211
        are vector spaces, 195
        bounded below by dimension, 206
        intrinsic characterization, 201, 204
    of quasiprojective varieties, 376
tangent spaces of affine varieties
    and products, 223
    in terms of local rings, 375
tangent vectors, 191, 194
tensor products, 215, 225
    and coordinate rings of products, 238
    extensions of scalars, 237
    of algebras, 229
        and bilinearity, 235
    of modules, 227

and bilinearity, 233
relations, 227
simple tensors, 228
Theorem on Fiber Dimensions
Part 1, 385
Part 2, 386
topological space, 315
transcendence bases, 157, 168
all have the same size, 175
exist, 173
transcendence degree, 175
bounds on, 176
transcendental numbers, 139
twisted cubic curve, 268, 269, 271, 290

unique factorization domains, 8
units in a ring, 7

vector spaces, 127
Veronese maps, 292
are embeddings, 294
images of, 293
Veronese varieties, 294
Veronese, Giuseppe, 292

Weak Nullstellensatz, 60, 152
well-ordering principle, 26
Whitney Embedding Theorem, 208

Zariski topology
on $\mathbb{A}^n$, 66
on $\mathbb{P}^n$, 315
on subsets of $\mathbb{P}^n$, 316
Zariski's Lemma, 151, 168
Zariski, Oscar, 66

Bibliography

[1] Renzo Cavalieri and Eric Miles. *Riemann Surfaces and Algebraic Curves: A First Course in Hurwitz Theory*. Cambridge University Press, 2016.

[2] David A. Cox, John Little, and Donal O'Shea. *Ideals, Varieties, and Algorithms: An Introduction to Computational Algebraic Geometry and Commutative Algebra*. Springer, 2015.

[3] David A. Cox, John B. Little, and Henry K. Schenck. *Toric Varieties*. American Mathematical Society, 2011.

[4] David Eisenbud and Joe Harris. *The Geometry of Schemes*. Springer, 2000.

[5] William Fulton. *Algebraic Curves*. Addison-Wesley, 1989.

[6] Robin Hartshorne. *Algebraic Geometry*. Springer, 1977.

[7] Rick Miranda. *Algebraic Curves and Riemann Surfaces*. American Mathematical Society, 1995.

[8] Igor R. Shafarevich. *Basic Algebraic Geometry 1: Varieties in Projective Space*. Springer, 2013.

[9] Joseph H. Silverman and John Tate. *Rational Points on Elliptic Curves*. Springer, 1992.

[10] Karen E. Smith, Lauri Kahanpää, Pekka Kekäläinen, and William Traves. *An Invitation to Algebraic Geometry*. Springer, 2000.

[11] Ravi Vakil. *The Rising Sea: Foundations of Algebraic Geometry*, 2017.

The manufacturer's authorised representative in the EU is Springer Nature Customer Service Centre GmbH, Europaplatz 3, 69115 Heidelberg, Germany. If you have any concerns regarding our products, please contact ProductSafety@springernature.com

Printed and bound by CPI Group (UK) Ltd, Croydon, CR0 4YY

26/03/2026

02078967-0004